Electrical Impedance Tomography

Series in Medical Physics and Biomedical Engineering

Series Editors: Kwan-Hoong Ng, E. Russell Ritenour, and Slavik Tabakov

Recent books in the series:

Ethics for Radiation Protection in Medicine
Jim Malone, Friedo Zölzer, Gaston Meskens, Christina Skourou

Introduction to Megavoltage X-Ray Dose Computation Algorithms
Jerry Battista

Problems and Solutions in Medical Physics: Nuclear Medicine Physics
Kwan Hoong Ng, Chai Hong Yeong, Alan Christopher Perkins

The Physics of CT Dosimetry: CTDI and Beyond
Robert L. Dixon

Advanced Radiation Protection Dosimetry
Shaheen Dewji, Nolan E. Hertel

On-Treatment Verification Imaging A Study Guide for IGRT
Mike Kirby, Kerrie-Anne Calder

Modelling Radiotherapy Side Effects Practical Applications for Planning Optimisation
Tiziana Rancati, Claudio Fiorino

Proton Therapy Physics, Second Edition
Harald Paganetti (Ed)

e-Learning in Medical Physics and Engineering: Building Educational Modules with Moodle
Vassilka Tabakova

Diagnostic Radiology Physics with MATLAB®: A Problem-Solving Approach
Johan Helmenkamp, Robert Bujila, Gavin Poludniowski (Eds)

Auto-Segmentation for Radiation Oncology: State of the Art
Jinzhong Yang, Gregory C. Sharp, Mark Gooding

Clinical Nuclear Medicine Physics with MATLAB: A Problem Solving Approach
Maria Lyra Georgosopoulou (Ed)

Handbook of Nuclear Medicine and Molecular Imaging for Physicists – Three Volume Set Volume I: Instrumentation and Imaging Procedures
Michael Ljungberg (Ed)

Practical Biomedical Signal Analysis Using MATLAB®
Katarzyna J. Blinowska, Jaroslaw Zygierewicz

Handbook of Nuclear Medicine and Molecular Imaging for Physicists – Three Volume Set Volume II: Modelling, Dosimetry and Radiation Protection
Michael Ljungberg (Ed)

Handbook of Nuclear Medicine and Molecular Imaging for Physicists – Three Volume Set Volume III: Radiopharmaceuticals and Clinical Applications
Michael Ljungberg (Ed)

Electrical Impedance Tomography: Methods, History and Applications, Second Edition
David Holder and Andy Adler (Eds)

Introduction to Medical Physics
Cornelius Lewis, Stephen Keevil, Anthony Greener, Slavik Tabakov, Renato Padovani

For more information about this series, please visit: https://www.routledge.com/
Series-in-Medical-Physics-and-Biomedical-Engineering/book-series/CHMEPHBIOENG

Electrical Impedance Tomography

Methods, History and Applications

Second Edition

Edited by
Andy Adler and David Holder

CRC Press
Taylor & Francis Group
Boca Raton London New York

CRC Press is an imprint of the
Taylor & Francis Group, an **informa** business

A CHAPMAN & HALL BOOK

Second edition published 2022
by CRC Press
6000 Broken Sound Parkway NW, Suite 300, Boca Raton, FL 33487-2742

and by CRC Press
2 Park Square, Milton Park, Abingdon, Oxon, OX14 4RN

© 2022 Taylor & Francis Group, LLC

First edition published by CRC Press 2004

CRC Press is an imprint of Taylor & Francis Group, LLC

Library of Congress Cataloging-in-Publication Data

Names: Adler, Andy, editor. | Holder, David, editor.
Title: Electrical impedance tomography : methods, history and applications
  / edited by Andy Adler and David Holder.
Description: Second edition. | Boca Raton : CRC Press, 2022. | Series:
  Series in medical physics and biomedical engineering | Includes
  bibliographical references and index.
Identifiers: LCCN 2021032150 | ISBN 9780367023782 (hardback) | ISBN
  9781032161174 (paperback) | ISBN 9780429399886 (ebook)
Subjects: LCSH: Electrical impedance tomography.
Classification: LCC RC78.7.E45 E44 2022 | DDC 616.07/57--dc23
LC record available at https://lccn.loc.gov/2021032150

ISBN: 978-0-7503-0952-3 (hbk)
ISBN: 978-1-032-16117-4 (pbk)
ISBN: 978-0-429-39988-6 (ebk)

DOI: 10.1201/9780429399886

Publisher's note: This book has been prepared from camera-ready copy provided by the authors.

Contents

SECTION II   EIT: Tissue Properties to Image Measures/Chapter

Andy Adler and William R. B. Lionheart

Chapter 6   The EIT Inverse Problem 109

William R. B. Lionheart and Andy Adler

Chapter 7   D-bar Methods for EIT 137

David Isaacson, Jennifer L. Mueller, and Samuli Siltanen

SECTION III *Applications*

Chapter 11   EIT Monitoring of Hemodynamics 207

Lisa Krukewitt, Fabian Müller-Graf, Daniel A. Reuter Stephan H. Böhm and Huaiwu He

Ryan Halter and David Holder

Martina Mosing and Yves Moens

SECTION IV Related Technologies

Chapter 18   Geophysical ERT...383

Alistair Boyle and Paul Wilkinson

Chapter 19   Industrial Process Tomography 403

Thomas Rodgers, William Lionheart and Trevor York

Section I

Section: Introduction

1 Electrical Impedance Tomography

Andy Adler
Systems and Computer Engineering, Carleton University, Ottawa, Canada

CONTENTS

1.1 INTRODUCTION

Electrical Impedance Tomography (EIT) is a technique to create tomographic images of the electrical properties of tissue within a body based on electrical impedance measurements at body-surface electrodes. Figure 1.1 illustrates the concept: electrodes are placed onto the chest, and images and waveforms of air and blood movement calculated. The term *Tomography* refers to imaging the volume within a body using penetrating energy from outside the body. Tomographic modalities are typically named after the energy used or its interaction with the body; EIT is thus named since it uses measurements of electrical *impedance*.

This is the second edition of this book, following up the 2004 edition[484] almost twenty years ago. The basics of the technology and its applications have not changed; what has changed is that EIT is finally being used, clinically, to make decisions about the patient therapy, such as for patients of the COVID-19 pandemic during the writing of this book[1069]. Commercial and research systems are available. EIT is accepted as a scientific tool, for example, to help understand the unique physiology of a baby's first cry[1042]. Technical developments now also focus on calculating robust and clinically useful parameters.

DOI: 10.1201/9780429399886-1

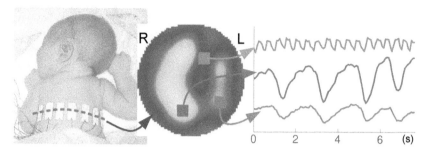

Figure 1.1   EIT images and waveforms from a healthy infant. (Left) A 10-day-old infant with EIT electrodes, from Heinrich *et al*[457], from which cross-sectional images are reconstructed (centre), with pixel waveforms showing the heart and left and right lung activity (right). For an infant with the head turned to the side, the contralateral lung receives most tidal ventilation. (color figure available in eBook)

1.2   OVERVIEW

Biomedical EIT is an inter-disciplinary field. Some clinical terminology and concepts in application chapters may be unfamiliar to readers with maths or physics backgrounds, and the reconstruction algorithms or instrumentation chapters may be difficult to follow for clinical readers. The next chapter is intended as a non-technical introduction to technology and concepts in biomedical electrical impedance tomography.

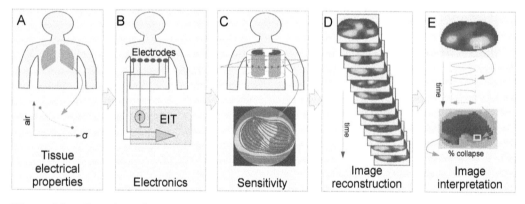

Figure 1.2   Overview of the steps in EIT image generation and interpretation. (A) physiologically interesting tissues have electrical properties which contrast from surrounding tissues or change over time; (B) EIT systems hardware applies electrical current and measures voltages on a sequence of body surface electrodes; (C) The pattern of electrical stimulation and measurements, and the body geometry and electrical properties, determine the sensitive region; (D) Images of the conductivity (absolute EIT) or the change in conductivity (difference EIT) are calculated using the sensitivity field using an image reconstruction algorithm; and (E) Physiologically relevant measures are calculated from image contrasts and their changes over time. (color figure available in eBook)

The second part of this book (Chapters 3–8) describes the story of an EIT measurement from tissue properties through measurement, imaging, and interpretation, as illustrated in figure 1.2. We have found it useful to categorize the various components of EIT analysis through this type of processing pathway[19,297]. The input is a physiological or anatomical feature of interest that leads to changes in tissue impedance, such as the movement of non-conductive air in the lungs or the presence of a tumour surrounded by a conductive network of blood vessels. The output is a functional image or measure which is scientifically or clinically relevant. If we wish to know if an intervention has improved air flow to a patient, we need to calculate flow from the EIT image and compare it before and after the treatment. After describing the tissue impedance properties (chapter 3), we review EIT electronics hardware and signal acquisition (chapter 4). We next need to understand how current propagates in the body (chapter 5) in order to build a model from which we can calculate tomographic images (chapter 6), as well as a new set of "D-bar" approaches for image reconstruction (chapter 7). Finally, from these images, clinically-relevant functional parameters and measures are calculated (chapter 8). These steps are reviewed in more detail below (§1.3).

The next part of this book (Chapters 9–14) reviews the clinical applications of EIT. EIT has seen the most widespread application for imaging of the lungs (chapter 9), and is currently seeing clinical application for monitoring of ventilation in intensive care and surgical patients (chapter 10). When making EIT recordings on the chest, useful information about blood flow (perfusion) is also available, and EIT shows promise for measuring hemodynamic parameters (chapter 11). Chapter 12 describes the application of EIT for imaging the brain and nerves, followed by a description of its use for imaging of cancerous regions (chapter 13). EIT has also been evaluated for use in many other applications (Chapter 14). In addition to human medicine, EIT is also seeing use in veterinary applications (chapter 15).

The next part describes many allied technologies which are relevant to biomedical EIT. Tissue electrical properties can be measured magnetically through Magnetic Induction Tomography (MIT, chapter 16). Very high-resolution images of tissue impedance can be created using magnetic resonance imaging in MREIT (chapter 17). The principles underlying medical EIT are also relevant to other disciplines. Geophysical prospecting with Electrical Resistivity Tomography (ERT) was the first tomographic imaging modality, starting in 1911[39], and, in many ways, geophysical ERT is more advanced than biomedical EIT (chapter 18). Another important use of ERT is in process tomography for monitoring of such processes as pipe flows and industrial mixing (chapter 19). Finally, we briefly describe the EIT devices available (both historically important and currently used systems) and details about conferences and proceedings in chapter 20.

1.3   EIT IMAGE GENERATION AND INTERPRETATION

EIT generates images with a relatively low resolution of about 10% of the diameter of the surrounding surface electrodes, but a high temporal resolution, of a few tens of ms. It uses physically small hardware with body surface electrodes that are less cumbersome than many other medical imaging devices. Figure 1.3 illustrates EIT data collection. These figures use data from a health newborn[457] with 16 ECG-type electrodes. A volume model of the tissue properties is used to reconstruct EIT images. We illustrate the heart activity, showing the electrocardiogram (ECG) signal (for reference only) and the ventricular volume. Blood is more conductive than most tissues, and the increased volume of the heart during diastole than systole, means the electrical current "prefers" to propagate through the blood. Next, we illustrate the process of EIT data collection. Current is applied between a pair of electrodes and then propagates diffusely throughout the body, creating distribution of voltage shown by the equipotential lines. Between systole and diastole, the changing blood volume in the heart affects the electric propagation thus the voltage measured at the electrodes. Figure 1.4 illustrates EIT image reconstruction and calculation of functional measures. Based on the data measured, a reference data set (frame) is chosen and images are reconstructed of the current data frame with respect to the reference, using a computational model to project and then filter the image. Reconstructed images of a transverse slice are calculated for each frame as a function of time. From these parameters, pixel waveforms can be extracted (showing activity in the heart and lungs) measures (such as the heart rate or tidal volume in each lung) calculated.

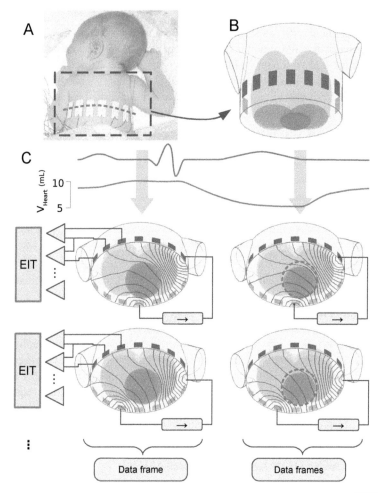

Figure 1.3 Illustation of raw data collection in EIT. (A) A healthy newborn[457]; (B) volume model of the tissue properties for reconstruction; (C) heart activity, and the increasing blood during diastole; (bottom) EIT data collection, applying current between alternating pairs of electrodes, creating a distribution of voltage in the body and changing the voltage at measurement electrodes. (color figure available in eBook)

1.3.1 TISSUE ELECTRICAL PROPERTIES

The electrical properties of biological tissue are measured by applying and measuring voltages and currents. EIT applies imperceptible currents which meet the safety standards for medical devices (ISO/IEC 60601). In order to improve electrical safety and to remain outside the range of the body's own electrical signals, EIT normally applies sinusoidal currents at frequencies above 50 kHz where the standards allow a maximum current of 5 mA (§3.5).

   At these small currents, Ohm's law applies: $V = IR$, a voltage V is required to send a current I through a resistance R. An analogy with fluid flow is useful: in order to force flow (current) through a narrowed pipe (resistance) a pressure (voltage) is required. Resistance (in Ω) describes the global properties of a body between electrodes. To describe the electrical properties of tissue, the relevant parameter is the resistivity ρ (normally measured in Ω·cm). For a uniform cylinder with electrodes at the ends, the resistance, $R = l\rho/A$, increases with length with the length, l, and decreases with area, A. It's often useful to think in terms of the inverse of resistance $G = R^{-1}$ measured in Siemens

(S) and conductivity $\sigma = \rho^{-1}$ in units of S/cm or S/m. In the body, electric current travels as ions in solution, and blood is one of the best conductors.

When our sinusoidal current $I = I_0 \cos 2\pi f$ is applied to a pure resistor, V is exactly in phase with I. In contrast, when a sinusoidal current flows into capacitor C, voltage is delayed by $90°$. When an electrical circuit combines resistances and capacitances, it is convenient to use complex numbers to express the in- and out of phase components at a single sinusoidal frequency. With complex values, impedance Z corresponds to resistance and admittance Y corresponds to conductance. For purely resistive circuits, $Z = R$ and $Y = G$. When a capacitor is in parallel to a resistor, the complex admittance $Y = G + i2\pi fC$.

In biological tissue at low frequency, tissue membranes act as capacitors which store electrical energy. As the frequency increases, a pathway through the membrane has a higher admittance, and current will begin to flow through the membrane (figure 3.2). The bulk properties of tissue are characterized by a complex conductivity (or admittivity) with a symbol σ^* (or γ). As the frequency increases, the change in σ^* can be used to characterize the tissue (figure 3.4). EIT systems typically measure both the amplitude and phase of the voltage at the electrodes, but most image reconstruction algorithms use only the amplitude. Thus conductivity and admittivity are treated as near-synonyms in this much of this book.

In EIT, a current at frequency f is applied to the body part of interest across a pair of electrodes. The current spreads out though the body to find the easiest (lowest impedance) path (figure 1.2C). Changes in the spatial distribution of σ^* of tissue will change the way the current flows and change the pattern of voltage. EIT is sensitive to contrasts in tissue admittivity. Since blood is more conductive than most other tissue, an increase of blood in the heart during diastole, or pooling of blood due to hemorrhage, create changes that can be measured by EIT. The most important conductivity contrast in the body is air. A large volume of non-conductive air moves with each breath, and, as a result, EIT imaging of breathing is easiest. Several other tissues give useful constrasts. Neural tissue changes conductivity when in an active state (chapter 12). Cancerous tissue and the new growth around tumours contrast with benign tissue (chapter 13). Additionally, conductivity contrasts can be induced using hypertonic saline injections (chapter 11), ingesting salty meals, or through thermal contrasts (chapter 14).

1.3.2 EIT ELECTRONICS

Most EIT systems apply currents across pairs of electrodes and measure the voltages across the remaining electrodes. This arrangement has an advantage in compensating for variable electrode contact quality. It is also typical to not use measurements made on the driven electrodes for the same reason. A *frame* of EIT data refers to a complete set of measurements. Most EIT systems apply, in sequence, current across each pair of electrodes and then measure voltage on all other electrode pairs, before switching the current to the next pair. Early systems treated adjacent electrodes as pairs, but many modern systems will use a *skip* pattern, where skip=2 means that electrode 1 is paired with electrode 4. The choice of current injection and voltages measurements is called the *simulation and measurement procotol*. Given N electrodes, there are a possible $N \times (N-3)/2$ independent measurements. Typical systems have 8, 16, or 32 electrodes, and thus have 20, 104, or 464 measurements per frame. Modern EIT systems function at impressive speeds, measuring 50 or more frames per second. For a 32 electrode system with a current injection at $100\,\text{kHz}$, this corresponds to 62.5 sinusoidal periods per pattern. An EIT system must calculate the amplitude and phase on all electrodes during this time.

There are numerous challenges to obtaining high quality EIT data. As mentioned, the high frame rates mean that it is difficult to perform lots of signal averaging. The hospital environment is electrically noisy and a patient generates their own electrical signals (heart, muscles and nerves). Since

EIT signals are in a narrow range of frequencies, precision filters are required to reject this interference. An additional challenge is the precision required, as seen in figure 1.3 where large changes of blood in the heart lead to small changes in the voltage equipotential lines on the body surface. A good EIT system should be able to accurately measure changes of 0.1%. Several further challenges arise. The dynamic range (ratio between the largest and smallest signals) is large, requiring precise data acquisition. At audio frequencies, circuits have numerous defects, non-zero common-mode rejection, leakage currents, and crosstalk between electrode channels. Precision electronics are sensitive to temperature and thus can drift over time. Finally, the contact quality of electrodes is variable: electrode contact often improves as a subject sweats, but electrodes can disconnect as posture changes. These challenges are difficult, but most modern EIT systems work surprisingly well most of the time.

For routine use, EIT electronics must be robust and user friendly. Disconnected electrodes must be detected and reported. Application of individual ECG-type electrodes is adequate for research, but too slow for clinical use. Instead, manufacturers have designed electrode belts which can be rapidly placed. Belts need to be of biocompatible material and either washable or cheap enough for one-time use. To improve contact quality a contact gel or saline water is normally applied to the skin. This leads to a further challenge that the saline/sweat mix is very corrosive and damages belts.

Time-difference EIT images typically use one measurement frequency. An alternative approach is to compare the difference between impedance images measured at different measurement frequencies, termed frequency-difference EIT (fdEIT). This technique exploits the different impedance characteristics of tissues at different measurement frequencies. An example of such a contrast would be the difference between cerebro-spinal fluid (CSF) and the grey matter of the brain. As the CSF is an acellular, ionic solution, it can be considered a pure resistance, so that its impedance is identical and equal to the resistance for all frequencies of applied current. However, the grey matter, which has a cellular structure, has a higher impedance at low frequencies than at high frequencies. This frequency difference could be exploited to provide a contrast in the impedance images obtained at different frequencies, and identifying different tissues in a multifrequency EIT image[1187].

1.3.3 MODELS OF SENSITIVITY

Sensitivity is the change in a measurement due to a change in an internal parameter. It is characterized by a Jacobian matrix, $\mathbf{J}_{i,j} = \partial \mathbf{v}_i / \partial \sigma_j^*$, which indicates how voxel j in the body affects measurement i. In order to reconstruct an image, the sensitivity explains what each measurement "means" and which parts of the body could be generating the measurements observed (chapter 5). At the frequencies used in EIT, voltages and currents distribute according to the Laplace equation: $\nabla \cdot \sigma^* \nabla V = 0$, with boundary conditions of the applied voltages. The Laplace equation means that EIT is non-local; every change in σ^* anywhere in the body affects voltages everywhere, and leads to the ill-conditioning of image reconstruction.

In EIT, two main approaches to sensitivity calculation have been used, analytical and finite element models (FEM). Analytical models are useful for regular shapes such as cylinders, and help understand the theoretical limits of sensitivity. For patient application, the FEM has been the main tool. An FEM splits a body into subdomains (normally tetrahedra) and solves the Laplace equation by constraints between the subdomain boundaries. Increased accuracy is achieved by a finer subdivison of the domain, especially in the high electric field regions near electrodes. The method is fairly rapid: a modern computer solves a one million element FEM in a few seconds.

Sensitivity matrix calculations allow an understanding of EIT system configurations. For example, where are electrodes best placed to achieve the best detectability of a given process?[347,376,391] An important question concerns the accuracy required of FEMs. Most algorithms use a *atlas* model based on an approximate shape (cylinder for early algorithms[83], and CT-based FEM models for

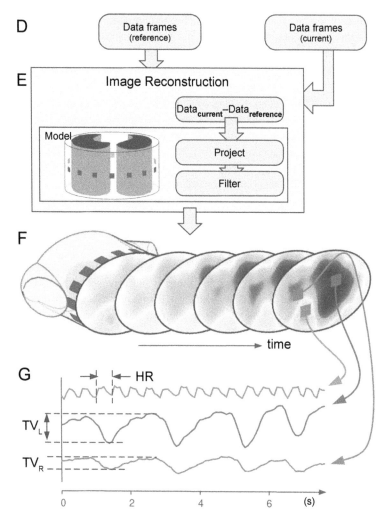

Figure 1.4 Based on the EIT signal (figure 1.3) a set of data frames are acquired which represents all the EIT measurements at each time point. A reconstruction algorithm calculates the image (of the volume or of a transverse slice) of the electrical impedivity, Z, of tissue. Almost all experimental and clinical EIT systems use time-difference reconstruction (tdEIT), in which the change in ΔZ is calculated between two data frames measured at two points in time, a reference time, t_{ref}, and the current time, t. (color figure available in eBook)

more recent ones). The quality of images certainly improves with accurate models (see figure 6.6), but these are rarely available. Even given patient-specific CT, the electrode positions would not be perfect, and posture changes[214] and breathing[27] would make changes to the shape. FEM meshing variability can be corrected through setting of image reconstruction parameters[29].

1.3.4 EIT IMAGE RECONSTRUCTION

Image reconstruction is the term used in tomographic imaging for calculation of an image from projection data. Image reconstruction is challenging as the underlying equations are ill-conditioned and often ill-posed. The ill-conditioning stems from the large difference in sensitivity between regions (the electrodes and the body centre in EIT). EIT is typically ill-posed because it is not possible to estimate a large number of image parameters from the limited number of measurements in each data frame. Image reconstruction calculates an estimate, $\hat{\mathbf{x}}$, of the distribution of internal properties, $\mathbf{x}$, which best explains the measurements, $\mathbf{y}$. A simplified schema for image reconstruction is shown in figure 6.1, which illustrates the process by which model parameters are iteratively adjusted to fit the measurements.

Image reconstruction in EIT has been divided into algorithms for difference EIT (§6.6) and absolute EIT (§6.7). Difference EIT is more stable and can be implemented as a matrix multiplication. Absolute EIT is a more difficult problem, and requires accurate models of body geometry and electrode positions, shapes and contact impedances. For difference EIT, shape and electronics modelling inaccuracies are less significant, as long as they remain the same between the difference measurements. While in geophysical ERT, absolute reconstruction is common, experimental and clinical EIT has almost exclusively used difference measurements and algorithms. Image reconstruction in EIT is a difficult problem and is highly sensitive to noise and outliers; §6.2 addresses the question "Why is EIT so hard?".

EIT image reconstruction is formulated to minimize a norm $\|\mathbf{y} - F(\hat{\mathbf{x}})\|^2_{\mathbf{W}} + \alpha^2 \|\mathbf{L}(\hat{\mathbf{x}} - \mathbf{x}_0)\|^2$, where the first term, $\mathbf{y} - F(\hat{\mathbf{x}})$, is the "data mismatch" between the measured data and their estimate via the forward model, $F(\cdot)$. $\mathbf{W}$ is a data weighting matrix, and represents the inverse covariance of measurements. The second term is penalty term on the mismatch between the reconstruction estimate, $\hat{\mathbf{x}}$, and an *a priori* estimate of its value, $\mathbf{x}_0$. The matrix $\mathbf{L}$ is normally chosen to measure the amplitude or roughness in the image. Thus this term penalizes solutions which would match the data but are large or non-smooth. The relative weighting between the data and prior mismatch terms is controlled by a hyperparameter, α. When α is large, solutions tend to be smooth and more similar to the prior; for small α, solutions better match the data, have higher spatial resolution, but are noisier and less well conditioned (figure 6.3).

Early work in EIT image reconstruction was motivated by the backprojection algorithms of X-ray CT. Interestingly, it is possible to interpret regularized linear inverse in terms of backprojection and filtering (see Box 6.10). A regularized inverse algorithm has many user-selected parameters, which control its performance. A standardization effort in 2007 lead to the GREIT algorithm[18], based on an consensus of the figures of merit required. Interestingly, high resolution was not a priority: clinical and experimental users want robust and predictable images.

Current work in algorithms is pursuing two directions. One seeks to increase the robustness of images given electrode movement[991] and detecting and compensating for failing electrodes[63]. The other pathway seeks novel techniques to improve image reconstruction, using innovative methods such as direct methods (chapter 7), level sets[851], and machine learning[908].

1.3.5 IMAGE INTERPRETATION

Medical images are valuable only if they can provide diagnostic information to usefully inform treatment. EIT images can thus be analyzed to determine relevant parameters: a rich literature of *functional EIT* (or fEIT) parameters has developed for analysis of individual images and time-sequences of images (chapter 8). Techniques for fEIT analysis were subdivided into *functional EIT images* and *EIT measures*[293]. In figure 1.4 the parameters calculated are the heart rate (HR) and an index of relative ventilation level between the right and left lungs. fEIT image analysis is illustrated in figure 8.6. Two basic categorizations are between measures which can be measured continuously, and examination-specific measures, which require specific interventions with patients. The literature on fEIT measures includes:

- *Distribution of lung volumes and flows*, relevant for managing ventilated patients and monitoring obstructive lung disease[1089].
- *Distribution of Aeration change*, due to an intervention or physiological activity.

- *Frequency analysis of impedance changes*, to separate breathing- and heart-related effects, using an ECG-gated or frequency-filtered signal, leading to the *perfusion-related* EIT signal[30,279].
- *Respiratory system mechanics*: Lung tissue can be characterized by a compliance, C and resistance, R. C is low at both low and high lung volumes due to alveolar collapse (atelectasis) or overdistension[213]. The time constant $\tau = RC$ of tissue introduces a ventilation which varies across the lung functional EIT measures of time constant[715].
- *Activation patterns*: Functional EIT of brain EIT images can characterize the pattern of activation within the cortex following stimulation[57].
- *Pulse transit time*, of blood through the vasculature is modulated by the blood pressure, and can measure pulmonary-[842] (transit time from heart to lungs) and systemic-arterial pressure (transit time from heart to aorta)[987].
- *Contrast agents*: Direct measures of blood flow can be measured with injected conductivity contrasts. EIT images show the blood flow through the heart and lungs[15,302] or brain[21].

1.4   EIT APPLICATIONS AND PERSPECTIVES

In the second part of this book, we review applications of EIT in many fields, from imaging lungs, heart and blood flow, brain and nerves, cancerous tissue and other applications. Good reviews of EIT applications have been written, to which we refer the interested reader[16,19,123,212,293,295,615,658,778]. Some applications are starting to become established in routine clinical use, such as monitoring of ventilation (Chapter 10), while others are still at an earlier stage. Overall EIT has several clear advantages: it is non-invasive and minimally cumbersome, requiring only electrodes and wires on the body; the electronics hardware is potentially fairly cheap; it is suitable for prolonged monitoring. At the same time, EIT images are of low resolution, EIT data are subject to many sources of interference, and we still don't know how to do absolute image reconstruction reliably.

The concepts in medical EIT are also used in geophysics and process tomography. Geophysical imaging with electrical measurements has a long history[39]; it can measure the metallic ores and ground water, and is used for imaging (e.g. archaeological surveys) and monitoring (bridges and embankments). Process tomography applications of ERT focus on monitoring of pipes and mixing vessels.

It is useful to review the two decades since the publication of the first edition of this book[484], perhaps in the context of a comment from our editorial[24]

> We are looking forward to the day when there is no longer a need to explain EIT; when there is no longer the need to call it a promising, "new" technology. Instead, we would like to write, "Based on evidence of improved patient outcomes and safety, EIT devices are increasingly used . . . ", and provide the references to prove it.

In the last decade, EIT has begun to see early clinical use. There are now several companies which sell clinical chest EIT systems. The COVID-19 pandemic has underlined the importance of intensive care medicine and the management of ventilated patients, a role for which EIT's advantages are clear[1069]. There is a strong and growing rate of citations of EIT in the medical literature. On the other hand, non-lung applications of EIT have not yet seen a transition to clinical use, but many are showing clear promise, such as monitoring of nerves[868].

To our readers: if you've read this far, you are interested in better understanding EIT, perhaps to conduct novel research or better treat patients. We eagerly await learning about your insights. We encourage you to interact with the EIT community, via our conferences (§20.1) software and mailing lists[14]. Our community is encouraging and supportive.

2 Introduction to EIT Concepts and Technology

David Holder
Medical Physics and Bioengineering, University College London, UK

CONTENTS

2.1 BIOMEDICAL ELECTRICAL IMPEDANCE TOMOGRAPHY

One of the attractions but also difficulties of biomedical Electrical Impedance Tomography (EIT) is that it is interdisciplinary. Topics which are second nature to one discipline may be incomprehensible to those with other backgrounds. Not all readers will be able to follow all the chapters in this book, but I hope that the majority will be comprehensible to most, especially those with a medical physics or bioengineering background. Nevertheless, the reconstruction algorithm or instrumentation chapters may be difficult to follow for clinical readers and some of the clinical terminology and

DOI: 10.1201/9780429399886-2

concepts in application chapters may be unfamiliar to readers with Maths or Physics backgrounds. This chapter is intended as a brief and non-technical introduction to biomedical EIT. It is didactic and explanatory, so that the more detailed chapters in the book which follow may be easier to follow for the general reader. It is intended to be comprehensible to readers with clinical or life sciences backgrounds but with the equivalent of high school physics. It commences with a non-technical introduction to the basics of bioimpedance which may be helpful for any reader wishing to refresh their understanding of the basics of electricity and its flow through biological tissues. A more detailed and technical introduction is in Chapter 3.

2.2 BRIEF INTRODUCTION TO BIOIMPEDANCE

Bioimpedance refers to the electrical properties of a biological tissue measured when current flows through it. This impedance varies with frequency and different tissue types and varies sensitively with the underlying histology.

2.2.1 RESISTANCE AND CAPACITANCE

The resistance and the capacitance of tissue are the two basic properties in bioimpedance.

Resistance (R) is a measure of the extent to which an element opposes the flow of electrons or, in aqueous solution as in living tissue, the flow of ions among its cells. The three fundamental properties governing the flow of electricity are voltage, current and resistance. The voltage may be thought of as the pressure exerted on a stream of charged particles to move down a wire or migrate through an ionized salt solution. This is analogous to the pressure in water flowing along a pipe. The current is the amount of charge flowing per unit time, and is analogous to water flow in a pipe. Resistance is the ease or difficulty with which the charged particles can flow, and is analogous to the width of a pipe through which water flows – the resistance is higher if the pipe is narrower (figure 2.1).

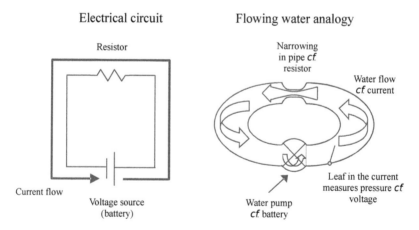

Figure 2.1   Basic concepts – current, voltage and resistance. Analogy to water flow. *Right*: Electrical circuit. *Left*: Flowing water analogy

They are related by Ohm's law:

$$V\,[\text{voltage, Volts}] = I\,[\text{current, Amps}] \times R\,[\text{resistance, Ohms } \Omega]. \qquad (2.1)$$

The above applies to steadily flowing, or "D.C." current – (direct current). Current may also flow backwards and forwards – "A.C." (alternating current). Resistance has the same effect on A.C current as D.C current. Capacitance (C) is an expression of the extent to which an electronic component, circuit, or system, stores and releases energy as the current and voltage fluctuate with each AC cycle. The capacitance physically corresponds to the ability of plates in a capacitor to store charge. With each cycle, charges accumulate and then discharge. Direct current cannot pass through a capacitor. AC can pass because of the rapidly reversing flux of charge. The capacitance is an unvarying property of a capacitive or more complex circuit. However, the effect in terms of the ease of current passage depends on the frequency of the applied current – charges pass backwards and forwards more rapidly if the applied frequency is higher.

For the purposes of bioimpedance, a useful concept for current travelling through a capacitance is "reactance" (X). The reactance is analogous to resistance – a higher reactance has a higher effective resistance to alternating current. Like resistance, its value is in Ohms, but it depends on the applied frequency, which should be specified (figure 2.2).

The relationship is :

$$\text{Reactance } [\Omega] = 1/(2 \times \pi \times \text{Frequency [Hz]} \times \text{Capacitance [Farads]}) \tag{2.2}$$

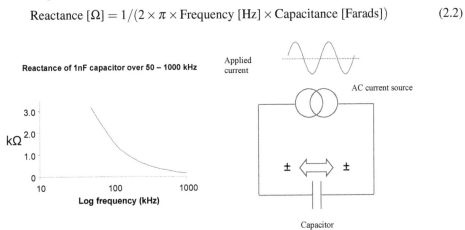

Figure 2.2   Capacitance, reactance and effect of frequency.

When a current is passing through a purely resistive circuit, the voltage recorded across the resistor will coincide exactly with the timing, or phase, of the applied alternating current, as one would expect. In the water flow analogy, an increase in pressure across a narrowing will be instantly followed by an increase in flow. When current flows across a capacitor, the voltage recorded across it lags behind the applied current. This is because the back and forth flow of current depends on repeated charging and discharging of the plates of the capacitor. This takes a little time to develop. To pursue the water analogy, a capacitor would be equivalent to a taut membrane stretched across the pipe. No continuous flow could pass. However, if the flow is constantly reversed, then for each new direction, a little water will flow as the membrane bulges, and then flow back the other way when the flow reverses. The development of pressure on the membrane will only build up after some water has flowed into the membrane to stretch it. In terms of a sine wave which has 360° in a full cycle, the lag is one quarter of a cycle, or 90°.

In practice, this is seen if an oscilloscope is set up as in figure 2.3. An ideal current AC source passes current across a resistor or capacitor. The current delivered by the source is displayed on the upper trace. The voltage measured over the components is displayed on the lower trace. When this is across a resistor, it is in phase – when across a capacitor, it lags by 90° and is said to be "out-of-phase". When the circuit contains a mixture of resistance and capacitance, the phase is intermediate between 0 and 90°, and depends on the relative contributions from resistance and capacitance. As a constant current is applied, the total combination of resistance or reactance, the impedance can be calculated by Ohm's law from the amplitude of the voltage at the peak of the sine wave.

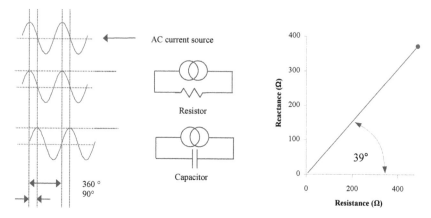

Figure 2.3   *Left:* The voltage that results from an applied current is in phase for a resistor and 90° out of phase for a capacitor. *Right:* Impedance data plotted in the complex plane.

Impedance is made of these two components, resistance or the real part of the data, and reactance – the out-of-phase data. These are usually displayed on a graph in which resistance is the x axis and reactance is the y axis. This is termed the "complex" impedance, and the graph as the "complex" plane. For mathematical reasons to do with solutions of the equations for the sine waves of the AC voltages, the in-phase resistive component is considered to be a "real", or normal number. The out-of phase, capacitive, component is considered to be "imaginary". This means that the amplitude of the capacitative voltage, a real number such as 3.2 V, is multiplied by "i" which is the square root of minus 1. Thus a typical complex impedance might be written as $450 + 370i\,\Omega$.

This would mean that the resistance is $450\,\Omega$ and the reactance is $370\,\Omega$, and would be displayed on the complex plane as in figure 2.3 (right), with the resistance on the x axis and reactance on the y axis. Another equivalent way is to calculate the length of the impedance line which passes from the origin of the graph to the complex impedance point. This is termed the "modulus" of the impedance (Z), and means its total amplitude, irrespective of whether it is resistance or reactance. In practice, this is identical to the amplitude of the sine wave of measured voltage, seen on an oscilloscope, as in figure 2.3, irrespective of the phase angle. The "phase angle" is calculated from the graph, and is given along with the modulus. The phase angle on the graph is exactly the same as the lag in phase of the measured voltage (figure 2.3 right). For the above example, $450 + 370i\,\Omega\,(R + iX)$ converts to $583\,\Omega$ at 39° ($Z\angle\theta$).

2.2.2   IMPEDANCE IN BIOLOGICAL TISSUE

Cells may be modelled as a group of electronic components. One of the simplest employs just three components: the extracellular space is represented as a resistor (R_e), and the intracellular space and the membrane is modelled as a resistor (R_i) and a capacitor (C_m) (figure 2.4a). Both the extracellular space and intracellular space are highly conductive, because they contain salt ions. The lipid membrane of cells is an insulator, which prevents current at low frequencies from entering the cells. At lower frequencies, almost all the current flows through the extracellular space only, so the total impedance is largely resistive and is equivalent to that of the extracellular space. As this is usually about 20% or less of the total tissue, the resulting impedance is relatively high. At higher frequencies, the current can cross the capacitance of the cell membrane and so enter the intracellular space as well. It then has access to the conductive ions in both the extra- and intra-cellular spaces, so the overall impedance is lower (figure 2.4b).

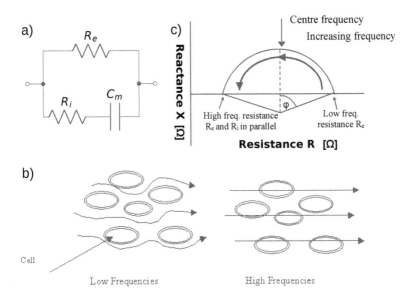

Figure 2.4   a) The cell modelled as basic electronic circuit: R_i and R_e are the resistances of the intracellular- and extracellular space, and C_m are is the membrane capacitance. b) The movement of current through cells at both low and high frequencies. c) Idealised Cole-Cole plot for tissue.

The movement of the current in the different compartments of the cellular spaces at different frequencies, and the related resistance and reactance values, are usefully displayed as a Cole-Cole plot. This is an extension of the resistance/reactance plot in the complex plane. Instead of the single point for a measurement at one frequency, as in figure 2.3c, the values for a range of frequencies are all superimposed. For simple electronic components, the arc will be a semicircle. At low frequencies, the measurement is only resistive, and corresponds to the extracellular resistance – no current passes through the intracellular path because it cannot cross the cell membrane capacitance. As the applied frequency increases, the phase angle gradually increases as more current is diverted away from the extracellular resistance and passes through the capacitance of the intracellular route. At high frequencies, the effective resistance of the intracellular capacitance (the reactance) becomes negligible, so current enters the parallel resistances of the intracellular and extracellular compartments. The cell membrane reactance is now nil, so the entire impedance again is just resistive and so returns to the x axis. Between these, the current passing through the capacitative path reaches a peak. The frequency at which this occurs is known as the centre frequency (f_c), and is a useful measure of the properties of an impedance. In real tissue, the Cole-Cole plot is not exactly semicircular, because the detailed situation is clearly much more complex; the plot is usually approximately semicircular, but the centre of the circle lies below the x-axis. Inspection of the Cole-Cole plot yields the high and low frequency resistances, as the intercept with the x axis, and the centre frequency is the point at which the phase angle is greatest. The angle of depression of the centre of the semicircle is another means of characterising the tissue (figure 2.4c).

Over the frequency ranges used for EIT and Magnetic induction Tomography (MIT), about 100 Hz to 100 MHz, the resistance and reactance of tissue gradually decreases. This is due to the simple effect of increased frequency passing more easily across capacitance, but also because cellular and biochemical mechanisms begin to operate which increase the ease of passage of the electrical current. A remarkable feature of live tissue is an extraordinarily high capacitance, which is up to

1000 times greater than inorganic materials, such as plastics used in capacitors. This is because capacitance is provided by the numerous and closely opposed cell membranes of cells, each of which behaves as a tiny capacitor. Over this frequency range, there are certain frequency bands where the phase angle increases, because mechanisms come into play which provide more capacitance. They may be seen as regions of increased decrease of resistance in a plot of resistance against frequency, and are termed "dispersions". At the low end of the frequency spectrum, the outer cell membrane of most cells is able to charge and discharge fully. This region is known as the alpha dispersion and is usually centred at about 100 Hz.

As the frequency increases, from 10 kHz to 10 MHz, the membrane only partially charges and the current charges the small intracellular space structures, which behave largely as capacitances. At these higher frequencies, the current can flow through the lipid cell membranes, introducing a capacitive component. This makes the higher frequencies sensitive to intracellular changes due to structural relaxation. This effect is largest around 100 kHz, and is termed the "beta dispersion". At the highest frequencies, dipolar reorientation of proteins and organelles can occur and affect the impedance measurements of extra- and intra-cellular environments. This is the gamma dispersion and is due to the relaxation of water molecules and is centred at 10 GHz. Most changes between normal and pathological tissues occur in the alpha and beta dispersion spectrum.

2.2.3 OTHER RELATED MEASURES OF IMPEDANCE

Resistance and reactance, as described above, are fixed measures of individual components or samples. It is useful to be able to describe the general properties of a material. The impedance of a sample increases, as one would expect, with increasing length of the sample between the measuring electrodes. Somewhat counter-intuitively, it decreases if the area contacting the measuring electrodes increases – this is because there is more conductive material to carry the current. The individual values for resistance or reactance can be converted to the general property – termed "resistivity" or "reactivity" by adjusting for these. Resistivity, ρ, is given in Ω·m and is the ability of a material to resist the passage of electrical current for a defined unit of tissue (figure 2.5). It is calculated as

$$\rho\,[\Omega{\cdot}m] = \text{Resistance}\,[\Omega] \times \text{Area/Length}. \qquad (2.3)$$

for an object such as a cylinder or cuboid which are uniform in the length direction.

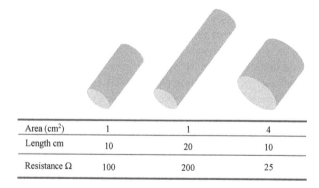

Area (cm²)	1	1	4
Length cm	10	20	10
Resistance Ω	100	200	25

Figure 2.5   The effect of changing the length or cross sectional area of the tissue sample measured. (color figure available in eBook)

The capacitative element of a material can be considered in the same way – the "reactivity" is also measured in Ω·m and is the general resistance property of a material, at a specified frequency.

The resistance and reactance fully describe the impedance of tissue, but there are several other related measures which are, sometimes confusingly, used in the EIT and Bioimpedance literature. These arise because different, reciprocal, terms may be used to describe the ease, as opposed to the difficulty, of passage of current. Secondly, with respect to capacitance, one can choose to use the effective resistance at a given frequency – the reactance, or the intrinsic property of the material, the capacitance, which is independent of frequency. Each of the different measures may be suffixed with "-ivity" to yield its general property. Finally, the measure given may refer to the complex impedance rather than the in or out of phase component.

Not all permutations, fortunately, are widely used. These are the most common: The conductivity σ is the inverse of resistivity and is given in Siemens/m or, more usually, S/m. The admittance is the inverse of impedance, and so is a combined measure of in and out of phase ease of passage of current through a tissue. The capacitance of the tissue is the capacity to store charge, and is given in Farads. The permittivity of a tissue is the property of a dielectric material that determines how much electrostatic energy can be stored per unit of volume when unit voltage is applied and is given in F/m. The relative dielectric constant ε_r is the permittivity relative to a vacuum, and indicates how much greater the capacitance of a capacitor would be if the sample was placed between the plates compared to a vacuum.

2.2.4   IMPEDANCE MEASUREMENT

The impedance of samples is usually recorded with Ag/Ag-Cl electrodes. The simplest arrangement is to place electrodes at either end of a cylindrical or cuboidal sample of the tissue. A constant current is passed and the impedance is calculated from the measured voltage (figure 2.6ab). The drawback of this method is that the impedance measured includes not only the tissue sample but also that of the electrodes. The method can be reliable, but requires that a calibration procedure is performed first to establish the electrode impedance. These then need to be subtracted from the overall impedance

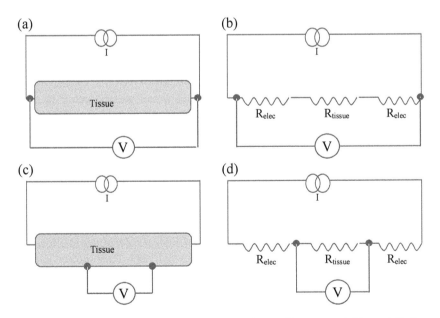

Figure 2.6   (a) The two-electrode measurement as a block diagram and (b) modelled as a simple electrical circuit. The two overlapping rings represent a constant current electrical source. (c) The four-electrode measurement as a block diagram and (b) modelled as a simple electrical circuit.

recorded, and it should be a fair assumption that the electrode impedances do not change between the calibration and test procedures.

Impedance is best measured using four electrodes, as this circumvents the error of inadvertent inclusion of the electrode impedance with two terminal recordings. The principle is that constant current is delivered to the electrodes through the two current electrodes; as it is constant, the correct current is independent of the electrode impedance. The voltage is recorded by high performance modern amplifiers, which are not significantly affected by the series electrodes impedance between the sample and amplifier (figure 2.6cd). As a result, the impedance is ideally unaffected by electrode impedance, although non-idealities in the electronics may cause inaccuracies in practice. The main drawback of this method is that the geometry of the sample is no longer clear cut, so that conversion to resistivity needs careful modelling of the path of current flow through the tissue.

2.2.5 RELEVANCE TO ELECTRICAL IMPEDANCE TOMOGRAPHY

Most systems which have been used to make clinical and human EIT measurements record both the in and out-of phase components of the impedance but only utilise the in-phase, resistive, component of the impedance. This is because unwanted capacitance in the leads and electronics introduce errors. Fortunately, these are all out-of-phase and so can be largely discounted by throwing away the out-of-phase data. For the same reason, images are generated of differences over time, as subtraction like this minimises errors. As a result, clinical EIT images are differences or a unitless ratio between the reference and test image data at a single frequency. Systems have also been constructed and tested which can measure at multiple frequencies, and provide absolute impedance data. As these are validated, and come into wider clinical use, then we may expect to see more absolute bioimpedance parameters, such as resistivity, admittivity, centre frequency, or ratio of extra- to intracellular resistivity, in EIT image data.

2.3 INTRODUCTION TO BIOMEDICAL EIT

2.3.1 HISTORICAL PERSPECTIVE

The first published impedance images appear to have been those of Henderson and Webster in 1976 and 1978. Using a rectangular array of 100 electrodes on one side of the chest earthed with a single large electrode on the other side, they were able to produce a transmission image of the tissues. Low conductivity areas in the image were claimed to correspond to the lungs. The first clinical impedance tomography system, then called Applied Potential Tomography (APT), was developed by Brian Brown and David Barber and colleagues in the Department of Medical Physics in Sheffield They produced a celebrated commercially available prototype, the Sheffield Mark 1 system, which was been widely used for performing clinical studies, and largely initiated interest in the field. It has been replaced by systems with more advanced electronics and image reconstruction but most systems in current use in clinical studies employ similar basic principles. This system made multiple impedance measurements of an object by a ring of 16 electrodes placed around the surface of the object. A constant current was applied to one pair of adjacent electrodes at a time. The first published tomographic images were from this group in 1982 and 1983. They showed images of the arm in which areas of increased resistance roughly corresponded to the bones and fat. As EIT was developed, images of gastric emptying, the cardiac cycle, and the lung ventilation cycle in the thorax were obtained and published. The Sheffield EIT system had the advantage that 10 images could be obtained a second, the system was portable and the system was relatively inexpensive compared to ultrasound, CT and MRI scanners. However, since the EIT images obtained were of low resolution compared to other clinical techniques such as cardiac ultrasound and X-ray contrast studies of the gut, EIT did not gain widespread clinical acceptance. Since the first flush of interest in the mid to late 1980's, about thirty groups and several commercial companies have developed their own EIT systems and reconstruction software. Initial interest in a wide range of applications at first has now settled into the main areas of imaging lung ventilation and perfusion, cardiac function, and brain and nerve function and pathology.

2.3.2 EIT INSTRUMENTATION

EIT systems are generally about the size of a shoe box, but some may be larger. They usually comprise a box of electronics and a PC. Connection to the subject is usually made by cables a metre or two long, and ECG type electrodes are placed in a belt or single on the body part of interest. All will sit on a movable trolley, so that recording can be made in a clinic or out-patient department. Typical systems are shown in Chapter 20.

2.3.2.1 Individual Impedance Measurements

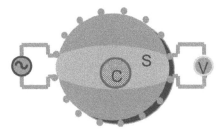

Figure 2.7 Typical single impedance measurement with EIT. Sixteen electrodes as placed on a circular body. The current source ($\sim$) applies a current of 5 mA at 100 kHz which is measured at V. Area of subject which is most sensitive, S; change at C affects voltages measured on the elecrodes. (color figure available in eBook)

A single impedance measurement forms the basis of the data set which is used to reconstruct an image. Most systems use a four electrode method, in which constant current is applied to two electrodes, and the resulting voltage is recorded at two others. This minimises the errors due to electrode impedance. The transfer impedance of the subject with this recording geometry is calculated using Ohm's law (figure 2.7). The current applied is approximately one tenth of the threshold for causing sensation on the skin. It is insensible and has no known ill effects. Most single frequency systems apply a current at about 50–200 kHz. At this frequency, the properties of tissue are similar to those at DC, in that the great majority of current travels in the extracellular space, but electrode impedance is much lower than at DC, so there are less instrumentation errors. At 50 kHz, a single measurement usually takes less than 1 msec.

The electronics for this four electrode arrangement comprises a current source, a voltage recording circuit, and a means to extract the impedance information from the acquired voltage. In the past, this was achieved using an electronic circuit but is now usually achieved by digital processing of the recorded converted voltage data. The phase of the injected current is known; the processing retrieves the value of the received waveform both in-phase with the applied current and with a phase delay of 90°. In this way, the resistance and reactance may be calculated (figure 2.8). Many systems discard the out-of-phase component, as it may be inaccurate due to effects of stray capacitance. Systems generally use multiple circuits for drive and receive. In principle, current may be injected into just one pair of electrodes at a time and then switched to other pairs, or it could be applied to multiple electrodes at the same time. Most current systems still apply current to just one pair of electrodes at a time.

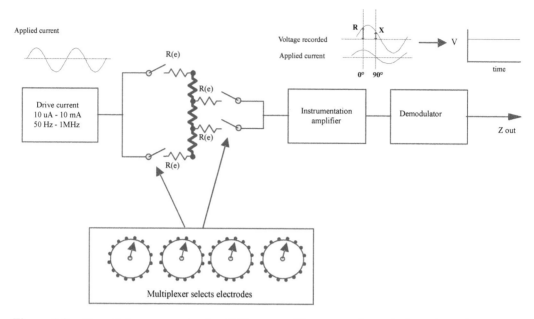

Figure 2.8 Essential components of an EIT system. The system shown is for a single impedance measuring circuit with connection to electrodes using a multiplexer. More complex systems may have multiple circuits attached directly to electrode pairs. The demodulator converts the AC recorded signal into a steady DC voltage for both resistance and reactance, although the reactance signal is discarded in many systems as the stray capacitance renders it inaccurate. The subject and electrode impedances (R(e)) are represented as resistances.

It will be seen below that EIT images in human subjects suffer from a relatively low resolution of about 10% of the diameter of the electrode ring. Part of this is due to an inherently limited resolution from the imaging procedure. Another cause is errors in individual measurements. The principal of these is a high skin-electrode impedance. In principle, measurement should be accurate with a four electrode system. Unfortunately, in practice, this is not the case. It may be necessary to abrade the skin of subjects to lower the impedance, and this can easily vary from site to site. Although leads are often coaxial, and may have driven screens to minimise stray capacitance, this is significant, especially at higher frequencies. The combination of variable skin impedance and stray capacitance conjoin to cause significant errors in recorded impedance values, especially in electrode combinations which are recording small voltages. Significant factors include fluctuations in current delivered, if skin impedances vary at different electrodes, and common mode errors on the recording side due to impaired common mode rejection as a result of stray capacitance (figure 2.9).

2.3.2.2 Data Collection

EIT systems may employ up to hundreds of electrodes but most used for human clinical studies still use a single ring of 16. Current systems may use several rings on the thorax or evenly distributed, for example, over the head. The following describes the procedure employed by the Sheffield Mark 1 system, which is still useful for explaining the basic principles. 16 electrodes are applied in a ring. A single measurement is made with four electrodes. A current of up to 5 mA at 50 kHz is applied between an adjacent pair of electrodes, and the voltage difference is recorded from two other adjacent electrodes. This yields a single transfer impedance measurement. Only the in-phase component of the voltage is recorded, so this is a recording of resistance, rather than impedance.

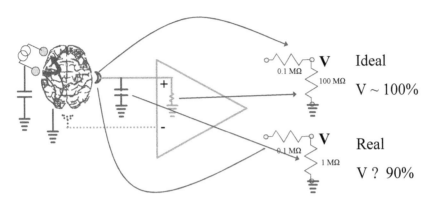

Figure 2.9   Sources of error in impedance measurements. There are two main sources of error. 1) A voltage divider exists, formed by the series impedance of the skin and input impedance of the recording instrumentation amplifier. Under ideal circumstances, the skin impedance is negligible compared to the input impedance of the amplifier, so that the voltage is very accurately recorded (upper example). In this example, skin impedance is 100 kΩ and input impedance is 100 MΩ, so the loss of signal is negligible. In practice, the stray capacitance in the leads, coupled to high skin impedances, may cause a significant attenuation of the voltage recorded – for example to 90%, if the input impedance reduces to 1 MΩ (lower example). In this diagram, only one side of a differential amplifier is shown, for clarity. This attenuating effect may be different for the two sides of the amplifier. This leads to a loss of common mode rejection ability, as well as absolute errors in the amplitude recorded. 2) The ideal current source is perfectly balanced, so that all current injected leaves by the sink of the circuit. The effect of stray capacitance and skin impedance may act to unbalance the current source. Some current then finds its way to ground, either by the ground, or by the high input impedance of the recording circuit. This causes a large common mode error. The common mode rejection ratio may be poor because of the effects in (1), so that the recorded voltage is inaccurate. (color figure available in eBook)

Voltage signals are measured on all other electrodes in turn (figure 2.10).

Sequential pairs are then successively used for injecting current until all possible combinations have been measured. Each individual measurement takes less than a millisecond, so a complete data set of 208 combinations could be collected in 80 mec, and 10 images per second acquired. Each data set comprised 104 measurements. Many different designs have been constructed and reported since. Most systems now record voltages simultaneously on all electrodes.

In theory, greater resolution within the image can be obtained if current is injected from many electrodes at once. This may be injected in different combinations to give fixed patterns of increasing spatial frequency, as in designs from groups at the Rensellaer Polytechnic (RPI), New York, USA, Oxford Brookes University, Oxford, UK, or Dartmouth, USA. Although these approaches are better in theory, this requires much greater precision as all the current sources have to be controlled accurately at once; it is not yet clear if, in practice, this confers an improvement in image quality over the simpler method of applying current only to two electrodes at a time. Other variations in hardware design include applying voltage and measuring current, using only two rather than four electrodes for individual measurements as in the RPI system, or recording many frequencies simultaneously – multifrequency EIT or EIT spectroscopy (EITS).

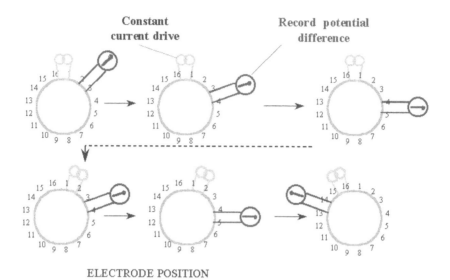

ELECTRODE POSITION

Figure 2.10   Data acquisition with the Sheffield Mark 1 system. A constant current is injected into the region between two adjacent electrodes, and the potential differences between all other pairs of adjacent electrodes are measured. The current drive is then moved to the next pair of adjacent electrodes and the measurements repeated, and so on for all possible current drive pairs. I, it is not possible to measure potential differences accurately at the pair of electrodes injecting current so there are 208 (13×16) measurements in a data set. (color figure available in eBook)

2.3.3   ELECTRODES

The great majority of clinical measurements have been made with ECG type adhesive electrodes attached to the chest or abdomen. Although the four electrode recording system should in theory be immune to electrode-skin impedance, in practice it is usually necessary to reduce the skin impedance first by abrasion. Similar EEG cup electrodes have been used for head recording. Errors in impedance measurement may occur from stray capacitance in leads connected to the electrodes. Some designs therefore employ electronic buffer amplifiers placed close to the recording electrodes – this reduces the length of lead prey to stray capacitance and are termed "active electrodes".

2.3.4   SETTING UP AND CALIBRATING MEASUREMENTS

Data collection in human subjects in EIT is sensitive to movement artefact and the skin-electrode impedance. It is therefore usually necessary to check signal quality before embarking on recordings. A simple widely used method is to check electrode impedance. Another method is to measure reciprocity. This principle is that the recorded transfer impedance should be the same, under ideal circumstances, if the recording and drive pair are reversed. A low reciprocity ratio – usually below 80% – generally indicates poor skin contact, which can be corrected by further skin abrasion or repositioning of the electrodes. Other systems, especially those using two, rather than four, electrodes may require special trimming before recording. Another potential problem lies in determining the correct zero phase setting for the impedance measuring circuit. The phase of the current produced by the electronics is, of course, accurately known, but stray capacitance and skin impedance may interact to alter the zero phase of the current delivered to the subject and similar effects on the recording side may also alter the phase of the signal delivered to the demodulator. Different approaches have been employed. One method is to calibrate the system on a saline filled tank. Others are to optimise the reciprocity, or to assume that the subject is primarily resistive at low frequencies, and adjust the phase detection accordingly.

For newly constructed systems, it is helpful to calibrate them on known test objects. Some employ agar test objects, impregnated with a saline solution, in a larger tank which contains saline of a different concentration. These can be accurate if images are made quickly, but the saline will diffuse into the bathing solution, so that the boundaries can become uncertain. Others have employed a porous test object such as a sponge, immersed in the bathing solution in a tank, so that the impedance contrast is produced by the presence of the insulator in the test object. Many tanks have been cylindrical; more realistic ones have simulated anatomy, such as the head, or used biological materials to produce multifrequency test objects. Typically, the spatial resolution of test objects in tanks is about 10% of the image diameter (figure 2.11).

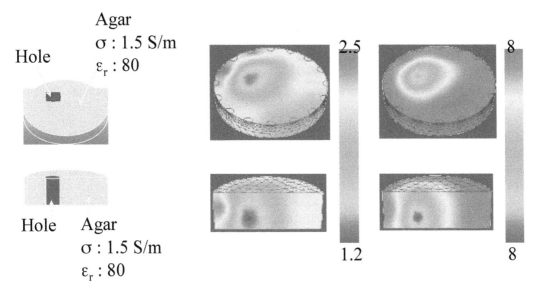

Figure 2.11   Example of image quality with multifrequency EIT system from Dartmouth, USA. (Courtesy of Prof. A. Hartov). (color figure available in eBook)

2.3.5   DATA COLLECTION STRATEGIES

Most EIT work has used EIT as a difference imaging method, in which images of the impedance change compared to a baseline condition are obtained. An example is EIT of gastric emptying. A reference baseline image is obtained at the start of the study when the stomach is empty. The stomach is then filled when the subject drinks a conductive saline solution. Subsequent EIT images are reconstructed with reference to the baseline image and demonstrate the impedance change as the stomach fills and then empties the conductive solution. A second example is of cardiac imaging: images are gated to the electrocardiogram (ECG) to demonstrate the change in impedance during systole, when the heart if full of blood in the cardiac cycle compared to a reference baseline image when the heart is emptied of blood in diastole. To image ventilation, a reference image is obtained when the lungs are partially emptied of air at the end of expiration and EIT images of the changes during normal ventilation are reconstructed with reference to the baseline image.

The main reason for imaging difference impedance changes is to eliminate or reduce errors that occur due to the instrumentation or differences between the model of the body part used in the reconstruction software and the actual object imaged. To reduce these, impedance changes are reconstructed with reference to a baseline condition; if the electrode placement errors in the baseline images and the impedance change images are the same, then these errors largely cancel if only impedance change is imaged. Although the difference imaging approach minimises reconstruction errors, it limits the application of EIT to experiments in which an impedance change occurs over a short experimental time course; otherwise electrode impedance drift may introduce artefacts in the data which cannot be predicted from the baseline condition. As difference imaging cannot be used to image objects present at the start of imaging and therefore in the baseline images, difference EIT cannot be used to obtain images of tumours or cysts. This contrasts with images obtained with CT, which can obtain static images of contrasting tissues such as tumours. Difference imaging has been used for almost all clinical studies to date in all areas of the body.

In principle, it should be possible to produce images of the absolute impedance. Unfortunately, image production is sensitive to errors in instrumentation and between the model used in reconstruction and the object imaged. Although there are some validated absolute EIT images in computer simulation and tanks, there are not yet any validated clinical data series, presumably because minor discrepancies between the actual anatomy and electrode placement are too great and image quality degrades.

Difference EIT images typically use one measurement frequency, usually at about 100 kHz for chest imaging, to make impedance measurements. An alternative approach is to compare the difference between impedance images measured at different measurement frequencies, termed EITS (EIT spectroscopy). This technique exploits the different impedance characteristics of tissues at different measurement frequencies. An example of such a contrast would be the difference between cerebrospinal fluid (CSF) and the grey matter of the brain. As the CSF is an acellular, ionic solution, it can be considered a pure resistance, so that its impedance is identical and equal to the resistance for all frequencies of applied current. However, the grey matter, which has a cellular structure, has a higher impedance at low frequencies than at high frequencies This frequency difference can theoretically be exploited to provide a contrast in the impedance images obtained at different frequencies, and provide a means of identifying different tissues in a multifrequency EIT image.

2.3.6   EIT IMAGE RECONSTRUCTION

2.3.6.1   Backprojection

The hardware described above produces a series of measurements of the transfer impedance of the subject. These may be transformed into a tomographic image using similar methods to X-ray CT. The earliest method, employed in the Sheffield Mark 1 system, is most clear intuitively. Each measurement may be conceived as similar to an X-Ray beam – it indicates the impedance of a volume between the recording and drive electrodes. Unfortunately, unlike X-rays, this is not a neat defined beam, but a diffuse volume which has graded edges. Nevertheless, a volume of maximum sensitivity may be defined. The change in impedance recorded with each electrode combination is then back projected into a computer simulation of the subject – a 2D circle for the Sheffield Mark 1. The back projected sets will overlap to produce to give a blurred reconstructed image, which can then be sharpened by the use of filters.

2.3.6.2   Sensitivity Matrix Approaches

Most EIT systems now employ a more powerful method, based on a "sensitivity matrix" (figure 2.12). This is based on a matrix, or table, which relates the resistivity of each voxel in the subject, and hence, images, to the recorded voltage measurements.

The method requires a mathematical model of the body part of interest. These may be modelled using mathematical formulae alone – these are termed "analytical" solutions. In general, these are only practical for simple shapes, such as a cylinder or sphere. More realistic shapes, such as the

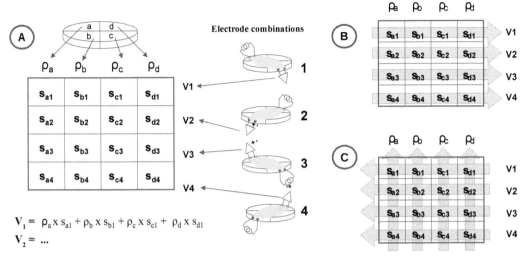

$$V_1 = \rho_a \times S_{a1} + \rho_b \times S_{b1} + \rho_c \times S_{c1} + \rho_d \times S_{d1}$$
$$V_2 = ...$$

Figure 2.12 Diagram of image reconstruction. (A) The sensitivity matrix. This is shown figuratively for a subject with four voxels and four electrode combinations. Each column represents the resistivity of one voxel in the subject. Each row represents the voltage measured for one electrode combination. The current from one current source flows throughout the subject, but the voltage electrodes are most sensitive to a particular volume, shown in grey. The resulting voltage is a sum of the resistivity in each of the voxels weighted by the factor S for each voxel, which indicates how much effect that voxel has on the total voltage. (B) The forward case. In a computer program, all the sensitivity factors are calculated in advance. Given all the resistivities for each voxel, the voltages from each electrode combination are easy to calculate. (C) The inverse. For EIT imaging, the reverse is the case – the voltages are known; the goal is to calculate all the voxel resistivities. This can be achieved by "inverting" the matrix. This is straightforward for the simple case of four unknowns shown here, but is not in a real imaging problem, where the voltages are noisy, and there may be many more unknown voxels than voltages measured.

thorax or head containing layers representing the internal anatomy, are achieved using imaginary meshes in the model, whose boundaries are determined by segmenting MRI or CT images. The equations of current flow are solved for each cell in the mesh; each cell's calculation is therefore simple, but solutions for the whole mesh, which may contain millions of cells, may be time consuming on even powerful computers, and may suffer from instability or hidden quantitative errors. These are termed "numerical" methods and common mesh types are FEM (finite element mesh) or BEM (boundary element mesh).

Using one of these models, the expected voltages at each electrode combination can be calculated. The principle is that the applied current actually flows everywhere in the subject, but, clearly, flows more in certain regions than others. Each voxel in the subject contributes to the voltage measured at a specified recording pair, but this depends on the resistance in the voxel, the amount of current which reaches it, and its distance from the recording electrodes. The total voltage at the recording pair is a sum of all these contributions from every voxel. Many of these, from voxels far away, may be negligible. This is illustrated in figure 2.12a for the case of a disc with just four voxels. In practice, for 16 or 32 electrode systems, several hundred electrode combinations are recorded, so the matrix will have several hundred rows. In principle, an image can only be accurately reconstructed if there is one independent measurement for each voxel. In practice, accurate anatomical meshes need to contain many more cells than a few hundred, especially if in 3D, so the matrix may contain many thousands or even millions of columns – one for each voxel – and a few hundred rows. If the resistivities of each voxel are given, then the expected voltages for each electrode combination may be easily calculated. This is termed the "forward" solution and is a simulation of the situation in reality (figure 2.12b). Its use is to generate a "sensitivity matrix". This is produced by, in a computer simulation, varying resistivity in each voxel, and recording the effect on different voltage recordings. This enables calculation of the sensitivity of a particular voltage recording to resistance change in a voxel – the "s" factor in figure 2.12.

To produce an image, it is necessary to reverse the forward solution. On collecting an image data set, the voltages for each electrode combination are known, and, by generating the sensitivity matrix, so is the factor relating each resistance to these. The unknown is the resistivity in each voxel. This is achieved by a process equivalent to inverting the matrix – which now takes measured impedances and yields all the resistivities (figure 2.12c). In principle, this can give a completely accurate answer, but this is only the case if the data is infinitely accurate, and that there are the same number of unknowns – i.e. voxels requiring resistance estimates, as electrode combinations. In general, none of these is true. In particular, in many of the voxels, very little current passes through, so the sensitivity factor for that cell in the table is near to zero. Just as dividing by zero is impossible, dividing by such very small numbers causes instabilities in the image. This is termed an "ill-conditioned" matrix inversion. There is a well established branch of mathematics which deals with these inverse problems, and matrix inversion is made possible by "regularising" the matrix. In principle, this is performed by undertaking a noise analysis of the data – noisy channels with little signal to noise are suppressed, so that the image production by inversion relies on electrode combinations with good quality data and so proceeds smoothly. Commonly used methods for this include truncated singular value decomposition or Tikhonov regularization. In practice, it is usually possible to produce EIT images with time difference regularised methods with a linear assumption with a spatial resolution of about 10% of the diameter of the ring or rings of recording electrodes (figure 2.13).

2.3.6.3 Other Developments in Algorithms

Initially, reconstruction was always performed with the assumption that the subject was a two dimensional circle. Although this actually worked quite well in practice, changes in impedance away from the plane of electrodes could be seen in the image, sometimes in an unpredictable way. 3D recording requires far more electrodes – such as four rings of 16 per ring around the chest. For simplicity, many continuing clinical studies still use a 2D approach. The first 3D images were of the chest in 1996 and in the head since 2003.

The sensitivity matrix approach described above requires an assumption that there is a direct unvarying, or "linear" relationship between the resistance of a voxel and its effect on recorded voltage. In practice, this is almost true for small changes in impedance below about 20%. However, it is not true for larger changes. This can be overcome by using more accurate non-linear approaches. This can be achieved by using a logical loop in the algorithm. A guess is made for the initial resistivities in the voxels. The forward solution is calculated to estimate the resulting electrode combination voltages. These are then compared with the original recorded voltage data. The resistances in the model are then adjusted, and the procedure is repeated continuously until the error between the calculated and recorded voltages is minimised to an acceptable level. In theory, this should give more accurate images, but it is time consuming in reconstruction and instabilities may creep in as the process is more sensitive to minor errors, such as anatomical differences between the mesh used and the subject's true anatomy, or the position of electrodes. Although there is interest in the development of non-linear approaches, there are not yet any validated studies using this approach in human subjects.

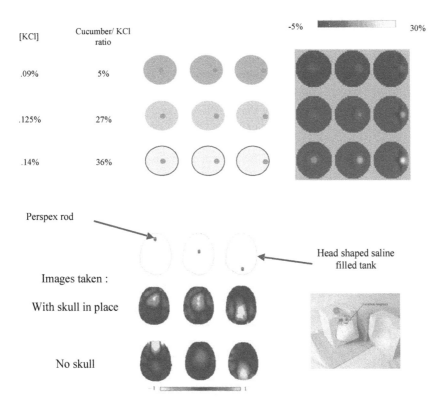

Figure 2.13 Calibration studies with a 16 electrode system in a saline filled tank. The tank was filled with saline, which was varied to give different contrasts with the test object of a cucumber. The cucumber may be seen in the correct location for all contrasts, but with more accuracy and greater change at near the edge. b) Images taken with 3D linear algorithm in a latex head-shaped tank, with or without the skull in place. The algorithm employed a geometrically accurate finite element mesh of the skull and tank. (color figure available in eBook)

2.3.7 CURRENT DEVELOPMENTS

This didactic overview has covered applications with conventional EIT. There are now well validated clinical data series demonstrating its quality and potential value in imaging lung ventilation and perfusion which are the main focus of current interest (Chapters 10,11). Research is currently active into its use for imaging cardiac output, cancer, and brain and nerve function and pathology but clinical data series have yet to be acquired (Chapters 11,12,13). Initial interest in gastric emptying, hyperthermia or pelvic structures have not evolved into active clinical research (Chapter 14). There are also related techniques of potential value, which are still in technical development, but have not yet been used for clinical studies. Magnetic induction tomography (MIT, Chapter 16) is similar in principle to EIT but injects and records magnetic fields from coils. It has the advantages that the position of the coils is accurately known, and there is no skin-electrode impedance, but the systems are bulkier and heavier than EIT. In general, higher frequencies have to be injected in order to gain a sufficient signal-to-noise ratio. Until now, spatial resolution has been the same or worse than EIT. The method could offer advantages in imaging brain pathology, as magnetic fields pass through the skull, and may in the thorax or abdomen if the method can be developed to demonstrate improved sensitivity over EIT. MR-EIT (magnetic resonance-EIT, Chapter 17) requires the use of an MRI scanner. Current is injected into the subject and generates a small magnetic field which alters the MRI signal. The pattern of resistivity in three dimensions may be extracted from the resulting changes in the MRI images. This therefore loses the advantage of portability in EIT but has the great advantage of the high spatial resolution of MRI. It could be used to generate accurate resistivity maps for use in models for reconstruction algorithms in EIT, especially for brain function, where prior knowledge of anisotropy is important.

Biomedical EIT has evolved substantially since the first edition of this book in 2005. Almost all clinical studies have been undertaken with variants of the 2D Sheffield Mark 1 system. The most promising applications appear to be in optimisation of ventilator settings in ventilated patients, and perhaps in imaging brain activity in normal function and epilepsy. Four companies are currently actively developing and marketing EIT for medial use in imaging lung function. However, EIT has yet to fulfil its promise in delivering a robust and widely accepted clinical application. Well funded clinical trials are in progress in the above applications, and there seems to be a reasonable chance that one or more, especially if using improved technology, may prove to be the breakthrough.

Section II

EIT: Tissue Properties to Image Measures/Chapter

3 Electromagnetic Properties of Tissues

Rosalind Sadleir
School of Biological and Health Systems Engineering, Arizona State
University, Tempe, AZ, USA

Camelia Gabriel
C. Gabriel Consultants, San Diego, CA, USA

CONTENTS

DOI: 10.1201/9780429399886-3

3.1 WHAT UNDERLIES TISSUE ELECTROMAGNETIC PROPERTIES?

The electrical properties of body tissues are characteristic of their cellular structure and composition (biological molecules and electrolytes in aqueous solution). These properties also depend on membrane characteristics. Tissue properties can be summarized using electrical conductivity and electrical permittivity values and their related spectra. The magnetic permeability of most biological materials is close to that of free space, which implies very weak interactions with the magnetic component of electromagnetic fields at low field strengths. The most significant magnetic contrasts in body tissues are related to presence of paramagnetic iron (in ferritin and deoxygenated hemoglobin) and calcium [1120].

The response of tissue to an incident sinusoidal electromagnetic field can be considered as either mostly conductive, or mostly capacitive and modelled from either perspective by summarizing its properties with complex quantities. Depending on the mixture, one approach may be more natural than another. For example, more solid tissues tend to have larger charge storage capability or polarizability and may be more appropriately considered using permittivity rather than conductivity measures. In either case, the property characterizing the tissue is a complex quantity. More resistive or dissipative (conductive) or energy storage (capacitive) properties may be measured using in-phase or quadrature measurements respectively.

The dependencies of real and quadrature (imaginary) tissue electrical properties on frequency are related via the Kramers-Kronig relations, therefore characteristics of conductivity spectra may be intuited from permittivity spectra, and vice versa. Tissue properties result from multiple influences including tissue heterogeneity, interfacial effects and directionality (anisotropy). The compound nature of these properties means that special electrical components must be introduced to best model tissue spectra.

At the frequencies most commonly used in EIT (below 250 kHz), phase shifts resulting from capacitive tissue characteristics are typically very small and signals are almost all related to conductive properties. The conductive properties of tissues are mostly due to the presence of electrolytes in body fluids, and those at higher frequencies are broadly related to the concentration of membrane interfaces and the presence of polar molecules.

As in most materials, passive tissue properties are linear (not dependent on the current or field applied to measure them) but only up to a limit, beyond which properties may start to be temporarily or irreversibly changed. Non-linear interactions may occur at high field intensities, the threshold for such effects is system and frequency dependent but known to be in excess of 10^6 V/m. Dielectrophoresis and electroporation are examples of non-linear electrical phenomena that are the focus of applications in biotechnology but are outside the scope of this chapter.

In the sections below we consider ionic contributions to tissue conductivities, followed by reactive and frequency-dependent characteristics. Commonly observed features of conductivity and permittivity spectra will be considered and properties of tissues important to EIT applications will be outlined. Methods and important considerations for making "benchtop" measurements of tissue properties for verification of EIT methods will also be summarized, as they are important for reconstruction validation.

3.1.1 IONIC CONDUCTIVITIES

Because the water content of tissues is significant, their properties are closely related to those of electrolytes in solution. The most common ions in the body are sodium and chloride, with other ions such as potassium, bicarbonate and calcium and magnesium also contributing to the intra- or extracellular environments. Properties of electrolytic solutions were investigated extensively in the nineteenth century.

The Kohlrausch law describes the dependence of the molar conductivity (conductivity per unit of concentration) on concentration. At low concentrations, this experimental data agrees very well with the resistivity values that may be calculated using the expression

$$\Lambda_m = \Lambda_m^0 - K\sqrt{c} \tag{3.1}$$

where Λ_M is the molar conductivity [S m^{-1} mol^{-1} L] and c is the electrolyte concentration in mol/L. K is the Kohlrausch coefficient, a value that depends on the solvent and the charge on ionic species. The molar conductivity of the salt at infinite dilution is Λ_m^0. The molar concentration of a solution and its conductivity κ are related via

$$\kappa = \Lambda_m c \tag{3.2}$$

Following (3.2) the conductivity of a solution broadly increases linearly with molar concentration of electrolyte, but this dependence progressively decreases as concentration increases. Figure 3.1 below shows the conductivity properties of NaCl in solution (saline) as a function of molar concentration at 25 C. Data from this plot were obtained from published experimental data[449]. In the case of NaCl dissolved in water, the value for K is $89.14 \times 10^{-4} \, \Omega^{-1} m^2 mol^{-1}/(molL^{-1})^{-0.5}$. The molar conductivity at infinite dilution at 25 C is approximately $126.4 \times 10^{-4} \, \Omega^{-1} m^2 mol^{-1}$. The predicted conductivity starts to underestimate the measured values at a molar concentration of around 0.3 mol/l (17 g/l) which is also about twice the NaCl concentration of normal (0.9% or 9 g/l) saline solution. Note that the graph indicates that at very high molarities the conductivity predicted by (3.1) and (3.2) actually begins to decrease, which is because this approximation breaks down as concentration becomes high.

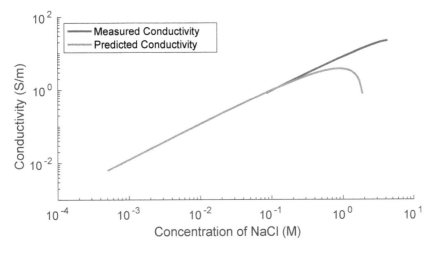

Figure 3.1 Dependence of conductivity of sodium chloride in aqueous medium on molar concentration at 25 C. The red line shows the predicted dependence of conductivity following (3.1) and (3.2). The blue line plots experimental measurements of saline conductivity. (color figure available in eBook)

The conductivity of an aqueous salt solution will also tend to increase with increasing temperature, as water viscosity decreases at higher temperatures. The decreasing viscosity leads to an increase in ionic mobility and hence in the resulting molar conductivity. This trend underlies the observed temperature dependence of tissue conductivity. Measurements of tissue properties, if not made in vivo, should therefore also specify the ambient temperature.

Conductivity measurements made using dielectric probes or conductivity cells may be calibrated using salt solutions. For example, Gabriel et al[322] present measurements of tissue conductivities and permittivities calibrated over the range 10 Hz–20 GHz based on dielectric probe calibrations performed using 5 mM saline solutions. Measurements of saline properties over a range of concentrations, frequencies and temperatures, as reported in Peyman[818] may also assist calibration.

3.1.2 MEMBRANES AND SOLID TISSUES

An illustration of the interaction of cells within tissue and a vertically applied electric field is shown in figure 3.2. As current flows through tissue, the electric properties measured vary because of the cell density and architecture, and ionic mobilities around this architecture. The relatively non-conducting membranes interrupt direct conduction, most significantly at low frequencies. The denser the cells are packed, the more important are these effects and solid tissues demonstrate more dielectric-like properties. The way that measurement currents are applied relative to the tissue architecture also influences the reported electric properties.

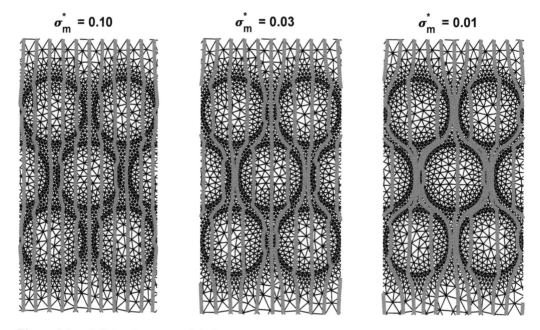

$$\sigma^*_m = 0.10 \qquad\qquad \sigma^*_m = 0.03 \qquad\qquad \sigma^*_m = 0.01$$

Figure 3.2 A finite element model of packed spherical cells in a uniform bath showing the current streamlines through a slide for vertically voltage gradient. The conductivity within cells is equal to the medium. The relative complex conductivity σ^*_r of the cell membrate is indicated for each case. (color figure available in eBook)

In our description of how properties may be represented, we follow the convention established by Grimnes and Martinsen[387]. The complex conductivity, $\boldsymbol{\sigma}$, consists of real and imaginary parts σ' and σ''

$$\boldsymbol{\sigma} = \sigma' + i\sigma''. \tag{3.3}$$

Similarly, the complex permittivity can be expressed as

$$\boldsymbol{\varepsilon} = \varepsilon' + i\varepsilon'', \tag{3.4}$$

where $\varepsilon' = \varepsilon'_r \varepsilon_0$, $\varepsilon'' = \varepsilon''_r \varepsilon_0$, and ε_0 is the permittivity of free space (8.8542×10^{-12} F/m). ε'_r and ε''_r are dimensionless relative parameters characteristic of the material.

If a measured electrical property is assumed to be capacitive in nature, it may be expressed as a complex admittance $\mathbf{Y} = i\omega\mathbf{C}$, where $\mathbf{C}$ is the complex capacitance. $\mathbf{Y}$ can alternatively be expressed as

$$\mathbf{Y} = G + i\omega C = \frac{A}{D}(\sigma' + i\omega\varepsilon'), \tag{3.5}$$

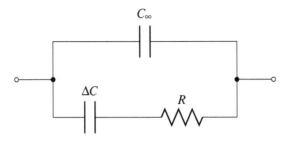

Figure 3.3 Example Debye dispersion circuit consisting of two capacitors and one resistor.

assuming the capacitor or electrodes have face area A, and thickness d. We assume that the ratio A/d is large and that there are negligible fringe field effects. The relationship between the complex conductivity and permittivity parameters can be found by substitution of (3.4) into (3.5),

$$\mathbf{Y} = i\omega\mathbf{C} = i\omega\frac{A}{d}\varepsilon = \omega\frac{A}{d}\varepsilon'' + i\omega\frac{A}{d}\varepsilon' \tag{3.6}$$

Comparing (3.5) and (3.6) leads to the identification

$$\varepsilon'' = \frac{\sigma'}{\omega}. \tag{3.7}$$

3.1.3 RELAXATION MODELS OF TISSUE PROPERTIES

A circuit that can be used to describe capacitive and conductive tissue properties over a small frequency range contains a capacitor in parallel with a series capacitor and resistor (figure 3.3).

It can be shown that if ω is the angular frequency and τ the time constant ($\tau = R\Delta C$), the complex capacitance of the circuit is given by

$$\mathbf{C} = C_\infty + \frac{\Delta C}{1 + i\omega\tau} \tag{3.8}$$

where C_∞ is the capacitance at high (∞) frequencies, or when $\omega\tau \gg 1$.

At low frequencies, when $\omega\tau \ll 1$, $C = \Delta C - C_\infty$, which makes $\Delta C = C_0 - C_\infty$ where the subscript (0) stands for the static (DC) value.

A parallel equation that can be written of terms of ε' and ε'' is

$$\boldsymbol{\varepsilon} = \frac{\varepsilon_0 - \varepsilon_\infty}{1 + i\omega\tau} = \varepsilon'(\omega) - \varepsilon''(\omega) \tag{3.9}$$

This relationship is known as the Debye equation; it applies whenever the decay of polarization is exponential and associated with a single relaxation time.

The frequency dependencies of the Debye parameters ε' and ε'' and measured impedances have the form illustrated in figure 3.4 below, for $R = 500\,\Omega$, $\varepsilon_{\infty,r} = 1000 = \varepsilon_\infty/\varepsilon_0$, $\Delta\varepsilon' = 10000 = \Delta Cd/A\varepsilon_0$, and a geometric factor (A/d) of 20 m.

The effective permittivity and conductivity of this model material varies as a function of frequency. The permittivity reduces as frequency increases and the conductivity increases, with the frequency of the transition occurring at the critical frequency at which $\omega\tau = 1$.

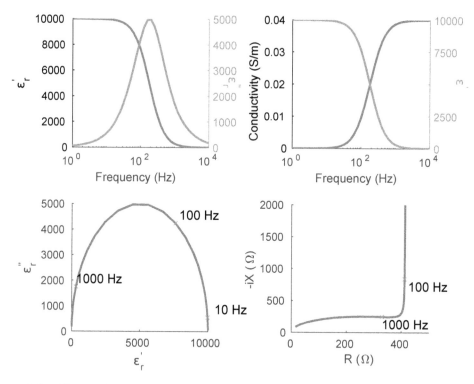

Figure 3.4   Characteristics of Debye circuit calculated with for $R = 500\,\Omega$, $\varepsilon_\infty, r = 1000$ $\Delta\varepsilon' = 10000 = \Delta Cd/A\varepsilon_0$, and (A/d) of 20 m. Part (A) shows characteristics of real (blue curve, left axis) and imaginary (red, right) components of relative permittivity as a function of frequency, (B) shows conductivity (blue, left) and real component of relative permittivity (red, right), (C) plots real and imaginary components of frequency, (D) plots real and imaginary components of effective impedance (R_z and $-X_z$). (color figure available in eBook)

Debye, or parallel or series simple resistor-capacitor models, involve ideal components as well as describing only a single dispersion. Naturally, properties of biological materials do not show the characteristics of pure resistors and capacitors; they experience multiple, overlapping polarizations better described in terms using a relaxation time distribution. Therefore, their properties are in turn better described with circuits containing components whose values all vary as a function of frequency. For most materials (at $\omega\tau \gg 1$) a power law dependence of the form $\frac{\omega^{m-1}}{\tau}$, with $m \neq 0$, applies for both $\varepsilon'(\omega)$ and $\varepsilon''(\omega)$, making the ratio $\varepsilon''(\omega)/\varepsilon'(\omega)$ frequency independent (constant phase). These dependencies are modelled with constant phase elements (CPEs), such that the overall phase of $\mathbf{Y}$ does not vary.

Instead of using a lumped resistor or capacitor, consider a parallel-component circuit where the admittance $\mathbf{Y}$ is described by

$$\mathbf{Y}_{cpe} = G_{cpe} + iB_{cpe} = (\omega\tau)^m(G_\boxed{1} + iB_\boxed{1}) = (\omega\tau)^m G_\boxed{1} + i\omega^m \tau^{m-1} C_\boxed{1} \tag{3.10}$$

where the susceptance, $B = \omega C$, m is a real number between 0 and 1 and τ is a frequency scaling factor. The notation $(\cdot)_\boxed{1}$ refers to value of G, B or C when $\omega\tau = 1$. Since both G and B scale with frequency in the same way, the constant phase value is

$$\phi_{cpe} = \tan^{-1}\frac{B}{G}. \tag{3.11}$$

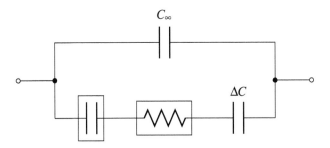

Figure 3.5 Cole-Cole Permittivity Model. The boxed resistor and capacitor symbols indicate constant phase element types arranged in series and having the combined property specified in (3.13). The capacitors C_∞ and ΔC describe capacitance values at high frequencies or changes in capacitance from low to high frequencies, respectively.

Using the CPE approach, both $G = (\omega\tau)^m G_{\square}$ and $B = (\omega\tau)^m B_{\square}$ increase with frequency. At 0 Hz, both Y and B are zero, therefore a DC admittance is not included in the model. Since $B_{\square} = \frac{1}{\tau} C_{\square}$ (recall $B = \omega C$ and $\omega = 1/\tau$ at $\square$), it also follows that $C = (\omega\tau)^{m-1} C_{\square}$ and that C decreases with frequency (since $m - 1 < 1$). As frequency increases, G tends toward ∞.

A special case of the CPE model was observed, in experimental data, by Fricke[314]; the frequency dependence of capacitance depends on a parameter, α, such that $C \propto f^{-\alpha}$, and that the constant phase angle is $\phi_{cpe} = \frac{1}{2}\alpha\pi$. In this case the model properties are described using the Fricke CPE as

$$Y_{cpe} = (\mathrm{i}\omega\tau)^\alpha G_{\square} = (\omega\tau)^\alpha G_{\square}\left(\cos\frac{\alpha\pi}{2} + \mathrm{i}\sin\frac{\alpha\pi}{2}\right) \tag{3.12}$$

As noted above, the general CPE of (3.10) has an infinite admittance at DC if $m \neq 0$. This is also true of the Fricke CPE in (3.12) if $\alpha \neq 0$. The parallel Fricke CPE can be modified to add a parallel frequency-independent conductance to modify this behaviour. Another behaviour can be derived if the two CPEs are combined in series. In this case the impedance of the CPE is

$$Z_{cpe} = (\mathrm{i}\omega\tau)^\alpha R_{\square} = (\omega\tau)^\alpha R_{\square}\left(\cos\frac{\alpha\pi}{2} - \mathrm{i}\sin\frac{\alpha\pi}{2}\right) \tag{3.13}$$

A special case of the Fricke CPE, the Cole-Cole model, involves a series Fricke CPE combination (3.13) with a fixed series capacitance and one parallel capacitance, as shown in figure 3.5 below. In this model, resistive and capacitive CPEs are shown in series in the lower arm of the parallel combination and are indicated with modified symbols. It is most appropriate for characterization of tissues with little DC conductance, or to describe higher frequency properties.

The admittance of this model is described by

$$\mathbf{Y} = \mathrm{i}\omega\mathbf{C} = \mathrm{i}\omega\frac{A}{d}\boldsymbol{\varepsilon} = \mathrm{i}\omega\frac{A}{d}\left(\varepsilon_\infty + \frac{\Delta\varepsilon}{1 + (\mathrm{i}\omega\tau)^{1-\alpha}}\right) \tag{3.14}$$

where we define $\Delta\varepsilon = \frac{d}{A}\Delta C$ and $\varepsilon_\infty = \frac{d}{A}C_\infty$.

The parameter α is a distribution parameter in the range $1 > \alpha \geq 0$. The factor ΔC describes the difference between capacitances of the of the CPE static element at low and high frequencies, that is $\Delta C = C_L - C_\infty$. From (3.4) and (3.5), the Cole-Cole dependence may written alternatively as

$$\boldsymbol{\varepsilon} = \varepsilon_\infty + \frac{\Delta\varepsilon}{1 + (\mathrm{i}\omega\tau)^{1-\alpha}}, \tag{3.15}$$

or

$$\mathbf{C} = C_\infty + \frac{\Delta C}{1 + (i\omega\tau)^{1-\alpha}}, \tag{3.16}$$

which is known as the Cole-Cole model; for $\alpha = 0$ it reverts to a Debye dispersion model.

The Cole-Cole models can be used to fit observed tissue measurements to the parameters α, $(\cdot)_\infty$, $\Delta(\cdot)$ and τ in (3.14), (3.15), or (3.16). This may be used to characterize tissue properties over a wide frequency range. A plot of the dependence of parameters on frequency for a tissue having a single dispersion characterized by a Cole-Cole model with $\Delta\varepsilon = 10000$, $\tau = 1$ ms and $\alpha = 0.3$ is shown in figure 3.6. It may be compared with matching parameters for the Debye dispersion in figure 3.4. Note that the low frequency properties shown in figure 3.6(D) are markedly different from those in figure 3.4(D).

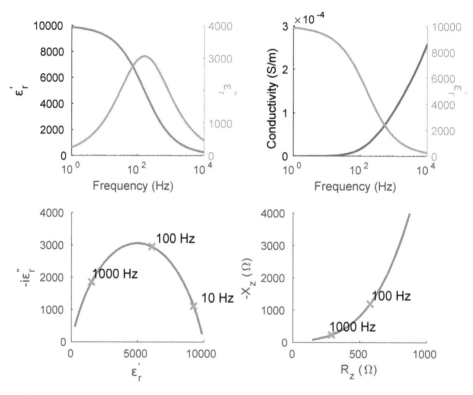

Figure 3.6   Cole-Cole dependency for tissue with $\Delta\varepsilon = 10000$, $\tau = 1$ ms, and $\alpha = 0.3$. (color figure available in eBook)

In real tissue, several characteristic relaxations are observed, with transitions (dispersions), typically in the Hz, kHz and MHz ranges. The processes involved in each transition are presumed to be related to interfacial processes at the different scales involved and are of course also each due to a mixture of possible relaxation times and processes. The transitions are labelled alpha, beta and gamma dispersions at lower, intermediate and higher frequency ranges, respectively. In addition, all tissues show a strong water-related dispersion in the GHz range.

To characterize individual dispersions, parameters in (3.14), (3.15), (3.16), may be fitted by considering frequencies around each dispersion and the relevant timescales and relaxation processes that apply, as in [322]. In this case, the parameters α, $(\cdot)_\infty$, $\Delta(\cdot)$ and τ_c (here, τ_c denotes the critical time and ω_c the critical frequency for an individual dispersion) are defined around the frequency range of each dispersion where $(\cdot)_\infty$, would be more correctly defined as $(\cdot)_H$, the parameter at the high-frequency end of the dispersion range. $\Delta(\cdot)$ denotes the change in ε or C that occurs

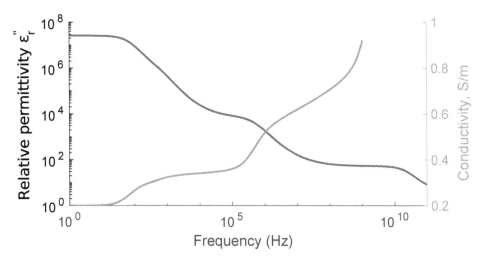

Figure 3.7 Alpha, beta and gamma dispersions illustrated in conductivity and permittivity spectra for muscle. Plots generated using parameters determined by Gabriel *et al.*[322]. (color figure available in eBook)

over the frequency range of the dispersion, and τ_c indicates the center frequency of the dispersion, $f_c = (2\pi\tau_c)^{-1}$. Each dispersion, characterized by its parameters, provides an indication of the distribution of relaxation times and scales of physical processes involved around its critical frequency.

3.2 OVERALL TISSUE CONDUCTIVITIES

As noted in 3.1.3, the overall observed frequency dependence of tissue electrical properties is a consequence of their specific mix of cell types, sizes and relaxation processes. A metanalysis by Faes[267] found that properties could be classified with reference to the bulk tissue water content. Properties of tissues have been measured in many different contexts, ranging from excised tissues or removed fluids; using multiple or single frequencies; using different measurement techniques; and in animal or human tissues in vivo. the database maintained by the Nello Carrara Institute of Applied Physics[1]. Another database has been established by the Foundation for Research on Information Technologies in Society[2]. Table 3.1 shows the conductivity of several key tissues at 50 kHz generated using the Italian database, which in turn was established using data collected by Gabriel *et al*[318].

The electrical characteristics of tissues, processes and pathologies of particular interest for EIT applications are discussed in the sections below.

---

[1] http://niremf.ifac.cnr.it/tissprop/

[2] https://itis.swiss/virtual-population/tissue-properties/database/

Table 3.1

Electrical Properties of Tissues at 50 kHz

Tissue	ε_r	σ (S/m)	Tissue	ε_r	σ (S/m)
Air (vacuum)	1.	0.	Lens	2626.5	0.33849
Aorta	1633.3	0.31686	Liver	690.	0.072042
Bladder	1912.4	0.21688	Lung (Deflated)	8531.4	0.26197
Blood	5197.7	0.7008	Lung (Inflated)	4272.5	0.10265
Bone (Cancellous)	613.18	0.083422	Muscle	10094.	0.35182
Bone (Cortical)	264.19	0.020642	Nerve	9587.5	0.069315
Breast Fat	117.75	0.024929	Ovary	3010.	0.33615
Cartilage	2762.1	0.17706	Skin (Dry)	1126.8	0.00027309
Cerebrospinal Fluid	109.	2.	Skin (Wet)	21876.	0.029369
Cervix	3150.7	0.54431	Small Intestine	17405.	0.58028
Colon	4160.6	0.24438	Spleen	5492.8	0.11789
Cornea	16970.	0.48145	Stomach	3551.2	0.53369
Dura	393.83	0.50168	Tendon	814.98	0.38779
Eye Sclera	5494.6	0.51475	Testis	6486.3	0.4344
Fat	172.42	0.024246	Thyroid	4023.1	0.53395
Gall Bladder	113.99	0.90012	Tongue	5496.	0.28422
Gall Bladder Bile	120.	1.4	Trachea	6912.4	0.32987
Gray Matter	5461.4	0.12752	Uterus	5669.9	0.52584
Heart	16982.	0.19543	Vitreous Humor	98.558	1.5
Kidney	11429.	0.15943	White Matter	3548.2	0.077584

3.2.1 PROPERTIES OF FLUIDS, CELL SUSPENSIONS AND BLOOD

3.2.1.1 Cerebrospinal Fluid

Cerebro-spinal fluid (CSF) conductivity is of interest in neural and spine applications. The influence of high-conductivity CSF surrounding brain tissue and spinal cord can significantly affect the passage of externally applied currents, and also has a great effect on EEG measurements because of its proximity to the cortical surface. The most-often cited work on CSF conductivity is by Baumann[91]. This involved low-frequency (10 Hz–10 kHz) measurements on CSF extracted from patients who had undergone brain surgery. Measurements of CSF conductivity at body temperature were found to be around 1.8 S/m. Recent reconstructions using Magnetic Resonance Electrical Impedance Tomography[186] (MREIT) and MR-Electric Properties Tomography[638] have found values of around 1.5 S/m at 10 Hz and 298 MHz respectively.

3.2.1.2 Blood

A simple model for blood is a suspension of cells (mostly erythrocytes) in plasma. The three main components of the model are the cell interior, cell membrane and suspending fluid. Each component has its own complex dielectric properties; together they determine the properties of the blood mixture. Fricke[311,313] derived the mixture equations for this relatively simple model; it predicts a frequency dependent permittivity and conductivity determined by the dielectric parameter of the components, the volume fraction, shape and size of the inclusions. The origin of this behaviour is due a transient accumulation of charges at the cell membrane, a phenomenon referred to as interfacial polarization which gives rise to the beta dispersion in tissue.

Measurement on dilute dispersed fluids such as milk, erythrocyte suspensions, and whole blood are some of the earliest reported electrical properties of tissues[209,314,315,320,323,470]. The general picture arising from these measurements is of a two-dispersion spectrum in which the permittivity of whole blood falls from a value of a few thousand at extremely low frequencies (ELF, <300 kHz) to around 60 at several hundred MHz, from which it then falls to a value between 4 and 5 at several hundred GHz. The first dispersion is the interfacial polarization predicted by Fricke[312] the second due to the aqueous component of the mixture. At ELF the conductivity of blood is of the order of 0.7 S/m, rising to about 70 S/m in the 100s of GHz region.

Visser[1086] reported the conductivity of blood at 100 kHz as function of the hematocrit (percentage volume fraction of RBC) at 37 C. The conductivity decreased monotonically from 1.6 to 0.15 S/m as the hematocrit increased from 0 to 80 (normal hematocrit is 40-50).

Healthy human erythrocytes have a biconcave disc shape with a flattened centre. When the cells are randomly oriented blood has isotropic dielectric properties. However, under flowing conditions erythrocytes tend to orient and stack up along their small axis. Complete orientation is known as rouleau formation; this gives directionality to blood and results in anisotropic dielectrical properties. This tends to happen to a greater or lesser extent in flowing blood.

In[1086], Visser developed a relationship for the conductivity of blood in terms of the conductivity of plasma, the hematocrit content and a function of the shape and dimension of the cells as well as their orientation with respect to the field. Orientation along and perpendicular to the field resulted in higher and lower values compared to random orientation.

Visser also measured the conductivity of flowing blood with the field along the direction of flow and observed an increase in conductivity with respect to average reduced velocity (velocity divided by the radius of the tube). A plateau is reached when almost all the cells are aligned. The conductivity can change by up 30% for normal hematocrit values and flow rates in the physiological range in tubes with dimensions of the main arterial branches. Hoetink[474] further expanded the modelling aspects of the work.

3.2.2 BONE

Bone electrical properties, and particularly those of the cranium, are central to many EIT-related applications such as reconstruction of cortical source location and strength using inverse EEG methods[794]. It is also of key importance when delivering transcranial AC or DC electrical stimulation[780] or measuring brain conductivity using MREIT[186]. Because the applications are characterised by low frequencies, it is critical to measure bone properties in the same frequency range, approximately DC–200 Hz. Low-frequency electrical properties of human body tissues are generally hard to measure because of the need to attach electrodes to samples (see 3.3.2) and measuring bone properties is particularly difficult because of bones' thinness and variable composition[607]. Bones may consist of hard tables of cortical bone, with spongiform or cancellous bone in their interior. Cancellous bone may be filled with varying amounts of red or yellow marrow, which also affects the overall conductivity. Cranial bone characteristics cannot be assumed to be the same as skeletal bone, because each bone type has distinct ontogenetic pathways and functional environments[684,912]. Skeletal bone may exhibit anisotropy at low frequencies because of variable loading[578], but there is little evidence for anisotropy in calvarial bone[912].

Because of its importance to EEG interpretation, many attempts have been made over the years to determine bone properties and scalp/skull conductivity ratios in the EEG frequency range[32,33,359,360,361,472,607,795,906]. Estimates for overall cranial conductivity estimated in vivo are as high as 15 mS/m[795] and those for cortical and cancellous bone have been found to approximate 5 mS/m and 20 mS/m respectively in live tissue[32]. Recently, the technique of bounded EIT (bEIT) has also been used to estimate skull conductivity, with estimates found to be around 5.5 mS/m at 42 Hz[276].

3.2.3 LIVER

Dielectric data for liver tissue were reported in several studies carried out under different conditions for a variety of reasons. For example, Riedel[881] developed a contact-free inductive measurement procedure and demonstrated the system by carrying out conductivity measurements on liver tissue between 50 kHz and 400 kHz as a function of time after death. Haemmerich[400] reported changes in the electrical resistivity of liver tissue during induced ischemia and postmortem. They observed increases in resistivity in vivo during occlusion. They analysed the data in terms of intra- and extra-cellular resistance and cell membrane capacitance.

Contributions by Raicu[853,854] have provided in vivo permittivity and conductivity data for rat liver tissue in the frequency range of 102–108 Hz. The measured data were corrected for electrode polarization and found to be in reasonable agreement with some previous studies (Foster and Schwan[285], Surowiec et al[1015]). Studies performed by Gabriel et al[318] were found to align more closely with the lower estimates; however, we note that Raicu's sample preparation involved flushing the surface with warm physiological saline, a practice that does not appreciably affect the spectral features except for the DC conductivity level of the conductivity spectrum[854].

3.2.4 LUNG

Wang et al[1113] showed lung conductivity properties in the frequency range of interest to EIT (down to 100 Hz) and reported conductivity values as function of air filling factor F (volume fraction of air to lung tissue). They found that lung conductivity decreased as the filling factor increased, ranging from 0.255 to 0.15 S/m for filling factors of 0 to 1.4. They also reported permittivity and conductivity data for cancerous lung tissue and found that the increase in low frequency conductivity was statistically significant ($p<0.05$).

An earlier paper by Nopp et al[786] reported data on lung electrical properties in the range 5–100 kHz, using an EIT system. They tested several electrode configurations and materials and observed that pressure on the lung tissue affected measured values. At 50 kHz, the measured conductivity values ranged from 50 to 100 mS/m when different electrode systems were used. They also observed a decrease in permittivity and conductivity with increasing air volume and attempted to quantify it using a model wherein air caused a thinning of alveolar walls and deformation of epithelial cells and blood vessels.

In a subsequent paper using a Sheffield EIT system on nine male subjects in vivo, Nopp et al[785] found that maximum lung resistivity reconstructed within a right lung region of interest increased linearly as inspiration volume increased and that the slope of increase increased with frequency. When the lung volume was kept constant, the maximum lung resistivity decreased as a function of frequency, with the approximate log-linear frequency slope of the curve becoming more shallow as lung volume increased. They further developed their model for lung tissue to attempt to explain the spectral changes with air volume.

For the use of EIT in pulmonary medicine, the most important variable is changes in lung air volume, and it is typical to assume that this is proportional to impedance change. Many papers have experimentally verified this linearity experimentally[17,469]. For example, Ngo et al found a >0.99 correlation between EIT images and spirometric data for slow breathing maneuvers[776]. The impedance of the lung is created by the distribution of the lung tissue and enclosed air volume. Models of this structure[787,902] also show a linear relationship over a large range. Even though the correlation is linear, in EIT images the ratio varies between patients due to the belt position[538], posture[214] and the amount of tissue between the body-surface electrodes and the lungs.

3.2.5 PATHOLOGY

The presence of abnormal cells or occult tissues and fluids also distinctly change baseline and spectral characteristics compared to healthy tissue regions. The sections below consider findings of changes in tissue electrical properties in tumors, lung tissue damage and ischemia.

3.2.5.1 Properties of Tumor Tissues

In general, it has been thought that cancerous tissues are overall more conductive than regular tissues below 1 MHz. This is believed to be a consequence of increased vascularization, or disruption of tissue layering facilitating lower resistance pathways through tissue at low frequencies[162]. Increased conductivity in tumors with respect to normal tissue was found in a study of breast cancer by Surowiec et al[1014], and in cervix[162]. However, several other studies, including in breast[522,524] and prostate tissue[404] reported increased conductivities in tumor tissue. Further tests may help identify spectral signatures for different cancer subtypes of specific tissues, and the relationship between observed electrical properties and tissue structural and biochemical composition.

3.2.5.2 Lung Injury

Changes in lung electrical characteristics caused by acute alveolar damage or chronic conditions such as congestive heart failure have also been measured. Acute lung injury (ALI) has been a focus of recent interest, and attempts have been made to mimic it by washing the lung with water, or administering endotoxins or oleic acid directly to alveoli[160,256]. An in vivo study performed by Brown et al[160] found that at 9.6 kHz resistivity in goat lungs measured via EIT decreased by 10–25% after oleic acid treatment. At 300 kHz, the reduction in lung resistivity was 20–35%. Freimark et al[292] studied resistivity of the lungs of 13 male human subjects as their congestive heart failure was treated with diuretics. After treatment, lung resistivity increased by an average of 8%.

3.2.5.3 Ischemia

Differences in electrical properties of blood and ischemic tissue may enable the identification and differentiation of cerebral stroke types[202,245,246,641,1183]. It could also potentially be used to localize effects of vessel blockages on heart tissue[182]. As noted in table 3.1, blood has a low-frequency conductivity of around 0.67 S/m, which is much larger than that of brain grey or white matter. Consequently, hemorrhagic strokes are considered more easily diagnosed by EIT. Experimental animal models of ischemia have been used to follow changes in electrical properties after a vessel has been mechanically obstructed using a suture. Measurements of the conductivity of ischemic myocardial tissue in pigs at frequencies between 100 Hz and 1 MHz[182] found that static conductivity decreased by a factor greater than two after 100 minutes of blood flow restriction. The same study demonstrated that ischemic processes initiated in the kidney, liver or skeletal muscle also resulted in a decrease in low-frequency conductivity, although these decreases were lesser.

Dowrick et al[245] summarized spectral properties of ischemic tissue measured in animals and compared these values to ones they measured in vivo in rats over the frequency range 10 Hz to 3 kHz. Impedance of healthy brain tissue was found to be around 40% lower at 3 kHz than at 10 Hz, while ischemic brain impedance decreased by only 30%. This was within the range found in two other studies[642,859], where decreases of around 30% were found.

3.2.6 ACTIVE MEMBRANE PROPERTIES

Conductivity changes are observed in active tissue membranes as a consequence of ionic flows during action potentials. This mechanism can be used for non-invasive neural source monitoring using EIT[58,915,1098]. While conductivity increases of around fortyfold are found in the membrane itself, effects on impedance measurements are only of the order of around 0.1%[346] because of their small effects on bulk current distributions. Models of EIT or MREIT measurements have been developed to verify the size of conductivity changes observed during active processes[632,914,1029,1030].

The Hodgkin-Huxley equations[471,676] describe the evolution of membrane conductance and voltage in neurons during activity. Membrane voltage is conventionally defined as the difference between intra and extracellular potentials

$$V_m = V_i - V_e. \tag{3.17}$$

Membrane conductance is governed by activity level of ion-specific channels within it. The simplest active membrane descriptions include probability states for sodium activation (m), inactivation (h) and potassium channels (n). The current flowing through the membrane is

$$I_m = C_m \frac{dV_m}{dt} + (V_m - V_{Na})G_{Na} + (V_m - V_K)G_K + (V_m - V_L)G_L \tag{3.18}$$

where I_m is the membrane current per unit area, and C_m is the membrane capacitance. V_{Na}, V_K and V_L are resting (Nernst) voltages for sodium, potassium and leakage ions, respectively, determined by the equilibrium concentrations of these species within or without the cell. The total membrane conductance per unit area is the sum of sodium and potassium conductances relative to their respective maxima $G_{Na,max}$ and $G_{K,max}$ and the constant leakage conductance G_L where

$$G_{Na} = m^3 h G_{Na,max} \tag{3.19}$$

and

$$G_K = n^4 h G_{K,max}. \tag{3.20}$$

The differential equations describing m, h and n channel states are

$$\frac{dm}{dt} = \frac{1}{\tau_m}(m_\infty - m), \tag{3.21}$$

$$\frac{dn}{dt} = \frac{1}{\tau_n}(n_\infty - n) \tag{3.22}$$

and

$$\frac{dh}{dt} = \frac{1}{\tau_h}(h_\infty - h) \tag{3.23}$$

where the variables m_∞, n_∞, h_∞, τ_m, τ_n and τ_h depend on channel-specific constants and resting membrane potential.

3.3 MEASUREMENT OF IMPEDANCE PROPERTIES

All measured impedance or apparent conductivity values are naturally influenced by the measurement method. For low-frequency measurements (below <50 MHz) electrodes may be attached to the sample boundary, and the properties of the electrode-tissue interface must be modelled, measured minimized or compensated prior to producing corrected data. Measurement of high frequency properties (>50 MHz) may be made using a dielectric probe. In this case the properties of the probe must also be considered.

A schematic diagram of a low-frequency impedance measurement is shown in figure 3.8. The process of measuring low-frequency electrical properties via electrodes is complex, because measured voltages also depend on electrode properties. Electrode-tissue interface properties are themselves complex and are influenced by the materials composing the electrode-tissue electrolyte interface, electrode area, measurement frequency and the amount of current flowing through the interface, amongst other factors. Four-terminal measurements greatly reduce the influence of electrode properties, but they should still be taken into account and included in models when performing EIT reconstructions[492].

3.3.1 ELECTRODE PROPERTIES

The connections between measurement source, wires, electrodes and tissues involve transitions in current flow from electronic to ionic. The interfaces between these entities and their consequent effects on measurements may be approximated with electrical elements. If a two-terminal measurement is made by applying a constant current into the circuit shown in figure 3.8(A), the voltage measured includes the interface properties. In its simplest form, an electrode-electrolyte interface comprises a half-cell potential, a resistance, and a capacitance contribution. Values for the half-cell potential, effective resistance and capacitance of the interface are characteristic of the electrode material, the interface area, and the type and concentrations of electrolytes in the electrode neighbourhood. An excellent basic summary of electrode properties can be found in the book by Geddes[334].

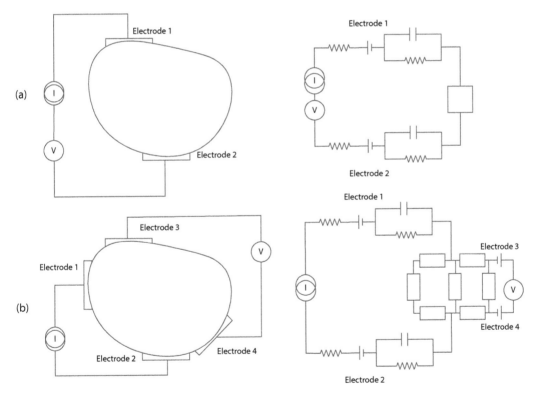

Figure 3.8 Impedance measurement configurations: (A) Two terminal measurement electrodes (B) four-terminal measurement circuit

A four-terminal measurement is illustrated in figure 3.8(B). In this idealized case it is assumed that no current flows through the voltage measurement circuitry (if the input impedance of the amplifiers involved are very large), and therefore the electrolyte interface impedance properties can be neglected. However differences in half-cell potentials, for example if the electrodes are of different materials, may produce a DC offset in the measured voltages.

3.3.2 CONDUCTIVITY CELL AND DEPENDENCE ON GEOMETRY

The relationship between a voltage difference measured on the periphery of an object and a current flowing through the object (a transfer impedance) is determined by both the geometry of the object and its electrical properties. For example, in the simple case of longitudinal current flow in a cylindrical object with length l, cross-sectional area A and conductivity σ, the measured resistance is

$$R = \frac{l}{\sigma A} \tag{3.24}$$

Here, the geometrical factor is $\frac{l}{A}$, measured in m^{-1}. Knowledge of the geometric factor combined with the measured resistance enables computation of the conductivity σ. If an object has a complicated shape, this geometrical factor can be retrieved using a substance of known conductivity: a conductivity standard. Such conductivity standards are conventionally used for low-frequency measurements and are usually potassium chloride solutions of different concentrations, chosen so as to have similar conductivities to materials to be tested. An impedance may be measured for a cell containing the standard and the geometric or shape-factor α may be derived using the relation

$$\alpha = R_{std}\sigma_{std} \tag{3.25}$$

where the shape factor is in units of m^{-1}. Subsequently, resistances measured with an unknown substance may be converted to conductivity using

$$\sigma_{meas} = \frac{\alpha}{R_{meas}} \tag{3.26}$$

A simple conductivity cell is shown in figure 3.9. This cell is cubic, and the geometric factor can be calculated or verified using a conductivity standard. If properties of a solid tissue are to be measured it may be easier to cut the sample to a simple geometric shape such as a cube and use the calculated shape factor instead of computing a shape factor using a standard. Most commercial conductivity cells also measure temperature so that compensation can be performed.

3.3.3 HIGH FREQUENCY (>50 MHz) PROPERTIES

The dielectric properties of biological materials are increasingly being performed using open-ended coaxial probes. The technique was described in numerous papers in the 1980s–90s, for example [319]. The probe in figure 3.10 has a ground plate but this is not necessary at low frequencies. Probes of different sizes are used for measurement across the frequency of interest, in general, the higher the frequency, the smaller the probe.

Dielectric probe measurements are made by placing the probe in contact with the sample and measuring its admittance or reflection coefficient. In the frequency range of interest to EIT (for example, below 100 kHz) the admittance of the probe in contact with the sample is

$$\mathbf{Y} = G + i\omega C = \frac{K\sigma}{\varepsilon_0} + K\varepsilon' \tag{3.27}$$

This admittance model is adequate when the dimensions of the probe are significantly smaller than the wavelength at the measurement frequency. In practice, standard liquids (e.g. water, dilute

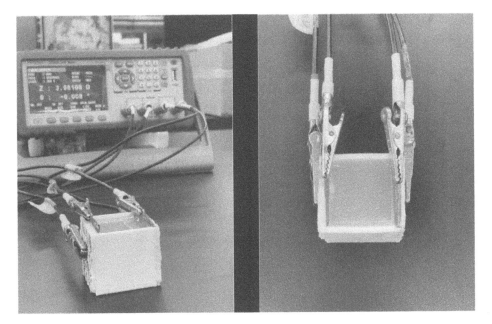

Figure 3.9 A commercial conductivity cell. (color figure available in eBook)

salt solutions) are used to obtain the cell constant and to eliminate effects of stray capacitance within the measurement system.

It is acceptable to use large probes at low frequencies, however large probes require large samples and are therefore not suitable for measuring most tissues. A 10 mm-diameter probe offers a good compromise between size and sensitivity and may be used at frequencies from Hz to MHz.

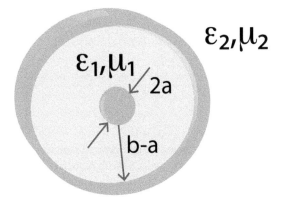

Figure 3.10 Dielectric probe, showing inner and outer radii of a and b respectively.

3.4 TISSUE ANISOTROPY

Anisotropy relates to the directionality of a property. Electrical anisotropy is manifested when the tissue conductivity is different when measured using current applied along different geometric directions. The best examples of tissue electrical anisotropy are observed in skeletal and cardiac muscle and in white matter. The directionality properties arise from the cellular structure of these tissues. In both white matter and skeletal muscle, the tissue is composed of long cell bundles aligned such that ion transport is along these cells and consequently conductivity measured using longitudinally-applied current is higher along bundles than across them. Shorter cardiac cells are connected in long chains to achieve similar properties.

It is important to emphasize that anisotropy is not the same as tissue inhomogeneity. Consider the case of measuring impedances across a cubic sample of tissue placed into the chamber shown in figure 3.9. While different impedances may be measured if the tissue were to be rotated into different orientations relative to the electrodes, this may be because the sample consists of mixed tissue types. In a simple case, the tissue might consist of two different substances layered as illustrated in figure 3.11. The apparent anisotropy σ_l/σ_t is the ratio of the conductivity measured on the object using current applied along the layers in the longitudinal (l) direction to that measured using current applied across them (t). The ratio is plotted as a function of the relative layer thicknesses of the two layer types as they vary away from $\alpha = 1$ ($t_1 = t_2$).

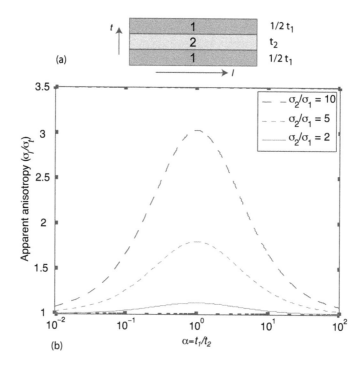

(a)

(b)

Figure 3.11   Measurement of anisotropy using simple layered structure. (A) Object composed of three layers of two different conductivities, σ_1 and σ_2 (with respective total thicknesses t_1 and t_2, such that $t_1 = t_2$ or $\alpha = 1$), having overall tangential length l, radial thickness T and width w. (B) Plot of apparent anisotropy σ_l/σ_t against α for conductivity ratios σ_2/σ_1 of 2, 3 and 10 for the brick shaped object shown in (A). (color figure available in eBook)

The apparent anisotropy depends on the relative thickness of the layers ($\alpha = t_1/t_2$) and their relative conductivities, with this quantity being much smaller than the actual conductivity contrast of the layers. As shown in figure 3.11, the maximum apparent anisotropy for alternating layers having a conductivity ratio of 10 occurs when both layers have the same thickness, and is only around 3 [912].

True anisotropy is an idealized case. The classification of real, inhomogeneous tissue as anisotropic relates to the geometric scale of its directionality or inhomogeneity with respect to the geometric scale of the measurement. In figure 3.12 below the apparent (measured) resistivity recovered by a four-terminal measurement configuration as the number of alternating conductivity layers increased is plotted as a function of the relative conductivity of the layered structure that would be measured by a uniformly applied measurement field. The measurements are compared with apparent resistivities characteristic of a truly anisotropic and inhomogeneous structure. It is clear from figure 3.12 that as the relative thickness of the layers decreases within the measurement configuration (i.e. become smaller relative to the measurement scale), the apparent resistivity becomes closer to the anisotropic curve.

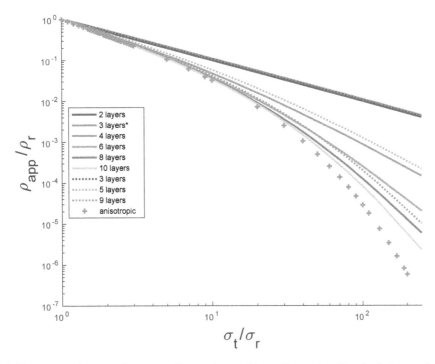

Figure 3.12 Transition to anisotropy. Comparison of (two-dimensional) calculations using a single four-terminal observation configuration for homogeneous anisotropic and layered structures. The graph plots apparent resistivities normalized with respect to the structure transverse resistivity as a function of relative conductivity as layer number n is increased. Layer thicknesses are equal in each case except where indicated by an asterisk (*) where the central layer had twice the thicknesses of the outer two layers. (color figure available in eBook)

3.5 ELECTRICAL SAFETY AND CURRENT LIMITATIONS

The subject of electrical safety of medical devices is complicated, and several good textbooks are available [856]. The key distinction has been made between devices where currents are applied on or near the heart (where currents of a few μA can be dangerous, and called a "microshock" hazard) and those where currents are applied far from the heart (such as most EIT systems) where larger currents are safe (called "macroshock" hazard). The relevant standard is ISO/IEC 60601-1 [504] and it's subparts. Most EIT system designers have based their safety assessments on the "patient auxiliary current" of the standard, defined as

"...current flowing in the patient in normal use between any patient connection and all other patient connections and not intended to produce a physiological effect."

In general, it has been assumed that the sum of all applied currents on all electrodes must be below the patient auxiliary current (I_{PA}) limit, since this current is defined to be measured "between a single part of the applied part and all other applied parts connected together". I_{PA} is the leakage current that would flow between parts applied to the patient under both normal and fault conditions (for EIT, the "normal" conditions are of relevance).

The standard establishes very low limits are established for I_{PA} at DC (0.01 mA), justified by the need to avoid tissue necrosis. For AC currents, the leakage currents specified for type "B" (i.e. non-cardiac) devices are 0.1 mA but are measured through a frequency-dependent "body model" which effectively allows higher values at higher frequencies. However, an additional rule limits the current: "irrespective of frequency no leakage current may be over 10 mA through 1 kΩ". Based on these two rules, the ISO60601-1 current limit for EIT current application at frequency f is

$$I_{PA} = \begin{cases} 0.1 \, \text{mA} & f < 1 \, \text{kHz} \\ f/(1 \, \text{kHz}) \, \text{mA} & 1 \, \text{kHz} < f < 100 \, \text{kHz} \\ 10.0 \, \text{mA} & f > 100 \, \text{kHz} \end{cases} \tag{3.28}$$

although, practically, values at about half this level have been used as a safety factor. To the authors' knowledge, there have been no reports of pain or injury from an EIT system functioning according to these specifications.

3.6 CONCLUSIONS AND PERSPECTIVE

Measured tissue electrical properties are used in formulating a multitude of electromagnetic models of biological tissue over a wide range of frequencies. For example, properties measured in the GHz range have been used to predict electromagnetic safety risks caused by mobile phones and microwave radiation; those at very low frequencies are used in forward models predicting field distributions caused by endogenous (EEG) or exogenous (transcranial DC stimulation) current sources, for use in source localization or treatment planning respectively. Measurement of properties at lower frequencies is much more difficult than in higher ranges, because of the need to attach electrodes, the geometry established by the electrode measurement array, and the need to compensate for electrode-electrolyte interfacial properties. Nevertheless, measurements at these frequencies (and those of interest to EIT) can be considered some of the most important for biomedicine, because these lower frequencies are those at which the body itself operates. This has great implications for further developments in understanding the biology underlying normal metabolism and disease.

Because of the difficulty of isolating and measuring low-frequency properties of individual human tissues in vivo, many values in the literature have been obtained invasively by performing measurements on animals in vivo[321,335,854,860,905] or in humans during surgical procedures[606], but the bulk of low frequency measurements have been made on excised (and therefore dead) tissues. The electrical properties of excised tissues change rapidly after they are extracted, and since properties also depend on temperature, this represents a further abstraction from living tissue[387]. To add to this, reported values have usually been obtained with different and often unspecified measurement geometries, and at different frequencies. Thus, the existing literature of low-frequency tissue electrical properties (below 1 MHz) is scarce and has larger than average uncertainties[321].

Use of EIT and related methods provide methods of determining relative or absolute electrical properties non-invasively, which may be of great value in understanding both dynamic and intrinsic aspects of tissue physiology.

4 Electronics and Hardware

Gary J. Saulnier
Electrical and Computer Engineering, University at Albany, State
University of New York, NY, USA

CONTENTS

4.1 HARDWARE CHALLENGES AND APPROACHES

Since the introduction of the first systems in the early 1980s, EIT instrumentation has continued to evolve in step with advances in analog and digital electronics. While early instruments were

DOI: 10.1201/9780429399886-4

designed using primarily analog techniques, newer instruments shift much of the processing to the digital domain, making extensive use of digital signal processors and programmable logic devices. Many of the most recent instruments use commercially-available computer-based instrumentation, minimizing the amount of custom hardware. Significant progress has also been made in implementing an EIT system on chip (SoCs) as well as parts of EIT systems on custom application specific integrated circuits (ASICs), resulting in miniaturization and the ability to move processing closer to the electrodes.

Along with advances in technology have come advances in system performance, particularly in the areas of system bandwidth and precision. While the original systems used relatively low frequency excitation – generally in the 10–50 kHz range—newer systems can apply waveforms up to the 1–10 MHz range. The ability to apply excitation signals over a significant range of frequencies makes it possible to perform impedance spectroscopy in which the variation of impedance with frequency can be used as a discriminating factor for imaging. Another significant change has been the increased use of "active electrodes" in which some of the electronics is placed at the electrodes, thereby eliminating some of the performance loss introduced by attaching the electrodes using cables. This chapter discusses some of the general issues involved in the design and implementation of the major functions required for EIT instrumentation.

4.1.1   SPEED AND PRECISION

The ill-posedness of the EIT inverse problem drives the design of the instrumentation for data collection. In an applied current system, large changes in the conductivity distribution away from the electrodes will result in small changes in the voltages produced at the electrodes, meaning that high measurement precision is needed to sense and, ultimately, image these changes. Distinguishability[500], defined as the ability to distinguish conductivity distributions σ_2 from σ_1, is one way to quantify the sensitivity of an EIT system. Distinguishability is a function of both the number of electrodes and the measurement precision. Due to this relationship, adding more electrodes to a system with insufficient precision or adding more precision to a system with an insufficient number of electrodes will not improve distinguishability (see also §5.3.3). The required precision is also affected by the large dynamic range of the measurements which, in some configurations, can be as large as 100 to 1000. Adjustable gain is often used to mitigate this large range of signal amplitudes and preserve measurement precision.

The time required to collect the data for a single image, the EIT system analog of the *shutter time* of the system, also impacts the design of the instrumentation. Clearly a shorter shutter time can create the ability to make more images in a given time interval. However, a short shutter time is important anytime that there are temporal changes in the conductivity distribution, even when only a single image is desired. At issue is the fact that changes in the conductivity distribution during the data collection interval can distort the reconstructed distribution in regions away from the location of the variation. The distortion is not confined to the region of the movement as it is in photography[1189]. Optimizing the order in which the data for an image is collected and/or using interpolation techniques to mitigate time skew can reduce distortion; however minimizing the time needed to collect a data set is the best way to improve performance.

Clearly, there is a need for both speed and precision in EIT instrumentation. Precision is generally characterized as a signal-to-noise ratio (SNR) or a number of bits. Many systems aim for 96 dB SNR or, equivalently, 16 bits of precision. An improvement in SNR can be obtained by lengthening the measurement time, effectively reducing noise by limiting signal bandwidth through integration, resulting in a trade-off between speed and precision. The optimal parameters for an EIT system, including the number of electrodes, speed, and precision, is highly dependent on the particular application.

4.1.2   APPLIED CURRENTS VS. VOLTAGES

Most EIT systems utilize the Neumann-to-Dirichlet map, applying currents and measuring voltages. Using applied currents is advantageous because the mapping of surface current density to voltage is smoothing while the mapping of surface voltage to current density is roughening. This smoothing effect is due to the fact that, with applied currents, the spatial pattern of voltages has one more derivative than the spatial pattern of applied currents. With applied voltages, the resulting spatial current patterns have one fewer derivatives. The net effect of the smoothing is that measurement errors, such as inaccuracy in electrode placement and noise in the measured voltages, will tend to be damped out when applying currents. There are patterns of currents that can be applied to optimize performance based on various criteria, such as those described in [500] and [235]. While applying currents and measuring voltages is theoretically optimal (see §5.5.3), implementing current sources that closely approximate ideal behaviour is difficult, particularly at higher frequencies. Consequently, some EIT systems use voltage sources, with some of these systems incorporating algorithms that enable the voltage sources to produce desired current patterns.

4.1.3   PAIR-DRIVE VS. PARALLEL-DRIVE SYSTEMS

EIT systems can be classified by the number of sources used to apply excitation. A pair-drive system, also called a single source system, applies currents between one pair of electrodes at a time and measures the voltages on the remaining electrodes. The voltage measurements can be done sequentially with a single voltmeter, though it is more common to measure the voltages in parallel using multiple voltmeters to reduce the data collection time. A parallel-drive system applies currents to, and measures voltages on, all electrodes simultaneously. A parallel-drive system has the advantage of being able to apply current patterns that are optimal for a specific geometry and conductivity distribution, yielding better data sets. However, this advantage has rarely been fully exploited due to the difficulty of rapidly finding and applying the optimal patterns when there is a time-varying conductivity distribution. Instead, patterns that are optimal for a pre-defined geometry and conductivity distribution, such as the canonical spatial sinusoids which are optimal for a circular region of homogeneous conductivity, are used in applications that approximate that geometry. Though not strictly optimal, these patterns are still significantly closer to the optimal patterns than those used in a pair-drive system. As shown below, the price of being able to approximate optimal patterns is a large increase in the complexity of the instrumentation.

The general structure of a pair-drive system is shown in figure 4.1 and figure 5.6. The waveform used in the system, in most cases a sinusoid, is produced by the waveform synthesis block. The waveform is fed to a dual current source or dual voltage-to-current converter, which produces a pair of currents having equal magnitude but opposite polarities. A 2-to-N multiplexer allows these sources to be applied to one pair of electrodes at a time. The currents are often supplied to the electrodes through shielded cables in which a driven shield is used to protect the signals from noise, as well as to minimize the cable capacitance and capacitance variation when the cables are flexed. Some systems, however, place the electronics near the electrodes to minimize the need for cables. Electrode voltages are measured using either single-ended or differential voltmeters. Differential voltage measurement, i.e. measurement of the voltage between pairs of electrodes, is often used to reduce the dynamic range requirements relative to single-ended (referenced to ground) voltage measurements. While a single voltmeter can be multiplexed to measure all electrode voltages, using more voltmeters (up to $N-2$) introduces parallelism that reduces measurement time at the expense of more hardware. The voltmetering process is performed synchronously, requiring a timing reference and/or reference waveform from the waveform synthesis block.

In the parallel-drive system shown in figure 4.2, the current source pair is replaced with N current sources, one for each electrode. This system is able to apply patterns of currents, where a pattern

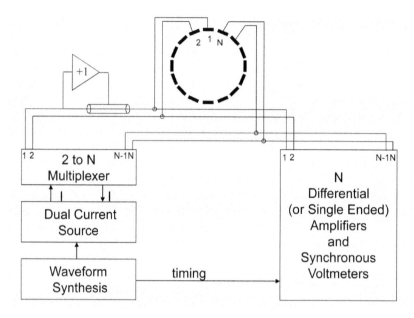

Figure 4.1   Pair-drive EIT system.

defines the current source value for each electrode. In all cases the sum of the currents applied to the electrodes must equal zero. The remainder of the system is the same as for the pair-drive system.

4.1.4   VOLTAGE MEASUREMENT ON CURRENT-CARRYING ELECTRODES

Pair-drive systems do not measure voltages on electrodes that are applying current while parallel-drive systems measure the voltages on all electrodes. The different approaches are used due to a combination of the differences between the systems and the nature of the electrode-body interface. The characteristics of the electrical interface between the electrode and the surface, generally called the contact impedance, depends on the materials involved and the quality of the contact. In general, this interface will have a non-zero impedance, meaning that current flow across the interface will produce some voltage drop. Of particular interest for EIT systems is the electrode-skin interface which has been extensively studied, characterized, and modelled[592,995].

The voltages measured on electrodes that are not injecting current are affected less by the contact impedance because of the high input impedance of the voltmeters. This high input impedance results in little current flow across the electrode-body interface and, consequently, little voltage drop even if the contact impedance is relatively high. Pair-drive systems use larger currents than parallel-drive systems to achieve greater sensitivity away from the electrodes such as, for instance, at the center of a circular region. This large current results in voltages that are both large and more dependent on the contact impedance than those on the other electrodes. Using these voltages would increase the dynamic range of the voltages being measured and introduce errors due to the contact impedance. These voltages are not used when reconstructing images, though not using them reduces the degrees of freedom in the measurements. Parallel-drive systems inject current and measure voltages on all electrodes simultaneously using patterns of current that are more optimal for the region being imaged. The use of optimal or near-optimal patterns and smaller currents enables the systems to work well with or without contact impedance[192]. For a parallel-drive system, the additional impedance introduced by contact impedance will be present at the periphery of reconstructed images of absolute impedivity at the locations of the electrodes.

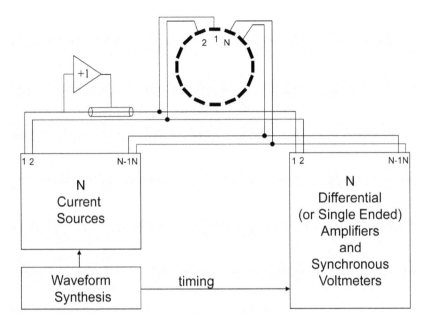

Figure 4.2 Parallel-drive EIT system.

4.2 ELECTRODE EXCITATION

Most EIT systems use sinusoidal excitation, enabling them to collect data at one frequency at a time. Direct digital synthesis (DDS) is generally used, either through a commercial DDS integrated circuit, programming a field-programmable gate array (FPGA) or digital signal processor (DSP), or using a computer. An analog waveform is produced by feeding the digital samples through a digital-to-analog converter (DAC). The performance of the synthesis is measured by the spectral purity and signal-to-noise ratio (SNR) of the resulting waveform.

Regardless of the implementation, a DDS system is constructed using either a sinusoid lookup table stored in read-only memory (ROM) as shown in figure 4.3 or a computational method for converting phases into sinusoid amplitudes. In the most common approach, a phase increment, $\Delta\Phi$, is fed into a phase accumulator that, in turn, provides addressing to the lookup table. The size of the phase increment along with the clock frequency sets the output frequency. The frequency can be

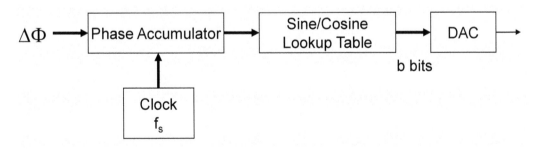

Figure 4.3 Direct digital synthesis.

adjusted by varying the size of the phase increment. However, the limited size of the lookup table requires rounding or truncation of the phase value that is used to access values in the table, resulting in periodic phase jitter that introduces line spectra (spurs) in the frequency spectrum of the resulting sinusoid[1047]. This phase jitter can be removed by restricting the choice of output frequency to those that use phase values that are in the lookup table. To help mitigate the spectral impurity introduced by the phase truncation, many DDS algorithms may utilize phase dithering to reduce the coupling between the phase error and the particular point in the sinusoid cycle.

The amount of noise present in the synthesized waveform after the DAC is a function of many things, including the resolution of the DAC, the sampling frequency and the noise present in the digital waveform itself. If we consider only the noise due to the digital-to-analog conversion using a voltage-output DAC, namely the quantization noise, the resulting voltage noise spectral density can be expressed as

$$v_{NQ} = \frac{A}{2^b \sqrt{12 f_s}} \ \mathrm{V}/\sqrt{\mathrm{Hz}}, \tag{4.1}$$

where A is the peak-to-peak voltage range of the waveform, b is the number of bits of resolution in the DAC and f_s is the sampling rate. This result is based on the common assumption that the quantization noise is white. Figure 4.4 shows the voltage noise spectral density as a function of the number of bits in the DAC and the sampling frequency when $A = 2$. Increasing the DAC resolution and/or increasing the sampling frequency results in a decrease in noise density. As a reference, typical low-noise operational amplifiers have a voltage noise spectral density in the range of 1 to $10\mathrm{nV}/\sqrt{\mathrm{Hz}}$.

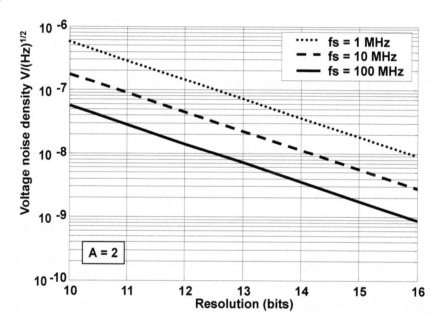

Figure 4.4   Voltage noise spectral density as a function of DAC resolution and sampling frequency.

4.2.1 CURRENT SOURCES

Most of the current sources used in EIT systems are more appropriately called voltage-to-current converters, since they produce an output current that is proportional to an input voltage. Ideally, a current source should have an infinite output shunt impedance, Z_0, resulting in the current delivered to the load being independent of the load voltage, V_L. Real current sources, however, have a finite Z_0 impedance that is usually characterized as the parallel equivalent of a resistance R_0 and capacitance

C_0. Figure 4.5A shows an ideal current source driving a load where the load current I_L equals the source current I_S. When a real current source drives a load, as shown in figure 4.5B, the current flowing in Z_0 varies with V_L, consequently the relationship between I_L and I_S varies with the value of the load impedance.

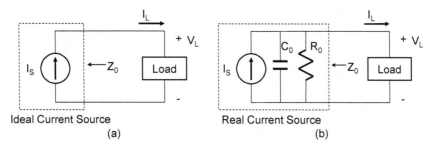

Figure 4.5   Ideal and real current sources.

The variation in I_L with V_L that occurs with finite current source output impedance is made worse by the presence of additional stray or parasitic capacitances. Though not associated with the current source itself but, rather, due to capacitance between wire and/or printed circuit board traces and ground, this capacitance provides an additional means for current to be shunted away from the load to ground, effectively reducing the output impedance of the source. In determining the required current source output impedance for a given application, it is essential to consider the impact of this. As will be discussed later, the cables to the electrodes can also introduce shunt capacitance and the use of a driven shield can help reduce the effect of this capacitance.

4.2.1.1   Floating and Single-Ended Current Sources

In a pair-drive EIT system that excites a pair of electrodes at a time, it is necessary to produce a current that flows into the body at one electrode and out of the body at another electrode. These currents can be produced using one "floating" current source that, as shown in figure 4.6A, makes a current that flows through a load without a reference to ground potential. The figure shows the presence of the current source output impedance, Z_0, as well as stray capacitance, C_S. In an idealized case, where Z_0 is infinite and C_S is zero, $I_1 = -I_2 = I_S$, as desired. With finite Z_0, the load currents will be equal and opposite but their relationship to I_S will vary with the load seen between the electrodes. The addition of the stray capacitance will make I_1 and I_2 dependent on the voltages between the corresponding electrode and ground, potentially producing a non-zero "common-mode" current of value $I_1 + I_2 \neq 0$.

Another way to produce the desired currents is to use a balanced pair of single-ended current sources, each of which produces a current that flows from a ground as shown in figure 4.6B. I_{S1} is set to equal $-I_{S2}$ so that, for infinite Z_0 and zero C_S, I_1 equals $-I_2$. The inclusion of finite Z_0 and non-zero C_S will result in $I_1 \neq I_{S1}$ and $I_2 \neq I_{S2}$. Additionally, the applied currents may not be equal and opposite, resulting in a common-mode current. The common-mode current issue has been addressed in many different ways, including feedback techniques that adjust the applied current to maintain a small common-mode voltage and the use of an additional electrode that provides a path for this common-mode current to ground.

Parallel EIT systems can be constructed using either floating or single-ended sources, though most use the latter. In both cases, the number of sources equals the number of electrodes. With a parallel system, common-mode current arises whenever the sum of the currents from all the sources does not equal zero. Keeping this common-mode current below a desired level with variations in the

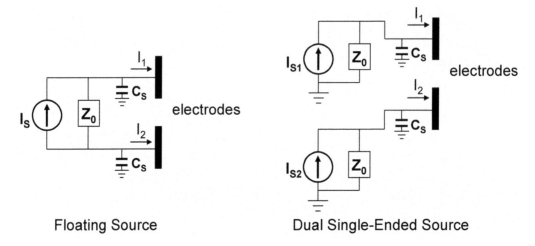

Figure 4.6   Floating and single-ended current sources.

load impedance seen by the electrodes requires a higher Z_0 and lower C_S as the number of electrodes increases.

4.2.1.2   Current Source Requirements

The current source in an EIT system must be able to deliver current with a desired precision over a specified frequency range to load impedances within an expected range of values. These requirements translate into specifications for the frequency response, output impedance and voltage compliance of the current source. Both the voltage compliance (maximum voltage a current source can supply into a load) and the output impedance requirements are a function of the expected load impedance. Since the voltage compliance of the source is the range of load voltages for which the current source continues to behave as a current source, it must exceed the voltage when the maximum current is sourced to (or sinked from) the load with the highest impedance.

In medical applications with single sinusoid excitation, maximum peak current values in the range of 0.1–5 mA are common, with smaller current values being used at lower frequencies due to safety concerns. Parallel-drive systems apply smaller maximum currents to individual electrodes than pair-drive systems since they apply current to all electrodes simultaneously. Load impedances, which are a function of electrode size, excitation frequency and the tissue being imaged, typically range from 100 Ω to 10,000 Ω with the lower values observed at higher frequencies. With these currents and impedances, voltage compliance in the range of a few volts to as much as 10–15 volts is required.

The required output impedance is also a function of the load impedance. However, there are two ways to look at the problem. In order to maintain a desired accuracy of the applied current, i.e. keeping I_L and I_S of figure 4.5B equal to within a given tolerance, it is necessary to consider the maximum load impedance that the current source will encounter. The error current equals the current through the output impedance of the source, I_{Z0}, which is given by

$$I_{Z0} = \frac{Z_{Lmax}}{Z_0 + Z_{Lmax}} I_S, \qquad (4.2)$$

where Z_{Lmax} is the maximum load impedance and Z_0 is the current source output impedance. For the I_L to be accurate to within b bits of precision requires that the current error be less than one least significant bit (LSB) or, equivalently, $1/2^b$. The output impedance requirement then becomes

$$Z_0 \geq (2^b - 1)Z_{Lmax}. \tag{4.3}$$

In this case, a system with 16 bit accuracy with a maximum load impedance of $10\,k\Omega$ requires a current source with an output impedance of over $655\,M\Omega$.

A second way to look at the problem is to consider the fact that, in general, EIT systems are more concerned with the precision of the current values rather than their accuracy. In other words, it is more important that the variation in load current between a minimum and maximum load impedance be within the desired tolerance than it is for the current be exactly equal to a desired value. This property is true for both pair- and parallel-drive systems. In a pair-drive system, the same source is applied to multiple loads (electrode pairs) to collect data for an image. In a parallel-drive system, different sources, each of which satisfies some minimum output impedance specification, are applied to the different loads. In both cases, the difference in load current with maximum and minimum load impedances of Z_{Lmax} and Z_{Lmin}, respectively, is given by

$$I_{Lmax} - I_{Lmin} = \left(\frac{Z_0}{Z_0 + Z_{Lmin}} - \frac{Z_0}{Z_0 + Z_{Lmax}} \right) I_S. \tag{4.4}$$

To determine the minimum Z_0 required to obtain b bits of precision, determine Z_0 such that $(I_{Lmax} - I_{Lmin})/I_S \leq 1/2^b$.

Figure 4.7 shows the output impedance in megohms that is needed to achieve a given number of bits of resolution for several ranges of load impedance. These results assume that all the impedances are real (resistive) whereas the impedances are generally complex. In a medical application, the larger load impedance values would generally be encountered at lower frequencies and the smaller values at higher frequencies. The figure includes results when the range of resistances goes from zero to some maximum value, indicated by plus sign markers. These results represent the case where the accuracy of the applied current is being maintained. Other results consider the case where the load impedance is expected to remain within 20% of a nominal value (box markers) and the case where load impedance remains within 10% of a nominal value (diamond markers). The plot demonstrates the benefit, in terms of reduced output impedance requirements, of considering the current precision over a restricted range of load impedances. However, high precision systems with relatively large load impedances still require high current source output impedance. For example, a 16 bit system with load impedances in the range of 9–$11\,k\Omega$ requires a current source output impedance in excess of $120\,M\Omega$.

While a higher level of precision is generally desired, current accuracy is also important. Higher accuracy can be obtained through current source calibration, where the current source is calibrated to deliver an accurate current to a test load having an impedance that is within the range of expected load impedances. Calibration is very important in a parallel system since it is necessary to account for gain differences between the sources in order to avoid problems with common-mode currents.

4.2.1.3 Stray Capacitance

Clearly, when implementing a high precision system that requires output impedances on the order of 10's of megohms, it is necessary to have extremely small stray capacitances – values much smaller than can be realistically achieved when building the circuit. There are two common approaches to this problem in EIT systems. One approach is to employ some type of capacitance cancellation technique to reduce the effective capacitance seen by the current source. A second approach, for use when the load impedance is resistive or nearly resistive, is to reduce the sensitivity to stray capacitance by measuring only the real part of the load voltage[161].

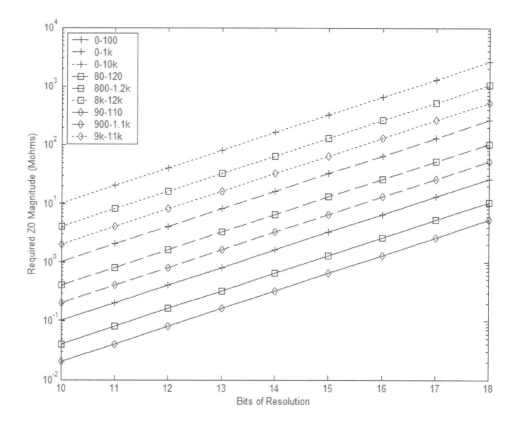

Figure 4.7   Required Z_0 as a function of desired precision and load impedance range.

To see how measuring the real voltage reduces the impact of stray capacitance, consider the circuit shown in figure 4.8. Here a current source drives a resistive load, R_L, which has a parallel capacitance, C. In the ideal case, where $C = 0$, the load voltage, V_L, is real and equal to IR_L. When the capacitor is present, V_L becomes complex due to the phase shift introduced by C. The normalized error equals $(IR_L - V_L)/IR_L$ and can be expressed as

$$\text{normalized error} = \frac{(2\pi fCR_L)^2}{1 + (2\pi fCR_L)^2} + j\frac{2\pi fCR_L}{1 + (2\pi fCR_L)^2}. \tag{4.5}$$

For the case where $2\pi fCR_L < 1$, the normalized imaginary (reactive) part of the error exceeds the real part. Consider, for example, the case where $C = 20\,\text{pF}$ and $R_L = 1\,\text{k}\Omega$ for which the real and reactive normalized error voltages are plotted in figure 4.9 as a function of frequency. For 16 bits of precision, the normalized error should be less than $2^{-16} \approx 15 \times 10^{-6}$. In considering the real voltage only, the system can operate up to approximately 10 kHz with an error below this level. The error in the reactive voltage is below this value only at very low frequencies. Note that, for these values of C and R_L, $2\pi fCR_L$ exceeds unity for frequencies of approximately 8 MHz and above where, on figure 4.9, the error for the real voltage moves above that for the reactive voltage.

By measuring only the real part of the load voltage, it is not possible to make images of the permittivity of the object. In order to achieve high precision while maintaining the ability to image both resistivity and permittivity, it is necessary to employ techniques that either cancel the stray capacitance or render it ineffective.

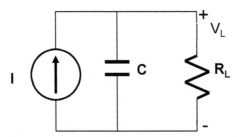

Figure 4.8 Current source with stray capacitance and a resistive load.

4.2.1.4 Current Source Compensation

Figure 4.10A illustrates the concept of using a negative capacitance[210] to cancel the positive capacitance that is present due to the current source output capacitance, C_0, and the stray capacitance, C_S. Since capacitors add in parallel, the compensating capacitance should equal the negative of the sum of the other capacitance present in the circuit. Figure 4.10B illustrates the second technique that uses an inductance to produce a parallel resonant circuit with the capacitance[901]. At resonance, the impedance of a parallel LC circuit goes to infinity, effectively canceling the much lower impedance presented by the capacitor itself. However, there are two drawbacks to the parallel resonant approach. First, the effect of the capacitance is cancelled at the resonant frequency only, making it unsuitable for systems that use an excitation other than a pure tone. For a system that employs variable frequency, the compensation must be tuned to accommodate any frequency change. The second disadvantage is that the resonant circuit has start-up and stop transients that depend on the quality factor (Q) of the circuit. This Q varies with the load and current source output resistances.

It is also possible to compensate for finite current source output impedance and additional stray capacitance by increasing the applied current by an appropriate amount. If the value of current source output impedance (including stray capacitance) and the load voltage are known, the amount of current that is shunted away from the load can be calculated. Increasing the applied current value to compensate for this current loss will result in the desired current being applied to the load[237]. While the output impedance and stray capacitance can be estimated using a calibration procedure, the current through this impedance is a function of the load voltage, which varies with the load impedance seen at the electrode as well as the applied current. This approach to compensating for output impedance has been implemented taking an iterative approach[704] and, more recently, using a direct solution that determines the needed current adjustment based on fixed output impedance measurements made during a calibration procedure and the time-varying measured load voltages[926].

4.2.1.5 Current Source and Compensation Circuits

Since they operate at relatively low frequencies, generally below 10 MHz, many EIT systems use current sources that are built using operational amplifiers. A smaller number of systems use sources based on operational transconductance amplifiers (OTAs). These types of current sources are particularly attractive for systems that are constructed using discrete ICs since they use few components and perform well. EIT systems that have been implemented using custom application specific integrated circuits (ASICs) have used a variety of different current source topologies.

Figure 4.11 shows a schematic diagram for a floating current source that uses a transformer to provide d.c. isolation between the source and load – an important feature for patient safety in medical applications – and allows the load voltage to float with respect to ground potential. The

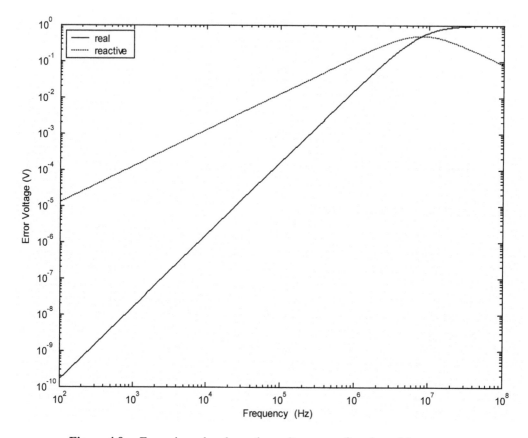

Figure 4.9 Errors in real and reactive voltages as a function of frequency.

voltage compliance and output impedance of the circuit are limited by the non-ideal behaviour of the operational amplifier and the transformer. As shown, the circuit includes a current sensing resistor R_S which enables direct measurement of the current on the load side of the transformer through the measurement of the voltage drop across the resistor. Measuring the current in this way, as opposed to relying on ideal behaviour of the operational amplifier and transformer, will enhance the precision of the source.

Current sources that use multiple operational amplifiers, such as those described in[1160], have been used in EIT systems. The Howland and improved Howland circuits, shown in figure 4.12 are single operational amplifier sources that have been used in a number of systems. These voltage-to-current converters offer both simplicity and good performance[290]. The topology of the current source has a forward path consisting of a non-inverting amplifier (the op amp along with R_1 and R_2) and positive feedback. Figure 4.12A shows the standard Howland source. Note that an instrumentation amplifier can be used in place of the non-inverting amplifier in the circuit as in[210]. For an ideal op amp, the output impedance of the Howland source is infinite when the resistors satisfy the relationship $R_4/R_3 = R_2/R_1$. At this 'balance' condition the load current can be expressed as $I_L = V_{in}/R_3$. Figure 4.12B shows the improved Howland current source that offers improved voltage compliance and lower power consumption. For this source, the balance condition is $(R_{4a} + R_{4b})/R_3 = R_2/R_1$ and the load current is $I_L = V_{in}(R_{4a} + R_{4b})/R_3R_{4b}$.

The primary advantages of the Howland sources are their simplicity and ability to produce a high output impedance with the appropriate trimming. In practice, it is possible to trim for an infinite

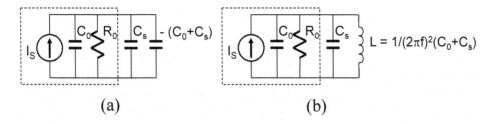

Figure 4.10   Current source compensation: (a) negative capacitance and (b) inductance.

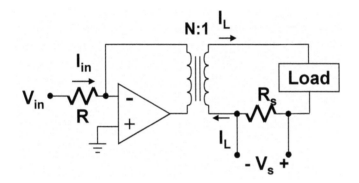

Figure 4.11   Floating current source with transformer coupling.

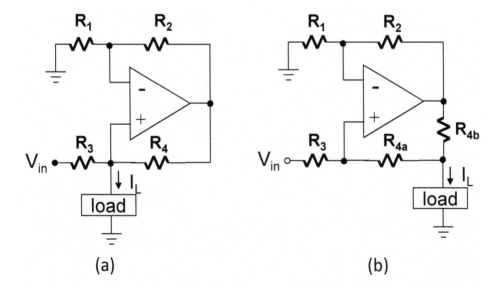

Figure 4.12   (a) Howland current source and (b) improved Howland current source.

output resistance by adjusting one resistor, but the non-ideal op-amp behaviour results in a non-zero output capacitance.

Operational transconductance amplifiers (OTAs) are also used to implement current sources for EIT. To understand an OTA, it is easiest to start with the current conveyor II (CCII)[858,952], shown in figure 4.13A. For an ideal CCII, $I_Y = 0$, $V_X = V_Y$, and $I_Z = \pm I_X$. In simple terms, Y presents an infinite input impedance, the voltage on X follows that on Y, and the current on X is reproduced on Z, either without an inversion for a CCII+ and with an inversion for a CCII-.

A simplified diagram of an OTA, which is essentially an implementation of a CCII, is shown in figure 4.13B. It is constructed around a high-input impedance/low output impedance unity gain amplifier (the input is CCII terminal Y and the output terminal X) driving a fixed load resistance R. Current mirrors on both the positive and negative voltage supplies of the unity gain amplifier reproduce the supply currents in the unknown load impedance (on terminal Z). If the unity gain amplifier has high input impedance, very little current flows into its input and, due to conservation of current, the current in R (from terminal X to ground) is nearly equal to the sum of the supply currents, $I_+ - I_-$, as indicated on the diagram. This current sum flows into the load.

CCIIs and commercially-available OTAs have been used to build current sources for EIT systems[146,444,866,1188]. An example device is the OPA860[224]. The OTA current source has the ad-

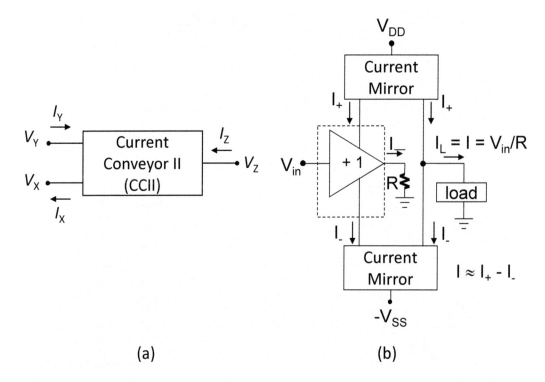

Figure 4.13   (a) Current Conveyor II and (b) Operational Transconductor Amplifier (OTA).

vantages of being adjustment-free and simple, consisting of a single IC. A disadvantage is a degradation in output impedance as frequency increases, a common problem with most current source implementations.

A variety of different EIT current sources have been implemented in custom ASICs as part of an integrated active electrode or front end circuit. A recent survey paper provides a useful summary of the evolution of these current sources, some types, and their performance[768]. Most of these sources are CMOS implementations and many use OTA-like structures, often with differential inputs and/or outputs. This is a rapidly developing area of investigation and, until recently, most of the published sources have had difficulty providing the combination of high SNR, low harmonic distortion, high output impedance, and large voltage compliance for needed for high-precision EIT systems. An integrated analog front end is presented in [865] that includes an op amp-based CCII current source, voltage buffering and 10 bit ADC. The active electrode ASIC in [1174] uses a differential difference transconductance amplifier (DDTA) followed by an OTA. This system implements a fully differential drive, creating a pair of current outputs with opposite sign, and uses a current sense resistor within a feedback network. Other recent approaches are described in [486,555,636].

4.2.1.6 Capacitance Mitigation Circuits

As discussed earlier, methods to compensate for excessive capacitance include placing a negative capacitance or an inductor in parallel with the current source output. A negative capacitance can be synthesized using a negative impedance converter (NIC) circuit, as shown in figure 4.14 [1067]. The impedance seen with respect to the ground when looking into the input terminal is given by

$$Z_{in} = -\frac{R_1}{R_2} Z. \tag{4.6}$$

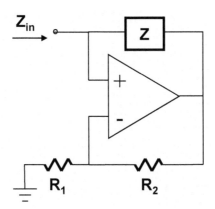

Figure 4.14   Negative impedance converter (NIC) circuit.

This impedance equals the impedance in the positive feedback path scaled by a negative value dependent on ratio of the resistors. By making Z a positive capacitor, a negative capacitance can be created having a value that is adjustable through R_1 and/or R_2. This adjustment can be made electronically using a digitally programmable potentiometer.

In theory, the NIC can create a relatively broadband negative capacitance, which would make it possible to cancel capacitance over a substantial frequency range. This behaviour is necessary for a multiple frequency EIT system in which the multiple frequencies are applied simultaneously. In the case where multiple frequencies are used one at a time, broadband compensation is desirable to avoid needing to retrim the source each time a new frequency is used. However, in practice, the usefulness of the NIC is limited by its tendency to oscillate. Stability can be improved by adding capacitance to the resistive feedback network, but only at the cost of reducing the frequency range over which the negative capacitance is produced.

A second compensation scheme is to create an LC parallel resonant circuit by introducing a parallel inductance [792,901]. This inductance can be synthesized using a generalized impedance converter (GIC) circuit such as that shown in figure 4.15 [1067]. This circuit is one of several implementations of the GIC. GICs are most commonly used to implement active filter equivalents of RLC ladder filters. The impedance seen looking into the GIC circuit is given by

$$Z_{\text{in}} = \frac{Z_1 Z_3 Z_5}{Z_2 Z_4}.$$ (4.7)

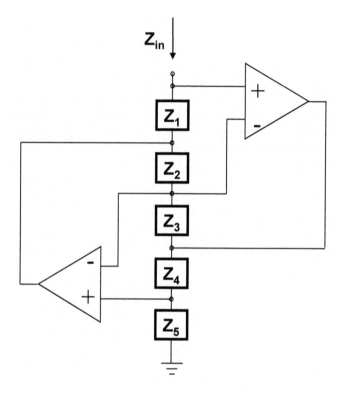

Figure 4.15   Generalized impedance converter (NIC) circuit.

By inserting a capacitor for Z_4 and resistors for the remaining impedances, the input impedance will be that of an inductance, i.e.

$$Z_{\text{in}} = s\frac{R_1 R_3 R_5 C_4}{R_2} = sL.$$ (4.8)

It is also possible to synthesize an inductance by inserting a capacitor for Z_2 and a resistor for the other impedances, but having the capacitance in the Z_4 location provides better performance. The GIC circuit exhibits good stability and component sensitivity properties. However, as described earlier, the effect of the capacitance is removed only at the LC resonant frequency, meaning that this compensation approach cannot be used in systems that apply multiple frequencies simultaneously, and retuning must occur whenever the frequency is changed in multi-frequency systems that apply a single frequency at a time.

4.2.2 VOLTAGE SOURCES

As discussed above, the precision requirements and, consequently, the output impedance requirements for the current source can be very large in order to avoid problems with common-mode currents, particularly for parallel systems. Implementing such high precision current sources requires relatively complex circuitry, including circuits for mitigating the impact of stray capacitance, and extensive calibration and/or tuning procedures. Some systems have avoided this issue by applying voltages instead of currents[444,927,1216]. While this approach can simplify the electronics, it is less desirable from a theoretical point of view and tends to increase the sensitivity to electrode placement and size errors[500]. When applying voltages, it is necessary to simultaneously measure the applied current. Figure 4.16 shows a voltage source circuit. The basic configuration is a non-inverting oper-

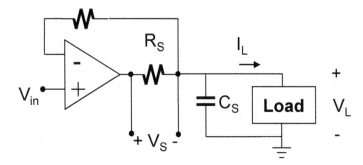

Figure 4.16   Voltage source with current measurement.

ational amplifier with a current sensing resistor R_S inserted to enable the measurement of the current leaving the voltage source. In the figure, R_S is contained within the feedback loop of the op amp which, for ideal behaviour, makes the load voltage V_L equal to the input voltage V_{in}. R_S can be placed outside the feedback loop, but the load voltage may differ significantly from V_{in} unless R_S is kept small.

While voltage sources are simpler to implement than current sources, they are not without problems. In practice, the limited open loop gain of the op amp and/or R_S outside the feedback loop will result in V_L being somewhat less than V_{in} in magnitude. This effect can also be viewed as a result of the nonzero output resistance of the voltage source. In either case, this voltage drop will result in errors in the applied voltages. To mitigate this problem the load voltage (the voltage at the negative terminal of V_S) can be measured directly, rather than assuming that the load voltage equals the input voltage. While this approach will not make the load voltage equal to the desired value, it at least enables precise knowledge of the actual load voltage. A bigger problem is inaccuracy in the measurement of the load current I_L. Figure 4.16 shows the presence of stray capacitance C_S in parallel with the load. A load-voltage-dependent current will flow in this stray capacitance, meaning that the current measured through R_S is not exactly equal to the load current. This problem is equivalent to the output capacitance/stray capacitance problem with a current source. Once again, techniques for cancelling the capacitance could be applied, although this would make the circuitry significantly more complex, removing one of the advantages of using voltage sources.

4.2.3   CONNECTING TO ELECTRODES

EIT systems commonly use cables to connect the electronics to the electrodes. For pair-drive systems, the need to multiplex currents from a single source to multiple electrodes makes using cables convenient. For parallel systems, the increase in hardware and, often, the need for a centralized calibration system, makes it useful to have the electronics in one place and connect to the the electrodes through cables. Despite their convenience, cables have complex electrical characteristics and their use can significantly degrade the performance of an EIT system. As a result, there has been a long-standing drive to move electronics closer to the electrodes and, to the degree possible, remove cables from between the current source, load, and voltage buffer. Making this shift to "active electrode" systems, while avoiding bulky electronics at the electrodes has helped motivate the implementation of ASICs for EIT. ASICs help minimize the size and power consumption of these active electrodes and, therefore, the inconvenience that they may present.

4.2.3.1   Cables

Generally, coaxial or triaxial cables are used between the electronics and electrodes as opposed to individual wires in order to minimize coupling between the signals to/from each electrode and reduce the susceptibility to noise and interference. To reduce the negative impact of these cables, some systems place some of the electronics at the end of the cable with the electrodes[194], moving the cable and its non-ideal electrical characteristics away from the source/electrode link to a place where driving and receiving ends of the cable can be controlled to minimize its impact. In putting electronics at the electrodes, the cost includes the need to deliver power and, possibly, control signals to the end of the cable as well as the inconvenience of having an electronics package, even if small, at the end of the cable.

Coaxial cables are transmission lines, meaning that they are distributed circuits in which voltage and current vary continuously along the length of the cable. A transmission line can be modelled as a cascade of lumped-parameter sections, one of which is shown in figure 4.17. A section is assumed to be of length Δz and R, L, G, and C are the series resistance, the series inductance, the shunting conductance, and the shunting capacitance, respectively, all *per unit length*. The greater the number of sections in the model, meaning the the smaller the Δz, the better the model and the accuracy of the predicted behaviour. However, when a transmission line is short relative the wavelength of the signal, the line output can be accurately approximated using a single section of figure 4.17 in which Δz is the total length of the transmission line. For example, a signal of $f = 1$ MHz in a coaxial cable with a propagation velocity of $v = 2 \times 10^8$ m/s has a wavelength of $\lambda = v/f = 200$ m. Consequently a 2 m length cable is approximately 0.01 times the wavelength, meaning that a single section should provide a good approximation to the cable behaviour.

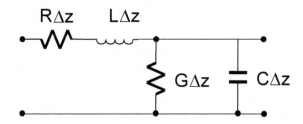

Figure 4.17   Lumped parameter transmission line model.

Due to their high output impedance, current source outputs are much more susceptible to noise and interference than voltage source outputs. While coaxial cables can provide the desired shielding, they typically present a significant distributed capacitance, on the order of 40–100 pF/m. As can be seen in figure 4.17, grounding the shield results in this capacitance acting as a shunt from the center conductor to ground, much like the stray capacitance and current source output capacitance.

The impact of the cable capacitance can be reduced by driving the shield with a voltage that is equal to that on the conductor as shown in figure 4.18. Since the voltage across the capacitance is now zero, the capacitance does not carry current and it is effectively removed from the circuit. In practice, a combination of the inability to reproduce the exact center conductor voltage on the shield and the impact of the series inductance results in a reduction in the effective value of the capacitance rather than its complete removal. When triaxial cables are used, a second grounded shield is positioned around the driven shield, providing added protection from external noise signals.

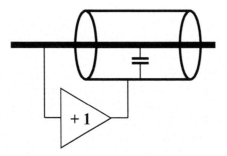

Figure 4.18   Driven cable shield.

The primary complication of using a driven shield is the potential for instability as the shield driver amplifier provides a positive feedback path. Additionally, the shield driver amplifier is typically presented with a highly capacitive load, which makes it less stable. Maintaining the gain of the shield driver somewhat less than unity reduces the risk of oscillation due to positive feedback through the signal conductor at the expense of increasing the residual cable capacitance. Many op amps are available that can drive large capacitive loads at unity gain and are suitable for shield drivers. If stability remains an issue, the circuit shown in figure 4.19 is commonly used to enhance the stability of the shield driver circuits. In this circuit, the combination of the 100 Ω series resistance and feedback capacitor allows negative feedback that is less sensitive to the phase shift introduced by the capacitive load[290].

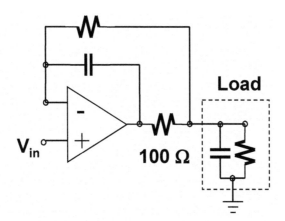

Figure 4.19   Driver circuit for capacitive loads.

4.2.3.2   Active Electrodes

Many new EIT systems have moved away from having cables linking the hardware directly to the electrodes. The issues with cables are not new, but they become more significant with the desire to operate at higher excitation frequencies while maintaining high precision. The idea of the active electrode is not new[886] but has become more attractive with the development of ASICs that both miniaturize and reduce the power requirements of the electronics placed at the electrodes.

The goal of an active electrode is to minimize the impact of any cabling on the performance of the system. A simple active electrode will only buffer the electrode voltages before they are returned for processing, providing an impedance transformation from the high impedance of the electrode to the low impedance of the buffer output. This impedance change will tend to reduce the coupling of noise and interference into the signal and reduce, though not completely remove, the impact of cable capacitance.

Adding buffers will improve the signal integrity but does not reduce the need for cables. Active electrodes can also be used to simplify the connection between electrodes and the system. The active electrode system described in[325] performs this buffering and also includes switches that can connect the buffered voltages to either of two buses, each of which connects to one side of a differential voltmeter. Designed for a pair-drive system, this approach eliminates the need for individual cables to each electrode while allowing the voltmeter to measure the voltage difference between any pair of electrodes. Similarly, this active electrode system includes switches that can connect an electrode to either of two current buses, one of which carries a positive current and the other a negative current. Again, this arrangement eliminates the need for individual current-carrying cables to each electrode while allowing any pair of electrodes to apply current.

Cable capacitance will result in some current from the source not being delivered to the electrode. To address this issue, the active electrode system described in[707] includes the ability to measure the current at the electrode. The current is applied to the electrode through a current sense resistor. The voltages at each end of the resistor are buffered and switches allow both of these voltages to be returned to the system, one at a time.

The most complex active electrodes include both current drive and voltage buffering, such as the system described in[1174]. Here, the active electrodes incorporate an ASIC that includes a current source and a voltage buffer, removing all cabling from directly interfacing with the electrodes.

4.2.4 MULTIPLEXERS VS. PARALLEL HARDWARE

Multiplexers are required to switch the applied current from one pair of electrodes to another in a pair-drive system. Parallel systems do not use multiplexers at the cost of fully parallel hardware. Multiplexers have many undesirable non-ideal properties, including a nonzero 'on' resistance that is somewhat dependent on the applied voltage, limited 'off' isolation, and charge injection during switching. Since the current in a pair-drive system is being applied using a current source with high output impedance, the impact of the 'on' resistance is insignificant. Limited 'off' isolation will produce the undesirable result of some current being redirected from the excitation electrodes to electrodes that should have zero current. The isolation degrades further as frequency increases. Probably the most significant non-ideal aspect of a multiplexer, however, is the relatively large capacitance of multiplexer devices. Typically the input capacitance is in the range 30–50 pF and the output capacitance on each line is in the range 5–10 pF. Multiplexers made using smaller devices will have lower capacitance values at the cost of higher 'on' resistance, a trade-off that is good for an EIT system. The impact of the unavoidable multiplexer capacitance can be lessened by only measuring real voltages as was discussed in Section 4.2.1.3.

Using a multiplexer to measure voltages on multiple electrodes using a single voltmeter can also introduce errors. In this case, buffering the electrode voltages before feeding them to the multiplexer will minimize the negative impact of the multiplexer capacitance because the low output impedance of the buffer amplifier can quickly charge and discharge the multiplexer capacitance. The cost is additional parallel hardware. Similarly, having a high input impedance at the input to the voltmeter will minimize the impact of the 'on' resistance.

4.3   VOLTAGE MEASUREMENT

EIT systems that image both the conductivity and permittivity in the body require phase-sensitive measurements, i.e. measurement of both the real and reactive voltages on the electrodes. Likewise, systems that assume that the load is resistive require phase-sensitive voltage measurements in order to extract the real part of the electrode voltage. As discussed earlier, measuring the magnitude of the electrode voltage would result in greater sensitivity to stray capacitance. These phase-sensitive measurements are generally made using a synchronous voltmeter that uses a coherent reference obtained from the system waveform generator. While early systems performed synchronous voltage measurement using analog circuitry, all modern EIT systems take a digital matched-filter approach. A discussion of both the analogue and digital approaches to phase-sensitive voltmetering is found in [984].

4.3.1   MATCHED FILTER

Figure 4.20 is a block diagram of a digital implementation of a matched filter phase-sensitive volt-

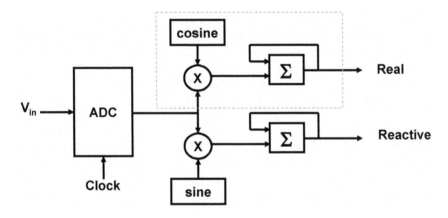

Figure 4.20   Digital synchronous voltmeter.

meter that produces both real and reactive measurements. The voltage is sampled and quantized by the ADC and the samples are multiplied by sine and cosine reference waveforms of exactly the same frequency. Note that an anti-aliasing filter may be needed before the ADC. The products are subsequently accumulated over an integral number of cycles of the signal frequency. For the system to work properly, the sampling clock for the ADC must have the necessary relationship to the signal frequency. It can be shown that the SNR of the measured voltages is optimal for a given ADC precision and integration period if the noise in the signal after the ADC is white, meaning that it has a flat (frequency independent) power spectral density. Real and reactive outputs in figure 4.20 are labelled, assuming that a real (resistive) load produces a voltage waveform that is a cosine having a phase angle of zero. It is necessary to integrate over an integral number of cycles of the signal in order to suppress the "double-frequency" components of the product of the ADC samples and the reference sine and cosine. Essentially, multiplying two sinusoids having the same frequency produces a result that consists of a d.c. signal, having an amplitude that is dependent on the amplitudes of the individual sinusoids and their relative phase, plus a sinusoid having double the original frequency. Integrating over an integral number of periods of the input signal frequency completely suppresses this double frequency and all other harmonics of the excitation frequency, because the integration 'filter' has a magnitude frequency response with a $|\sin x/x|$ shape centered at d.c. and nulls at frequencies kT, where T is the integration period and k is any integer not equal to zero. When $T = N/f$, where f equals the signal frequency, the nulls are at kf/N.

4.3.1.1 Noise

The quantization noise from an ADC is generally assumed to be white with power

$$\sigma_Q^2 = \frac{\Delta^2}{12} \qquad (4.9)$$

where Δ is the ADC quantization step size. Increasing the precision of the ADC by one bit results in a reduction of Δ by a factor of 2 and a corresponding decrease in the quantization noise power by a factor of 4. Using the assumption that this noise is white, the power is uniformly distributed over a bandwidth of f_S Hz, where f_S is the sampling frequency, resulting in a one-sided noise power spectral density of

$$PSD = \frac{\Delta^2}{6 f_S}. \qquad (4.10)$$

Consequently, increasing f_S for a given ADC resolution results in a decrease in the PSD of the quantization noise.

The length of the integration period also affects the SNR at the voltmeter output. We can assume that the signal that is input to the voltmeter has some additive white noise that results from various noise sources, including thermal noise in the electronic components. Integrating over a larger number of cycles of the signal, i.e. oversampling, results in an improvement in the SNR of the voltage measurements, where the noise consists of noise at the ADC input plus the quantization noise of the ADC itself. If it is assumed that the noise samples are uncorrelated with each other, and that the noise is uncorrelated with the sinusoidal signal being measured, integrating the signal results in SNR improvement by a factor that is equal to the number of samples being accumulated. There are two ways to view how this improvement occurs. One way is to consider the fact that the bandwidth of the integrator is inversely proportional to the integration period. Integrating over N samples results in a decrease in bandwidth by a factor of N and a corresponding reduction in the output noise power by a factor of N or, equivalently, a reduction in root mean square (RMS) noise voltage by a factor of $\sqrt{N}$. Since the signal itself has zero bandwidth, reducing the filter bandwidth does not reduce the signal power and the result is an increase in SNR by a factor of N. The second view is that when summing N samples in the integrator the signal samples (all the same d.c. value) add coherently, resulting in a voltage increase by a factor of N and a power increase by a factor of N^2. The noise samples are uncorrelated and add non-coherently, resulting in an increase in power by a factor of N. SNR increases, then, by a factor of $N^2/N = N$. Since an additional bit of precision corresponding to a factor of 4 decreases in noise power, every increase in integration period by a factor of 4 produces an additional bit of effective resolution. Therefore, the resolution of the voltmeter is not strictly limited by the resolution of the ADC itself, but can be increased by integrating over multiple samples.

4.3.2 DIFFERENTIAL VS. SINGLE-ENDED VOLTAGE MEASUREMENT

Some EIT instruments measure differential voltages, i.e. voltages between a pair of electrodes, while others measure single-ended voltages, where the measurement is made with respect to ground potential. Each approach has its advantages and disadvantages. The primary advantage of performing differential measurements is the fact that the voltage between a pair of electrodes may be significantly smaller than the voltage between each individual electrode and ground potential, particularly when the electrodes are located near each other on the body. This smaller voltage may result in a reduction in the dynamic range of the voltage signals being measured, which, in turn, reduces the dynamic range requirements for the ADC. Differential voltage measurements are used extensively in pair-drive systems in which the voltages are measured only on non-current carrying electrodes, and differential voltages between adjacent electrodes can be much smaller than the single-ended voltages. In practice, the voltage difference between a pair of electrodes is generally converted to a single-ended voltage by an instrumentation amplifier for processing by the voltage measurement system. In parallel systems, particularly those that measure voltages on current carrying electrodes, the fact that adjacent electrodes may be carrying large currents with opposite polarity makes using differential measurements less advantageous.

The primary disadvantage of differential voltage measurements is a loss of precision due to nonzero common-mode amplifier gain. Figure 4.21(a) shows an instrumentation amplifier and its inputs and outputs. These inputs can be expressed in terms of a differential signal, $V_D = V_1 - V_2$, and

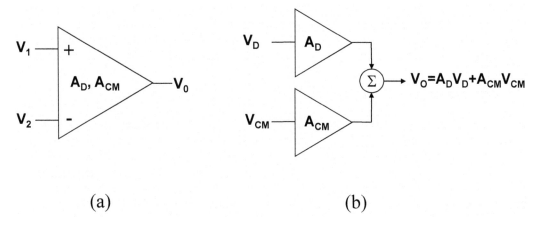

(a) (b)

Figure 4.21 Behaviour of an instrumentation amplifier: (a) amplifier showing actual inputs; (b) block diagram showing how the output is produced from differential and common-mode inputs.

a common-mode signal, $V_{CM} = (V_1 + V_2)/2$. If the instrumentation amplifier is ideal, the common-mode gain is zero and the output is determined solely by the differential gain A_D and the difference between the input voltages

$$V_0 = A_D V_D = A_D(V_1 - V_2). \tag{4.11}$$

A real instrumentation amplifier, however, will respond to both V_D and V_{CM}, and its output is given by

$$V_0 = A_D V_D + A_{CM} V_{CM}, \tag{4.12}$$

where A_{CM} is the common-mode gain. Figure 4.21(b) is a block diagram that illustrates the behaviour of the instrumentation amplifier. A figure of merit for an instrumentation amplifier is its common-mode rejection ratio (CMRR) given by

$$CMRR = 20 \log_{10}(A_D/A_{CM}). \tag{4.13}$$

While an ideal differential amplifier has a CMRR of infinity, real instrumentation amplifiers generally have a CMRR that is large at d.c. and drops with increasing frequency. Typical CMRR values at d.c. are in the range 100–120 dB, while values at 1 MHz that are in the range 0–60 dB are common. The common-mode rejection of an instrumentation amplifier is degraded when there is an imbalance between the driving impedances for each input. Figure 4.22 shows an instrumentation amplifier with capacitors C_i representing its input capacitance. A common-mode voltage is applied through unequal resistances, R_1 and R_2. The impact of the unequal driving resistances is that the common

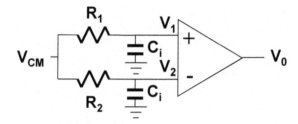

Figure 4.22　Instrumentation amplifier with input capacitance and driving impedances.

mode input signal produces a differential voltage between the inputs to the instrumentation amplifier. This differential voltage is then multiplied by the differential gain of the amplifier to produce and output, even if the common-mode gain of the instrumentation amplifier itself is zero. As discussed in [218] the degradation in common-mode rejection due to mismatches in driving impedance impacts the reactive part of the voltage more severely than the real part. Therefore, as with the case of stray capacitance impacting the application of current, using only the real part voltage from the output of the instrumentation amplifier mitigates the performance loss that this effect produces.

4.3.3　COMMON-MODE VOLTAGE FEEDBACK

Since it is difficult to achieve sufficient insensitivity to common-mode voltage, particularly at higher frequencies, some systems employ a voltage feedback system to reduce the common-mode voltage presented to the instrumentation amplifier [897,983]. Since an ideal current source will produce a current that is independent of its load voltage, it is possible, in principle, to vary the load voltage in a way that minimizes the common-mode voltage seen by the differential voltage amplifier without affecting the applied current. In practice, however, the finite output impedance and/or stray capacitance will produce some variation in current with changes in load voltage, and the load voltage must be kept within the voltage compliance of the current source. The compensation systems apply a voltage to an additional electrode, typically located away from the electrodes being used for imaging, that minimizes the common-mode voltage seen by the instrumentation amplifier.

4.4　EIT SYSTEMS

Many EIT instruments have been built during the 40 years that investigators have been developing the technology for medical applications. Early on, a small number of investigators built relatively simple prototypes that demonstrated the feasibility of EIT and began the evolution towards better, more capable systems. Some of the advances made through the years have been described in this chapter. In recent years, the community of investigators building and using EIT hardware as expanded significantly. This growth is likely fueled by an increase in interest in EIT and EIT applications as well as the availability of modular hardware that simplifies building an instrument. The availability of commercial EIT systems is also part of this expansion. Details of many research and commercial EIT systems are provided in chapter 20, based on the literature and requests to the designers.

EIT systems can be broadly classified by whether their use a pair-drive or parallel-drive architecture. The choice of which approach to use is fundamentally one of complexity versus performance, with a single-source system having much simpler hardware and a multiple-source system having better performance. Both approaches were developed early and are still used today, though pair-drive systems are much more common. Table 4.1 lists several EIT systems, with links to detailed descriptions. There is a concentration on some of the earlier systems that made large advances in the way that EIT systems are constructed.

Table 4.1
Research EIT systems

Device	Details	Pair/Parallel
Sheffield Mk 2	20.3.2.1	Pair
Goettingen Goe MF II	20.3.2.2	Pair
École Polytechnique de Montréal Sigmatôme	20.3.2.3	Pair
Russian Academy of Sciences: Breast Imaging system	20.3.2.4	Pair
Middlesex University: CRADL system	20.3.2.5	Pair
Rensselaer Polytechnic Institute: ACT3	20.3.2.6	Parallel
Kyung Hee University KHU Mark 2.5	20.3.2.7	Parallel
Dartmouth University Broadband High Frequency System	20.3.2.8	Parallel

A number of commercial EIT systems are now available, enabling investigators interested in developing EIT applications to avoid having to build custom hardware. The commercial systems all use a serial architecture, though information about the specifics of their designs are not readily available. Table 4.2 summarizes the operating parameters for several of these commercial systems.

Table 4.2
Commercially available EIT systems

Device	#Electrodes	Frames/s	Frequency	Details	Pair/Parallel
Drager PulmoVista 500[1]	16	50	80-130 kHz	20.3.1.1	Pair
Sentec[2] BB^2	32	50	<150 kHz	20.3.1.2	Pair
Timpel Enlight 1800[3]	32	50	<125 kHz	–	Parallel

4.5 CONCLUSION

This chapter has reviewed various approaches for implementing the major components of an EIT system and discussed some of the advantages and disadvantages of each approach. A few example systems were presented to show how these components have been combined to produce EIT instruments. The design of a particular system is driven by the application, required precision, and tolerance for cost and complexity.

The future EIT hardware will likely be with ASIC- and SoC-based, active electrode systems. The integration of EIT circuit functions is an active area of research, with progress being made in improving all aspects of performance. While current integrated systems provide a significant advantage in terms of form factor, they still lag behind discrete hardware systems in performance. This gap will certainly disappear as we move forward.

5 The EIT Forward Problem

Andy Adler
Systems and Computer Engineering, Carleton University, Ottawa,
Canada

William R. B. Lionheart
Department of Mathematics, University of Manchester, UK

CONTENTS

5.1 INTRODUCTION

The story so far is shown in figure 5.1: an EIT system has been attached to a body containing tissues with interesting electrical properties; it makes voltage measurements at body-surface electrodes, as described in §4.3, and our next task is to reconstruct images from these measurements (chapter 6). But in order to reconstruct images, we need to know what each measurement "means": this is the *forward problem*. This chapter describes how EIT measurements are related to the spatial distribution of impedance within a body.

DOI: 10.1201/9780429399886-5

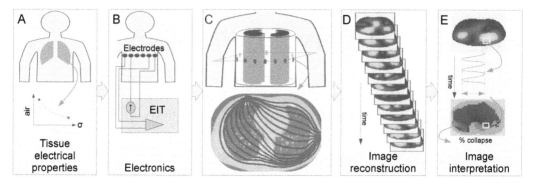

Figure 5.1 Overview of the steps in EIT image generation and interpretation, in the context of the *forward problem*. (A) tissue electrical properties; (B) EIT systems hardware; (C) The pattern of electrical stimulation and measurements, and the body geometry and electrical properties, determine the sensitive region; (D) image reconstruction; and (E) image interpretation. In (C) the body is modelled with a FEM in the chest region and the streamlines of current between two electrodes calculated. (color figure available in eBook)

The concept of a *forward problem* is contrasted with that of an *inverse problem* (see Box 6.8). Image reconstruction (the "inverse") describes the process of beginning with the effects (measurements) and calculating the causes (parameters). We thus need a model of the forward process, describing how the causes (parameters) lead to effects (measurements).

In EIT, two main approaches to the foward problem calculation have been used, analytical and finite element models (FEM). Analytical models are useful for regular shapes such as cylinders, and to help understand the theoretical limits. For irregular geometries, such as for a patient's body, the FEM has been the main tool. The forward model calculations allow an understanding of EIT system configurations: where are electrodes best placed to achieve the best detectability of a given process? how accurate do models of shape and electrodes need to be? and how big is the effect of electrode movement (due to breathing[27] and posture change[214])?

5.2 MATHEMATICAL SETTING

Our starting point for consideration of EIT should be Maxwell's equations. But for simplicity let us assume direct current or sufficiently low a frequency current that the magnetic field can be neglected.

We have a given body Ω a closed and bounded subset of three-dimensional space with a smooth (or smooth enough) boundary $\partial\Omega$. The body has a conductivity σ which is a function of the spatial variable $\mathbf{x}$ (although we will not always make this dependence explicit for simplicity of notation). The scalar potential is ϕ and the electric field is $\mathbf{E} = -\nabla\phi$. The current density is $\mathbf{J} = -\sigma\nabla\phi$, which is a continuum version of Ohm's law. In the absence of interior current sources, we have the continuum Kirchhoff's law[1]

$$\nabla \cdot \sigma\nabla\phi = 0 \qquad (5.1)$$

The current density on the boundary is

$$j_\mathbf{n} = -\mathbf{J} \cdot \mathbf{n} = \sigma\nabla\phi \cdot \mathbf{n}$$

where $\mathbf{n}$ is the outward unit normal to $\partial\Omega$. Given σ, specification of the potential $\phi|_{\partial\Omega}$ on the boundary (Dirichlet boundary condition) is sufficient to uniquely determine a solution ϕ to (5.1).

---

[1]There is a recurring error in the EIT literature of calling this Poisson's equation, it however a natural generalisation of the Laplace equation.

Similarly specification of boundary current density $j_{\mathbf{n}}$ (Neumann boundary conditions) determines ϕ up to an additive constant, which is equivalent to choosing an earth point. From Gauss' theorem, or conservation of current, the sum of all currents into and out of the body must be zero, and the boundary current density must satisfy the consistency condition $\int_{\partial\Omega} j_{\mathbf{n}} = 0$.

The ideal complete data in the EIT reconstruction problem is to know all possible pairs of Dirichlet and Neumann data $\phi|_{\partial\Omega}, j_{\mathbf{n}}$ (pairs of of voltage measurements and current stimulation). As any Dirichlet data determines unique Neumann data we have a Dirichlet-to-Neumann operator $\Lambda_\sigma : \phi|_{\partial\Omega} \mapsto j$.

In electrical terms, Λ_σ is the admittance matrix at the boundary, and can be regarded as the response of the system we are electrically interrogating at the boundary. Given a body with a large number of electrodes, and voltages $\mathbf{v}$ and currents $\mathbf{i}$ on the electrodes, we have an $\mathbf{i} = \mathbf{Y}\mathbf{v}$, where the admittance matrix $\mathbf{Y}$ is a discretization of Λ_σ (see Box 6.9).

5.2.1 QUASI-STATIC APPROXIMATION

Practical EIT systems use sinusoidal currents at fixed frequency, $f = \omega/(2\pi)$. The electric field, current density and potential are all represented by complex phasors multiplied by $e^{i\omega}$ (or sometimes $e^{j\omega}$). See Box 5.1 on phasor notation.

Box 5.1: Phasor Notation

Phasor notation is convenient way to express sinusoidally varying quantities as a complex number. A voltage, $\tilde{V}(t)$ applied to a resistor, R, and capacitor, C, in parallel, leads to a current

$$\tilde{I}(t) = R^{-1}\tilde{V}(t) + C\frac{d}{dt}\tilde{V}(t).$$

Using phasor notation, $\tilde{V}(t) = V\cos\omega t = \mathrm{Re}[Ve^{i\omega t}]$. The most useful aspect of this notation is that the derivative functions as a multiplication by $i\omega$ and

$$I = R^{-1}V + i\omega CV = (R^{-1} + i\omega C)V = YV$$

for an admittance Y. Here, the angle $\angle Y$ represents the phase offset between I and V.

Ignoring magnetic effects (see Box 5.2) we replace the conductivity σ in 5.1 by the complex *admittivity* $\gamma = \sigma + i\omega\varepsilon$ where ε is the permittivity. Elsewhere in this book, we use notation $\sigma^* = \gamma$. In biological tissue one can expect ε to be frequency dependent (§3.1.2) which becomes important in a multi-frequency system (§4.3).

Box 5.2: Maxwell's Equations

The basic field quantities in Maxwell's equations are the electric field, $\vec{E}(\mathbf{x},t) = \mathrm{Re}[\mathbf{E}(\mathbf{x})e^{i\omega t}]$ and the magnetic field, $\vec{B}(\mathbf{x},t) = \mathrm{Re}[\mathbf{B}(\mathbf{x})e^{i\omega t}]$, vector valued functions of space varying at the simulation frequency and time.

We will assume that there is no relative motion in our system. The fields, when applied to a material or indeed a vacuum, produce fluxes — electric displacement $\mathbf{D}$ and magnetic flux $\mathbf{B}$. The charge density is $\nabla \cdot \mathbf{E} = \rho$, and $\mathbf{J}$ is the electric current density. There are no magnetic monopoles, $\nabla \cdot \mathbf{B} = 0$. The material properties appear as relations between fields and fluxes. The simplest case is of non-dispersive, local, linear, isotropic media. The magnetic permeability is then a scalar function $\mu > 0$ of space and the material response is $\mathbf{B} = \mu \mathbf{H}$. Similarly the permittivity $\mathbf{D} = \varepsilon \mathbf{E}$ with $\varepsilon > 0$. In a conductive medium we have the continuum counterpart to Ohm's law where the conduction current density $\mathbf{J_c} = \sigma \mathbf{E}$. The total current is then $\mathbf{J} = \mathbf{J_c} + \mathbf{J_s}$ the sum of the conduction and source currents.

Using phasor notation, the spatial and temporal variations of the fields and fluxes are linked by Faraday's Law of induction,

$$\nabla \times \mathbf{E} = -\frac{\partial \mathbf{B}}{\partial t} = -i\omega \mathbf{B} = -i\omega \mu \mathbf{H}$$

and Coulomb's law

$$\nabla \times \mathbf{H} = \frac{\partial \mathbf{D}}{\partial t} + \mathbf{J} = i\omega \mathbf{D} + \mathbf{J} = i\omega \varepsilon \mathbf{E} + \mathbf{J}. \tag{5.2}$$

We can combine conductivity and permittivity as a complex admittivity $\sigma + i\omega\varepsilon = \gamma$ and write (5.2) as

$$\nabla \times \mathbf{H} = (\sigma + i\omega\varepsilon)\mathbf{E} + \mathbf{J_s} = \gamma \mathbf{E} + \mathbf{J_s}$$

and $\nabla \cdot (\gamma \mathbf{E} + \mathbf{J_s}) = 0$ by the identity $\nabla \cdot (\nabla \times \mathbf{H}) = 0$. In EIT the source term $\mathbf{J_s}$ is typically zero at frequency ω. The quasi-static approximation usually employed in EIT is to assume $\omega\mu\mathbf{H}$ is negligible so that $\nabla \times \mathbf{E} = 0$ and hence on a simply connected domain $\mathbf{E} = -\nabla\phi$ for a scalar potential distribution ϕ, and thus

$$\nabla \cdot \gamma \nabla \phi = 0. \tag{5.3}$$

The quasi-static approximation is valid for bodies of moderate conductivity, if the dimentions of interest are much smaller than the electromagnetic wavelength, $\lambda = c/f$ and the skin depth, $\delta = 1/\sqrt{\pi f \mu \sigma}$ (see Box 5.3). The skin depth characterizes the outer layer of a conductor where most electric current flows; it is is most pronounced with very conductive materials in which significant eddy currents flow. If we use 200 kHz electrical stimulation on a patient filled with conductive blood $\sigma = 0.7\,\mathrm{S/m}$, $\lambda = 1500\,\mathrm{m}$ and $\delta = 4.8\,\mathrm{m}$. These values are greater than the size of patient organs under EIT investigation.

Box 5.3: Limits of the Quasi-static Approximation

The general formulation for the electric field is

$$\mathbf{E} = -(\nabla\phi + \frac{\partial}{\partial t}\mathbf{A}) = -(\nabla\phi + i\omega\mathbf{A}) \tag{5.4}$$

where $\mathbf{A}$ is the magnetic vector potential, and

$$\nabla^2\mathbf{A} = \mu\mathbf{J} = \mu\gamma\mathbf{E}.$$

The second term can be ignored if $\omega\mathbf{A}$ contributes a small part of $\mathbf{E}$. Following [603], we define a smallest length scale l over which fields vary and approximate $[\nabla^2\mathbf{A}]_{max} \sim \frac{1}{l^2}\mathbf{A}$ and thus require $[\omega\mathbf{A}]_{max} \sim \omega l^2\mu\gamma\mathbf{E} \ll \mathbf{E}$. Here

$$\omega l^2\mu(\sigma + i\omega\varepsilon) = \frac{1}{2}\left(\frac{l}{\delta}\right)^2 + 4\pi^2 i\left(\frac{l}{\lambda}\right)^2 \ll 1 \tag{5.5}$$

where $\lambda = c/f$ is the wavelength, $c = 1/\sqrt{\mu\varepsilon}$ is the speed of light in the medium, and $\delta = 1/\sqrt{\omega\sigma\mu/2}$ is the skin depth. Thus, the quasi-static approximation is valid if the length scales over which fields change are smaller than both the electromagnetic wavelength and the skin depth.

5.3 CURRENT PROPAGATION IN CONDUCTIVE BODIES

In order to understand EIT imaging, it is useful to have an appreciation for how current propagates in a conductive body. We start with the case where current flows in one direction and the body is uniform except for a small region which is more or less conductive than the background. Figure 5.2 shows uniform current flow around some simple conductivity contrasts. Conductive regions "pull" inward the currents, while non-conductive regions "push" them out. As the magnitude of the contrast ratio increases, the effect becomes more pronounced. However there is a limit, or saturation. Little further change occurs when the conductivity ratio goes from 100 to 1000. Next, we notice a shape effect. In 2D, a circular region creates the same magnitude of perturbation whether it is conductive or non-conductive. An ellipsoidal region, by contrast, a more complex effect. Non-conductive ellipsoidal regions have most effect when across (blocking) the flow, and conductive regions have most effect parallel to the flow. In this case, current is drawn toward the conductor so it can take an "easier" path.

Next, we look at this non-linearity in more detail. The sensitivity $S(\sigma_c)$ is the change in voltages measured for a given change in the conductivity of a contrasting region. Later §5.6 we will study sensitivity in more detail and formulate it as a Jacobian, $\mathbf{J}$. Figure 5.3 shows the relative sensitivity as a function of the contrast ratio σ_c of a cylindrical region to the background of the body. The sensitivity for a very conductive contrast is normalized to one. Note the saturation: sensitivity is linear for small changes in σ_c, but stops increasing at ratios of about 100:1. Next, we note that EIT is less sensitive to non-conductive changes than it is to conductive ones, and that this difference depends on the shape of the object. When the inclusion is tall, a non-conductive object has more similar signal to a conductive one. A full hight object is vertically uniform, and thus more similar to the 2D case, where conductive and non-conductive cylinders far from the boundary have the same sensitivity magnitude.

In a body, non-uniform tissues lead to complex patterns of current flow. Figure 5.4 shows a simulation of current injection between two electrodes in a FEM of the body based on a thoracic CT.

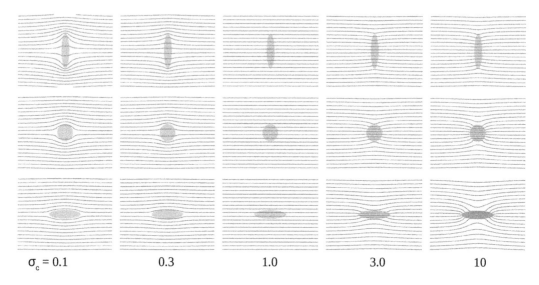

$\sigma_c = 0.1$ 0.3   1.0   3.0   10

Figure 5.2   Streamlines in a uniform conductor perturbed with an elliptical contrast region. The relative conductivity of the region to the background σ_c varies from 0.1 (left) to 10 (right). A conductive contrast has the most effect when it is along the streamline (bottom right), while a non-conductive contrast has most effect perpendicular to the streamline. (color figure available in eBook)

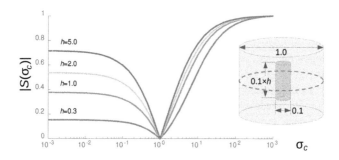

Figure 5.3   Normalized relative absolute EIT sensitivity, $|S(\sigma_c)|$, as a function of the shape and conductivity of a cylindrical inclusion in the centre. The stimulation configuration (subfigure at right) has 16 electrodes in a central plane (dotted), and has a contrasting cylindrical ROI with a height/diameter, h, and conductivity σ_c while elsewhere $\sigma = 1$. The graph shows the normalized EIT signal, $S(\sigma_c)$, as a function of σ_c for four values of h. (color figure available in eBook)

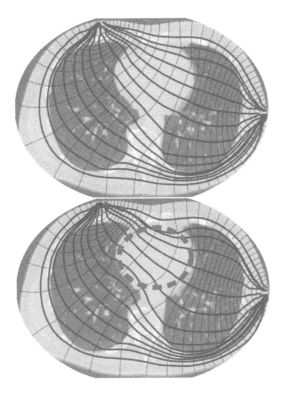

Figure 5.4   Illustration of the propagation of current in a body. In each image, a FEM of a volumetric model of the thorax is used to simulate the propagation of electric current from a pair of surface electrodes with the indicated current source. Blue lines show current streamlines while the black lines are isopotential surfaces. From left to right, an increase of conductivity in the heart (dotted red lines) is simulated. (color figure available in eBook)

A set of equipotential lines and current streamlines are shown. It is interesting to see how the current "avoids" travelling through non-conductive lungs, and takes a longer path to travel through tissue. When the conductivity changes, the voltage distribution and current flow changes, as illustrated by the difference made by a doubling of the conductivity of the heart region (to simulate the increased blood volume during diastole). Two effects can be seen, first current streamlines are drawn to the conductive heart, and the equipotential lines space further apart (given the higher conductivity). It is worth noting how little the equipotential lines on the body surface change even with a large internal change. This is one of the reasons that EIT is difficult (§6.2).

5.3.1   ANALYTICAL SOLUTIONS

Analytic solutions are possible for regular shapes. For arbitrary shapes we require numerical methods (§5.7). However, an understanding of analytical solutions is useful to understand limits of EIT sensitivity.

   To enclude discontinuities in conductivity and boundary conditions, we will need to include weak solutions, but still have a measure of how smooth these solutions are and a way to measure convergence of our numerical approximations. We do this using Sobolev spaces and norms, see Box 5.4.

Box 5.4: Sobolev Spaces

In the mathematical literature you will often see the assumption that ϕ lies in the Sobolev Space $H^1(\Omega)$, which can look intimidating to the uninitiated. Actually these spaces are easily understood on an intuitive level and have a natural physical meaning. For mathematical details see Folland[281]. A (generalized) function f is in $H^k(\Omega)$ for integer k if the square kth derivative has a finite integral over Ω. For non-integer and negative powers Sobolev spaces are defined by taking the Fourier transform, multiplying by a power of frequency and demanding that the result is square integrable. For the potential we are simply demanding that $\int_\Omega |\nabla\phi|^2 dV < \infty$ which is equivalent, provided the conductivity is bounded, to demanding that the ohmic power dissipated is finite. An obviously necessary physical constraint. Sobolev spaces are useful as a measure of the smoothness of a function, and are also convenient as they have an inner product (they are Hilbert spaces). To be consistent with this finite power condition the Dirichlet boundary data $\phi|_{\partial\Omega}$ must be in $H^{1/2}(\partial\Omega)$ and the Neumann data $j \in H^{-1/2}(\partial\Omega)$. Note that the current density is one derivative less smooth than the potential on the boundary as one might expect.

5.3.2 CIRCULAR ANOMALY IN A UNIT DISK

To understand this problem further it is best to use a simple example. Let us consider a unit disk in two dimensions with conductivity σ_b and a concentric circular anomaly in the conductivity, σ_i

$$\sigma(\mathbf{x}) = \left\{ \begin{array}{ll} \sigma_i & \rho < |\mathbf{x}| < 1 \\ \sigma_b & |\mathbf{x}| \le \rho \end{array} \right. .$$

Although this is a two dimensional example, it is equivalent to a three dimensional cylinder with a central cylindrical anomaly, electrodes the full hight of the cylinder, and insulating faces.

The forward problem can be solved by separation of variables giving

$$\Lambda_\sigma[\cos k\theta] = k\frac{1+\mu\rho^{2k}}{1-\mu\rho^{2k}}\cos k\theta \tag{5.6}$$

and similarly for $\sin k\theta$, where

$$\mu = \frac{\sigma_i - \sigma_b}{\sigma_i + \sigma_b},$$

which has a symmetric sigmoid shape: $\mu = -1$ when $\sigma_i = 0$, $\mu = 1$ if $\sigma_i = \infty$ and $\mu = 0$ when $\sigma_i = \sigma_b$. We can now express any arbitrary Dirichlet boundary data as a Fourier series

$$\phi(1,\theta) = \sum_{k}^{\infty} a_k \cos k\theta + b_k \sin k\theta$$

and notice that the Fourier coefficients of the current density are $k(1+\mu\rho^{2k})/(1-\mu\rho^{2k})a_k$ and similarly b_k. The lowest frequency component is clearly most sensitive to the variation in the conductivity of the anomaly. This shows that patterns of voltage (or current) with large low frequency components are best able to detect an object near the centre of the domain. This might be achieved for example by covering a large proportion of the surface with driven electrodes and exciting a voltage or current pattern with low spatial frequency. For a pair-drive system, separating the stimulation pair will improve this sensitivity at depth. We will explore this further in section 5.5.3.

We also see again the non-linearity of EIT from this simple example – saturation. Fixing the radius of the anomaly and varying the conductivity we see that for high contrasts from the effect on the voltage of further varying the conductivity is reduced. A detailed analysis of the circular anomaly was performed by Seagar[949] using conformal mappings, including offset anomalies. It is found of course that a central anomaly produces the least change in boundary data. This illustrates the positional dependence of the ability of EIT to detect an object. By analogy to conventional imaging problems one could say that the "point spread function" is position dependent.

In three dimensions, many of these lessons from the circular disk remain. The sensitivity to anomalies increases with increased low-order Fourier components (and the separation of the electrodes). However, sensitivities in 3D are much lower than 2D. This can be seen looking at the voltage distribution at a radius r from a point current source, $V_{3D} \sim 1/r$ while $V_{2D} \sim -\log(r)$. The 3D voltage decreases much more rapidly.

In 3D, when then boundary conditions have axial symmetry (no dependence on an azimuthal angle), a solution to Laplace's equation can be written in terms of a sum of Legendre polynomials, $P_n(\cos\theta)$. For a spherical anomaly, the perturbation has a factor[700].

$$\mu_{\text{sphere}} = \frac{n(\sigma_i - \sigma_b)}{n(\sigma_i + \sigma_b) + \sigma_b} = \frac{\mu}{1 + \frac{\sigma_b}{n(\sigma_i + \sigma_b)}}$$

and μ_{sphere} is less than μ for the 2D system if $\sigma_i < \infty$. This effect explains the decreased saturation at the non-conductive limit (figure 5.3).

To be linear, a system must be scale invariant, $f(\alpha x) = \alpha f(x)$, and respect superposition, $f(x + y) = f(x) + f(y)$. We have seen that the saturation of voltages for large conductivity contrasts means that EIT is not homogeneous. EIT is also not additive. When two contrasting regions are close, the current perturbed by the one will affect the other, and the overall behaviour will not be the same as sum from the regions individually.

5.3.3   DETECTABILITY

Detectability or *Distinguishability* is a measure of the reliability with which a system can detect changes of interest. We want to distinguish between two scenarios: (A) H_0, from which we measure, on average, vector $\mathbf{v}_0$, from (B) H_1 and $\mathbf{v}_1$. Under all circumstances we have additive noise $\mathbf{n}$. We image the object and select the average in a region of interest, a process which can be described by a norm $\|\cdot\|_{\mathbf{R}}$. We now have a statistical detection problem: how reliably can we reject H_0 with measurement variance $\mathbb{E}[\|\mathbf{n}\|_{\mathbf{R}}^2]$.[23,500,629]

Our central circular anomaly also demonstrates the ill-posed nature of the problem. For a given level of measurement precision, we can construct a circular anomaly undetectable at that precision. We can make the change in conductivity arbitrarily large and yet by reducing the radius we are still not be able to detect the anomaly.

While still on the topic of a single anomaly, it is worth pointing out that finding the location of a single localised object is comparatively easy, and with practise one can do it crudely by eye from the voltage data. Box 5.5 describes the disturbance to the voltage caused by a small object and explains why, to first order, this is the potential for a dipole source. This idea can be made rigorous and Ammari[50] and Seo[518] show how this could be applied locating the position and depth of a breast tumour using data from a T-Scan measurement system.

Box 5.5: Sensitivity to a Localised Change in Conductivity

Studying the change in voltage from a small localised change in conductivity is a useful illustration of EIT. Suppose we fix a current pattern, and a background conductivity of γ, which results in a potential ϕ. Now consider a perturbed conductivity $\gamma + \delta\gamma$ which results in a potential, with the same current drive, $\phi + \delta\phi$. From $\nabla \cdot (\gamma + \delta\gamma)\nabla\phi = 0$ we see that

$$\nabla \cdot \delta\gamma\nabla\phi + \nabla \cdot \gamma\nabla\delta\phi + \nabla \cdot \delta\gamma\nabla\delta\phi = 0$$

The same procedure used to calculate the Jacobian can be used to show that the last term is $O(\delta\gamma^2)$ so that to first order

$$\nabla \cdot \gamma\nabla\delta\phi \approx -\nabla \cdot \delta\gamma\nabla\phi$$

now for simplicity take $\gamma = 1$ and we have the Poisson equation for $\delta\phi$

$$\nabla^2\delta\phi = \nabla\delta\gamma \cdot \nabla\phi$$

If we now take $\delta\gamma$ to be a small change, constant on a small ball near some point p, then the source term in this Poisson equation approximates a dipole at p whose strength and direction is given by $\nabla\phi$.

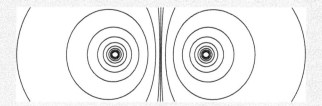

Figure 5.5   Equipotential lines around a dipole in 3D.

Observing $\delta\phi$ at the boundary we see it as a dipole field from which a line through p can be estimated by eye. This goes some way to explain the ease with which one small object can be located, even with only a small number of current patterns. It also illustrates the depth dependence of the sensitivity as the dipole field decays with distance even if the electric field is relatively uniform. Typically the electric field strength is also less away from the boundary. This approach was further developed by Hancke[428], to locate small objects from the perturbed field. Is had been further hypothesized that electro-locating fish may use a similar method[550].

5.4   MEASUREMENTS AND ELECTRODES

An electrical imaging system uses conducting electrodes attached to the surface of the body under investigation. One can apply current or voltage to these electrodes, and a sequence of measurements, y_i, are made using different *stimulation and measurement patterns*, illustrated in figure 5.6.

In a current-drive system, for each measurement i, a *drive pattern* is applied, with current $I_{l,i}$ into each electrode l. This current leads to voltages $V_{l,i}$ on electrodes, and depends on the admittivity of the body. Measurement y_i is generated by a *measurement pattern*, of (complex) voltage gains at each electrode, $G_{l,i}$, and

$$y_i = \sum_l V_{l,i}\overline{G}_{l,i}. \tag{5.7}$$

The complex conjugate of the gains corresponds to demodulation: when sinusoidal voltages are demodulated, the phase of the demodulation clock subtracts from the phase of the signal. This detail is rarely important; normally $G_{l,i}$ is real.

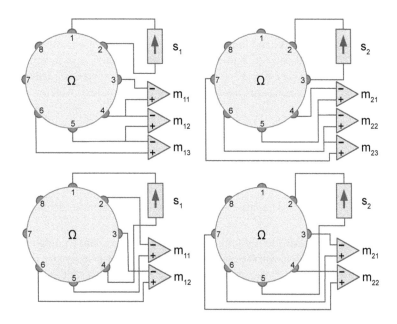

Figure 5.6 Stimulation and measurement patterns using an adjacent pattern (*top row*) and a "skip 2" pattern (*bottom row*) for an eight electrode EIT system, in which measurements are not made on driven electrodes. The first column shows the measurements $m_{11} \ldots m_{1n}$ made during the first stimulation s_1, and the second column shows the measurements $m_{21} \ldots m_{2n}$ during the second stimulation. (color figure available in eBook)

A complete set of measurements is a *frame* of EIT data, and the rate at which sets of measurements are acquired is the *frame rate*. In figure 5.6, the top row shows an adjacent stimulation and measurement pattern. Current is injected between electrodes 1 and 2, and voltage is measured on pairs (3,4), (4,5), ..., (7,8). Next current injection is moved to electrodes 2 and 3, and the next set of voltages read. Thus, measurement #1, with stimulation on electrodes $1 \to 2$ and measurement $3 \to 4$, has $I_{1\ldots8,1} = [+1,-1,0,0,0,0,0,0]$ and $G_{1\ldots8,1} = [0,0,-1,+1,0,0,0,0]$.

Adjacent stimulation and measurement patterns have the limitation that current does not penetrate deeply into the body, and separating the electrode pairs increases detectability (§5.5.3). Pair drive patterns with increased electrode spacing are defined in terms of a *skip* parameter, such that electrode 1 is paired with electrode 2+*skip*. The bottom row of figure 5.6 shows stimulation and measurement patterns with *skip*=1.

Electromagnetic measurements are characterized by *reciprocity*, which means that the measured value remains the same if current drive and voltage measurement patterns are interchanged. In the EIT literature, the reciprocity principle is commonly cited as[340], although it was known much much earlier[652]. Reciprocity reduces the number of independent measurements available from EIT data, but can also be used to test for data consistency and electrode errors[442].

Let us suppose that the subset of the boundary in contact with the l-th electrode is E_l, and $1 \leq l \leq L$. For one particular measurement the voltages (with respect to some arbitrary reference) and currents are arranged in vectors as $\mathbf{v}$ and $\mathbf{i} \in \mathbb{C}^L$. The discrete equivalent of the Dirichlet-to-Neumann Λ map is the transfer admittance, or mutual admittance, matrix $\mathbf{Y}$ which is defined by $\mathbf{i} = \mathbf{Y}\mathbf{v}$.

Assuming that the electrodes are perfect conductors for each l we have that $\phi|_{E_l} = V_l$, a constant. Away from the electrodes where no current flows $\partial\phi/\partial\mathbf{n} = 0$. This mixed boundary value problem is well-posed, and the resulting currents are $I_l = \int_{E_l} \sigma \partial \phi / \partial \mathbf{n}$. It is easy to see that the vector $\mathbf{1} = (1, 1, \ldots, 1)^T$ is in the null space of $\mathbf{Y}$, and that the range of $\mathbf{Y}$ is orthogonal to the same vector. Let S be the subspace of $\mathbb{C}^L$ perpendicular to $\mathbf{1}$ then it can be shown that $\mathbf{Y}|_S$ is invertible from S to S. The generalized inverse (see §6.4) $\mathbf{Z} = \mathbf{Y}^\dagger$ is called the transfer impedance. This follows from uniqueness of solution of the so called *shunt model* boundary value problem, which is (5.1) together with the boundary conditions

$$\int_{E_l} \sigma \partial \phi / \partial \mathbf{n} = I_l \text{ for } 0 \leq l \leq L \tag{5.8}$$

$$\partial \phi / \partial \mathbf{n} = 0, \text{ on } \Gamma' \tag{5.9}$$

$$\nabla \phi \times \mathbf{n} = 0 \text{ on } \Gamma \tag{5.10}$$

where $\Gamma = \bigcup_l E_l$ and $\Gamma' = \partial \Omega - \Gamma$. The last condition (5.10) is equivalent to demanding that ϕ is constant on electrodes.

The transfer admittance, or equivalently transfer impedance, represents a complete set of data which can be collected from the L electrodes at a single frequency for a stationary linear medium. From reciprocity we have that $\mathbf{Y}$ and $\mathbf{Z}$ are symmetric (but for $\omega \neq 0$ *not* Hermitian). The dimension of the space of possible transfer admittance matrices is clearly no bigger than $L(L-1)/2$, and so it is unrealistic to expect to recover more unknown parameters than this. In the case of planar resistor networks the possible transfer admittance matrices can be characterized completely[204], a characterization which is known at least partly to hold in the planar continuum case[499]. A typical electrical imaging system applies current or voltage patterns which form a basis of the space S, and measures some subset of the resulting voltages which as they are only defined up to an additive constant can be taken to be in S.

The shunt model with its idealization of perfectly conducting electrodes predicts that the current density on the electrode has a singularity of the form $O(r^{-1/2})$ where r is the distance from the edge of the electrode. The potential ϕ while still continuous near the electrode has the asymptotics $O(r^{1/2})$. Although some electrodes may have total current $I_l = 0$ as they are not actively driven the shunting effect means that their current density is not only non-zero but infinite at the edges.

In a medical application with electrodes applied to the skin, and in phantom tanks with ionic solutions in contact with metal electrodes, a contact impedance layer exists between the solution or skin and the electrode. This modifies the shunting effect so that the voltage under the electrode is no longer constant. The voltage on the electrode is still a constant V_l so now on E_l there is a voltage drop across the contact impedance layer

$$\phi + z_l \sigma \frac{\partial \phi}{\partial \mathbf{n}} = V_l \tag{5.11}$$

where the contact impedance z_l could vary over the electrode material E_l but is generally assumed constant. This new boundary condition together with (5.8) and (5.9) form the *Complete Electrode Model* or CEM. For experimental validation of this model see[592], theory[995] and numerical calculations[811,1075]. A nonzero contact impedance removes the singularity in the current density, although high current densities still occur at the edges of electrodes (fig 5.7). For asymptotics of ϕ with the CEM see[200]. The contact impedance, z_l properties of an electrode are measured in impedance/areas; thus, a right circular cylinder of hight h and area A, made of material with admittivity γ, with electrodes of contact impedance z_c at each end, will have an impedance $Z = h/(\gamma A) + 2z_c/A^2$.

The singular values (§6.4.3) of $\mathbf{Z}$ sometimes called *characteristic impedances* are sensitive to the electrode model used and this was used by[592] to validate the CEM. With no modelling of electrodes and a rotationally symmetric conductivity in a cylindrical tank, the characteristic impedances tend toward a $1/k$ decay, as expected from (5.6) with sinusoidal singular vectors of frequency k, as the number of electrodes increases.

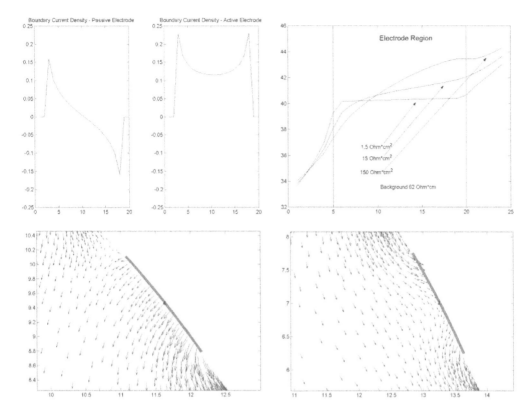

Figure 5.7 The current density on the boundary with the CEM is greatest at the edge of the electrodes[484]. For passive electrodes, current takes a "short cut" through a conductive pathway. This effect is reduced as the contact impedance increases. (A) Current density on the boundary for passive and active electrodes. (B) The effect of contact impedance on the potential beneath an electrode. (C) Interior current flux near an active electrode. (D) Interior current flux near a passive electrode. (color figure available in eBook)

5.5 MEASUREMENT STRATEGY

In EIT we seek to measure some discrete version of Λ_γ or Λ_γ^{-1}. We can choose the geometry of the system of electrodes, the excitation pattern and the measurements that are made. We have to strike a balance between the competing requirements of accuracy, speed and simplicity of hardware.

For a system of L electrodes, the complete relationship between current and voltage at the given frequency is summarized by the transfer impedance matrix $\mathbf{Z} \in \mathbb{C}^{L \times L}$. Each measurement, $y_i = (\bar{\mathbf{g}}_i)^T \mathbf{Z} \mathbf{i}_i$ in terms of the vectors of drive, $\mathbf{i}_i$, and measurement, $\mathbf{g}_i$, patterns. The null space of $\mathbf{Z}$ is spanned by the constant vector $\mathbf{1}$, and for simplicity we set the sum of voltages also to be zero $\mathbf{1}\mathbf{Z} = 0$ so that $\mathbf{Z}$ is symmetric $\mathbf{Z} = \mathbf{Z}^T$ (transpose not conjugate).

The space of contact impedances is a subset of the vector space of symmetric $L \times L$ matrices with column and row sums zero, which has dimension $\frac{1}{2}L(L-1)$. In addition the real part of $\mathbf{Z}|_S$ is positive definite, otherwise there would be direct current patterns which dissipate no power. There are other conditions on $\mathbf{Z}$, given in the planar case by[204], associated with Ω being connected, and its is shown in the planar case that the set of feasible $\mathbf{Z}$ is an open subset of the vector space descried above. This confirms that we can measure up to $\frac{1}{2}L(L-1)$ independent parameters. Some systems however measure fewer than this, primarily to avoid measuring voltage on actively driven electrodes.

This limit, $\frac{1}{2}L(L-1)$, is the maximum number of independent image parameters which can be calculated with L electrodes, and helps explain the low resolution of EIT. For a 16 electrode system $\frac{1}{2}16 \times 15 = 120$, less than the unknowns in a 11×11 pixel image.

5.5.1 LINEAR REGRESSION

We will illustrate the ideas mainly using the assumption that currents are prescribed and the voltages measured, although there are systems which do the opposite. In this approach we regard the matrix of voltage measurements to be contaminated by noise, while the currents are known accurately. This should be compared with the familiar problem of linear regression where we aim to fit a straight line to experimental observations, $y = ax$. For will assume an intercept of zero and mean $\bar{x}$ of zero. The abscissae x_i are assumed accurate and the y_i contaminated with noise. Assembling the x_i and y_i into *row* vectors **x** and **y** we estimate the slope a by

$$\hat{a} = \arg\min_a \|\mathbf{y} - a\mathbf{x}\|^2. \tag{5.12}$$

The solution is the Moore-Penrose generalized inverse (6.1) $a = \mathbf{y}\mathbf{x}^\dagger$, another way of expressing the usual regression formulae. The least squares approach can be justified statistically[731]. Assuming the errors in **y** have zero correlation $\hat{a}$ is an unbiased estimator for a. Under the stronger assumption that the y_i are independently normally distributed with identical variance, $\hat{a}$ is the maximum likelihood estimate of a, and is normally distributed with mean a. Under these assumptions we can derive confidence intervals and hypothesis testing for a[731,p14].

Although less well known, linear regression for several independent variables follows a similar pattern. Now X and Y are matrices and we seek a linear relation of the form $Y = AX$. The estimate $\hat{A} = YX^\dagger$ has the same desirable statistical properties as the single variable case[731].

Given a system of K current patterns assembled in a matrix $\mathbf{I} \in \mathbb{C}^{L \times K}$ (with column sums zero) we measure the corresponding voltages $\mathbf{V} = \mathbf{ZI}$. Assuming the currents are accurate but the voltages contain error we then obtain our estimate $\hat{\mathbf{Z}} = \mathbf{VI}^\dagger$. If we have two few linearly independent currents, rank $\mathbf{I} < L - 1$, then this will be an estimate of a projection of $\mathbf{Z}$ on to a subspace; if we have more than $L - 1$ current patterns then the generalized inverse averages over the redundancy, reducing the variance of $\hat{\mathbf{Z}}$. Similarly we can make redundant measurements: let $\mathbf{M} \in \mathbb{R}^{M \times L}$ be a matrix containing the measurement patterns used (for simplicity the same for each current pattern), so that we measure $\mathbf{V}_m = \mathbf{MV}$. For simplicity we will assume that separate electrodes are used for drive and measurement, so there is no reciprocity in the data. Our estimate for $\mathbf{Z}$ is now $\mathbf{M}^\dagger \mathbf{V}_m \mathbf{I}^\dagger$. For a throrough treatment of the more complicated problem of estimating Z for data with reciprocity see [233]. In both cases redundant measurements will reduce variance. Of course it is common practice to take multiple measurements of each voltage (§4.1.1), and the averaging of these may be performed within the data acquisition system before it reaches the reconstruction program. In this case the effect is identical to using the generalized inverse. The benefit in using the generalized inverse is that it automatically averages over redundancy where there are multiple linearly dependent measurements. If quantization in the analog to digital converter (ADC) is the dominant source of error, averaging over different measurements reduces the error, in a similar fashion to *dithering* (adding a random signal and averaging) to improve the accuracy of an ADC. Some EIT systems use variable gain amplifiers before voltage measurements are passed to the ADC. In this case the absolute precision varies between measurements and a weighting must be introduced in the norms used to define the least squares problem.

For the case where the voltage is accurately controlled and the current measured an exactly similar argument holds for estimating the transfer admittance matrix. However where there are errors in both current and voltage, for example caused by imperfect current sources, a different estimation procedure is required. What we need is *multiple correlation analysis*[731] rather than multiple regression.

One widely used class of EIT systems which use voltage drive and current measurement are ECT systems used in industrial process monitoring[171]. Here each electrode is excited in turn with a positive voltage while the others are at ground potential. The current flowing to ground through the non-driven electrode is measured. Once the voltages are adjusted to have zero mean this is equivalent to using the basis (5.41) for $\mathbf{Y}|_S$.

We know that feasible transfer impedance matrices are symmetric and so employ the orthogonal projection on to the feasible set and replace $\hat{\mathbf{Z}}$ by $\operatorname{sym}\hat{\mathbf{Z}}$ where $\operatorname{sym}\mathbf{A} = \frac{1}{2}(\mathbf{A} + \mathbf{A}^T)$. This is called *averaging over reciprocity error*. The skew-symmetric component of the estimated $\mathbf{Z}$ gives an indication of errors in the EIT instrumentation.

5.5.2 ADJACENT MEASUREMENT PROTOCOL

The Sheffield Mark I and II systems[161] and the many systems inspired by this design use a protocol with $L = 16$ electrodes which are typically arranged in a circular pattern on the subject. Adjacent pairs E_l, E_{l+1} are excited with equal and opposite currents. These can be assembled in to a matrix $I_P \in \mathbb{R}^{L \times (L-1)}$ with $\delta_{l,k} - \delta_{l,k+1}$ in the l, k position. Clearly the columns of I_P span S. Measurements are made similarly between adjacent pairs and $\mathbf{i}_P^T$ gives the measurement patterns so that the matrix of all possible voltages measured is $\mathbf{Z}_P = \mathbf{i}_P^T \mathbf{Z} \mathbf{i}_P$, a symmetric $(L-1) \times (L-1)$ matrix of full rank. However when the l-th electrode pair is excited, the measurement pairs $l - 1, l$ and $l + 1$ are omitted (indices are assumed to wrap around when out of range). The subset of $\mathbf{Z}_P$ which is actually measured by the Sheffield system is shown in (figure 5.6) and a counting argument shows that the number of independent measurements is $\frac{1}{2}(L-2)(L-1) - 1 = \frac{1}{2}L(L-3)$, or 104 for $L = 16$.

A pair drive system has the advantage that only one current source is needed, which can then be switched to each electrode pair. With a more complex switching network other pairs can be driven at the expense of higher system cost and possibly a loss of accuracy. A study of the dependence of the SVD of the Jacobian for different separations between driven electrodes can be found in[154]

One feature of the Sheffield protocol is that on a two dimensional domain the adjacent voltage measurements are all positive. This follows as the potential itself is monotonically decreasing from source to sink. The measurements also have a 'U' shaped graph for each drive. This provides an additional feasibility check on the measurements. Indeed if another protocol is used Sheffield data $\mathbf{Z}_P$ can be synthesized to employ this check.

5.5.3 OPTIMAL DRIVE PATTERNS

Detectability (§5.3.3) increases with applied current, but these currents are limited by safety norms (§3.5), so we may wish to choose drive patterns with the largest detectability. The optimization of drive patterns in EIT was first considered by Seagar[949] who calculated the optimal placing of a pair of point drive electrodes on a disk to maximize the voltage differences between the measurement of a homogeneous background and an offset circular anomaly. Isaacson and colleagues[349,500] argued that one should choose a single current pattern to maximize the L^2 norm of the voltage difference between the measured $\mathbf{V}_m$ and calculated $\mathbf{V}_c$ voltages constraining the L^2 norm of the current patterns in a multiple drive system. This is a simple quadratic optimization problem

$$I_{\text{opt}} = \arg \min_{I \in S} \frac{\|(\mathbf{V}_m - \mathbf{V}_c)I\|}{\|I\|} \tag{5.13}$$

to which the answer is that I_{opt} is the eigenvector of $|\mathbf{Z}_m - \mathbf{Z}_c|$ corresponding to the largest eigenvalue (here $|\mathbf{A}| = (\mathbf{A}^*\mathbf{A})^{1/2}$). One can understand this eigenvector to be a current pattern which focuses the dissipated power in the regions where actual and predicted conductivity differs most. If one is to apply only one current pattern then in a particular sense this is best. The eigenvectors for smaller eigenvalues are increasingly less useful for telling these two conductivities apart and one could argue that eigenvectors for eigenvalues which are smaller than the error in measurement contain no useful information. Gisser et al[349] argued that the eigenvector for this eigenvalue can be found experimentally using the *power method* a classical fixed point algorithm for numerically finding an eigenvector.

Later Gisser et al[350] used a constraint on the maximum dissipated power in the test object which results in the quadratic optimization problem

$$I_{\text{opt}} = \arg\min_{I \in S} \frac{\|(\mathbf{V}_m - \mathbf{V}_c)I\|}{\|\mathbf{V}_m^* I\|} \tag{5.14}$$

which is a generalized eigenvalue problem. The argument here is that the dissipated (and stored) power should be limited in a medical application, rather than the rather artificial constraint of sum of squares of current.

Optimal current patterns can be incorporated in iterative reconstruction algorithms, at each iteration the optimal current pattern to distinguish between the actual and conductivity and the latest approximation can be applied, and the voltage data from this pattern used in the next iterative update. As the current pattern used will change at each iteration eventually all the information in $\mathbf{Z}_m$ will be used. Alternatively more than one of the eigenvectors of $|\mathbf{Z}_m - \mathbf{Z}_c|$ can be used, provided the resulting voltages differences are above the noise level. In practice this method is an improvement over pair drives even for simulated data[155].

Driving current patterns in eigenvectors requires multiple programmable current sources with consequent increase in cost and complexity. There is also the possibility that a pair drive system could be made with sufficiently better accuracy that it counteracts the advantage of a multiple drive system with optimal patterns. Even neglecting the errors in measurement, there is numerical evidence[153] that using optimal currents produces better reconstructions on synthetic data. In this respect one can also use synthetic optimal voltage patterns[812].

Kaipio et al[530] suggest choosing current patterns that minimize the total variance of the posterior. In this Bayesian framework the choice of optimal current patterns depends on the prior and a good choise will result in a 'tighter' posterior. Demidenoko et al[234] consider optimal current patterns in the framework conventional optimal design of experiments, and define an optimal set of current patterns as one that minimizes the total variance of $\mathbf{Z}$.

Eyöboğlu and Pilkington[261] argued that medical safety legislation demanded that one restrict the maximum total current entering the body, and if this constraint was used the distinguishability is maximized by pair drives. Cheney and Isaacson[190] study a concentric anomaly in a disk, using the 'gap' model for electrodes. They compare trigonometric, Walsh, and opposite and adjacent pair drives for this case giving the dissipated power as well as the L^2 and power distinguishability Köksal and Eyöboğlu[569] investigate the concentric and offset anomaly in a disk using continuum currents. Further study of optimization of current patterns with respect to constraints can be found in[23,629].

To get a numerical sense of the affect of current patterns, we calculated the distinguishability for a small circular anomaly in a 2D circular domain for pair-drive patterns with various skip levels, as well as the low-frequency Walsh pattern, and compared that to the distinguishability from the first sinusoidal pattern. If we assume the total current into each electrode is limited, we get 1.2% (adjacent), 12.3% (opposite) and 128% (Walsh). On the other hand, if we assume total current into the body is limited, we get 12.4% (adjacent), 124% (opposite) and 80.9% (Walsh). Overall, optimal current patterns appear to offer small or no benefit depending on how safety constraints are interpreted, given the increased electronics complexity required.

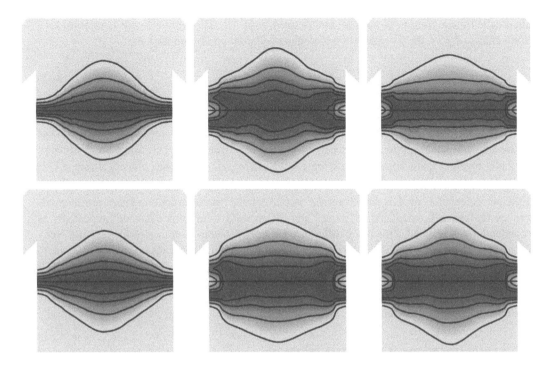

Figure 5.8   Vertical sensitivity of an EIT configuration with 32 electrodes with different electrode placements and stimulations and measurement patterns. *left*: one plane of 32 electrodes, *centre*, two planes of 16 electrodes configured in a planar pattern and two (*right*, two planes of 16 electrodes configured in a square pattern[377]. Data are simulated with a homogeneous elliptic model, and sensitivity shown on frontal plane through the centre. Sensitivity in each vertical row is normalized to the maximum value. Contours at 90%, 75%, 50%, and 25% are shown. For each column, the *top* image uses an adjacent stimulation and measurement pattern, while the *bottom* uses "skip=4". (color figure available in eBook)

5.6   EIT SENSITIVITY

The sensitivity of a measurement system is the change in measured quantity (voltage) for a small change in the underlying quantity interest (tissue electrical properties). Consider a perturbation $\gamma \to \gamma + \delta\gamma$ leading to changes of body potential distribution $\phi \to \phi + \delta\phi$ and electrode voltages $V_l \to V_l + \delta V_l$, with the current in each electrode I_l held constant.

If we parameterize the body into discretized voxel or simplex regions, then $\delta\gamma_j$ is a small uniform change in region j. We can calculate the change in measurement δy_i, based on its associated drive and measurement patterns $I_{l,i}$ and $G_{l,i}$. The complete matrix of partial derivatives of voltages with respect to conductivity parameters is the *Jacobian* matrix, **J**, where element i, j of $\mathbf{J}_{i,j}$ is the sensitivity of measurement y_i to changes in conductivity parameter γ_j.

In the medical and industrial EIT literature **J** is also called the *sensitivity matrix*. The Jacobian matrix is useful to understand the distribution of sensitivity in the body of an EIT configuration. Figure 5.8 shows an example of a vertical slice through a thorax, in which different electrode placements and stimulation and measurement configurations are evaluated[390].

Conceptually, the simplest way to calculate **J** is using a small perturbation in each region Ω_j and then calculating the change in estimated voltages. Thus calculates a *perturbation* Jacobian, but is inefficient. A method for efficiently calculating **J** with a minimal number of forward solutions is often called the *adjoint field method*, and derived in §5.6.1. The simplest case is to consider a uniform change $\delta\gamma_j$ on the jth region of a parameterized domain Ω_j and we have

$$J_{i,j} = \frac{\partial y_i}{\partial \gamma_j} = -\int_{\Omega_j} \nabla\phi(I_{l,i}) \cdot \nabla\phi(\overline{G}_{l,i}) \, dV \qquad (5.15)$$

where $\phi(I_{l,i})$ is the (actual) potential distribution in the body created by the applied current pattern $I_{l,i}$. $\phi(\overline{G}_{l,i})$ is the potential distribution in the body which *would have occured* if the conjugate of the measurement pattern gains $\overline{G}_{l,i}$ were applied as if they were a current. Since the Neumann-to-Dirichlet operator takes voltages to currents (division by Ω), $\phi(\overline{G}_{l,i})$ has units Ω^{-1}, since G is unitless. This formula is derived in the next section.

Efficient calculation of **J** is important for iterative image reconstruction. There are methods where the derivative is calculated only once and the forward solution is calculated repeatedly as the conductivity is updated. This is the difference between *Newton-Kantorovich* method and Newton's method. There are also *Quasi-Newton* methods in which the Jacobian is update approximately from the forward solutions that have been made. Indeed this has been used in geophysics [645]. It also worth pointing out that where the conductivity is parameterized in a non-linear way for example using shapes of an anatomical model, that the Jacobian with respect to those new parameters can be calculated using the chain rule.

5.6.1 STANDARD FORMULA FOR THE JACOBIAN

Using the adjoint-field formulation (Box 5.6), we restrict the perturbation to a scalar $\delta\gamma_j$ in body region Ω_j and calculate from (5.23),

$$\oint_{\partial\Omega} \delta\phi \, j_{\mathbf{n}}^w \, dS = -\int_\Omega \gamma\nabla\phi \cdot \nabla w \, dV = -\delta\gamma_j \int_{\Omega_j} \nabla\phi \cdot \nabla w \, dV \qquad (5.16)$$

where $j_{\mathbf{n}}^w = \gamma\nabla w \cdot d\mathbf{n}$ is the normal current at the boundary associated with potential distribution, w.

From potential w, the current into electrode l is $I_l^w = \int_{E_l} j_{\mathbf{n}}^w \, dS$ over the surface area E_l of the electrode, and V_l is the potential ϕ in the conductive electrode connection. We have $\sum_l \delta V_l I_l^w = \oint_{\partial\Omega} \delta\phi \, j_{\mathbf{n}}^w \, dS$ since normal current is zero where the body surface is not connected to an electrode. For each measurement, y_i, (5.7), we see this expression corresponds to the Jacobian $\mathbf{J}_{i,j} = \delta y_i/\delta\gamma_j$, where

$$\delta y_i = \sum_l \delta V_{l,i}\overline{G}_{l,i} = -\delta\gamma_j \int_{\Omega_j} \nabla\phi(V_{l,i}) \cdot \nabla\phi(\overline{G}_{l,i}) \, dV \qquad (5.17)$$

where, as for (5.15), $\phi(I_{l,i})$ is potential distribution in the body due to currents $I_{l,i}$, and $w = \phi(\overline{G}_{l,i})$ is the potential which *would occur* if the currents I_l^w were replaced by the conjugate of the gains G. The latter has been referred to as a "lead field" — the potential field if currents were applied to the measurement leads. Note that, in general $\overline{\phi(G)} \neq \phi(\overline{G})$ if γ is complex.

Some EIT and capacitance tomography systems use a voltage source and in this case the Jacobian has the opposite sign to the applied current case. A common variation in the case of real conductivity is to use the resistivity $\rho = 1/\sigma$ as the primary variable or more commonly to use $\log\sigma$ [85,153,1075], which has the advantage that it does not need to be constrained to be positive (see also §18.5). With a simple parameterization of conductivity as constant on voxels, parameter $g(\gamma)$ is constant on voxels as well. In this case, we simply use the chain rule, dividing the k-th column of Jacobian we have calculated by $g'(\gamma_k)$. The regularization will also be affected by the change of variables.

While this formula gives the Fréchet derivative for $\delta\gamma \in L^\infty(\Omega)$, care is needed to show that the voltage data is Fréchet differentiable in other norms, such as those needed to show that the total variation regularization scheme works[1102]. For a finite dimensional subspace of $L^\infty(\Omega)$ a proof of differentiability is given in[529].

For full time-harmonic Maxwell's equations, an analysis yeilds the same sensitivity to a perturbation of admittivity, but the electric field **E** is no longer a gradient and sensitivity to a change in the magnetic permeability is given by $\mathbf{H} \cdot \mathbf{H}$[995].

Box 5.6: Adjoint-field formulation of the Sensitivity

For a small change $\gamma \to \gamma + \delta\gamma$, the voltage $\phi \to \phi + \delta\phi$, and the Laplace equation, $\nabla \cdot (\gamma + \delta\gamma)\nabla(\phi + \delta\phi) = 0$, yeilds

$$\nabla \cdot \gamma\nabla\phi + \nabla \cdot \gamma\nabla\delta\phi = -(\nabla \cdot \delta\gamma\nabla\phi + \nabla \cdot \delta\gamma\nabla\delta\phi). \tag{5.18}$$

The first term is the Laplace equation (and thus zero); the last represents higher order terms (*HOT*), which we ignore for small $\delta\gamma$, following Calderon[172] (for $\gamma = 1$) and Breckon[153].

Boundary conditions specify the normal current $j_n = \gamma\frac{d}{dn}\phi = (\gamma + \delta\gamma)\frac{d}{dn}(\phi + \delta\phi)$. We assume that γ does not change on the boudary. Since, using a current source, j_n is constant, $\delta\gamma|_{\partial\Omega} = 0$, and $\delta\phi$ must also be zero on the boundary.

We multiply (5.18) by test function, w, and integrate. For clarity, the spatial or volume differentials are omitted below.

$$\int_\Omega w\nabla \cdot \gamma\nabla\delta\phi = -\int_\Omega w\nabla \cdot \delta\gamma\nabla\phi + HOT \tag{5.19}$$

For scalar a and vector **b**, we have $\int_\Omega \nabla \cdot (a\mathbf{b}) = \oint_{\partial\Omega} a\mathbf{b} \cdot \mathbf{n}$ (divergence theorem) and $\nabla \cdot (a\mathbf{b}) = \nabla a \cdot \mathbf{b} + a\nabla \cdot \mathbf{b}$ (separation of variables). For the first term in (5.19), use $\int_\Omega a\nabla \cdot \mathbf{b} = \oint_{\partial\Omega} a\mathbf{b} \cdot d\mathbf{n} - \int_\Omega \nabla a \cdot \mathbf{b}$, with $a = v$, $\mathbf{b} = \gamma\nabla\delta\phi$,

$$\int_\Omega w\nabla \cdot \gamma\nabla\delta\phi = \oint_{\partial\Omega} w\gamma\nabla\delta\phi - \int_\Omega (\nabla w) \cdot (\gamma\nabla\delta\phi) = -\int_\Omega \nabla\delta\phi \cdot \gamma\nabla w \tag{5.20}$$

since $\delta\phi = 0$ on the boundary and the dot product commutes. We apply integration by parts again to (5.20), using $\int_\Omega \nabla a \cdot \mathbf{b} = \oint_{\partial\Omega} a\mathbf{b} \cdot d\mathbf{n} - \int_\Omega a\nabla \cdot \mathbf{b}$, with $a = \delta\phi$, $\mathbf{b} = \gamma\nabla w$,

$$\int_\Omega \nabla\delta\phi \cdot \gamma\nabla w = \oint_{\partial\Omega} \delta\phi\gamma\nabla w \cdot d\mathbf{n} - \int_\Omega \delta\phi\nabla \cdot \gamma\nabla w \tag{5.21}$$

By choosing w which solves the Laplace equation, $\nabla \cdot \gamma\nabla w = 0$, the second term is zero.

For second term in (5.19), use $a = v$, $\mathbf{b} = \delta\gamma\nabla\phi$,

$$\int_\Omega w\nabla \cdot \delta\gamma\nabla\phi = \oint_{\partial\Omega} w\delta\gamma\nabla\phi \cdot d\mathbf{n} - \int_\Omega \nabla w \cdot \delta\gamma\nabla\phi = -\int_\Omega \delta\gamma\nabla\phi \cdot \nabla w \tag{5.22}$$

using $\delta\phi = 0$ on the boundary and commuting the dot product. From (5.21) and (5.22),

$$\oint_{\partial\Omega} \delta\phi \, \gamma\nabla w \cdot d\mathbf{n} = -\int_\Omega \delta\gamma\nabla\phi \cdot \nabla w \tag{5.23}$$

which describes the change in boundary voltage $\delta\phi$ due to $\delta\gamma$.

Some iterative nonlinear reconstruction algorithms, such as nonlinear Landweber, or non-linear conjugate gradient (see section 5.7.3 and[1087]) require the evaluation of transpose (or adjoint) of the Jacobian multiplied by a vector $\mathbf{J}^*z$. For problems where the Jacobian is very large it may be undesirable to store the Jacobian and then apply its transpose to z. Instead the block of z_i corresponding to the ith current drive is written as distributed source on the measurement electrodes. A forward solution is performed with this as the boundary current pattern so that when this measurement field is combined with the field for the drive pattern as 5.15, and this block accumulated to give $\mathbf{J}^*z$. For details of this applied to diffuse optical tomography see[60], and for a general theory of adjoint sources see[1087]

For fast, calculation of the Jacobian using (5.15) one can precompute the integrals of products of finite element basis functions over elements. If non-constant basis functions are used on elements, or higher order elements used one could calculate the product of gradients of FE basis functions at quadrature points in each element. As this depends only on the geometry of the mesh and not the conductivity this can be precomputed unless one is using an adaptive meshing strategy. The same data is used in assembling the FE system matrix efficiently when the conductivity has changed but not the geometry. It is these factors particularly which make current commercial FEM software unsuitable for use in an efficient EIT solver.

5.7   SOLVING THE FORWARD PROBLEM: THE FINITE ELEMENT METHOD

To solve the inverse problem one needs to solve the forward problem for some assumed conductivity so that the predicted voltages can be compared with the measured data. In addition the interior electric fields are usually needed for the calculation of Jacobian. Only in cases of very simple geometry, and homgeneous or at least very simple conductivity, can the forward problem be solved analytically. These can sometimes be useful for linear reconstruction algorithms on highly symmetric domains. Numerical methods for general geometry and arbitrary conductivity require the discretization of both the domain and the conductivity. In the Finite Element Method (FEM), the three dimensional domain is decomposed in to (possibly irregular) polyhedra (for example tetrahedra, prisms or hexahedra) called *elements*, and on each element the unknown potential is represented by a polynomial of fixed order. Where the elements intersect they are are required to intesect only in whole faces or edges or at vertices, and the potential is assumed continuous (or derivatives up to a certain order continuous), across faces. The finite element method converges to the solution (or at least the weak solution) of the partial differential equation it represents, as the elements become more numerous (provided their interior angles remain bounded) or as the order of the polynomial is increased[1010].

The Finite Difference Method and Finite Volume Method, are close relatives of the FEM, which use regular grids. These have the advantage that more efficient solvers can be used at the expense of the difficulty in accurately representing curved boundaries or smooth interior structures. In the Boundary Element Method (BEM) only surfaces of regions are discretized, and an analytical expression for the Green function used within enclosed volumes that are assumed to be homogeneous. BEM is useful for EIT forward modelling provided one assumes piecewise constant conductivity on regions with smooth boundaries (organs for example). BEM results in a dense rather than a sparse linear system to solve and its computational advantage over FEM diminishes as the number of regions in the model increases. BEM has the advantage of being able to repressent unbounded domains. A hybrid method where some regions assumed homogeneous are repressented by BEM, and inhomogeneous regions by FEM may be computationally efficient for some applications of EIT[505].

In addition to the close integration of the Jacobian calculation and the FEM forward solver, another factor which leads those working on EIT reconstruction to write their own FEM program that the Complete Electrode Model is a non-standard type of boundary condition not included in commercial FEM software. It is not hard to implement and there are freely available codes[28,838,1078], but it is worth covering the basic theory here for completeness. A good introduction to FEM in electromagnetics is[975], and details of implementation of the CEM can be found especially in the theses[836,1075].

5.7.1 BASIC FEM FORMULATION

Our starting point is to approximate the domain Ω as union of a finite number of *elements*, which for simplicity we will take to be *simplices*. In two dimensions a simplex is a triangle and in three dimensions a tetrahedron. A collection of such simplices is called a finite element mesh, and we will suppose that there are K simplices with N vertices. We will approximate the potential using this mesh by functions which are linear on each simplex, and continuous across the faces. These functions have the appealing feature that they are completely determined by their values at the vertices. A natural basis is the set of functions w_i that are one on vertex i and zero at the other vertices and we can represent the potential by the approximation

$$\phi_{\mathrm{FEM}}(\mathbf{x}) = \sum_{i=1}^{N} \phi_i w_i(\mathbf{x}).$$ (5.24)

so that $\Phi = (\phi_1, \ldots, \phi_n)^T \in \mathbb{C}^N$ is vector which represents our discrete approximation to the potential.

As our basis functions w_i are not differentiable we cannot directly satisfy (5.1). Instead we derive the weak form of the equation. Multiplying (5.1) by some function v and integrating over Ω,

$$\int_{\Omega} v \nabla \cdot (\gamma \nabla \phi) \, dV = 0 \qquad \text{in } \Omega$$ (5.25)

and we demand that this vanishes for all functions v in a certain class. Clearly this is weaker than assuming directly that $\nabla \cdot (\gamma \nabla \phi) = 0$.

Using Green's second identity and the vector identity

$$\nabla \cdot (v \gamma \nabla \phi) = \gamma \nabla \phi \cdot \nabla v + v \nabla \cdot (\gamma \nabla \phi)$$ (5.26)

the equation (5.25) is changed to

$$\int_{\Omega} \nabla \cdot (v \gamma \nabla \phi) \, dV - \int_{\Omega} \gamma \nabla \phi \cdot \nabla v \, dV = 0$$ (5.27)

Invoking the divergence theorem

$$\int_{\Omega} \nabla \cdot (v \gamma \nabla \phi) \, dV = \int_{\partial \Omega} v \gamma \nabla \phi \cdot \mathbf{n} \, dS$$ (5.28)

gives

$$\int_{\Omega} \gamma \nabla \phi \cdot \nabla v \, dV = \int_{\partial \Omega} \gamma \nabla \phi \cdot \mathbf{n} v \, dS$$
$$= \int_{\Gamma} \gamma \nabla \phi \cdot \mathbf{n} v \, dS$$ (5.29)

where $\Gamma = \bigcup_l E_l$ is the union of the electrodes, and we have used the fact that the current density is zero off the electrodes. For a given set of test functions v (5.29) is the weak formulation of the boundary value problem for 5.1 with current density specified on the electrodes.

Rearranging the boundary condition (5.11) as

$$\gamma \nabla \phi \cdot \mathbf{n} = \frac{1}{z_l} (V_l - \phi)$$ (5.30)

on E_l for $z_l \neq 0$ and incorporating it into (5.29) gives

$$\int_\Omega \gamma \nabla \phi \cdot \nabla v \, dV = \sum_{l=1}^{L} \int_{E_l} \frac{1}{z_l} (V_l - \phi) \, v \, dS \qquad (5.31)$$

In the finite element method we use test functions from the same family as used to approximate potentials $v = \sum_{i=0}^{N} v_i w_i$, substitution of this and ϕ_{FEM} for ϕ gives for each i

$$\sum_{j=1}^{N} \left\{ \int_\Omega \gamma \nabla w_i \cdot \nabla w_j \, dV \right\} \phi_j + \sum_{l=1}^{L} \int_{E_l} \frac{1}{z_l} w_i w_j \, dS \left. \right\} \phi_j - \sum_{l=1}^{L} \left\{ \int_{E_l} \frac{1}{z_l} w_i \, dS \right\} V_l = 0 \qquad (5.32)$$

Together with the known total current

$$I_l = \int_{E_l} \frac{1}{z_l} (V_l - \phi) \, dS$$

$$= \int_{E_l} \frac{1}{z_l} V_l - \sum_{i}^{N} \left\{ \int_{E_l} \frac{1}{z_l} w_i \, dS \right\} \phi_i \qquad (5.33)$$

and if we assume z_l is constant on E_l this reduces to

$$I_l = \frac{1}{z_l} |E_l| V_l - \frac{1}{z_l} \sum_{i}^{N} \left\{ \int_{E_l} w_i \, dS \right\} \phi_i \qquad (5.34)$$

where $|E_l|$ is the area (or in 2D length) of the l-th electrode.

We now need to choose how to approximate γ, and a simple method is to choose γ to be constant on each simplex (piece-wise constant). The characteristic function χ_j is one on the j-th simplex and zero elsewhere so we have an approximation to γ

$$\gamma_{\text{PWC}} = \sum_{j=1}^{k} \gamma_j \chi^j \qquad (5.35)$$

which has the advantage that the γ_j can be taken outside of an integral over each simplex. If a more elaborate choice of basis is used it would be wise to use a higher order quadrature rule.

Our FE system equations now take the form

$$\begin{bmatrix} A_M + A_Z & A_W \\ A_W^T & A_D \end{bmatrix} \begin{bmatrix} \Phi \\ V \end{bmatrix} = \begin{bmatrix} 0 \\ I \end{bmatrix} \qquad (5.36)$$

were A_M is an $N \times N$ symmetric matrix

$$A_{Mij} = \int_\Omega \gamma \nabla w_i \cdot$$

$$= \sum_{k=1}^{K} \gamma_k \int_{\Omega_k} \nabla w_i \nabla w_j \, dV \qquad (5.37)$$

which is the usual system matrix for (5.1) without boundary conditions, while

$$A_{Zij} = \sum_{l=1}^{L} \int_{E_l} \frac{1}{z_l} w_i w_j \, dS, \qquad (5.38)$$

$$A_{Wli} = -\frac{1}{z_l} \int_{E_l} w_i \, dS, \qquad (5.39)$$

and

$$A_D = \text{diag}\left(\frac{|E_l|}{z_l}\right) \tag{5.40}$$

implement the CEM boundary conditions. One additional constraint is required as potentials are only defined up to an added constant. One elegant choice is to change the basis used for the vectors V and I to a basis for the subspace S orthogonal to constants, for example the vectors

$$\left[\tfrac{1}{L-1}, \dots, \tfrac{1}{L-1}, 1, \tfrac{1}{L-1}, \dots, \tfrac{1}{L-1}\right]^T. \tag{5.41}$$

while another choice is to 'ground' an arbitrary vertex i by setting $\phi_i = 0$. The resulting solution Φ can then have any constant added to produce a different grounded point.

As the contact impedance decreases the system (5.36) becomes ill-conditioned. In this case, (5.11) in the CEM can be replaced by the *shunt model* which simply means the potential ϕ is constrained to be constant on each electrode. This constraint can be enforced directly replacing all nodal voltages on electrode E_l by one unknown V_l.

It is important for EIT to notice that the conductivity only enters in the system matrix as linear multipliers of

$$s_{ijk} = \int_{\Omega_k} \nabla w_i \cdot \nabla w_j \, dV = |\Omega_k| \nabla w_i \cdot \nabla w_j$$

which depend only on the FE mesh and not on γ. These coefficients can be pre-calculated during the mesh generation saving considerable time in the system assembly. An alternative is to define a discrete gradient operator $D : \mathbb{C}^N \to \mathbb{C}^{3K}$ which takes the representation as a vector of vertex values of a piecewise linear function ϕ to the vector of $\nabla\phi$ on each simplex (on which of course the gradient is constant). On each simplex define $\Sigma_k = (\gamma_k/|\Omega_k|)\mathbf{I}_3$ where $\mathbf{I}_3$ is the 3×3 identity matrix, or for the anisotropic case simply the conductivity matrix on that simplex divided by its volume, and $\Sigma = \text{diag}(\Sigma_k) \otimes \mathbf{I}_K$. We can now use

$$A_M = D^T \Sigma D \tag{5.42}$$

to assemble the main block of the system matrix. It useful to think of a FEM as a resistor network, see Box 5.7. Figure 5.9 illustrates assembly of the system matrix through the parallel combination of resistors.

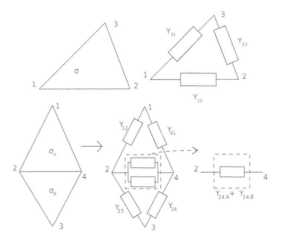

Figure 5.9   Equivalence between FEM models and resistor networks. *Top:* Triangular element and equivalent circuit. *Bottom:* Connection of two elements and the parallel combination of $Y_{24,A}$ and $Y_{24,B}$.

Box 5.7: FEM as a Resistor Network

It may help to think of the finite element method in terms of resistor networks. For the case we have chosen with piecewise linear potentials on simplicial cells and conductivity constant on cells there is an exact equivalence[975]. To construct a resistor network equivalent to such a FEM model replace each edge by a resistor (figure 5.9). To determine the conductance of that resistor consider first a triangle (in the two dimensional case), and number the angles θ_j opposite the j-th side. The resistor on side j has a conductance $\sigma\cot\theta_j$. When the triangles are assembled in to a mesh the conductances add in parallel summing the contribution from triangles both sides of an edge. In the three dimensional case θ_j is the angle between the two faces meeting at the edge opposite edge j, and of course several tetrahedra can meet at one edge. In 3D, each edge in a tetrahedron is replaced with a resistor of conductance $6\sigma\cot\theta$ where θ is the opposing dihedral angle[630].

With a resistor mesh assembled in this way, voltages ϕ_i at vertex i are governed by Ohms law and Kirchhoff's law, and the resulting system of equations is identical to that derived from the FEM. The situation is not reversible as not all resistor networks are the graphs of edges of two or three dimensional a FE mesh. Also some allocation of resistances do not correspond to a piecewise constant isotropic conductivity. For example there may be no consistent allocation of angles θ_j so that around any given vertex (or edge in 3D) they sum to 2π.

The question of uniqueness of solution, as well as the structure of the transconductance matrix for real planar resistor networks well is understood[204,217].

5.7.2 SOLVING THE LINEAR SYSTEM

We now consider the solution of the system (5.36). The system has the following special features. The matrix is sparse: the number of non-zeros in each row of the main block depends on the number of neighbouring verticies connected to any given vertex by an edge. It is symmetric (for complex conductivity and contact impedance that means real and imaginary parts are symmetric), and the real part is positive definite. In addition, we have multiple righthand sides for the same conductivity, and we wish to solve the system repeatedly for similar conductivities.

A simple approach to solving $\mathbf{Ax} = \mathbf{b}$ is LU-factorization[358], where an upper triangular matrix $\mathbf{U}$ and lower triangular matrix $\mathbf{L}$ are found such that $\mathbf{A} = \mathbf{LU}$. As solving a system with a diagonal matrix is trivial, one can solve $\mathbf{Lu} = \mathbf{b}$ (forward substitution) and then $\mathbf{Ux} = \mathbf{u}$ (backward substitution). The factorization process is essentially Gaussian elimination and has a computational cost $O(n^3)$ while the backward and forward substitute have a cost $O(n^2k)$ for k righthand sides. An advantage of a factorization method such as this is that one can apply the factorization to multiple right hand sides, in our case for each current pattern. Although the system matrix is sparse, the factors are in general less so. Each time a row is used to eliminate the non-zero elements below the diagonal it can create more non-zeros above the diagonal. As a general rule it is better to reorder the variables so that rows with more non-zeros are further down the matrix. This reduces the *fill in* of non-zeros in the factors. For a real symmetric or Hermitian matrix the Symmetric Multiple Minimum Degree Algorithm[338] reduces fill in, whereas the Column Multiple Minimum Degree algorithm is designed for the general case. For an example see figure 5.10. The renumbering should be calculated when the mesh is generated so that it is done only once.

For large 3D systems direct methods can be expensive and iterative methods may prove more efficient. A typical iterative scheme has a cost of $O(n^2k)$ per iteration and requires fewer than n iterations to converge. In fact the number of iterations required needs to be less than Cn/k for some C depending on the algorithm to win over direct methods. Often the number of current patterns

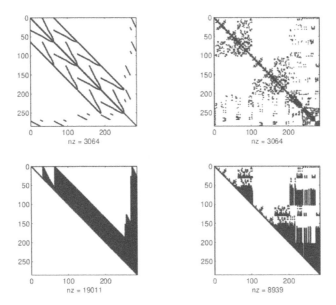

Figure 5.10   Top left the sparsity pattern of a system matrix which is badly ordered for fill-in. Bottom left sparsity pattern for the U factor. On the right the same after reordering with `colmmd`.

driven is limited by hardware to be small while the number of verticies in a 3D mesh needs to be very large to accurately model the electric fields, and consequently iterative methods are often preferred in practical 3D systems. The potential for each current pattern can be used as a starting value for each iteration. As the adjustments in the conductivity become smaller this reduces the number of iterations required for forward solution. Finally it is not necessary to predict the voltages to full floating point accuracy when the measurements system itself is far less accurate than this, again reducing the number of iterations required.

The convergence of iterative algorithms, such as conjugate gradient method (see §5.7.3, can be improved by replacing the original system by $\P\mathbf{Ax}=\P\mathbf{b}$ for some matrix $\P$ which is an approximation to the inverse of $\mathbf{A}$. A favourite choice is to use an approximate LU- factorization to derive $\P$. In EIT one can use the same preconditioner over a range of conductivity values.

5.7.3   CONJUGATE GRADIENT AND KRYLOV SUBSPACE METHODS

The conjugate gradient (CG) method[89,358] is a fast and efficient method for solving $\mathbf{Ax}=\mathbf{b}$ for real symmetric matrices $\mathbf{A}$ or Hermitian complex matrices. It can also be modified for complex symmetric matrices[168]. The method generates a sequence $\mathbf{x}_i$ (iterates) of successive approximations to the solution and residuals $\mathbf{r}_i=\mathbf{b}-\mathbf{Ax}_i$ and search directions $\mathbf{p}_i$ and $\mathbf{q}_i=\mathbf{Ap}_i$ used to update the iterates and residuals. The update to the iterate is

$$\mathbf{x}_i=\mathbf{x}_{i-1}+\alpha_i\mathbf{p}_i \tag{5.43}$$

where the scalar α_i is chosen to minimize

$$\mathbf{r}(\alpha)^*\mathbf{A}^{-1}\mathbf{r}(\alpha) \tag{5.44}$$

where $\mathbf{r}(\alpha)=\mathbf{r}_{i-1}-\alpha\mathbf{r}_{i-1}$ explicitly

$$\alpha_i=\frac{\|\mathbf{r}_{i-1}\|^2}{\mathbf{p}_i^*\mathbf{Ap}_i}. \tag{5.45}$$

The search directions are updated by

$$\mathbf{p}_i = \mathbf{r}_i + \beta_{i-1}\mathbf{p}_{i-1} \tag{5.46}$$

where using

$$\beta_i = \frac{\|\mathbf{r}_i\|^2}{\|\mathbf{r}_{i-1}\|^2} \tag{5.47}$$

ensures that $\mathbf{p}_i$ are orthogonal to all $\mathbf{Ap}_j$ and $\mathbf{r}_I$ are orthogonal to all $\mathbf{r}_j$, for $j < i$. The iteration can be terminated when the norm of the residual falls below a predetermined level.

Conjugate Gradient Least Squares (CGLS) method solves the least squares problem (6.1) $\mathbf{A}^T\mathbf{Ax} = \mathbf{A}^T\mathbf{b}$ without forming the product $\mathbf{A}^T\mathbf{A}$ (also called CGNR or CGNE Conjugate Gradient Normal Equations[89,174]) and is a particular case of the non-linear conjugate gradient (NCG) algorithm of Fletcher and Reeves[280] (see also[1087, Ch 3]). The NCG method seeks a minimum of a cost functions $f(\mathbf{x}) = \frac{1}{2}\|\mathbf{b} - F(\mathbf{x})\|^2$, which in the case of CGLS is simply the quadratic $\frac{1}{2}\|\mathbf{b} - \mathbf{Ax}\|^2$. The direction for the update in (5.43) is now

$$\mathbf{p}_i = -\nabla f(\mathbf{x}_i) = \mathbf{J}_i^*(\mathbf{b} - F(\mathbf{x}_i)) \tag{5.48}$$

where $\mathbf{J}_i = F'(\mathbf{x}_i)$ is the Jacobian. How far along this direction to go is determined by

$$\alpha_i = \arg \min_{\alpha > 0} f(\mathbf{x}_{i-1} + \alpha\mathbf{p}_i) \tag{5.49}$$

which for non-quadratic f requires a line search.

CG can be used for solving the EIT forward problem for real conductivity, and has the advantage that it is easily implemented on parallel processors. Faster convergence can be used using a preconditioner, such as an incomplete Cholesky factorization, chosen to work well with some predefined range of conductivities. For the non Hermitian complex EIT forward problem, and the linear step in the inverse problem, other methods are needed. The property of orthogonal residuals for some inner product (Krylov subspace property) of CG is shared by a range of iterative methods. Relatives of CG for non-symmetric matrices include Generalised Minimal Residual (GMRES)[909], Bi-Conjugate Gradient BiCG, Quasi Minimal Residual (QMR) and Bi-Conjugate Gradient Stabilized (Bi-CGSTAB). All have their own merits[89] and as implementations are readily available have been tried to some extent in EIT forward or inverse solutions, not much is published[399,654] applications of CG itself to EIT include[723,804,832,837] and to optical tomography[60,62]. The application of Krylov subspace methods to solving elliptic PDEs as well as linear inverse problems[174,427] are active areas of research and we invite the reader to seek out and use the latest developments.

5.7.4 MESH GENERATION

Mesh generation is a major research area in itself, and posses particular challenges in medical EIT. The mesh must be fine enough to represent the potential with sufficient accuracy to predict the measured voltages as a function of conductivity. In medical EIT this means we must adequately represent the surface shape of the region to be images, and the geometry of the electrodes. The mesh needs to be finer in areas of high field strength and this means in particular near the edges of electrodes. Typically there will be no gain in accuracy from using a mesh in the interior which is as fine. As we are usually not interested so much in conductivity changes near the electrodes, and in any case we cannot hope to resolve conductivity on a scale smaller than the electrodes, our parameterization of the conductivity will inevitably be coarser than the potential. One easy option is to choose groups of tetrahedra as voxels for conductivity, another is to use basis functions interpolated down to the FE mesh. If there are regions of known conductivity, or regions where the conductivity is known to be constant, the mesh should respect these regions. Clearly the electric field strengths will vary with the current pattern used, and it is common practice to use a mesh which is suitable for all current patters, which can mean that it would be unnecessarily fine away from excited electrodes. The trade-off is that the same system matrix is used for each current pattern.

Any mesh generator needs to have a data structure to represent the geometry of the region to be meshed. This includes the external boundary shape, the area where the electrodes are in contact with the surface and any internal structures. Surfaces are can represented either as a triangularization, or more by more general polygons, or by spline patches. The relationship between named volumes, surfaces curves and points must also be maintained usually as a tree or incidence matrix. Simple geometric objects can be constructed from basic primitive shapes, either with a graphical user interface or from a series of commands in a scripting language. Set theoretic operations such as union, intersection can be performed together with geometric operations such as extrusion (for example of a circle into a cylinder).

As each object is added consistency checks are performed and incidence data structures maintained. For general objects these operations require difficult and time consuming computational geometry.

For examples of representations of geometry and scripting languages see the documentation for QMG[719], Netgen[943] and FEMLAB[77]. Figure 5.11 illustrates a large FEM of a body shape.

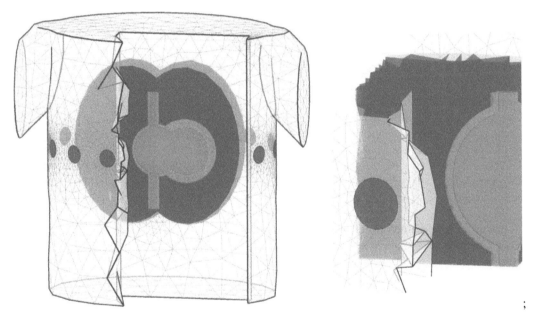

Figure 5.11   FEM model of a human thorax including elliptical lungs and a spherical heart with veins, created with Netgen[943] and EIDORS[28]. Right insert illustrates mesh refinement near electrodes and around internal structures. (color figure available in eBook)

Commercial Finite Element software can often import geometric models from Computer Aided Design programs, which makes life easier for industrial applications. Unfortunately human bodies are not supplied with blueprints from their designer. The problem of creating good FE meshes of the human body remains a significant barrier to progress in EIT, and of course such progress would also benefit other areas of biomedical electromagnetic research. One approach[92] is to segment Nuclear Magnetic Resonance, or X-ray CT images and use these to develop a FE mesh specific to an individual subject. Another is to warp a general anatomical mesh to fit the external shape of the subject[343], measured by some simpler optical or mechanical device.

Once the geometry is defined, one needs to create a mesh. Mesh generation software generally use a combination of techniques such as advancing front, octtree[719], bubble-meshing[967]. In a convex region given a collection of verticies, a tetrhedral mesh of their convex hull can be found with the Delaunay property that no tetrahedron contains any vertex in the interior of its circumsphere, using the QuickHull algorithm[81].

The standard convergence results for the FEM[1010], require that as the size of the tetrahedra tend to zero the ratio of the circumscribing sphere to inscribing sphere is bounded away from zero. In practice this means that for an isotropic medium without *a priori* knowledge of the field strengths tetrahedra which are close to equilateral are good and those with a high aspect ratio are bad. Mesh generators typically include methods to *smooth* the mesh. This simplest is *jiggling* in which each interior vertex in turn is moved to the center of mass of the polyhedron defined by the verticies with which it shares an edge (its neighbours). This can be repeated for some fixed number of iterations or until the shape of the elements ceases to improve. Jiggling can be combined with removal of edges and swapping faces which divide polyhedra into two tetrahedra. In EIT where the edges of electrodes and internal surfaces need to be preserved this process is more involved. It is also possible to incorporate a term into image reconstruction to meshing variability[29].

5.8 FURTHER COMMENTS ON THE FORWARD MODEL

This chapter has discussed the basic features of the EIT forward problem, sensitivity calculations, and selection of stimulation and measurement patterns. There are some common questions about forward modelling specific to applications of EIT which we address below.

- **Interpretation of sensitivity matrix:** $\mathbf{J}$ is a useful matrix for understanding a system. Each column $\mathbf{J}_{[:,j]}$ corresponds to the change in measurements δV for a small change in parameter j. Given a set of measurements, it is often useful to plot measurements against parts of $\mathbf{J}$ as a first test of the accuracy of the model.
  Each row $\mathbf{J}_{[i,:]}$ corresponds to the image sensitivity map of measurement i, and shows which image parameters contributed in which proportion. It is often useful to show the image of this sensitivity map, however, the rows of $\mathbf{J}$ must be normalized by the area (volume) of each voxel or parameter.
- **Perturbation approaches to calculating J:** It is often useful to know the system sensitivity with respect to some model parameter, like electrode movement or contact impedance. It is possible to estimate the value of each row using a perturbation $\mathbf{J}[:,p] = \frac{1}{\varepsilon}\left(F(\sigma + \varepsilon_p) - F(\sigma)\right)$, for a small change ε in model parameter p. Some care is required to find a value of ε which is small but doesn't lead to numerical errors. Perturbation Jacobian calculations are also an excellent way to validate more complicated numerical software.
- **Parametrizing the forward problem:** Often each FEM element of the forward model is taken as a parameter of $\mathbf{J}$ and thus the inverse calculation. However this is not necessary: a fine discretization of the FEM near the electrodes is not needed in the image. Instead a parameterization of the image with fewer degrees of freedom is made which maps to the FEM. This is often called a "dual model", and an example is shown in figure 6.1. If we have a coarse (inverse) parameterization, x_c, and a fine (forward) parameterization, x_f, we calculate a coarse-to-fine map, $\mathbf{M}$ such that $\mathbf{x}_f = \mathbf{M}\mathbf{x}_c$, where $\mathbf{M}_{i,j}$ is the fraction of fine parameter region i inside coarse parameter region j. The Jacobian on the coarse parameters (used in image reconstruction), $\mathbf{J}_c = \mathbf{J}_f\mathbf{M}$, where $\mathbf{J}_f$ is the fine Jacobian of the forward model. It is also possible to accelerate this computation in the FEM.
- **Modelling hardware and cable properties:** One important aspect of EIT systems that has seen little work is the modelling of hardware imperfections[142,324]. Systems have offsets, mismatched gains and crosstalk between channels. Many of these defects can be expressed as a

linear correction to the sensitivity. Once a good model of the specific hardware is available the corrected **J** can be directly used for reconstructing images[441].

- **Off-plane sensitivity of EIT** The Laplace equation, and thus EIT, is non-local. Changes in γ anywhere affect voltage everywhere. Because of this, we should create a forward model which contains areas of the body far from the electrodes. See[391] for a discussion of this approach.
- **How accurate should FEMs be?** In EIT, a FEM must estimate both the measured voltages and the sensitivity. Accurate FEMs are computationally expensive: a model of the thorax to 1 mm could require 100 million elements and be computationally almost infeasible. Surprisingly, the EIT community doesn't have a good answer to the accuracy requirements. Difference EIT can work well at fairly low accuracy, while absolute EIT requires more accuracy[23,388].
- **How to model EIT in the time domain** EIT circuits and forward models have been analyzed in the frequency domain using phasors. Sometimes there is interest in modelling time domain transient responses of circuits, or questions about the behaviour of and EIT system when multiple frequencies are simultaneously used. Given that Ohm's law is linear, it is possible use a Fourier analysis to decompose the time-domain stimulations into frequency components, model each through the forward (or inverse) model, and then reconstruct the temporal behaviour in the image using an inverse Fourier transform.

As advice to the reader, there are now many excellent software packages for meshing and EIT imaging (including our own, EIDORS[28] [2]). We also recommend the EIDORS mailing list as a place where many questions have been discussed. As we mention above there are several topics which are still not well understood by the EIT community. In our opinion, the most pressing need is to design forward models which account for the many real-world sources of variability. These are variability in shape through breathing and posture, electrode contact quality, and inaccuracies in the hardware. This kind of modelling is often the "secret sauce" that successful imaging equipment vendors use to enhance their systems.

---

[2]http://www.eidors.org

6 The EIT Inverse Problem

William R. B. Lionheart
Department of Mathematics, University of Manchester, UK

Andy Adler
Systems and Computer Engineering, Carleton University, Ottawa, Canada

CONTENTS

DOI: 10.1201/9780429399886-6

6.1   INTRODUCTION

The story of EIT so far: we have a conductive body with interesting tissue properties (chapter 3) and have put electrodes on it used an EIT system (chapter 4) to make measurements. Next, we built a model (chapter 5) of the body, electrodes and the stimulation and measurement patterns, which allows us to understand and predict our measurements; this is the *forward* model. In this chapter, we reconstruct images of the volumetric distribution of tissue properties, the *inverse* problem (see Box 6.8). For a more mathematical treatment, see [22].

In this chapter, we consider parameter estimation methods, illustrated in figure 6.1. Our body has an internal conductivity distribution described by a vector, $\boldsymbol{\sigma}$. Our goal is to estimate a vector of parameters, $\mathbf{x}$, which describes (via some mapping) our $\boldsymbol{\sigma}$. The parameters, $\mathbf{x}$, will typically be at a lower spatial resolution than $\boldsymbol{\sigma}$ especially in regions far from the electrodes. Our forward model $F(\boldsymbol{\sigma})$ predicts the measurements for a given value of parameters. We start at an initial guess $\mathbf{x}_0$

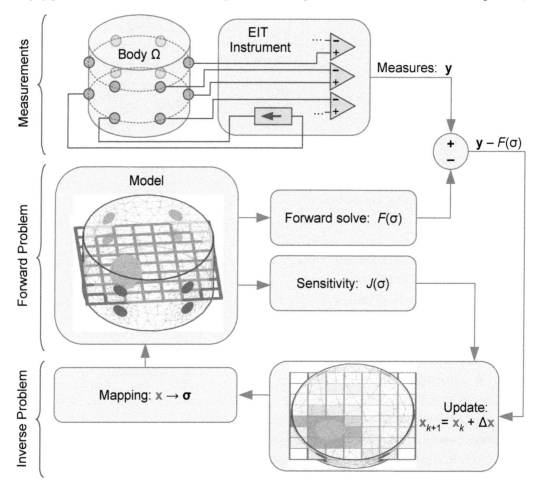

Figure 6.1   Schema for image reconstruction. Top: EIT data, $\mathbf{y}$, are measured with an instrument from body Ω. Middle: EIT forward problem, for a finite-element model of the body. Bottom: inverse solution, where, iteratively, a model $\mathbf{x}_k$ is improved by updates, $\Delta \mathbf{x}$, calculated from the mismatch between the current forward estimate and sensitivity, and a prior model. The mapping between image parameters, $\mathbf{x}$, of a planar slice, and the model are illustrated. (color figure available in eBook)

which is unlikely to accurately predict our measurements, and so the error or data misfit $\mathbf{y} - F(\mathbf{x})$ is large. Using this error and an estimate of the sensitivity we calculate an update $\Delta\mathbf{x}$ to improve our estimate $\mathbf{x}_1 = \mathbf{x}_0 + \Delta\mathbf{x}$. In many cases, EIT algorithms accept the first update as a solution. We can also continue updating our estimate until some stopping criteria is met.

All experimental and clinical results so far published use these types of techniques. However, there are other mathematically interesting techniques, known as direct methods, which are described in chapter 7. See [420] for a comparison of direct and regularized approaches.

Box 6.8: What is an Inverse Problem?

The term "inverse problem" refers to calculating the internal parameters that describe a system from a set of observations or measurements. The classical inverse problem in medicine is X-ray computed tomography. The terminology of "inverse" describes the process of beginning with the effects (measurements) and calculating the causes (parameters). In order to solve an inverse problem, we need a model of the forward process, describing how the causes (parameters) lead to effects (measurements). This forward process can be called the "foward model" or "forward problem".

Not every parameter estimation is called an inverse problem. Term term is usually applied to problems that difficult because the solution is sensitive to the noise and inaccuracies in the data. Inverse problems have been a focus of research for a century and sophisticated mathematical methods techniques have been developed to overcome the difficulties.

6.2 WHY IS EIT SO HARD?

In conventional medical imaging modalities, such as X-ray computerized tomography, a collimated beam of radiation passes through the object in a straight line, and the attenuation of this beam is affected only by the matter which lies along its path. In this sense X-ray CT is *local*, and it means that the pixels or voxels of our image affect only some (in fact a very small proportion) of the measurements. If the radiation were at lower frequency (*softer* X-rays) the effect of scattering would have to be taken into account and the effect of a change of material in a voxel would no longer be local. As the frequency decreases this nonlocal effect becomes more pronounced until we reach the case of direct current, in which a change in conductivity would have some effect on any measurement of surface voltage when any current pattern is applied. This *non-local* property of conductivity imaging, which still applies at the moderate frequencies used in EIT, is one of the principal reasons that EIT is difficult. It means that to find the conductivity image one must solve a system of simultaneous equations relating every voxel to every measurement.

In addition to non-locality, EIT is non-linear. The change in measured voltages as a conductivity changes will saturate for large contrasts (see §5.3.2). A conductivity increase or decrease by a large factor will produce a much smaller change in measurements. Another aspect of the non-linearity of EIT is the lack of superpostion. Two nearby contrasting regions do not lead to the same measurements as the combination of the measurements from the regions separately.

Non-locality in itself is not such a big problem provided we attempt to recover a modest number of unknown conductivity parameters from a modest number of measurements. Worse than that is the ill-posed nature of the problem. According to Hadamard a mathematical model of a physical problem is well posed if

1) For all admissible data, a solution exists.
2) For all admissible data, the solution is unique.
3) The solution depends continuously on the data.

The problem of recovering an unknown conductivity from boundary data is severely ill-posed, and it is the third criterion which gives us the most trouble. In practice that means for any given measurement precision, there are arbitrarily large changes in the conductivity distribution which are undetectable by boundary voltage measurements at that precision. This is clearly bad news for practical low frequency electrical imaging. Before we give up EIT altogether and take up market gardening, there is a partial answer to this problem – we need some additional information about the conductivity distribution. If we know enough *a priori* (that is in advance) information, it constrains the solution so that the wild variations causing the instability are ruled out.

The other two criteria can be phrased in a more practical way for our problem. Existence of a solution is not really in question: we know the body can be characterized by a conductivity. The issue is more whether the data are sufficiently accurate to be consistent with a conductivity distribution. Small errors in measurement can violate consistency conditions, such as reciprocity. One way around this is to project our infeasible data on to the closest feasible set. The mathematician's problem of uniqueness of solution is better understood in experimental terms as sufficiency of data. In the mathematical literature the conductivity inverse boundary value problem (or Calderón problem) is to show that a complete knowledge of the relationship between voltage and current at the boundary determines the conductivity uniquely. This has been proved under a variety of assumptions about the smoothness of the conductivity distribution[503]. This is only a partial answer to the practical problem as we have only finitely many measurements from a fixed system of electrodes, the electrodes typically cover only a portion of the surface of the body and in many cases voltage are not measured on electrodes driving currents. In the practical case the number of degrees of freedom of a parameterized conductivity we can recover is limited by the number of independent measurements made and the accuracy of those measurements.

This introductory section has deliberately avoided mathematical treatment, but a further understanding of why the reconstruction problem of EIT is difficult, and how it might be done requires some mathematical prerequisites. The minimum required for the following is reasonably thorough understanding of matrices[1009], and little multi-variable calculus, such as are generally taught to engineering undergraduates. For those desirous of a deeper knowledge of EIT reconstruction, for example those wishing to implement reconstruction software, an undergraduate course in the finite element method[975] and another in inverses problems[71,105,435] would be advantageous.

6.3 INVERSE PROBLEM

The inverse problem, as formulated by Calderón[172], is to recover σ from Λ_σ (Box 6.9). The uniqueness of solution, or if you like the sufficiency of the data, has been shown under a variety of assumptions, notably in the work of Kohn and Vogelius[567] and Sylvester and Uhlmann[1016]. For a summary of results see Isakov[503]. Astala and Paivarinta[68] have shown uniqueness for the 2D case without smoothness assumptions. There is very little theoretical work on what can be determined from incomplete data, but knowing the Dirichlet-to-Neumann mapping on an open subset of the boundary is enough[605]. It is also known that one set of Dirichlet and Neumann data, provided it contains enough frequency components, is enough to determine the boundary between two homogeneous materials with differing conductivities[38]. These results show that the second of Hadamard's conditions is not the problem, at least in the limiting, 'infinitely many electrodes' case. As for the first of Hadamard's condition, the difficulty is characterising 'admissible data' and there is very little work characterising what operators are valid Dirichlet-to-Neumann operators. The real problem however is in the third of Hadamard's conditions. In the absence of *a priori* information about the conductivity, the inverse problem $\Lambda_\sigma \mapsto \sigma$ is extremely unstable in the presence of noise.

Box 6.9: The Dirichlet-to-Neumann Map: Λ_σ

The Dirichlet-to-Neumann (DtoN) map, Λ_σ, is the Voltage-to-Current map; it describes what current will flow into a body if we impose a pattern of boundary currents. It depends on the body's internal conductivity distribution, σ. The Neumann-to-Dirichlet (NtoD) map describes the converse, how a pattern of applied current maps to a distribution of voltage. The NtoD is not unique, since adding a constant voltage everywhere corresponds to the same currents.

Unfortunately the terminology is often difficult for engineers: it's easiest to understand DtoN as a continuous extension of an admittance matrix, $\mathbf{Y}$, and the NtoD as an impedance matrix, $\mathbf{Z}$. Given a body with a large number of electrodes, and voltages $\mathbf{V}$ and currents $\mathbf{I}$ on the electrodes, we have $\mathbf{V} = \mathbf{ZI}$ and $\mathbf{I} = \mathbf{YV}$. Given a finite-element model of a body, matrices each row of the matrices can be assembled by applying the corresponding voltage or current pattern.

6.4 REGULARIZING LINEAR ILL-POSED PROBLEMS

In this section we consider the general problem of solving a linear ill-posed problem, before applying this specifically to EIT in the next section. Detailed theory and examples of linear ill-posed problems can be found in [71,258,435,1027,1087]. We assume a background in basic linear algebra [1009]. For complex vectors $\mathbf{x} \in \mathbb{C}^n$ and $\mathbf{b} \in \mathbb{C}^m$ and a complex matrix $\mathbf{A} \in \mathbb{C}^{m \times n}$ we wish to find $\mathbf{x}$ given $\mathbf{Ax} = \mathbf{b}$. Of course in our case $\mathbf{A}$ is the Jacobian while $\mathbf{x}$ will be a conductivity change and $\mathbf{b}$ a voltage change. In practical measurement problems it is usual to have more data than unknowns. If the surfeit of data were are only problem, the natural solution would be to use the Moore-Penrose generalized inverse

$$\mathbf{x}_{\text{MP}} = \mathbf{A}^\dagger \mathbf{b} = (\mathbf{A}^*\mathbf{A})^{-1}\mathbf{A}^*\mathbf{b} \tag{6.1}$$

which is the *least squares solution* in that

$$\mathbf{x}_{\text{MP}} = \arg \min_{\mathbf{x}} \|\mathbf{Ax} - \mathbf{b}\| \tag{6.2}$$

(here $\arg \min_\mathbf{x}$ means the argument $\mathbf{x}$ which minimizes what follows). In Matlab[1] the backslash (left division) operator can be used to calculate the least squares solution, for example $x = A\backslash b$.

6.4.1 ILL-CONDITIONING

It is the third of Hadamard's conditions, instability, which causes us problems. To understand this first we define the operator norm of a matrix

$$\|A\| = \max_{\mathbf{x} \neq 0} \frac{\|\mathbf{Ax}\|}{\|\mathbf{x}\|}.$$

This can be calculated as the square root of the largest eigenvalue of $\mathbf{A}^*\mathbf{A}$. There is another norm on matrices in $\mathbb{C}^{m \times n}$ the *Frobenius norm* which is simply

$$\|\mathbf{A}\|_F^2 = \sum_{i=1}^{m} \sum_{j=1}^{n} |a_{ij}|^2 = \text{trace}\,\mathbf{A}^*\mathbf{A}$$

---

[1]Matlab®is a matrix oriented interpreted programming language for numerical calculation. (The MathWorks Inc, Natick, MA, USA.) While we write Matlab for brevity we include its free software relative Octave.

which treats the matrix as simply a vector rather than an operator. We also define the condition number

$$\kappa(\mathbf{A}) = \|\mathbf{A}\| \cdot \|\mathbf{A}^{-1}\|.$$

for $\mathbf{A}$ invertible. Assuming that $\mathbf{A}$ is known accurately, $\kappa(\mathbf{A})$ measures the amplification of relative error in the solution.

Specifically if

$$\mathbf{A}\mathbf{x} = \mathbf{b} \quad \text{and} \quad \mathbf{A}(\mathbf{x} + \delta\mathbf{x}) = \mathbf{b} + \delta\mathbf{b}$$

then the relative error in solution and data are related by

$$\frac{\|\delta\mathbf{x}\|}{\|\mathbf{x}\|} \leq \kappa(\mathbf{A})\frac{\|\delta\mathbf{b}\|}{\|\mathbf{b}\|}$$

as can be easily shown from the definition of operator norm. Note that this is a 'worst case' error bound, often the error is less. With infinite precision, any finite $\kappa(\mathbf{A})$ shows that $\mathbf{A}^{-1}$ is continuous, but in practice error in data could be amplified so much the solution is useless. Even if the data $\mathbf{b}$ were reasonably accurate, numerical errors mean that, effectively $\mathbf{A}$ has error, and

$$\frac{\|\delta\mathbf{x}\|}{\|\mathbf{x}\|} \leq \kappa(\mathbf{A})\frac{\|\delta\mathbf{A}\|}{\|\mathbf{A}\|}.$$

(Actually this is not quite honest, it should be a 'perturbation bound' see[466]) So in practice we can regard linear problems with large $\kappa(A)$ as 'ill-posed' although the term *ill-conditioned* is better for the discrete case.

6.4.2 TIKHONOV REGULARIZATION

The method commonly known as Tikhonov Regularization was introduced to solve integral equations by Phillips[826] and Tikhonov[1039] and for finite dimensional problems by Hoerl[473]. In the statistical literature, following Hoerl, the technique is known as *ridge regression*. We will explain it here for the finite dimensional case. The least squares approach fails for a badly conditioned $\mathbf{A}$ but one strategy is to replace the least squares solution by

$$\mathbf{x}_\alpha = \arg\min_{\mathbf{x}} \quad \|\mathbf{A}\mathbf{x} - \mathbf{b}\|^2 + \alpha^2\|\mathbf{x}\|^2. \tag{6.3}$$

Here we trade off actually getting a solution to $\mathbf{A}\mathbf{x} = \mathbf{b}$ and not letting $\|\mathbf{x}\|$ get too big. The term $\|\mathbf{x}\|$ can be considered a "penalty" on solutions with large $\mathbf{x}$ values, and the number α controls this trade-off and is called a *regularization parameter*. Notice that as $\alpha \to 0$, $\mathbf{x}_\alpha$ tends to a generalized solution $\mathbf{A}^\dagger\mathbf{b}$.

There are two classic solutions for (6.3). We can expand the norm and set its derivative with respect to $\mathbf{x}$ at the solution $\mathbf{x}_\alpha$ to zero.

$$\frac{d}{d\mathbf{x}}(\mathbf{A}\mathbf{x} - \mathbf{b})^*(\mathbf{A}\mathbf{x} - \mathbf{b}) + \alpha^2(\mathbf{x}^*\mathbf{x}) = 2\left(\mathbf{A}^*\mathbf{A}\mathbf{x} - \mathbf{A}^*\mathbf{b} + \alpha^2\mathbf{x}\right) = 0$$

The other approach is represent (6.3) as the Moore-Penrose inverse of the block matrices

$$\begin{bmatrix} \mathbf{A} \\ \alpha\mathbf{I} \end{bmatrix}\mathbf{x} = \begin{bmatrix} \mathbf{b} \\ 0 \end{bmatrix}.$$

The explicit formula for the minimum is

$$\mathbf{x}_\alpha = (\mathbf{A}^*\mathbf{A} + \alpha^2\mathbf{I})^{-1}\mathbf{A}^*\mathbf{b}, \tag{6.4}$$

and can also be represented as

$$\mathbf{x}_\alpha = \mathbf{A}^*(\mathbf{A}\mathbf{A}^* + \alpha^2\mathbf{I})^{-1}\mathbf{b}. \tag{6.5}$$

Given more unknowns than measurements, the inverse in (6.5) is generally smaller and faster to calculate. See Box 6.10 for an interpretation in terms of CT backprojection.

The condition number $\kappa(\mathbf{A}^*\mathbf{A} + \alpha^2\mathbf{I})^{-1}$ is $\frac{\lambda_1 + \alpha^2}{\lambda_n + \alpha^2}$ where λ_i are the eigenvalues of $\mathbf{A}^*\mathbf{A}$, which for λ_n small is close to $\frac{\lambda_1}{\alpha^2} + 1$, so for big α it is well conditioned. Notice also that even if $\mathbf{A}$ does not have full rank ($\lambda_n = 0$), $\mathbf{A}^*\mathbf{A} + \alpha^2\mathbf{I}$ does.

Box 6.10: Tikhonov Regularization and Filtered Backprojection

The Tikhonov regularised solutions (6.4) and (6.5) can be usefully understood with an analogy to the filtered backprojection algorithms used X-ray CT[437].

First, we understand the operation $\mathbf{b} = \mathbf{A}\mathbf{x}$ as "projection", where the internal image $\mathbf{x}$ is projected onto the data $\mathbf{b}$. The adjoint matrix $\mathbf{A}^*$ is thus "backprojection" of the data onto an image $\mathbf{x}_b = \mathbf{A}^*\mathbf{b}$. While having some structure, image $\mathbf{x}_b$ is not very good; it will be blurry and reflect any uneven sensitivity in the projection.

To improve the image, a filter is required. We have two (equivalent) options

- first backproject then filter:
  $\mathbf{x} = \mathscr{F}_i\mathbf{A}^*\mathbf{b}$, where $\mathscr{F}_i = (\mathbf{A}^*\mathbf{A} + \alpha\mathbf{I})^{-1}$ or (6.4)
- first filter then backproject:
  $\mathbf{x} = \mathbf{A}^*\mathscr{F}_d\mathbf{b}$, where $\mathscr{F}_d = (\mathbf{A}\mathbf{A}^* + \alpha\mathbf{I})^{-1}$ or (6.5)

6.4.3 THE SINGULAR VALUE DECOMPOSITION

The singular value decomposition (SVD) is the generalization to non-square matrices of orthogonal diagonalization of Hermitian matrices. We describe the SVD in some detail here due to its importance in EIT. Although the topic is often neglected in elementary linear algebra courses and texts ([1009] is an exception), it is described well in texts on Inverse Problems, eg[71]. For some problems, such as X-ray CT, an explicit analytical form for the SVD is known[437], while for EIT it is calculated numerically.

For $\mathbf{A} \in \mathbb{C}^{m \times n}$, we recall that $\mathbf{A}^*\mathbf{A}$ is a non-negative definite Hermitian so has a complete set of orthogonal eigenvectors $\mathbf{v}_i$ with real eigenvalues $\lambda_1 \geq \lambda_2 \geq \cdots \geq 0$. These are normalized so that $\mathbf{V} = [\mathbf{v}_1 \mid \mathbf{v}_2 \mid \cdots \mid \mathbf{v}_n]$ is a unitary matrix $\mathbf{V}^* = \mathbf{V}^{-1}$. We define $\sigma_i = \sqrt{\lambda_i}$ and for $\sigma_i \neq 0$, $\mathbf{u}_i = \sigma_i^{-1}\mathbf{A}\mathbf{v}_i \in \mathbb{C}^m$. Now notice that $\mathbf{A}^*\mathbf{A}\mathbf{v}_i = \lambda_i\mathbf{v}_i = \sigma_i^2\mathbf{v}_i$. And $\mathbf{A}^*\mathbf{u}_i = \sigma_i^{-1}\mathbf{A}^*\mathbf{A}\mathbf{u}_i = \sigma_i\mathbf{u}_i$. Also $\mathbf{A}\mathbf{A}^*\mathbf{u}_i = \sigma_i^2\mathbf{u}_i$, where σ_i are called *singular values*[2] $\mathbf{v}_i$ and $\mathbf{u}_i$ right and left *singular vectors* respectively.

We see that the $\mathbf{u}_i$ are the eigenvectors of the Hermitian matrix $\mathbf{A}\mathbf{A}^*$, so they too are orthogonal. For a non-square matrix $\mathbf{A}$, there are more eigenvectors of either $\mathbf{A}^*\mathbf{A}$ or $\mathbf{A}\mathbf{A}^*$, depending on which is bigger, but only $\min(m,n)$ singular values. If rank $\mathbf{A} < \min(m,n)$ some of the σ_i will be zero. It is conventional to organize the singular values to be in decreasing order $\sigma_1 \geq \sigma_2 \geq \cdots \geq \sigma_{\min(m,n)} \geq 0$.

If $\text{rank}(\mathbf{A}) = k < n$ then the singular vectors $\mathbf{v}_{k+1}, \ldots, \mathbf{v}_n$ form an orthonormal basis for null $(\mathbf{A})$, whereas $\mathbf{u}_1, \ldots, \mathbf{u}_k$ form a basis for range$(\mathbf{A})$. On the other hand, if $k = \text{rank}(\mathbf{A}) < m$, then $\mathbf{v}_1, \ldots, \mathbf{v}_k$

---

[2]The use of σ for singular values is conventional in linear algebra, and should cause no confusion with the use of this symbol for conductivity, which is the accepted symbol for conductivity.

form a basis for range$(\mathbf{A}^*)$, and $\mathbf{u}_{k+1}, \ldots, \mathbf{u}_m$ form an orthonormal basis for null $(\mathbf{A}^*)$. In summary

$$\mathbf{A}\mathbf{v}_i = \sigma_i \mathbf{u}_i \quad i \leq \min(m,n)$$
$$\mathbf{A}^*\mathbf{u}_i = \sigma_i \mathbf{v}_i \quad i \leq \min(m,n)$$
$$\mathbf{A}\mathbf{v}_i = 0 \quad \text{rank}(\mathbf{A}) < i \leq n$$
$$\mathbf{A}^*\mathbf{u}_i = 0 \quad \text{rank}(\mathbf{A}) < i \leq m$$
$$\mathbf{u}_i^*\mathbf{u}_j = \delta_{ij}, \mathbf{v}_i^*\mathbf{v}_j = \delta_{ij}$$
$$\sigma_1 \geq \sigma_2 \geq \cdots \geq 0.$$

It is clear from the definition that for any matrix $\mathbf{A}$ $\|\mathbf{A}\| = \sigma_1$ while the Frobenius norm is $\|\mathbf{A}\|_F = \sqrt{\sum_i \sigma_i^2}$. If $\mathbf{A}$ is invertible $\|\mathbf{A}^{-1}\| = 1/\sigma_n$.

The singular value decomposition, SVD, allows us to diagonalize $\mathbf{A}$ using orthogonal transformations. Let $\mathbf{U} = [\mathbf{u}_1 \mid \cdots \mid \mathbf{u}_m]$ then $\mathbf{A}\mathbf{V} = \mathbf{U}\boldsymbol{\Sigma}$, where $\boldsymbol{\Sigma}$ is the diagonal matrix of singular values padded with zeros to make an $m \times n$ matrix. The nearest thing to diagonalization for non-square $\mathbf{A}$ is

$$\mathbf{U}^*\mathbf{A}\mathbf{V} = \boldsymbol{\Sigma}, \text{ and } \mathbf{A} = \mathbf{U}\boldsymbol{\Sigma}\mathbf{V}^*.$$

Although the SVD is a very important tool for understanding the ill-conditioning of matrices, it is rather expensive to calculate numerically and the cost is prohibitive for large matrices.

In Matlab the command s=svd(A) returns the singular values, [U,S,V]=svd(A) gives you the whole singular value decomposition. There are special forms if $\mathbf{A}$ is sparse, or if you only want some of the singular values and vectors.

Once the SVD is known, it can be used to rapidly calculate the Moore-Penrose generalized inverse from

$$\mathbf{A}^\dagger = \mathbf{V}\boldsymbol{\Sigma}^\dagger\mathbf{U}^*$$

where $\boldsymbol{\Sigma}^\dagger$ is simply $\boldsymbol{\Sigma}^T$ with the non-zero σ_i replaced by $1/\sigma_i$. This formula is valid whatever the rank of $\mathbf{A}$ and gives the minimum norm least squares solution. Similarly the Tikhonov solution is

$$\mathbf{x}_\alpha = \mathbf{V}\mathbf{T}_\alpha\mathbf{U}^*\mathbf{b}$$

where $\mathbf{T}$ is $\boldsymbol{\Sigma}^T$ with the non-zero σ_i replaced by $\sigma_i/(\sigma_i^2 + \alpha^2)$. As only $\mathbf{T}_\alpha$ varies with α one can rapidly recalculate $\mathbf{x}_\alpha$ for a range of α once the SVD is known. A truncated SVD (TSVD) solution

$$\mathbf{A}_{TSVD} = \mathbf{V}\boldsymbol{\Sigma}_{TSVD}\mathbf{U}^*$$

can be created by setting each element i to $1/\sigma_i$ for σ_i above some user-defined treshold, and setting other values to zero. The SVD was one of the first approaches used in EIT image reconstruction[754]; however the abrupt threshold is generally worse than the smoother transition in singular values provided by Tikhonov regularization[436].

6.4.4 STUDYING ILL-CONDITIONING WITH THE SVD

The singular value decomposition is a valuable tool in studying the ill-conditioning of a problem. Typically we calculate numerically the SVD of a matrix which is a discrete approximation to a continuum problem, and the decay of the singular values gives us an insight into the extent of the instability of the inverse problem. In a simple example[435], calculating k-th derivatives numerically is an ill-posed problem, in that taking differences of nearby values of a function is sensitive to error in the function values. Our operator $\mathbf{A}$ is a discrete version of integrating trigonometric polynomials k-times. The singular vectors of $\mathbf{A}$ are a discrete Fourier basis and the singular value for the i-th frequency proportional to i^{-k}. Problems such as this where $\sigma_i = O(i^{-k})$ for some $k > 0$ are called *mildly ill-posed*. If we assume sufficient *a priori* smoothness on the function the problem becomes well-posed. By contrast problems such as the inverse Laplace transform, the backward heat equation[435], and linearized EIT the singular values decay faster than any power i^{-k}, and we term them *severely ill-posed*. This degree of ill-posedness technically applies to the continuum problem, but a discrete approximation to the operator will have singular values that approach this behaviour as the accuracy of the approximation increases.

In linearized EIT we can interpret the singular vectors $\mathbf{v}_i$ as telling us that the components $\mathbf{v}_i^*\mathbf{x}$ of a conductivity image $\mathbf{x}$ are increasingly hard to determine as i increases, as they produce voltage changes $\sigma_i\mathbf{u}_i^*x$. With a relative error of ε in the data $\mathbf{b}$ we can only expect to reliably recover the components $\mathbf{v}_i^*\mathbf{x}$ of the image when $\sigma_i/\sigma_1 > \varepsilon$. A graph of the singular values (for EIT we typically plot σ_i/σ_0 on a logarithmic scale), gives a guide to the number of degrees of freedom in the image we can expect to recover with measurement at a given accuracy. See figure 6.2.

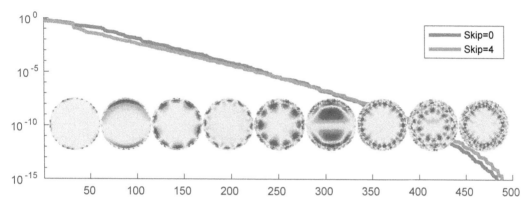

Figure 6.2   Singular values on a logarithmic scale of the Jacobian matrix for a 32 electrode model, for two different stimulation and measurement patterns: adjacent (skip=0) and skip=5. Illustrated are images of the singular vectors from 10 to 250. The first singular vectors represent the boundary, later ones represent low spatial frequency patterns in the interior, and the rightmost represent higher spatial frequency interior patterns. (color figure available in eBook)

Another use of the graph of the singular values is determination of rank. Suppose we collect a redundant set of measurements, for example some of the voltages we measure could be determined by reciprocity. As the linear relations between the measurements will transfer to dependencies in the rows of the Jacobian, if n is greater than the number of independent measurements k, the matrix $\mathbf{A}$ will be rank deficient. In numerical linear algebra linear relations are typically not exact due to rounding error, and rather than having zero singular values we will find that after σ_k the singular values will fall abruptly by several decades. For an example of this in EIT see[154].

The singular values themselves do not tell the whole story. For example two EIT drive configurations may have similar singular values, but if the singular vectors $\mathbf{v}_i$ differ then they will be able to reliably reconstruct different conductivities. To test how easy it is to detect a certain (small as we have linearized) conductivity change $\mathbf{x}$, we look at the singular spectrum $\mathbf{V}^*\mathbf{x}$. If most of the large components are near the top of this vector the change is easy to detect, where as if they are all bellow the l-th row they are invisible with relative error worse than σ_l/σ_0. The singular spectrum $\mathbf{U}^*\mathbf{b}$ of a set of measurements $\mathbf{b}$, gives a guide to how useful that set of measurements will be at a given error level.

6.4.5 MORE GENERAL REGULARIZATION

In practical situations the standard Tikhonov regularization is rarely useful unless the variables $\mathbf{x}$ represents coefficients with respect to some well chosen basis for the underlying function. In imaging problems it is natural to take our vector of unknowns as pixel or voxel values, and in EIT one often takes the values of conductivity on each cell (eg triangle or tetrahedron) of some decomposition of the domain, and assumes the conductivity to be constant on that cell. The penalty term $\|\mathbf{x}\|$ in standard Tikhonov prevents extreme values of conductivity but does not enforce smoothness, nor constrain nearby cells to have similar conductivities. As an alternative we choose a positive definite (and without loss of generality Hermitian) matrix $\mathbf{P} \in \mathbb{C}^{n \times n}$ and the norm $\|\mathbf{x}\|_{\mathbf{P}}^2 = \mathbf{x}^* \mathbf{P} \mathbf{x}$. A common choice is to use an approximation to a differential operator $\mathbf{L}$ and set $\mathbf{P} = \mathbf{L}^* \mathbf{L}$.

There are two further refinements which can be included, the first is that we penalise differences from some background value $\mathbf{x}_0$, which can include some known non-smooth behaviour and penalise $\|\mathbf{x} - \mathbf{x}_0\|_{\mathbf{P}}$. The second is to allow for the possibility that we may not wish to fit all measurements to the same accuracy, in particular as some may have larger errors than others. This leads to consideration of the term $\|\mathbf{A}\mathbf{x} - \mathbf{b}\|_{\mathbf{Q}}$ for some diagonal weighting matrix $\mathbf{Q}$. If the errors in $\mathbf{b}$ are correlated, one can consider a non-diagonal $\mathbf{Q}$ so that the errors in $\mathbf{Q}^{1/2}\mathbf{b}$ are not correlated. The probabilistic interpretation of Tikhonov regularization in Box 6.11 makes this more explicit. Our generalized Tikhonov procedure is now

$$\mathbf{x}_{GT} = \arg\min_{\mathbf{x}} \|\mathbf{A}\mathbf{x} - \mathbf{b}\|_{\mathbf{Q}}^2 + \|\mathbf{x} - \mathbf{x}_0\|_{\mathbf{P}}^2.$$

which reduces to the standard Tikhonov procedure for $\mathbf{P} = \mathbf{I}$, $\mathbf{Q} = \alpha^2 \mathbf{I}$, $\mathbf{x}_0 = 0$. We can find the solution by noting that for $\tilde{\mathbf{x}} = \mathbf{P}^{1/2}(\mathbf{x} - \mathbf{x}_0)$, $\tilde{\mathbf{A}} = \mathbf{Q}^{1/2}\mathbf{A}\mathbf{P}^{-1/2}$, and $\tilde{b} = \mathbf{Q}^{1/2}(\mathbf{b} - \mathbf{A}\mathbf{x}_0)$

$$\mathbf{x}_{GT} = \mathbf{x}_0 + \mathbf{P}^{-1/2} \arg\min_{\tilde{\mathbf{x}}} \left(\|\tilde{\mathbf{A}}\tilde{\mathbf{x}} - \tilde{b}\|^2 + \|\tilde{\mathbf{x}}\|^2 \right)$$

which can be written explicitly as

$$\mathbf{x}_{GT} = \mathbf{x}_0 + \mathbf{P}^{-1/2} \left(\tilde{\mathbf{A}}^* \tilde{\mathbf{A}} + \mathbf{I} \right)^{-1} \tilde{\mathbf{A}}^* \tilde{b} \qquad (6.6)$$

$$= \mathbf{x}_0 + \left(\mathbf{A}^* \mathbf{Q} \mathbf{A} + \mathbf{P} \right)^{-1} \mathbf{A}^* \mathbf{Q} (\mathbf{b} - \mathbf{A}\mathbf{x}_0) \qquad (6.7)$$

or in the alternative form

$$\mathbf{x}_{GT} = \left(\mathbf{A}^* \mathbf{Q} \mathbf{A} + \mathbf{P} \right)^{-1} \left(\mathbf{A}^* \mathbf{Q} \mathbf{b} + \mathbf{P} \mathbf{x}_0 \right) \qquad (6.8)$$

$$= \mathbf{x}_0 + \mathbf{P} \mathbf{A}^* \left(\mathbf{A} \mathbf{P}^{-1} \mathbf{A}^* + \mathbf{Q}^{-1} \right) (\mathbf{b} - \mathbf{A}\mathbf{x}_0). \qquad (6.9)$$

As in the standard Tikhonov case, generalized Tikhonov can be explained in terms of the SVD of $\tilde{\mathbf{A}}$ which can be regarded as the SVD of the operator $\mathbf{A}$ with respect to the $\mathbf{P}$ and $\mathbf{Q}$ norms. Sometimes it is useful to consider a non-invertible $\mathbf{P}$, for example if $\mathbf{L}$ is a first order difference operator $\mathbf{L}^* \mathbf{L}$ has a non-trivial null space. Provided the null space can be expressed as a basis of singular vectors of $\mathbf{A}$ with large σ_i the regularization procedure will still be successful. This situation can be studied using the Generalized Singular Value Decomposition, GSVD[435].

Box 6.11: Probabilistic Interpretation of Regularization

The statistical approach to regularization[1087] gives an alternative justification of generalized Tikhonov regularization. For a detailed treatment of the application of this approach to EIT see[529]. Bayes' theorem relates conditional probabilities of random variables.

$$P(\mathbf{x}|\mathbf{b}) = \frac{P(\mathbf{b}|\mathbf{x})P(\mathbf{x})}{P(\mathbf{b})}.$$

The probability of $\mathbf{x}$ given $\mathbf{b}$ is the probability of $\mathbf{b}$ given $\mathbf{x}$ times $P(\mathbf{x})/P(\mathbf{b})$.

We now want the most likely $\mathbf{x}$, so we maximize the *posterior* $P(\mathbf{x}|\mathbf{b})$, obtaining the so called *Maximum A-Posteriori* (MAP) estimate.

This is easy to do if we assume $\mathbf{x}$ is multivariate Gaussian with mean $\mathbf{x}_0$ and covariance $\mathrm{Cov}[\mathbf{x}] = P^{-1}$, and $\mathbf{e}$ has mean zero and $\mathrm{Cov}[\mathbf{e}] = \mathbf{Q}^{-1}$

$$P(\mathbf{x}|\mathbf{b}) = \frac{1}{P(\mathbf{b})} \exp\left(-\frac{1}{2}\|\mathbf{Ax} - \mathbf{b}\|_{\mathbf{Q}}^2\right) \cdot \exp\left(-\frac{1}{2}\|x - x_0\|_{\mathbf{P}}^2\right)$$

where we have used that $\mathbf{x}$ and $\mathbf{e}$ are independent so $P_{\mathbf{b}}(\mathbf{b}|\mathbf{x}) = P_{\mathbf{e}}(\mathbf{b} - \mathbf{Ax})$. We notice that $P(\mathbf{x}|\mathbf{b})$ is maximized by minimizing

$$\|\mathbf{Ax} - \mathbf{b}\|_{\mathbf{Q}}^2 + \|\mathbf{x} - \mathbf{x}_0\|_{\mathbf{P}}^2.$$

This is called the *maximum a posteriori* or MAP estimate. It is also possible to derive many other useful estimates from this distribution, including error estimates[529].

6.5   REGULARIZING EIT

Image reconstruction in EIT has traditionally been divided into algorithms for difference EIT (§6.6) and absolute EIT (§6.7). Difference EIT is more stable and can be feasibly reconstructed with linear techniques, while absolute EIT is a more difficult problem and requires advanced methods. In contrast to geophysical ERT (chapter 18), in which absolute reconstructions are common, experimental and clinical EIT has almost exclusively used difference measurements and algorithms. For successful absolute reconstructions, it is necessary for the model to be accurate in terms of body geometry and electrode positions, shapes and contact impedances, as well as hardware and electronics imperfections. For difference EIT, shape and electronics modelling inaccuracies are less significant, as long as they remain the same between the difference measurements.

We define a forward operator F by $F(\mathbf{x}) = \mathbf{V}$ which takes the vector of parameters $\mathbf{x}$ to the measured voltages at the boundary $\mathbf{V}$. In simple cases, each image element can be a parameter, but often this leads to an unnecessarily large degrees of freedom; instead parameters can have a dense representation in high-sensitivity areas, and a less dense representation elsewhere. $\mathbf{V}$ can represent either the matrix of measured voltages at each stimulation pattern, or, in the case where voltages are not measured on driven electrodes, it can represent a vector of pair-difference measurements made by the system.

We will leave aside the adaptive current approach (§5.5.3) where the measurements taken depend on the conductivity. The goal is to fit the actual measured voltages $\mathbf{V}_{\mathrm{m}}$ and the simplest approach, as in the case of a linear problem, is to minimize the sum of squares error

$$\|\mathbf{V}_{\mathrm{m}} - F(\mathbf{x})\|_F^2$$

the so called *output least squares* approach. We have emphasized the Frobenius norm here as $\mathbf{V}_m$ is a matrix, however in this section we will use the notational convenience of using the same symbol when the matrix of measurements is arranged as a column vector. In practice it is not usual to use the raw least squares approach, but at least a weighted sum of squares which reflects the reliability of each voltage. More generally (Box 6.11) we use a norm weighted by the inverse of the error covariance. Such approaches are common both in optimization and the statistical approach to inverse problems. To simplify the presentation we will use the standard norm on voltages, or equivalently, assume they have already been suitably scaled. The more general case is easily deduced from the last section.

Minimization of the voltage error (for simple parameterizations of γ) is doomed to failure as the problem is ill-posed. In practise the minimum lies in a long narrow valley of the objective function[153]. For a unique solution one must include additional information about the conductivity, and this is typically expressed as a penalty $G(\mathbf{x})$, just as in the case of a linear ill-posed problem. We seek to minimize

$$f(\mathbf{x}) = \|\mathbf{V}_m - F(\mathbf{x})\|^2 + G(\mathbf{x}). \tag{6.10}$$

In EIT a typical simple choice[1075] is

$$G(\mathbf{x}) = \|\alpha \mathbf{L}(\mathbf{x} - \mathbf{x}_{\text{ref}})\|^2 \tag{6.11}$$

where $\mathbf{L}$ is a matrix approximation to some partial differential operator and $\mathbf{x}_{\text{ref}}$ a reference conductivity (for example including known anatomical features). The minimization of f represents a trade-off between fitting the data exactly and constraining spatial derivatives of the image $\mathbf{x}$, the trade off being controlled by the regularization parameter α.

6.6 DIFFERENCE EIT

Difference EIT reconstructs the change in image parameters $\delta \mathbf{x}$ from between two sets of measurements $\delta \mathbf{V} = \mathbf{V}_t - \mathbf{V}_{\text{ref}}$, where $\mathbf{V}_t$ is the "current" data set, and $\mathbf{V}_{\text{ref}}$ is reference data calculated at some stable reference time (or potentially an average over multiple time in order to improve stability). Note that, for most EIT systems, data are not actually available at the same time, but are sequentially acquired by varying the simulation pattern. For rapid changes, it is necessary to correct for this non-simultaneous sampling[1189]. When the two measurements are separated in time we refer to it as time-difference EIT (tdEIT), and when the measurements are simultaneous, but at different frequencies, it is frequency-difference EIT (fdEIT).

The key advantage of difference EIT is that it compensates for errors and offsets in the data or reconstruction model which do not vary with time[27]. Difference EIT will typically work well even if the electrode or boundary shape (or even dimention: 2D vs 3D) is incorrect.

One special case of difference EIT is normalized-difference, in which the a value $\delta \mathbf{V} = (\mathbf{V}_t - \mathbf{V}_{\text{ref}}) \oslash \mathbf{V}_{\text{ref}}$, where $\oslash$ denotes element by element (or Hadamard) division. Normalized-difference EIT can also normalize differences between measurement channels, for example if the channels have different gain. To reconstruct normalized-difference EIT, the sensitivity must also be normalized, $\mathbf{J}_{i,j} = \tilde{\mathbf{J}}_{i,j}/\mathbf{V}_{\text{ref},i}$, where $\tilde{\mathbf{J}}_{i,j}$ is the original model sensitivity.

In this chapter, we describe only regularization across an image at a single time. Clearly, data are correlated across time as there is a maximum rate at which a given body can change. Filtering of the data can improve the signal to noise ratio. Frequency filtering can be done before reconstruction or simultaneously with it[143]. Traditional temporal filtering schemes such as Kalman filtering can be incorporated into the image reconstruction[1077].

For fdEIT, $\mathbf{V}_t$ becomes the frequency of interest data set, and $\mathbf{V}_{\text{ref}}$ is reference frequency. It is not sufficient to simply subtract these values, since impedance decreases with frequency. Instead, a weighted frequency difference is calculated, $\delta\mathbf{V} = \mathbf{V}_t - k\mathbf{V}_{\text{ref}}$, for a (complex) constant, k, which best matches the data[957].

6.6.1 LINEARIZED EIT RECONSTRUCTION

For linearized difference EIT Reconstruction, we consider $F(\mathbf{x})$ replaced by a linear approximation

$$F(\mathbf{x}) \approx F(\mathbf{x}_0) + \mathbf{J}(\mathbf{x} - \mathbf{x}_0) \tag{6.12}$$

where $\mathbf{J}$ is the Jacobian matrix of F calculated at some initial conductivity estimate $\mathbf{x}_0$ (not necessarily the same as $\mathbf{x}_{\text{ref}}$). Defining $\delta\mathbf{x} = \mathbf{x} - \mathbf{x}_0$ and $\delta\mathbf{V} = \mathbf{V}_m - F(\mathbf{x}_0)$, our solution $\delta\mathbf{x}$ minimizes $\|\delta\mathbf{V}_m - \mathbf{J}\delta\mathbf{x}\|_{\mathbf{W}}^2 + \alpha^2\|\mathbf{L}(\delta\mathbf{x} - (\mathbf{x}_{\text{ref}} - \mathbf{x}_0))\|^2$ using (6.7). Here $\mathbf{W}$ can be understood as a measurement error weighting matrix. If all measurements have equal error $\mathbf{W} = \mathbf{I}$. This parameter can be varied to account for practical scenarios. If some electrodes are poorly connected the faculty measurements using these electrodes can be removed by setting the corresponding diagonal elements in $\mathbf{W}$ to zero[13]. Erroneous electrodes can be detected autonomically[63], or $\mathbf{W}$ can be weighted by the reciprocity error[442]. Off-diagonal terms in $\mathbf{W}$ can help compensate for meshing errors and electrode movement[29].

The solution to the linearized regularization problem for the choice of regularization in (6.11) (now a quadratic minimization problem) is given by

$$\begin{aligned}
\delta\mathbf{x} &= \mathbf{x}_{\text{ref}} - \mathbf{x}_0 + (\mathbf{J}^*\mathbf{W}\mathbf{J} + \alpha^2\mathbf{L}^*\mathbf{L})^{-1}\mathbf{J}^*\mathbf{W}(\delta\mathbf{V} - \mathbf{J}(\mathbf{x}_{\text{ref}} - \mathbf{x}_0)) \\
&= (\mathbf{J}^*\mathbf{W}\mathbf{J} + \alpha^2\mathbf{L}^*\mathbf{L})^{-1}\left(\mathbf{J}^*\mathbf{W}\delta\mathbf{V} + \alpha^2\mathbf{L}^*\mathbf{L}(\mathbf{x}_{\text{ref}} - \mathbf{x}_0)\right).
\end{aligned} \tag{6.13}$$

or any of the equivalent forms[1027]. In most applications of difference EIT, $\mathbf{x}_{\text{ref}} = \mathbf{x}_0$, and

$$\delta\mathbf{x} = \left[(\mathbf{J}^*\mathbf{W}\mathbf{J} + \alpha^2\mathbf{L}^*\mathbf{L})^{-1}\mathbf{J}^*\mathbf{W}\right]\delta\mathbf{V} = \mathbf{R}_M\delta\mathbf{V} \tag{6.14}$$

and a reconstruction matrix $\mathbf{R}_M$ calculated. Most experimental and clinical EIT systems precalculate $\mathbf{R}_M$ so an image can be rapidly calculated through a matrix multiplation and a mapping of parameters $\delta\mathbf{x}$ to the image.

While there are many other forms of regularization possible for a linear ill-conditioned problem, this generalised Tikhonov regularization has the benefit that (see Box 6.11) the *a priori* information it incorporates is made explicit and that under Gaussian assumptions it is the statistically defensible MAP estimate. When only a linearised solution is to be used with a fixed Jacobian $\mathbf{J}$, the regularized solution is calculated through a factorization of $(\mathbf{J}^*\mathbf{W}\mathbf{J} + \alpha^2\mathbf{L}^*\mathbf{L})$ with complexity $O(N^2)$ for N degrees of freedom in the conductivity (which should be smaller than the number of independent measurements). Although LU factorization would be one alternative, perhaps a better choice is to use the Generalized Singular Value Decomposition (GSVD)[435], which allows the regularized solution to be calculated efficiently for any value of α. The GSVD is now a standard tool for understanding the effect of the choice of the regularization matrix $\mathbf{L}$ in a linear ill-conditioned problem, and has been applied to linearized EIT[136,1079]. The use of a single linearized Tikhonov regularized solution is widespread in medical industrial and geophysical EIT, the NOSER algorithm[191] being a well known example.

6.6.2 SELECTION OF HYPERPARAMETER, α

Regularization imposes a trade-off between fidelity to the data and fidelity to the prior; the trade-off "dial" is α, which is often called the "hyperparameter" since it affects other parameters. In many algorithms, another parameter takes a similar role to α; for example in a truncated SVD inverse the number is singular values retained controls the amount of regularization.

In general when the hyperparameter is small, images are reconstructed with higher spatial resolution, but the images are much more sensitive to data noise. To illustrate the effect of the hyperparameter, different images are reconstructed with values of α with and without noise in figure 6.3. Clearly, the best choice of α varies depending on the noise level.

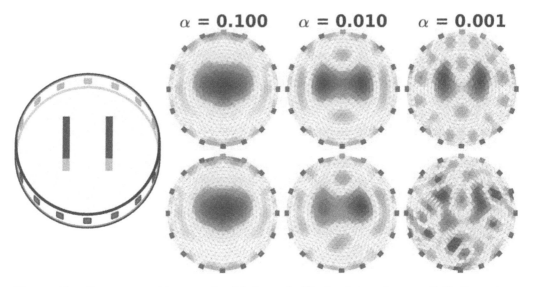

Figure 6.3 Reconstructed images of a 16 electrode 3D simulation phantom (*left*), for various values of the hyperparameter, α. *Top:* simulated data with no noise; *Bottom:* simulated data with added Gaussian noise. (color figure available in eBook)

An appropriate choice of α is vital for regularized image reconstructions. However, the meaning of a value of α depends on the details of the inverse problem. Simply changing the units of measure (Volts to mV) will completely change the meaning of α. Instead of directly chosing the parameter, a number of strategies have been used in the EIT literature to select an appropriate hyperparameter value.

The most widely used technique is the *L-curve*[435,436], which plots the data misfit against the prior as a function of α. The recommended value is the corner of the L. Several other classic techniques are also widely studied, such as generalized cross validation (GCV). Unfortunately, the experience in the EIT community with these algorithms is not good[376]: at low noise levels, the L-curve works well, but at experimental or clinical noise, there is no clear corner. Experience shows the GCV typically under-regularizes.

Given this experience, another approach has been pursued in the EIT community. Instead of seeing α as a property of the data (i.e. choose the best α to reconstruct this measurement) it is seen as a property of an EIT system (i.e. choose the best α over the range of likely data to be measured in this particular clinical application). This approach guides a manufacturer in their setting of the value that would work for their customers. Two EIT-specific approaches have been widely used: Noise Figure[26] and Image SNR[149]. In each case a target value is chosen, for example to achieve an adequate probability of detection of a particular contrast, and α is adjusted to achive the value, typically by bisection search.

6.6.3 REGULARIZATION PARAMETERS

Tikhonov regularization with $\mathbf{L} = \mathbf{I}$ is rarely used in EIT, because the sensitivity varies so widely from the centre of the body to the side. One early scheme has come to be called the NOSER prior[191] in which the diagonal is matched to the sum squared sensitivity from the Jacobian, so $\mathbf{L}^*\mathbf{L} = \operatorname{diag}\mathbf{J}^*\mathbf{J}$. Another common choice[838,1078] is to use a discrete approximation to the Laplacian on piecewise constant functions on the mesh. For each element a sum of the neighbouring element values is taken, weighted by the area (or length in 2D) of the shared faces and the total area (perimeter length) of the element multiplied by the element value subtracted. This is analogous to the common five point difference approximation to the Laplacian on a square mesh. Where elements have faces on the boundary, there are no neighbours and the scheme is equivalent to assuming an extension outside the body with the same value. This enforces a homogeneous Neumann boundary condition so that the null space of $\mathbf{L}$ is just constants. As constant conductivity values are easily obtained in EIT the null space does not diminish the regularizing properties of this choice of G. Similarly one could choose a first order differential operator for $\mathbf{L}$[1079]. Figure 6.4 illustrates this choice of $\mathbf{L}$, and also compares to a non-linear approach (§6.8).

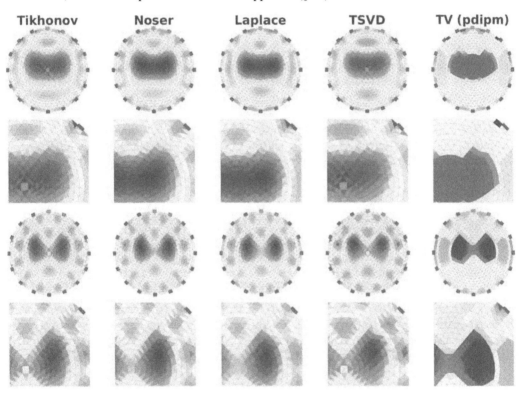

Figure 6.4   Reconstructed images of a simulation phantom (figure 6.3) for various values of the structural prior, $\mathbf{L}$, in each column. The penalty function, $G(\mathbf{x})$ is from left to right: Tikhonov ($\|\alpha\mathbf{Ix}\|^2$), Noser ($\alpha^2\mathbf{x}^*\operatorname{diag}(\mathbf{J}^*\mathbf{J})\mathbf{x}$)[191], Laplace ($\|\alpha\mathbf{Lx}\|^2$, where $\mathbf{L}^*\mathbf{L}$ is the discrete Laplacian on the FEM grid), TSVD (§6.4.3), and Total Variation (§6.8). Each row shows A reconstructed image and a zoomed-in version below. Parameters are chosen so that all images on each row are calculated at the same noise figure[26]. (color figure available in eBook)

Other smooth choices of G include the inverse of a Gaussian smoothing filter[26,136], effectively an infinite order differential operator. In these cases where G is smooth and for α large enough the Hessian of f will be positive definite, we can then deduce that f is a *convex* function[1087], so that a critical point will be a strict local minimum, guaranteeing the success of smooth optimization methods. Such regularization however will prevent us from reconstructing conductivities with a sharp transition, such as an organ boundary. However the advantage of using a smooth objective function f is that it can be minimized using smooth optimization techniques.

Another option is to include in G the *Total Variation*, that is the integral of $|\nabla\gamma|$. This still rules out wild fluctuations in conductivity while allowing step changes. We study this in more detail in §6.8

6.6.4   BACKPROJECTION

The earliest EIT reconstruction algorithms were motivated by the filtered backprojection algorithms used in X-ray CT and, in fact, many pictures of EIT reconstruction show equipotential lines. Although this approach is quite different from the filtered back projection along equipotential lines of Barber and Brown[82,925] it is sometimes confused with this in the literature. The *Sheffield Backprojection* algorithm is of historical importance because it was used in the Sheffield Mark II and (with minor variations) in the Goe MF II EIT systems, and thus in the majority of clinical and experimental publications before 2010. A comparison between Sheffield Backprojection and regularized algorithms showed that it performed reasonably well on experimental data[389]. For historical interest, the algorithm has been made available[3].

It is an interesting historical observation that in the medical and industrial applications of EIT numerous authors have calculated $\mathbf{J}$ and then proceeded to use *ad hoc* regularized inversion methods to calculate an approximate solution. Often these are variations on standard iterative methods which, if continued would for a well posed problem converge to the Moore-Penrose generalised solution. It is a standard method in inverse problems to use an iterative method but stop short of convergence (Morozov's discrepancy principle tells us to stop when the output error first falls below the measurement noise). Many linear iterative schemes can be represented as a filter on the singular values. However they have the weakness that the *a priori* information included is not as explicit as in Tikhonov regularization. One extreme example of the use of an *ad hoc* method is the method described by Kotre[579] in which the normalized transpose of the Jacobian is applied to the voltage difference data. In the Radon transform used in X-ray CT[767], the formal adjoint of the Radon transform is called the *back projection* operator. It produces at a point in the domain the sum of all the values measured along rays through that point. Although not an inverse to the Radon transform itself, a smooth image can be obtained by backprojecting smoothed data, or equivalently by backprojecting then smoothing the resulting image.

Kotre's back projection was until recently widely used in the process tomography community for both resistivity (ERT) and permittivity (ECT) imaging[250]. Often supported by the fallacious arguments, in particular that it is fast (it is no faster than the application of any precomputed regularized inverse) and that it is commonly used (only by those who know no better). In an interesting development the application of a normalized adjoint to the residual voltage error for the linearised problem was suggested for ECT, and later recognized as yet another reinvention of the well known Landweber iterative method[1184]. Although there is no good reason to use pure linear iteration schemes directly on problems with such small a number of parameters as they can be applied much faster using the SVD, an interesting variation is to use such a slowly converging linear solution together with projection on to a constraint set; a method which has been shown to work well in ECT[171].

6.6.5   GREIT

Given the still-widespread use of Sheffield backprojection in the mid 2000s, an effort was created to help move the clinical and experimental users of EIT to mathematically well-understood regularized techniques. A committee to design a consensus algorithm was established at the 2007 conference

---

[3]http://eidors.org/data_contrib/db_backproj_matrix/db_backproj_matrix.shtml

in Graz, Austria, and came to be known as the "Graz consensus Reconstruction algorithm for EIT" or GREIT[18]. The approach was then extended to 3D[390,391].

Work was done in two parts. First, clinical and experimental users of EIT were asked to identify the most important features (figures of merit) in EIT images, and subsequently, the parameters of a generalized Tikhonov scheme were selected to best match these features. Figure 6.5 illustrates the features itentified. The requirements, in order of importance were identified as: 1) uniform amplitude response, 2) small and uniform position error, 3) small ringing artefacts, 4) uniform resolution, 5) limited shape deformation, and 6) high resolution. The relative low importance for resolution can be surprising for image reconstruction specialists. The consensus of the experts was thus that it was most important to be able to trust the images resported, rather than specific image quality items such as high resolution.

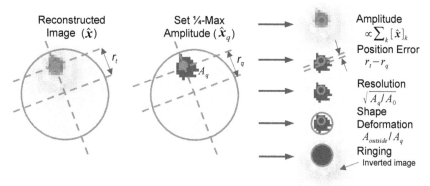

Figure 6.5 Performance figures of merit for evaluation of GREIT images. Based on a reconstructed image ($\hat{\mathbf{x}}$, *left*), an image ($\hat{\mathbf{x}}_q$, *centre*) is constructed of all image pixels which exceed $\frac{1}{4}$ of the maximum amplitude. From these images, figures of merit at *right* are calculated as described. (color figure available in eBook)

GREIT and its variants are used in many commercial EIT systems and in many clinical and experimental papers. It is formulated as a training problem, in which the calculated linear reconstruction matrix needs to produce "desired images" for training targets. In most cases, these training targets were generated from a finite element model, however they can be calculated experimentally, in which case the reconstruction matrix reflects any inaccuracies in the hardware[324].

As described, this training scheme results in a generalized Tikhonov norm and thus a linear reconstruction matrix, $\mathbf{R}_M$, such that $\hat{\mathbf{x}} = \mathbf{R}_M \mathbf{y}$, where $\hat{\mathbf{x}}$ is a reconstructed voxel image, from from difference data, $\mathbf{y} = \Delta \mathbf{V}_m$. The GREIT reconstruction matrix minimizes an error $\varepsilon^2(\mathbf{R}_M) = \mathrm{E}_w \left[\|\mathbf{x} - \mathbf{R}_M \mathbf{y}\|^2 \right]$ where w is the weight assigned to each target, $\mathbf{t}$, to represent the importance of its contribution. The expectation, $\mathrm{E}_w[\cdot]$ is over a distribution of "training" targets, $\mathbf{t}^{(i)}$, for which the corresponding data, $\mathbf{y}^{(i)}$, and a "desired" image, $\mathbf{x}^{(i)}$, are calculated. The reconstruction matrix which minimizes the error is $\mathbf{R}_M = \mathrm{E}_w \left[\mathbf{xy}^T \right] \left(\mathrm{E}_w \left[\mathbf{yy}^T \right] \right)^{-1}$.

Given a distribution $\mathbf{t} \sim \mathcal{N}(0, \Sigma_t)$ of training targets, $\mathbf{x} = \mathbf{Dt}$ and $\mathbf{y} = \mathbf{Jt} + \mathbf{n}$ are calculated, where $\mathbf{D}$ is the "desired image" matrix, which maps each training sample location onto the larger desired image region as discussed above. Noise, $\mathbf{n}$, is distributed as $\mathbf{n} \sim \mathcal{N}(0, \Sigma_n)$. Using these values

$$\mathbf{R}_M = \mathbf{D}\Sigma_t^* \mathbf{J}^T \left(\mathbf{J}\Sigma_t^* \mathbf{J}^T + \alpha \Sigma_n \right)^{-1} \tag{6.15}$$

where Σ_t^* is the effective covariance of the training targets when weighted by w. The parameter α is selected so that noise performance of the reconstruction matrix matches a noise figure of 0.5[26].

For robust 3D image reconstruction, it is important to introduce targets above and below the layers which form part the reconstructed image. This requirement may be understood as "allowing" the image reconstruction to "explain" measurements via out-of-field contrasts. If not allowed to do this, the reconstruction will be "forced" to create artifacts in the image plane to account for these off-plane contrasts. We separate targets $\mathbf{t} = [\mathbf{t}_i | \mathbf{t}_o]$ into those inside, $\mathbf{t}_i$, and outside, $\mathbf{t}_o$, the reconstructed layers. The sensitivity matrix is similarly partitioned $\mathbf{J} = [\mathbf{J}_i | \mathbf{J}_o]$, and $\mathbf{D} = \mathrm{diag}(\mathbf{D}_i, \mathbf{D}_o)$. From (6.15), we calculate

$$\begin{bmatrix} \mathbf{R}_i \\ \mathbf{R}_o \end{bmatrix} = \begin{bmatrix} \mathbf{D}_i & 0 \\ 0 & \mathbf{D}_o \end{bmatrix} \begin{bmatrix} A & B^T \\ B & C \end{bmatrix} \begin{bmatrix} \mathbf{J}_i^T \\ \mathbf{J}_o^T \end{bmatrix} \left(\begin{bmatrix} \mathbf{J}_i & \mathbf{J}_o \end{bmatrix} \begin{bmatrix} A & B^T \\ B & C \end{bmatrix} \begin{bmatrix} \mathbf{J}_i^T \\ \mathbf{J}_o^T \end{bmatrix} + \lambda \Sigma_n \right)^{-1} \tag{6.16}$$

where A and C are the covariances of in- and outside plane targets, while B represents the covariance between them, and is typically small. Here the reconstruction matrix for the inside region,

$$\mathbf{R}_i = \mathbf{D}_i A \mathbf{J}_i^T \left(\mathbf{J}_i A \mathbf{J}_i^T + N \right)^{-1}, \tag{6.17}$$

where N represents data noise as well as the influence of out-of-plane contrasts, and $N = \lambda \Sigma_n + \mathbf{J}_o B \mathbf{J}_i^T + \mathbf{J}_i B^T \mathbf{J}_o^T + \mathbf{J}_o C \mathbf{J}_o^T$.

Figure 6.6 shows volumetric images of ventilation in 3D on several subjects. Image resolution between the electrode planes is good, but resolution decreases above and below. This is a natural consequence of the reduced sensitivity in those regions.

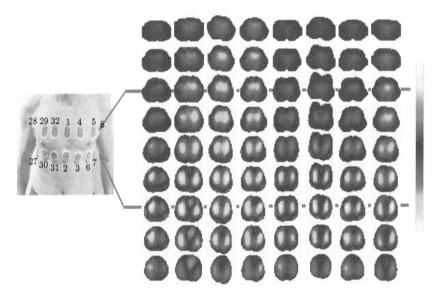

Figure 6.6 Tidal ventilation in healthy subjects in supine position from[147]. EIT data were reconstructed with two planes of 16 electrodes located at the 3rd and 7th levels in the images. (Left) Subject with electrode positions. (Right) Transverse slices of images reconstructed with GREIT[391] using a subject-specific finite element model. (color figure available in eBook)

6.7 ABSOLUTE EIT

In contrast to difference EIT (§6.6), absolute EIT (aEIT) reconstructs an image of the distribution of conductivity at a single instant in time. Unfortunately, aEIT has proven to be much more difficult. It is sensitive to all sorts of parameters in the hardware and the forward model that can be ignored in tdEIT. Electrode shapes and positions must be accurate, and data channels must be calibrated.

The use of linear approximation is only valid for small deviations from the reference conductivity. In medical problems conductivity contrasts can be large, but there is a good case for using the linearized method to calculate a change in admittivity between two states, measured either at different times or with different frequencies. Although this has been called "dynamic imaging" in EIT the term *difference imaging* is now preferred (*dynamic imaging* is a better used to describe statistical time series methods such as[1077]). In industrial ECT modest variations of permittivity are commonplace. In industrial problems and in phantom tanks it is possible to measure a reference data set using a homogeneous tank. This can be used to calibrate the forward model, in particular the contact impedance can be estimated[453]. In an *in vivo* measurement there is no such possibility and it may be that the mismatch between the measured data and the predictions from the forward model is dominated by the errors in electrode position, boundary shape and contact impedance rather than interior conductivity. Until these problems are overcome it is unlikely, in our opinion, to be worth using iterative non-linear methods *in vivo* using individual surface electrodes. Note however that such methods are in routine use in geophysical problems[644,645]. One difference is that geophysical ERT seeks to image objects at a shallower depth with respect to the electrode spacing than is typical in biomedical EIT.

6.7.1   ITERATIVE NONLINEAR SOLUTION

The essence of non-linear solution methods is to repeat the process of calculating the Jacobian and solving a regularized linear approximation. However a common way to explain this is to start with the problem of minimizing f, which for a well chosen G will have a critical point which is the minimum. At this minimum $\nabla f(\mathbf{x}) = \mathbf{0}$ which is a system of N equations in N unknowns which can be solved by multi-variable Newton-Raphson method. The Gauss-Newton approximation to this, which neglects terms involving second derivatives of F, is a familiar Tikhonov formula updating the n th approximation to the conductivity parameters $\mathbf{x}_n$

$$\mathbf{x}_{n+1} = \mathbf{x}_n + \Delta \mathbf{X}_n \tag{6.18}$$

using a step at each iteration

$$\Delta \mathbf{X}_n = (\mathbf{J}_n^* \mathbf{W} \mathbf{J}_n + \alpha^2 \mathbf{L}^* \mathbf{L})^{-1} \left(\mathbf{J}_n^* \mathbf{W} (\mathbf{V}_m - F(\mathbf{x}_n)) + \alpha^2 \mathbf{L}^* \mathbf{L} (\mathbf{x}_{ref} - \mathbf{x}_n) \right) \tag{6.19}$$

where $\mathbf{J}_n$ is the Jacobian evaluated at $\mathbf{x}_n$, and care has to be taken with signs. Figure 6.7 illustrates iterative reconstruction using an Agar tank phantom from[502]. For the first few iterations, the image improves nicely, and the data mismatch decreases dramatically, after which point little further improvement occurs. The figure also illustrates the range of possible images from highly noisy to oversmoothed, given various choices of α.

There are many variations to (6.18) which improve the stability of aEIT reconstructions. The estimate $\mathbf{x}_0$ must be a reasonably good estimate of the homogeneous "background" conductivity. Often this value can be calculated by a single-parameter fit to the data. Next, it is normally better not to take a single step of $\Delta \mathbf{X}_n$, but rather to do a *line search* in the direction of $\Delta \mathbf{X}_n$ to find the step $s\Delta \mathbf{X}_n$ which best minimizes the f at each iteration. Iteration is typically stopped after f is no longer improving significantly, such as shown in figure 6.7.

Notice that in this formula the Tikhonov parameter is held constant throughout the iteration, by contrast the Levenberg-Marquardt[685] method applied to $\nabla f = 0$ would add a diagonal matrix λD in addition to the regularization term $\alpha^2 L^* L$ but would reduce λ to zero as a solution was approached. For an interpretation of λ as a Lagrangian multiplier for an optimization constrained by a *trust region*. Both methods are described in[1087].

The parameterization of the conductivity can be much more specific than voxel values or coefficients of smooth basis functions. One example is to assume that the conductivity is piecewise constant on smooth domains and reconstruct the shapes parameterized by Fourier series[465,552,570,572] or by level sets[243,851]. Level sets can be extended to 3D[908,1199], as well as to regions with several

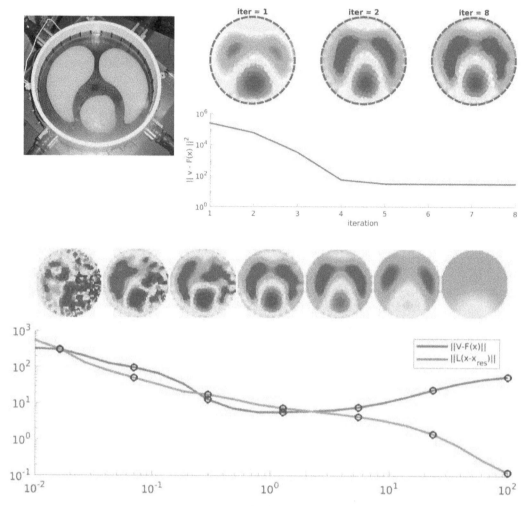

Figure 6.7   Left: experimental phantom from[502]. Right: non-linear iterative reconstructions as a function of iteration number (top) and data mismatch vs iteration number (bottom). Bottom: iterative reconstructions for various values of α (top) and data mismatch and image prior as a function of α. (color figure available in eBook)

different constant values[637]. For an overview of level set methods in inverse problems see[244]. For this and other model based approaches the same family of smooth optimization techniques can be used as for simpler parameterizations, although the Jacobian calculation may be more involved. For inclusions of known conductivities there are a range of direct techniques we will briefly survey in section 6.10.2.

6.8   TOTAL VARIATION REGULARIZATION

Total variation (TV) regularization offers image reconstructions with desirable properties in terms of sensitivity to noise and reduced image bluring. The term TV refers to the use of total variation (or ℓ_1) norms $\|\mathbf{x}\|_1^1 = \Sigma_i |x_i|$ instead of the ℓ_2-norms used in generalized Tikhonov reconstruction, $\|\mathbf{x}\|_2^2 = \Sigma x_i^2$. The reconstructed EIT image $\hat{\mathbf{x}}$ minimizes $f(\hat{\mathbf{x}})$ defined in (6.10) and (6.11)

$$f(\mathbf{x}) = \|\mathbf{V}_m - F(\mathbf{x})\|_p^p + \|\alpha \mathbf{L}(\mathbf{x} - \mathbf{x}_{\text{ref}})\|_q^q \qquad (6.20)$$

using norms p on the data mismatch and q on the prior.

To illustrate the effect of p, imagine our data contains as single measurement with a mismatch, $\varepsilon = \mathbf{V}_m - F(\mathbf{x})$, of $\varepsilon_1 = 10$ and ten measurements with $\varepsilon_2 = 1$. Consider first using $p = 2$ (the ℓ_2 norm): the first error leads to $\|\varepsilon_1\|^2 = 100$ while the ten values lead to to $10 \times \|\varepsilon_2\|^2 = 10$ and thus the minimization will "try" 10 times "harder" to match the first voltage mismatch. Unfortunately, single large mismatch values are often outliers or errors, and it is not desired that they have such a large impact. Instead, using $p = 1$ both groups of errors have equal weight and the single large error does not unduly bias the reconstructed image. Thus $p = 1$ has the desirable property of being less sensitive to outliers and is called a "robust error norm".

Using a ℓ_1 norm for q also has desirable properties: here it promotes sharp edges in images and reduces the tendency of image reconstruction to create blurred objects. First, we set $\mathbf{L}$ to a discrete approximation of the spatial gradient, and it thus measures the variations of a function over its domain. The important difference between ℓ_1 and ℓ_2 is that functions with bounded total variation includes discontinuous functions, which makes TV particularly attractive for the regularization of non–smooth profiles. The following one-dimensional example illustrates the advantage of using the TV against a quadratic functional in non-smooth contexts

Let $F = \{f : [0,1] \rightarrow \mathbb{R}, | f(0) = a, f(1) = b\}$, we have

- $\min\limits_{f \in F} \int_0^1 |f'(x)| dx$ is achieved by any monotonic function, including discontinuous ones.
- $\min\limits_{f \in F} \int_0^1 (f'(x))^2 dx$ is achieved only by the straight line connecting the points $(0,a)$ $(1,b)$.

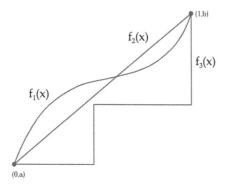

Figure 6.8   Three possible functions: $f_1, f_2, f_3 \in F$. All of them have the same TV, but only f_2 minimises the H^1 semi-norm.

Figure 6.8 shows three possible functions f_1, f_2, f_3 in F. All of them have the same total variation, including f_3 which is discontinuous. Only f_2 however minimises the H^1 semi-norm

$$|f|_{H_1} = \left(\int_0^1 \left(\frac{\partial f}{\partial x} \right)^2 dx \right)^{1/2}. \tag{6.21}$$

The quadratic functional, if used as a penalty, would therefore bias the inversion toward the linear solution and the function f_3 would not be admitted in the solution set as its H^1 semi-norm is infinite. (To see that (6.21) is a semi-norm, consider that $|f|_{H_1} = 0$ for constant f.)

We thus have four possible combinations of norms (p, q) on the data mismatch and image priors. In order to visualize the effect, figure 6.9 shows the four combinations on experimental data in which known electrode errors are present[17]. Images were calculated using the PD-IPM method of[133]. Note the different appearance of TV prior ($q = 1$), and the much sharper image edges. While visually impressive, it is important to note that the appearance of sharp edges can lead to misinterpretations, with the viewer assuming that the edges correspond exactly to anatomical structes.

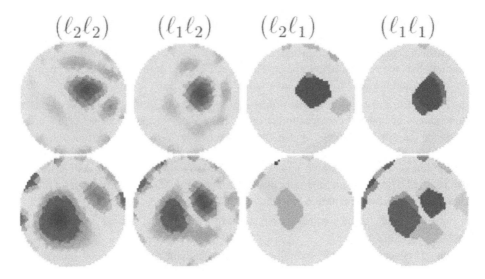

Figure 6.9   Reconstructed images of anesthetized and ventilated dog from[17], which had known electrode contact quality problems in the top left electrode. The data before fluid instillation was used as reference data for the subsequent measurements right after fluid injection (top row) and 60 minutes after the fluid injection (bottom row). The data were known to contain electrode errors. Each column shows images reconstructed with (p,q) norm pairs for the data mismatch and image prior, respectively[681]. (color figure available in eBook)

In the general image reconstruction literature, there has been significant recent developments in optimization methods including those for TV, see[52] and[242]. Within the EIT community several different approaches have been used for application of TV to EIT. The main challenge of TV algorithms is the numerical problems because the TV functional is non-differentiable at zero. This problem can be addressed using a slightly larger norm $p > 1$, but the convergence is slow. The first approach was by Dobson and Santosa[240] formulated TV for difference EIT with stabilized linear contraints. Somersalo *et al*[996] and Kolehmainen *et al*[571] applied MCMC methods to solve the TV regularized inverse problem, which do not suffer from the numerical problems involved with non-differentiability of the TV functional. Borsic *et al* introduced the use of the primal-dual interior point method (PD-IPM) for TV applied to EIT[132,134] and extended to use on data and image norms[133].

Several authors have evaluated TV reconstruction on experimental and clinical data[367,681,1215]. Recently, newer TV approaches have been proposed including Alternating Direction Method of Multipliers (ADMM)[527] and first-iterative shrinkage-thresholding algorithm (FISTA)[511]. The latter, FISTA, has shown particularly good performance, and is now used to accelerate X-ray CT image reconstructions.

6.9   COMMON PITFALLS AND BEST PRACTICE

The ill-posed nature of inverse problems means that any reconstruction algorithm will have limitations on what images it can accurately reconstruct and that the images degrade with noise in the data. When developing a reconstruction algorithm it is usual to test it initially on simulated data. Since the reconstruction algorithms typically incorporate a forward solver. A solver, a natural first test is to use the same forward solver to generate simulated data with no simulated noise and to then find to one's delight that the simulated conductivity can be recovered fairly well, the only difficulties being if it violates the *a priori* assumptions built into the reconstruction and the limitations of floating point arithmetic. Failure of this basic test is used as a diagnostic procedure for the program. On the other hand, claiming victory for one's reconstruction algorithm using this data is what is slightly jokingly called an "inverse crime"[206] (by analogy with the "variational crimes" in FEM perhaps). We list a few guidelines to avoid being accused of an inverse crime and to lay out what we believe to be best practice. For slightly more details see[28,1168].

A first resource for reconstruction algorithms in EIT is the Electrical Impedance and Diffuse Optical Tomography Reconstruction Software (EIDORS) project. This project began as software for 2D reconstruction[1078] and was subsequently extended to 3D[838]. Starting with version 3, EIDORS has developed into a complete software suite offering forward modelling, finite element meshing, and image reconstruction tools[28]. EIDORS also hosts a large number of sample experimental and clinical data sets. Currently on version 3.10,[14] EIDORS has over 120 k lines of code and – as of writing – 1516 citations to all references. The reader is encouraged to take a look at EIDORS and also at the mailing list as a starting point for many questions about image reconstruction.

We give the following advice for readers developing reconstruction approaches using simulated data:

- **Use a different mesh.** If you do not have access to a data collection system and phantom tank, or if your reconstruction code is at an early stage of development, you will want to test with simulated data. To simulate the data use a finer mesh than is used in the forward solution part of the reconstruction algorithm. But not a strict refinement. The shape of any conductivity anomalies in the simulated data should not exactly conform with the reconstruction mesh, unless you can assume the shape is known *a priori*.
- **Simulating noise.** If you are simulating data you must also simulate the errors in experimental measurement. At the least there is quantisation error in the analogue to digital converter. Other sources of error include stray capacitance, gain errors, inaccurate electrode position, inaccurately known boundary shape, and contact impedance errors. To simulate errors sensibly it is necessary to understand the basics of the data collection system, especially when the gain on each measurement channel before the ADC is variable. When the distribution of the voltage measurement errors is decided this is usually simulated with a pseudo random number generator.
- **Pseudo random numbers.** A random number generator models a draw from a population with a given probability density function. To test the robustness of your reconstruction algorithm with respect to the magnitude of the errors it is necessary to make repeated draws, or calls to the random number generator, and to study the distribution of reconstruction errors. As our inverse problem is non-linear even a Gaussian distribution of error will not produce a (multivariate) Gaussian distribution of reconstruction errors. Even if the errors are small and the linear approximation good, at least the mean and variance should be considered.
- **Not tweaking.** Reconstruction programs have a number of adjustable parameters such as Tikhonov factors and stopping criteria for iteration, as well as levels of smoothing, basis constraints and small variations of an algorithms. While there are rational ways of choosing reconstruction parameters based on the data (such as generalized cross validation and L-curve), and on an estimate of the data error (Morotzov's stopping criterion). In practice on often finds acceptable values empirically which work for a collection of conductivities one expects to encounter. There will always be other cases for which those parameter choices do not work well. What one should avoid is *tweaking* the reconstruction parameters for each set of data until one obtains an image which one knows is close to the real one. By contrast an honest policy is to show examples of where a certain algorithm and parameters performs poorly as well as the best examples.

We give the following advice for readers developing reconstruction approaches using experimental and clinical data:

- **Understand the electronics hardware and the measurement setup.** Where were the electrodes connected? What stimulation and measurement pattens were used? What experimental protocols were used. If at all possible watch the data collection or speak to those who did it. It is easy for details of the experiment to dramatically impact image analysis: for example if the subject changed posture between measurements, then the electrode positions will be slightly different.
- **Look at the raw data** compared to the simulations before running the reconstruction algorithm, or whenever the results are uncertain. A common problem is that the number of electrodes or the stimulation and measurement patterns in the data do not match the experimental conditions. "Eye-balling" the data in this way will help to set expectations. If the match is close, then a high-resolution reconstruction is possible. If not, then work to identify and reject noisy measurements if required [63,442].
- **Real data errors are not pseudo-random** zero-mean Gaussian noise. There are many sources of systematic noise. EIT hardware has non-zero common-mode rejection, crosstalk between cables and drift over time (chapter 4). Electrodes have contact impedance that varies between electrodes and over time as the patient sweats or the electrode dries. Electrodes move with breathing [27] and with posture [214] which introduces complicated patterns of noise into the images. Note that deforming the finite element geometry does not match electrode movement errors, but this can be addressed by incorporating meshing noise into the reconstruction [29]. Measurements in hospitals are electrically noisier than the lab [303]. Frustratingly patient data has proven much more challenging than data from healthy volunteers; this could be because many patients have more (non-conductive) body fat and have organs further from the electrodes.
- **High resolution is not useful by itself** if it means artifacts will be misinterpreted. What we mean by this is that the clinicians who interpret EIT images will be looking for features in the data. Noise leading to a small contrast in a particular anatomical location may correspond to a expected feature, and the image will confirm this. Such misinterpretations is more likely with edge preserving approaches like TV. The overall advice is that it is better to show nothing (e.g. a blurry blob) than invalid results.
- **Focus on the end use of the images.** The only reason to make medical EIT recordings is if they will somehow influence clinical decisions. A knowledge of the use of the images helps clarify the requirements in terms of robustness, resolution and other features. Refer to chapter 8 on image interpretation for for an understanding of the ways EIT images are processed.

6.10   FURTHER DEVELOPMENTS IN RECONSTRUCTION ALGORITHMS

In this chapter there is not space to describe in any detail many of the exciting current development in reconstruction algorithms. Before highlighting some of these developments it is worth emphasizing that for ill-posed problem *a priori* information is essential for a stable reconstruction algorithms, and it is better that this information is incorporated in the algorithm in a systematic and transparent way. Another general principle of inverse problems is to think carefully what information is required by the end user. Rather than attempting to produce an accurate image what is often required in medical (and indeed most other) applications is an estimate of a much smaller number of parameters which can be used for diagnosis. For example we may know that a patient has two lungs as well as other anatomical features but we might want to estimate their water content to diagnose pulmonary edema. A sensible strategy would be devise an anatomical model of the thorax and fit a few parameters of shape and conductivity rather than pixel conductivity values. The disadvantage of this approach is that each application of EIT gives rise to its own specialised reconstruction method, which must be carefully designed for the purpose. In the authors' opinion the future development of EIT systems, including electrode arrays and data acquisition systems as well as reconstruction software, should focus increasingly on specific applications, although of course such systems will share many common components.

6.10.1   BEYOND TIKHONOV REGULARIZATION

We have also discussed the use of more general regularization functionals including total variation. For smooth G traditional smooth optimization techniques can be used, whereas for Total Variation the PD-IPM is promising. Other functionals can be used to penalize deviation from the *a priori* information, one such choice is the addition of the Mumford-Shah functional which penalizes the Hausdorf measure of the set of discontinuities[895]. In general there is a trade-off between incorporating accurate *a priori* information and speed of reconstruction. Where the regularization matrix **L** is discretized partial differential operator, the solution of the linearized problem is a compact perturbation of a partial differential equation. This suggests that multigrid methods may be used in the solution of the inverse problem as well. For a single linearized step this has been done for the EIT problem by McCormick and Wade[698], and for the non-linear problem by Borcea[127]. In the same vein adaptive meshing can be used for the inverse problem as well as the forward problem[657,722,723]. In both cases there is the interesting possibility to explore the interaction between the meshes used for forward and inverse solution.

At the extreme end of this spectrum we would like to describe the prior probability distribution and for a known distribution of measurement noise and calculate the entire posterior distribution. Rather than giving one image, such as the MAP estimate gives a complete description of the probability of any image. If the probability is bimodal for example, one could present the two local maximum probability images. If one needed a diagnosis, say of a tumour, the posterior probability distribution could be used to calculate the probability that a tumour like feature was there. The computational complexity of calculating the posterior distribution for all but the simplest distributions is enormous, however the posterior distribution can be explored using the Markov Chain Monte Carlo Method (MCMC) which has been applied to two dimensional EIT[529]. This was applied to simulated EIT data[286], and more recently to tank data, for example[1137]. For this to be a viable technique for the 3D problem highly efficient forward solution will be required.

6.10.2   DIRECT NON-LINEAR METHODS

Iterative methods which use optimization methods to solve a regularized problem are necessarily time consuming. The forward problem must be solved repeatedly and the calculation of an updated conductivity is also expensive. The first direct method to be proposed was the Layer Stripping algorithm[994] however this is yet to be shown to work well on noisy data.

An exciting development is the implementation of a Scattering Transform ($\bar{\partial}$ or d-bar) algorithm (see chapter 7 for a detailed description). Siltanen *et al*[973] showed that this can be implemented stably and applied to *in vitro* data[747]. The main limitation of this technique is that is inherently two dimensional and no one has found a way to extend it to three dimensions, also in contrast to the more explicit forms of regularization it is not clear what *a priori* information is incorporated in this method as the smoothing is applied by filtering the data. Hamilton *et al*[420] compared a d-bar to some regularized approaches and showed that d-bar calculated an effectively spatially invariant prior. A strength of the method is its ability to accurately predict absolute conductivity levels. In some cases where long electrodes can be used and the conductivity varies slowly in the direction in which the electrodes are oriented a two dimensional reconstruction may be a useful approximation. This is perhaps more so in industrial problems such as monitoring flow in pipes with ECT or ERT. In some situations a direct solution for a two dimensional approximation could be used as a starting point for an iterative three dimensional algorithm.

Two further direct methods show considerable promise for specific applications. The monotonicity method of Tamburrino and Rubinacci[1019] relies on the monotonicity of the map $\rho \mapsto R_\rho$ where ρ is the real resistivity and R_ρ the transfer resistance matrix. This method, which is extremely fast, relies on the resistivity of the body to be known to be one of two values. It works equally well in two and three dimensions and is robust in the presence of noise. The time complexity scales linearly with the number of voxels (which can be any shape) and scales cubically in the number of electrodes. It works for purely real or imaginary admittivity, (ERT or ECT), and for Magnetic Induction Tomography for real conductivity. It is not known if it can be applied to the complex case and it requires the voltage on current carrying electrodes.

Linear sampling methods[164,427,929] have a similar time complexity and advantages as the monotonicity method. While still applied to piecewise constant conductivities, linear sampling methods can handle any number of discrete conductivity values provided the anomalies separated from each other by the background. The method does not give an indication of the conductivity level but rather locates the jump discontinuities in conductivity. Considerable progress has been made on practical and theoretical aspects of the monotonicity method for conductivity[438,439,1214]. The extension to complex admittivity is open but compare with the monotonicity approach to MIT[1020]. Both monotonicity and linear sampling methods are likely to find application in situations where a small anomaly is to be detected and located, for example breast tumours.

Finally a challenge remains to recover anisotropic conductivity which arises in applications from fibrous or stratified media (such as muscle), flow of non-spherical particles (such as red blood cells), or from compression (for example in soil). Tissue anisotropy is frequency dependent (see §3.4) and decreases with frequency. The inverse conductivity problem at low frequency is known to suffer from insufficiency of data, but with sufficient *a priori* knowledge (for example[1167]) the uniqueness of solution can be restored. One has to take care that the imposition of a finite element mesh does not predetermine which of the family of consistent solutions is found[816]. Numerical reconstructions of anisotropic conductivity in a geophysical context include[804], although there the problem of non-uniqueness of solution (diffeomorphism invariance) has been ignored. Another approach is to assume piece-wise constant conductivity with the discontinuities known, for example from an MRI image, and seek to recover the constant anisotropic conductivity in each region[352,362].

6.11   MACHINE LEARNING AND INVERSE PROBLEMS

We will first consider the idea of *training data*. Suppose we have examples of pairs typical parameter vectors for conductivity that we expect to encounter and their simulated or experimental inverse voltage measurements $(\mathbf{x}_i, \mathbf{V}_i)$, for $i = 1...N$. The idea is to then find an approximate nonlinear inverse that works well for those examples, and hopefully for nearby conductivity vectors as well. We choose a smooth parameterized family of functions $G(\mathbf{V}, \mathbf{p})$ where $\mathbf{p}$ is a vector of parameters and then use an optimization method to find some $\mathbf{p}$ that minimizes the fitting error

$$\sum_{i=1...N} ||G(\mathbf{V}_i, \mathbf{p}) - \mathbf{x}_i||^2.$$

This minimization is called *training* and conceptually, from a Machine Learning ('ML') perspective G is a model of interconnected neurons in the brain and the process of finding the parameters (often called *weights*) is analogous to a learning in an animal. Obviously the inverse problem of EIT is unstable, but the key factor here is that the training data are all contained in some ball of finite radius. If we assume we are only interested in a constrained set of parametrized conductivities there are *conditional stability results* for the EIT inverse problem that say, roughly that once the conductivities are sufficiently constrained the inverse problem has a certain stability[37]. Two key problems are the choice of a sufficiently rich training set that is representative of the sort of conductivities one might expect to encounter in this application and the choice of G so it has sufficient flexibility to accurately represent some regularized inverse of the forward problem, while doing a reasonable job of interpolating between points in the training data. A family of functions inspired originally by a model of interconnected neurons is known as an Artificial Neural Net. These are constructed by compositions of multiple linear functions of several variables and sigmoid functions performing approximate thresholding. These have been found to have good approximation properties and an early example of their use for EIT reconstruction, along with more details of how ANNs are constructed and trained are given in[25]. As highly efficient non-linear optimization routines are available the training phase, finding a good $\mathbf{p}$ can be performed offline in a reasonable time, and the reconstruction is then just an evaluation of $G(\mathbf{V_m}, \mathbf{p})$ for that $\mathbf{p}$ and can be very quick and done in real time, with a time complexity higher than using a precomputed inverse to a regularized linear approximation but could be lower than an iterative optimization based solver, although[907] found an ANN to be slower in their implementation.

The Machine Learning approach has several disadvantages, in common with any *empirical* rather than *mechanistic* modelling. A mechanistic model, such as deriving the equations for the EIT forward problems, has the advantage that it generalizes to a wide class of problems. For example we can change the size, shape, number and position of electrodes, contact impedances in our finite element model and we still know how to solve the inverse problem. With an empirical model, such as the machine learning approach, we have to acquire a new training set and fit the parameters for any slightly different problem. On the other hand an empirical model has the possibility to fit physics in the problem we did not build in to a mechanistic model. For example stray capacitance between electrode leads, external capacitive coupling between electrodes, different gains in measurement channels, non-ideal current sources[324]. A training set derived from a realistic experimental phantom would include all the physical effects and the machine learning approach would mean that we found the best fit we could given our choice of G. Even with a mechanistic model we would typically do some calibration using undetermined parameters in the model, for example the contact impedances on electrodes. This suggest that a hybrid between empirical and mechanistic model is usually a good approach. While a purely empirical one may fit physics that we did not anticipate it is quite difficult to "interrogate" and ANN: as it were to ask it what it has learned. One has to investigate further to find what the physics is that we neglected in our mechanistic model.

New developments in ML applied to inverse problems focus on hybrid approaches where the mechanistic, and usually well understood, part of the problem is used while the parts that are not well understood are fitted or "learnt". For example there is considerable success using ML to learn a regularization term or prior which is then applied to an inverse problem. For a recent survey see[61].

6.12   PRACTICAL APPLICATIONS

For time difference EIT for thoracic application, linearized algorithms work quite well. EIT systems are manufactured and used clinically. EIT algorithms in such systems are often based on the GREIT approach (§6.6.5) and quite robust to the noise, interference and variability of the clinical reality and are able to calculate functional parameters with adequate accuracy[389]. Especially in chest EIT, there has been little innovation in EIT reconstruction itself with work focusing on calculating robust parameters[293]. In most applications, electrodes are placed in in single plane around the thorax, with little change from the first EIT systems of the 1980s. There is surprisingly little push toward 3D EIT images, perhaps because there has been little work to make recommendations of the practical details (exact electrode location and spacing of the planes, stimulation and measurement patterns)[391]. For chest monitoring, there remains work to improve the robustness and accuracy of the EIT images. For example, how important is it to get the exact body shape and electrode position? Is it possible to use an atlas shape and update it to better fit the data[29]? How can we compensate the reconstructed for defects in the electronics hardware and for the specific inferference in the clinical scenario. For a given EIT application, in which limitations are imposed, such as electrodes excluded from a surgical field, where should electrodes be best placed? Is it possibly to usefully quantify the uncertainty in EIT images and functional measures?

For more advanced image reconstruction, there are many unsolved questions. The major non-linear algorithms have all been tested on tank data. Yorkey[1194] compared Tikhonov regularized Gauss-Newton with *ad hoc* algorithms on two dimensional tanks, Goble[355,356] and Metherall[712,713] applied one step regularized Gauss-Newton to 3D tanks. P Vauhkonen[1080,1081] applied a fully iterative regularized Gauss-Newton method to 3D tank data using the complete electrode model. More recently the linear sampling method[929] and the scattering transform method[747] have been applied to tank data. However there is a paucity of application of non-linear reconstruction algorithms to *in vivo* human data.

The EIT problem is inherently non-linear (§5.3.2). However, to use a non-linear algorithm the forward model used must be able to fit the data accurately when the correct conductivity is found. This means that the shape, electrode position and electrode model must all be correct. Until an accurate model is used, including a method of constructing accurate body shaped meshes and locating electrodes is perfected, it will not be possible to do justice to the EIT hardware by giving the reconstruction algorithms the best chance of succeeding.

Medical EIT reconstruction algorithms are approximately fourty years old[949] and are now being used to treat patients. At the same time, many advanced concepts are still active research areas and show great promise to help improve the state of the art.

7 D-bar Methods for EIT

David Isaacson
Department of Mathematical Sciences, Rensselaer Polytechnic
Institute, Troy, NY, USA

Jennifer L. Mueller
Department of Mathematics and School of Biomedical Engineering,
Colorado State University, Fort Collins, CO, USA

Samuli Siltanen
Department of Mathematics and Statistics, University of Helsinki,
Finland

CONTENTS

7.1   INTRODUCTION

This chapter focuses on the D-bar method for EIT, which is really one method in a family of direct reconstruction methods based on *complex geometrical optics* (CGO), which generally lead to a D-bar equation to be solved as part of the method. By "direct" reconstruction methods, we refer to algorithms that are *not* based on improving a reconstructed conductivity iteratively step by step, as is done for example in variational regularization, but rather calculate a regularized reconstruction directly from the data. One of the advantages of direct methods is that there is often no need for a forward problem solver, or it is only invoked once to compute a simulated reference state. Furthermore, many of the methods are trivially parallelizable.

This chapter begins with Calderón's method in §7.2, and continues with an exposition of the breakthrough approach of Sylvester and Uhlmann, based on *complex geometric optics* (CGO) solutions, in §7.3. Our main focus is on the 2-D D-bar method based on Nachman's 1996 uniqueness proof[764], discussed in §7.4. We have chosen to focus on this D-bar method since it is arguably the most developed of the direct EIT reconstruction algorithms. We present the mathematical equations

DOI: 10.1201/9780429399886-7

to be solved, a method for their numerical solution, and several examples of reconstructions. New directions of the D-bar methodology is found in §7.5.

To establish notation, we state here the inverse conductivity problem. Given a domain Ω in R^n, $n \geq 2$ with Lipschitz boundary $\partial\Omega$, the unknown conductivity appears as a (possibly complex) coefficient $\gamma(x) = \sigma(x) + i\omega\varepsilon(x)$ in the generalized Laplace equation

$$\nabla \cdot (\gamma(x)\nabla u(x)) = 0, \quad x \in \Omega, \qquad u|_{\partial\Omega} = f, \tag{7.1}$$

where u is the electric potential, $\gamma = \sigma + i\omega\varepsilon$ where σ is the conductivity of the medium, ε is the permittivity, and ω is the temporal angular frequency of the applied electromagnetic wave. The data is the Dirichlet-to-Neumann, or voltage-to-current density map defined by

$$\Lambda_\gamma(u|_{\partial\Omega}) = \gamma\frac{\partial u}{\partial v}\bigg|_{\partial\Omega}. \tag{7.2}$$

The linear operator Λ_γ is a mathematical model for the set of all infinite-precision electric boundary measurements.

The Fourier transform of a function g is defined by

$$\hat{g}(z) = \int_{\mathbb{R}^n} g(x)e^{-2\pi ix \cdot z}\, dx$$

where $x \cdot z$ is the usual dot product in $\mathbb{R}^n$, and the inverse Fourier transform is

$$g(x) = \int_{\mathbb{R}^n} \hat{g}(z)e^{2\pi ix \cdot z}\, dx.$$

7.2 CALDERON'S METHOD

In [172] Calderón proved uniqueness for the linearized inverse problem in EIT in dimension $n \geq 2$. He linearized the forward map $\gamma \mapsto \Lambda_\gamma$ about a constant conductivity and showed that it is one-to-one, and therefore invertible. He also proposed a direct reconstruction method for approximating conductivities that are a small perturbation from a constant. The method relies on the use of certain exponentially growing harmonic functions, which as a family from a dense subset of L^2. This was the first use of complex geometrical optics solutions for the solution of the inverse conductivity problem, and his work inspired a new direction of research on the global uniqueness of the inverse conductivity problem under various conditions on σ. The problem also hereafter became known as the Calderón Problem.

In this section, let us adopt the same notation used by Calderón and denote the solution to (7.1) by w with $w|_{\partial\Omega} = \Phi$, and $w = u + v$ where $\Delta u = 0$ and $u|_{\partial\Omega} = \Phi$. Note that since $w|_{\partial\Omega} = \Phi$, we know $v|_{\partial\Omega} = 0$. Calderón defines the quadratic form of the Dirichlet-to-Neumann map by

$$Q_\gamma(\Phi) = \int_\Omega \gamma|\nabla w|^2\, dx, \quad w|_{\partial\Omega} = \Phi. \tag{7.3}$$

Then since w satisfies (7.1),

$$\begin{aligned} Q_\gamma(\Phi) &= \int_\Omega \gamma|\nabla w|^2\, dx = \int_{\partial\Omega} \gamma w\nabla w \cdot v\, ds - \int_\Omega (\nabla \cdot \gamma\nabla w)w\, dx \\ &= \int_{\partial\Omega} \gamma\frac{\partial w}{\partial v}w\, ds. \end{aligned} \tag{7.4}$$

Physically, Q_γ represents the power required to maintain the voltage Φ on the boundary. Calderón posed the question: Is γ uniquely determined by Q_γ, and if so, how can one calculate γ in terms of Q_γ?

For two solutions w_1 and w_2 of (7.1) with $w_1|_{\partial\Omega} = \Phi_1$ and $w_2|_{\partial\Omega} = \Phi_2$, and $w_1 = u_1 + v_1$, $w_2 = u_2 + v_2$, Calderón introduces the bilinear form

$$B(\Phi_1, \Phi_2) = \frac{1}{2}[Q_\sigma(w_1 + w_2) - Q_\sigma(w_1) - Q_\sigma(w_2)]. \tag{7.5}$$

Then the data is related to this bilinear form as follows:

$$
\begin{aligned}
B(\Phi_1, \Phi_2) &= \frac{1}{2}\left(Q_\gamma(\Phi_1 + \Phi_2) - Q_\gamma(\Phi_1) - Q_\gamma(\Phi_2)\right) \\
&= \frac{1}{2}\int_\Omega \gamma|\nabla(w_1 + w_2)|^2 - \gamma|\nabla w_1|^2 - \gamma|\nabla w_2|^2 dx \\
&= \frac{1}{2}\int_\Omega 2\gamma(\nabla w_1 \cdot \nabla w_2) dx \\
&= \int_{\partial\Omega} w_1 \gamma \frac{\partial w_2}{\partial \nu} ds = \int_{\partial\Omega} w_1 \Lambda_\gamma w_2 ds.
\end{aligned}
\tag{7.6}
$$

The linearization assumption is that γ is a constant plus a perturbation $\delta(x)$, and for convenience he sets the constant to 1 so that $\gamma(x) = 1 + \delta(x)$. To reconstruct γ from the data, Calderón shows that the bilinear form can be expressed in terms of the Fourier transform of γ plus a remainder term. To do this, using the definition of Q, write

$$
\begin{aligned}
B(\Phi_1, \Phi_2) &= \int_\Omega (1 + \delta(x))(\nabla u_1 \cdot \nabla u_2) + \delta(x)(\nabla u_1 \cdot \nabla v_2 + \nabla v_1 \cdot \nabla u_2) \\
&+ (1 + \delta(x))\nabla v_1 \cdot \nabla v_2 \, dx.
\end{aligned}
\tag{7.7}
$$

Now, choosing u_1 and u_2 to be the exponentially growing harmonic functions

$$u_1 = e^{\pi i(z \cdot x) + \pi(a \cdot x)} \tag{7.8}$$
$$u_2 = e^{\pi i(z \cdot x) - \pi(a \cdot x)}, \tag{7.9}$$

where $a, z \in \mathbb{R}^2$ with $z \cdot a = 0$ and $|z| = |a|$, we see that $\nabla u_1 \cdot \nabla u_2 = -2\pi^2|z|^2 e^{2\pi(z \cdot x)}$ and

$$\frac{B(\Phi_1, \Phi_2)}{-2\pi^2|\xi|^2} = \int_\Omega (1 + \delta)e^{2\pi(z \cdot x)} dx - R(z), \tag{7.10}$$

where $R(z)$ is a remainder term given by

$$R(z) = \frac{1}{2\pi^2|z|^2}\int_\Omega \delta(x)(\nabla u_1 \cdot \nabla v_2 + \nabla v_1 \cdot \nabla u_2) + (1 + \delta(x))\nabla v_1 \cdot \nabla v_2 \, dx.$$

If γ is extended to be zero outside Ω, then

$$\hat{\gamma}(z) = \int_{\mathbb{R}^n} \gamma(z)e^{2\pi i(z \cdot x)} dx$$

can be interpreted as the Fourier transform of γ. Defining

$$\hat{F}(z) = -\frac{1}{2\pi^2|z|^2}B(u_1, u_2)$$

equation (7.10) is of the form

$$\hat{F}(z) = \hat{\sigma}(z) + R(z). \tag{7.11}$$

The linearized reconstruction method then is to neglect the remainder and compute the inverse Fourier transform of $\hat{F}(z)$ to obtain an approximation to $\gamma(x)$. Since $\gamma(x)$ may be a discontinuous function, a mollifier may be applied to avoid a Gibbs' phenomenon, as Calderón suggests. The effect of mollifying is studied in[110]. Denoting the mollifier by η, in[108] η was chosen to be $\eta_\beta(x) = \beta^n e^{-\pi|\beta x|^2}$ where n is the dimension. Equation (7.11) becomes

$$\hat{\gamma}(z)\hat{\eta}\left(\frac{z}{\sigma}\right) = \hat{F}(z)\hat{\eta}\left(\frac{z}{\sigma}\right) + R(z)\hat{\eta}\left(\frac{z}{\sigma}\right). \tag{7.12}$$

Taking the inverse Fourier transform of (7.12) results in

$$(\gamma * \eta_\sigma)(x) \approx (F * \eta_\sigma)(x) \tag{7.13}$$

where $*$ denotes convolution.

To calculate $B(u_1, u_2)$, one can expand u_1 and u_2 in terms of basis functions suitable for the domain and applied current patterns. For reconstructions on a circular 2-D domain of radius R with trigonometric current patterns, the basis functions $e^{in\theta}$ are a good choice.

In this case, let $x|_{\partial\Omega} = Re^{i\theta}$, $z = |z|e^{i\phi}$ and $a = |z|e^{i(\phi \pm \pi/2)}$ so that $a \cdot z = 0$ and $|z| = |a|$. Then one can show $\pi(a \cdot x + i(z \cdot x))|_{\partial\Omega} = |z|\pi Rie^{\mp i(\theta - \phi)}$ and we can expand $u_1|_{\partial\Omega}$ by $u_1|_{\partial\Omega} = \sum_{j=0}^\infty a_j(z)e^{\mp ij\theta}$, where $a_j(z) = \frac{(|z|\pi Rie^{\pm i\phi})^j}{j!}$ and $u_2|_{\partial\Omega} = \sum_{j=0}^\infty b_j(z)e^{\pm ij\theta}$, where $b_j(z) = \frac{(|z|\pi Rie^{\mp i\phi})^j}{j!}$. (See[110] for further details.)

Using these expansions in equation (7.6) results in

$$Q_\gamma(\phi_1, \phi_2) = R\sum_{j=0}^\infty \sum_{k=0}^\infty a_j b_k \int_0^{2\pi} e^{\mp ij\theta} \Lambda_\gamma e^{\pm ik\theta} d\theta,$$

which gives an equation for $\hat{F}$:

$$\hat{F}(z) = -\frac{1}{2\pi^2|z|^2} B(\phi_1, \phi_2)$$

$$= \frac{-R}{2\pi^2|z|^2} \sum_{j=0}^\infty \sum_{k=0}^\infty a_j(z)b_k(z) \int_0^{2\pi} e^{\mp ij\theta} \Lambda_\gamma e^{\pm ik\theta} d\theta$$

Calderón's method has been implemented on a circular domain and shown to be effective on experimental data in[110], on elliptical domains in[750] where the effect of domain-shape error was analyzed, and on subject-specific domains in[751] with a real-time implementation for human subject data. A method of introducing a spatial prior in Calderón's method has been proposed in[968]. See Figure 7.1 for an example of absolute images computed with Calderón's method with and without spatial priors applied to experimental data consisting of three cucumber targets in a saline-filled tank measured with adjacent current patterns of amplitude 3.3 mA using the ACE 1 EIT system[707].

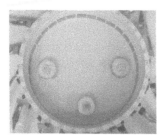 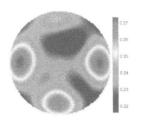 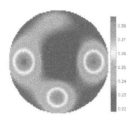

Figure 7.1   Left: Saline-filled tank with three cucumber targets. Center: Absolute image computed using Calderòn's method no spatial prior. Right: Absolute image computed using Calderòn's method with a spatial prior. The positions of the cucumber targets were estimated from the photo, and approximate values of the conductivity were chosen for the prior. (color figure available in eBook)

7.3   THE RISE OF THE CGO SOLUTIONS

The history of multidimensional CGO solutions goes back to the 1960s, when Faddeev introduced them in the context of inverse quantum scattering. CGO solutions were also used in the study of two-dimensional nonlinear evolution equations in the early 1980s by Beals and Coifman. This is the origin of the D-bar methodology. CGO solutions were first used in the analysis of inverse boundary value problems by Sylvester and Uhlmann [1016,1017] where they were used to prove uniqueness for the Calderón problem in dimensions three and higher.

The D-bar approach for reconstructing PDE coefficients from boundary measurements was first formally suggested by R.G. Novikov for $n \geq 2$ in [788], and rigorously analysed by Nachman in 1988 ($n \geq 3$) and 1996 ($n = 2$).

The first numerical inversion method based on CGO solutions was published in 2000 by Siltanen, Mueller and Isaacson [973] for two-dimensional EIT. The method was later applied successfully applied to real data in 2006 [501] and shown in 2009 to provide a formal regularization strategy for the highly nonlinear inverse problem of EIT [562].

7.4   A 2-D D-BAR METHOD

In this section we present the 2-D D-bar method based on the 1996 uniqueness proof by Nachman [764] for real-valued conductivities σ. The underlying theory for the numerical method is well developed, and the reader is referred to [745] for a thorough discussion. It should be noted that the regularization of this method has a solid foundation [562]. The regularization is accomplished via low-pass filtering in the nonlinear Fourier Transforms with a cutoff frequency dependent upon the noise level.

The 2-D D-bar method based on the 1996 uniqueness proof by Nachman [764] for real-valued conductivities σ is arguably the most developed D-bar method as it has a proven regularization strategy [562], has a real-time implementation [241], has been demonstrated to be robust in the presence of electrode placement errors, incorrect boundary shape assumptions, and noise [422,755], and has been used with clinical data to compute EIT-derived measures of pulmonary function [752] and to identify regions of air trapping in a small number of cystic fibrosis patients [743]. For this reason, we will focus on the 2-D D-bar method based on [562,764,973] and the "texp" approximation that simplifies and speeds up the method [973].

7.4.1 EQUATIONS OF THE D-BAR METHOD

Throughout this section, we will assume that σ is constant in a neighbourhood of the boundary of Ω. This assumption is sufficiently realistic for human data since the skin provides a nearly constant resistive layer near the boundary of the domain. A deeper investigation of this assumption and an implementation of the extension of the conductivity and the DN map to a larger domain proposed in [764] to handle non-constant conductivities on the boundary can be found in [974]. For further simplicity, assume the conductivity is scaled so that it has the value 1 in a neighbourhood of $\partial\Omega$.

For a twice differentiable conductivity, there is a standard transformation of the generalized Laplace equation to the Schrödinger equation. Let $\tilde{u}(x,y) = \sqrt{\sigma(x,y)}u(x,y)$ and $q(x,y) = \Delta\sqrt{\sigma(x,y)}/\sqrt{\sigma(x,y)}$. Then

$$-\Delta\tilde{u} + q(x,y)\tilde{u} = 0, \quad (x,y) \in \Omega. \tag{7.14}$$

Since σ is constant in a neighbourhood of $\partial\Omega$, extending q to be 0 outside Ω allows us to pose equation (7.14) in the whole plane $\mathbb{R}^2$. Next, introduce a complex parameter $k = k_1 + ik_2$, identify (x,y) with the point $z = x + iy$ in the complex plane, and seek a special solution $\psi(z,k)$ to the Schrödinger equation in the complex plane with the asymptotic behaviour $\psi(z,k) \sim e^{ikz}$ for large $|z|$ or large $|k|$. The function $\psi(z,k)$ is a CGO solution and its existence is proved in [764], where it is shown to satisfy

$$-\Delta\psi(z,k) + q(z)\psi(z,k) = 0, \quad z \in \mathbb{R}^2 \tag{7.15}$$
$$e^{-ikz}\psi(z,k) - 1 \in W^{1,p}(\mathbb{R}^2), \quad p > 2. \tag{7.16}$$

Note that (7.16) provides the mathematical formulation of the asymptotic condition on ψ, where $W^{1,p}(\mathbb{R}^2)$ is the Sobolev space of functions with one weak derivative in L^p (see, for example,[11]) Also note that multiplication between z and k and the conjugates is in the complex plane.

An important closely related CGO solution to ψ is

$$\mu(z,k) \equiv e^{-ikz}\psi(z,k).$$

These solutions are the keys to the reconstruction since the conductivity can be obtained directly from μ or ψ through the formula [764]

$$\sigma(z) = \mu^2(z,0), \quad z \in \Omega. \tag{7.17}$$

Based on the critical relationship shown in equation (7.17), the goal of the D-bar method is to compute the CGO solution $\mu(z,k)$ at any point of interest z in the domain Ω, and evaluate at $k = 0$ to obtain the conductivity at that point, $\sigma(x,y)$. We will see that the equations allow for independent computation of the conductivity at each point in the region of interest, which can be a subset of the entire domain – even just a single point!

Clearly, a connection between the measured data and the CGO solutions is required for success of this method. This link comes from an intermediate function known as the scattering transform $\mathbf{t}(k)$, defined by

$$\mathbf{t}(k) = \int_\Omega e^{i k \bar{z}} q(z)\psi(z,k)dz. \tag{7.18}$$

The scattering transform can be regarded as a nonlinear Fourier transform of q due to the asymptotic behaviour of ψ. Recalling that the data takes the form of the DN map Λ_σ in this method, we will now make use of data corresponding to a homogeneous conductivity of 1, which is then denoted by Λ_1. The scattering transform is related to the DN data through the equation

$$\mathbf{t}(k) = \int_{\partial\Omega} e^{i k \bar{z}}(\Lambda_\sigma - \Lambda_1)\psi(z,k)ds. \tag{7.19}$$

Note that (7.19) requires knowledge of ψ on the boundary of Ω, which can be computed by solving the boundary integral equation[764]

$$\psi(z,k)|_{\partial\Omega} = e^{ikz}|_{\partial\Omega} - \int_{\mathbb{R}^2} G_k(z-\zeta)(\Lambda_\gamma - \Lambda_1)\psi(\cdot,k)|_{\partial\Omega}, \qquad (7.20)$$

where G_k is a special Green's function for the Laplacian known as the *Faddeev Green's function*[266] given by

$$G_k(z) = e^{ikz}\int_{\mathbb{R}^2}\frac{e^{iz\cdot\xi}}{\xi(\bar\xi + 2k)}d\xi, \qquad -\Delta G_k(z) = \delta(z).$$

It is shown in[662] that equation (7.20) is uniquely solvable even for discontinuous conductivity distributions, validating the existence of the scattering transform even if the twice-differentiable conductivity assumption is violated.

The final step of the method is to solve the D-bar equation at each point z in the region of interest. The D-bar equation is derived by differentiating the Lippman-Schwinger equation satisfied by $\mu(z,k)$ with respect to the conjugate of the complex parameter k. Since $\frac{\partial}{\partial\bar k}$ is often denoted $\bar\partial_k$, and is called a $\bar\partial$ (D-bar) operator, this is the origin of the method's name. The D-bar equation satisfied by μ is

$$\frac{\partial\mu}{\partial\bar k} = \frac{\mathbf{t}(k)}{4\pi\bar k}e_{-z}(k)\overline{\mu(z,k)}, \qquad (7.21)$$

where $e_z(k) \equiv e^{i(kz+\bar k\bar z)}$. The generalized Cauchy integral formula facilitates writing equation (7.21) in integral form as

$$\mu(z,s) = 1 + \frac{1}{(2\pi)^2}\int_{\mathbb{R}^2}\frac{\mathbf{t}(k)}{(s-k)\bar k}e_{-z}(k)\overline{\mu(z,k)}dk_1 dk_2, \qquad (7.22)$$

which is practical for the numerical solution of the D-bar equation.

For fast implementation, we recommend a linearized approximation to the scattering transform, denoted by $\mathbf{t}^{\text{exp}}$, which is defined by replacing $\psi|_{\partial\Omega}$ by its asymptotic behaviour in formula (7.19)

$$\mathbf{t}^{\text{exp}}(k) \equiv \int_{\partial\Omega} e^{i\bar k\bar z}(\Lambda_\sigma - \Lambda_1)e^{ikz}ds. \qquad (7.23)$$

This approximation was first introduced in[973] and was later studied in[561] where it was shown that the D-bar equation (7.21) with $\mathbf{t}(k)$ replaced by $\mathbf{t}^{\text{exp}}$ truncated to a disk of radius R in the k-plane has a unique solution which is smooth with respect to z, and the reconstruction is smooth and stable. Further, it was shown that no systematic artifacts are introduced when the method with $\mathbf{t}^{\text{exp}}$ is applied to piecewise continuous conductivities.

We will now describe the numerical implementation of the D-bar method with the $\mathbf{t}^{\text{exp}}$ approximation.

7.4.2 NUMERICAL SOLUTION OF THE EQUATIONS

7.4.2.1 Computing the DN Map from Measured Data

The direct reconstruction methods in this chapter all require a matrix approximation to the DN map; that is, a matrix that approximates the action of the DN map on any given voltage pattern. The dimension of the approximation depends on the number of linearly independent current patterns being applied. For example, for L electrodes, the trigonometric current patterns result in $L-1$ linearly independent patterns, and the DN matrix will be of size $L-1$ by $L-1$. Bipolar injection patterns that skip α electrodes between the injection electrodes result in $L - \gcd(L, \alpha + 1)$ linearly independent patterns. Denote the basis of orthonormal current patterns by ϕ^n, where $n = 1, \ldots, N$, $N < L$, and denote the vector of voltages on the electrodes arising from the nth current pattern by U^n.

Denoting the DN matrix by $\mathbf{L}_\sigma$, we have that $\mathbf{L}_\sigma$ is the inverse of the matrix approximation $\mathbf{R}_\sigma$ to the Neumann-to-Dirichlet map, provided that the voltage measurements sum to zero for each current pattern. Since the usual data collection scenario is to apply current and measure voltage, one can form $\mathbf{R}_\sigma$ from the discrete inner product, and invert the matrix to obtain $\mathbf{L}_\sigma$. As long as the matrix of current patterns is full rank, $\mathbf{R}_\sigma$ will be a well-conditioned matrix. Thus, setting

$$\mathbf{R}_\sigma(m,n) = (s_\ell \phi^m, U^n)_L, \tag{7.24}$$

where s_ℓ is the arc length of the boundary segment connecting the centers of electrodes ℓ and $\ell + 1$, we have

$$\mathbf{L}_\sigma = \mathbf{R}_\sigma^{-1}. \tag{7.25}$$

Let $\delta\mathbf{L}$ denote the matrix approximation to $\delta\Lambda = \Lambda_\sigma - \Lambda_1$.

Other options for approximating DN or ND maps from practical measurements are given in [173,495].

7.4.2.2 Computation of the Scattering Transform

Denoting the area of an electrode region along $\partial\Omega$ by e_ℓ, (7.19)

$$\mathbf{t}^{\mathrm{exp}}(k) = \sum_{\ell=1}^{L} \int_{e_\ell} e^{i\bar{k}\bar{z}} \delta\Lambda e^{ikz} ds(z). \tag{7.26}$$

In light of equation (7.24), we wish to express the action of the DN map on e^{ikz} in equation (7.19) in terms of the orthonormal basis of current patterns. We can write

$$e^{ik\zeta_\ell} = 1 + \sum_{n=1}^{N} b_n(k)\phi^n(\zeta_\ell), \tag{7.27}$$

where the 0th order term in the expansion, 1, accounts for the fact that no basis function ϕ^n can be a constant function due to Kirchhoff's law, and

$$b_n(k) = (e^{ik\zeta_\ell}, \phi^n(\zeta_\ell))_L.$$

However, since $\Lambda_\sigma 1 = 0$ for any σ, we see that in the formula for computing $\mathbf{t}^{\mathrm{exp}}$ the 0th order term is annihilated, and we have

$$\delta\Lambda e^{ikz} = \sum_{n=1}^{N} b_n(k)\delta\Lambda\phi^n(z).$$

Define Φ to be the matrix of current patterns formed by setting the nth column of Φ to be $[\phi^n(z_1), \ldots, \phi^n(z_L)]^T$. Then $\delta\Lambda\phi^n(\zeta_\ell) \approx (\Phi\delta\Lambda)(\ell,n)$.

We can now compute the scattering transform by discretizing the integral in (7.26) as follows.

$$\mathbf{t}^{\mathrm{exp}}(k) \approx \sum_{\ell=1}^{L} \int_{e_\ell} e^{i\bar{k}\bar{z}} \sum_{n=1}^{N} b_n(k)\delta\Lambda\phi^n(z) ds \approx \sum_{\ell=1}^{L} A_\ell \sum_{n=1}^{N} e^{i\bar{k}\bar{z}_\ell} b_n(k)(\Phi\delta L)(\ell,n),$$

where A_ℓ is the area of the ℓth electrode.

In the presence of noise, the scattering transform or its $\mathbf{t}^{\mathrm{exp}}$ approximation will blow up, and therefore the computations must be restricted to a finite region of the k-plane. For simplicity, we will take this region to be a disk of radius R, and will refer to R as the *truncation radius*, setting $\mathbf{t}^{\mathrm{exp}}(k) = 0$ for $|k| > R$, and including the subscript R as a reminder of the truncation, $\mathbf{t}_R^{\mathrm{exp}}$.

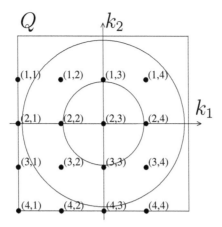

Figure 7.2   Computational 4×4 grid $\mathscr{G}_2$ (black dots) in the square Q. The two circles shown have radii R and $2R$.

7.4.2.3   Numerical Solution of the D-bar Equation

A numerical method for solving equations of the form

$$\overline{\partial}_k v(k) = T(k)\overline{v(k)} \tag{7.28}$$

based on the fast method[1066] that uses FFT's for solving integral equations with weakly singular kernels was developed in [564] for the inverse conductivity problem. This method and its adaptation to the D-bar equation have computational complexity $M^2 \log M$ when the solutions are computed on a $M \times M$ grid. An alternative method of solution based on finite difference methods is found in[749]. See also the spectral method introduced in [558].

Since we have truncated the scattering transform to $\mathbf{t}_R^{\text{exp}}$, we denote the solution to the integral form of the D-bar equation with truncated $\mathbf{t}^{\text{exp}}$ by

$$\mu_R(z,s) = 1 + \frac{1}{(2\pi)^2} \int_{|k| \le R} \frac{\mathbf{t}_R^{\text{exp}}(k)}{(s-k)\overline{k}} e_{-z}(k)\overline{\mu_R(z,k)}dk_1 dk_2. \tag{7.29}$$

It is shown in[745] that for fixed $z \in \mathbb{R}^2$, if we know the restriction $\mu_R(z,k)$ to the disc $|k| < R$, then we know $\mu_R(z,k)$ for any $k \in \mathbb{C}$ by substituting $\mu_R(z,k)|_{D(0,R)}$ to the right hand side of equation (7.29). The reduction to a periodic equation on a torus is given explicitly in[745].

Having chosen a truncation radius $R > 2$ in the k-plane, let Q be the square $Q := [-R, R]^2$, choose a positive integer m, let $M = 2^m$, and set $h = 2R/(M-1)$. Define an $M \times M$ grid $\mathscr{G}_m \subset Q$ by

$$\mathscr{G}_m = \{(hj_1, hj_2) \mid j_\ell \in \mathbb{Z}, \ -2^{m-1} \le j_\ell < 2^{m-1}, \ \ell = 1, 2\}. \tag{7.30}$$

See Figure 7.2 for an illustration of the grid. Note that the grid does not contain the point $(0,0)$ and that the grid can be easily created using Matlab's meshgrid command.

For a function $\varphi : Q \to \mathbb{C}$, define the matrix of values of ϕ on the grid $\mathscr{G}_m$ by a complex $M \times M$ matrix ϕ_h where $\phi_h(1,1)$ is the value of ϕ on the upper left grid point, $\phi_h(1,2)$ is the value of ϕ on the second grid point in the top row, and so on, following matrix notation.

Equation (7.29) is of the form

$$\mu_R = 1 - PT_R \rho(\mu_R), \tag{7.31}$$

where ρ is the conjugation operator, T_R is the multiplicative operator

$$T_R = \frac{\mathbf{t}_R^{\text{exp}}(k)}{4\pi\overline{k}} e_{-z}(k),$$

and P is convolution with the Green's function for the $\overline{\partial}_k$ operator, $\tilde{g} = \frac{1}{\pi k}$.

We use Fast Fourier Transforms (FFTs) to implement convolution in (7.31) by multiplication in the frequency domain. Given a periodic function φ, $P\varphi$ can be approximately computed by

$$(\widetilde{P}\varphi)_h \approx h^2 \, \mathrm{IFFT}\big(\mathrm{FFT}(\tilde{g}_h) \cdot \mathrm{FFT}(\varphi_h)\big),$$

where FFT and IFFT stand for the two-dimensional direct and inverse fast Fourier transform, respectively.

To solve equation (7.31) numerically, there are a number of choices, but GMRES provides a matrix-free method. That is, no intermediate storage of the matrices is necessary during the GMRES iterations. Due to the complex conjugate in equation (7.22), the real and imaginary parts must be separated, creating a linear system of dimension $2M^2$.

Consider stacking the real and imaginary parts of the $M \times M$ matrix φ_h as a vector and calling the result $\vec{\phi}_h$:

$$\mathrm{vec}(\varphi_h) := \begin{bmatrix} \mathrm{Re}\,\varphi_h(1,1) \\ \vdots \\ \mathrm{Re}\,\varphi_h(M,M) \\ \mathrm{Im}\,\varphi_h(1,1) \\ \vdots \\ \mathrm{Im}\,\varphi_h(M,M) \end{bmatrix} = \vec{\phi}_h. \tag{7.32}$$

Also, denote the inverse operation by $\mathrm{mat}(\vec{\phi}_h) = \varphi_h$. We now vectorize the operator $I - PT_R\rho$ appearing in equation (7.31) as

$$\vec{\phi}_h - \mathrm{vec}\left(h^2 \, \mathrm{IFFT}\left(\mathrm{FFT}(\tilde{g}_h) \cdot \mathrm{FFT}\left(\big(\frac{\mathbf{t}_R^{\exp}(k)}{4\pi \bar{k}} e^{-i(kz + \bar{k}\bar{z})}\big)_h \cdot \overline{\mathrm{mat}(\vec{\phi}_h)} \right) \right) \right).$$

This is in a form suitable for feeding into real-valued iterative GMRES solver. The solution to (7.31) is then $\vec{\phi}_h$, and the value of μ_R at $k = 0$ can be approximated by interpolation on the k-grid to the point $k = 0$, and the conductivity can be computed from $\mu_R^2(z,0) = \sigma(z)$. This can be solved for each point z in the region of interest in parallel.

7.4.3 EXAMPLES OF RECONSTRUCTIONS

In this section we show several examples of reconstructions using the D-bar method on experimental data, and many more can be found in the references cited. Here we include reconstructions from data collected on healthy human subjects with the ACE 1 the pairwise current injection system[707] at Colorado State University. Figure 7.3 contains six difference images from a sequence of 500 frames of perfusion in a human chest with data collected on a healthy subject during breath-holding reconstructed using the D-bar method with the $\mathbf{t}^{\exp}$ approximation, and figure 7.4 contains six frames from a sequence of 500 frames of difference images of ventilation in a human chest with data collected on a healthy subject also reconstructed using the D-bar method with the $\mathbf{t}^{\exp}$ approximation.

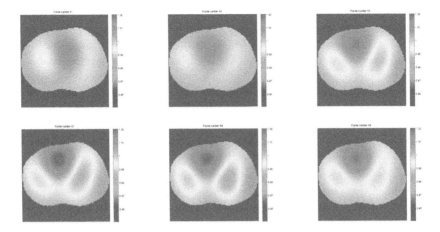

Figure 7.3 Difference images of perfusion in a human chest with data collected on a healthy subject during breath-holding and reconstructed using the D-bar method with the $\mathbf{t}^{exp}$ approximation. The heart is at the top of the images, and red represents regions of high conductivity relative to the reference frame, and blue low conductivity. Computations by Melody Alsaker. (color figure available in eBook)

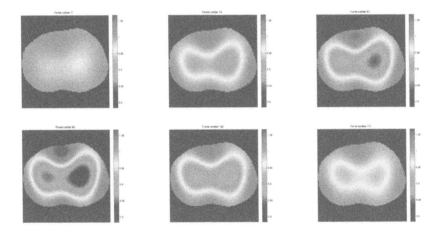

Figure 7.4 Difference images of ventilation in a human chest with data collected on a healthy subject reconstructed using the D-bar method with the $\mathbf{t}^{exp}$ approximation. The heart is at the top of the images, and red represents regions of high conductivity relative to the reference frame, and blue low conductivity. Computations by Melody Alsaker. (color figure available in eBook)

7.5 NEW AND RECENT DIRECTIONS IN D-BAR METHODS

While the D-bar method described in the previous sections is the most developed and tested, there are many directions still to be explored. In this section we briefly describe some recent progress in D-bar methods that take into account other medically relevant aspects of the EIT problem. For comparison of images reconstructed with the D-bar method to least-squares-based iterative methods, the reader is referred to [420,422,502,923].

The nonlinear low-pass filter used in two-dimensional D-bar method results in a Gibbs-type phenomenon analogously to the linear Fourier transform [70]. Also, the computational results in [70] suggest that the D-bar reconstructions of real-valued conductivities from truncated scattering data are almost the same regardless of the variety of the D-bar method. There are three main varieties. The first are the methods based on [764], see [501,562,564,744,973]. The smoothness assumptions in the proof in [764] were relaxed to permit once-differentiable conductivities in [163]. The D-bar problem that arises in this context is a first-order elliptic system of equations, which adds some complexity to the implementation. The reader is referred to [416,421,565] for results and reconstructions based on this approach. Finally, all assumptions on smoothness were removed in [69], and a Beltrami equation serves as the PDE for the CGO solutions. For methods based on [67,69], see [65,66,87].

The nonlinear low-pass filter used in two-dimensional D-bar method results in a Gibbs-type phenomenon analogously to the linear Fourier transform [70]. Also, the computational results in [70] suggest that the D-bar reconstructions of real-valued conductivities from truncated scattering data are almost the same regardless of the variety of the D-bar method. There are three main varieties. The first are the methods based on [764], see [501,562,564,744,973]. The smoothness assumptions in the proof in [764] were relaxed to permit once-differentiable conductivities in [163]. The D-bar problem that arises in this context is a first-order elliptic system of equations, which adds some complexity to the implementation. The reader is referred to [416,421,565] for results and reconstructions based on this approach. Finally, all assumptions on smoothness were removed in [69], and a Beltrami equation serves as the PDE for the CGO solutions. For methods based on [67,69], see [65,66,86].

A D-bar method for reconstructing complex-valued conductivities (that is, conductivity and permittivity with $\gamma = \sigma + i\omega\varepsilon$) was developed in [416], with the theoretical proof of unique determination in [289] and Lipschitz stability proved in [104]. The method relies on the solution of an elliptic system of D-bar equations, as in the proof [163] for the real-valued case. Reconstructions from numerically simulated data representing pleural effusion, hyperinflation, and pneumothorax are found in [421]. Another implementation differing from [421] in the computation of the scattering transform is given in [462] with reconstructions of ventilation and perfusion in a healthy human subject.

Progress in extending the D-bar method to partial-boundary data can be found in [419,446,447]. Also, the assumption of constant conductivity near the boundary can be avoided by a computational extension technique [974] based on [764,Ch6]. An example applied to simulated data is shown in Figure 7.5.

If the conductivity is anisotropic, the EIT imaging problem suffers from non-uniqueness. However, one can still use the D-bar method to recover an isotropic conductivity that is a deformation of the anisotropic one. This was first shown in [460] and later re-analysed via a different route and computationally implemented in [417].

The D-bar method is typically regularized using a nonlinear low-pass filter, as explained in [562]. However, the conductivity often has sharp features such as crisp boundaries between areas of smooth conductivity. In such cases it is not advisable to replace the high frequency information by zero, as

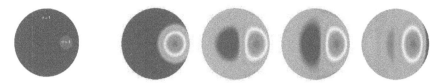

Figure 7.5 Left: The ideal discontinuous conductivity. Right: Reconstructions from partial boundary data using radius of truncation $|k| \leq 3$ in the scattering transform and (from left) DN data for 100%, 75%, 50%, and 25% of the boundary. Results from [419]. (color figure available in eBook)

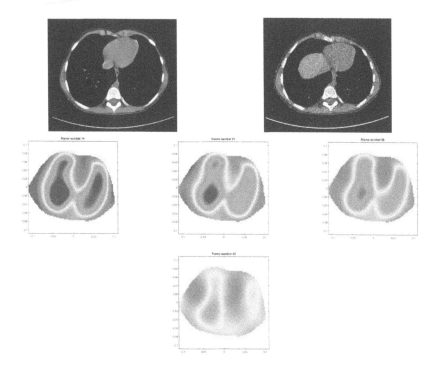

Figure 7.6 Top left: Inspiratory CT scan. Top right: Expiratory CT scan. The diaphragm is visible in the right lung (viewer's left) in the expiratory CT scan. Bottom row: Reconstructions of conductivity changes due to ventilation at 6 frames of 600 frame sequence using the D-bar method with a dynamic prior, which does NOT include a diaphragm, and conductivity values of the prior chosen by the optimization algorithm. All images are shown on the same scale, with blue corresponding to regions of low relative conductivity and red to regions of high relative conductivity. The first image corresponds to a maximal inhalation, and the sequence progresses to nearly maximal exhalation. The reader is referred to [45] for further details. (color figure available in eBook)

is done in the low-pass filter. This issue can be addressed by re-introducing edges to the blurred D-bar reconstruction in a data-driven way [417] or with a segmentation process [423].

Another approach is to introduce spatial priors in the reconstructions, for example by appending the scattering transform computed from the data with the scattering transform computed from a priori in a region of k values outside the original truncation radius. Since larger values of k correspond to high frequency information in the reconstruction, this improves resolution in the images. This approach was introduced in [42] and applied to cases including partial boundary data [41] and complex conductivities in [418]. An optimization method for choosing the conductivity values in the prior was introduced in [45] to accommodate challenges arising when using experimental data, and the method was further developed for dynamic reconstructions of human subject data in [43,44]. See Figure 7.6 for an example of the effect of a dynamic prior in a ventilatory image sequence. Another approach to including *a priori* information in the D-bar method is to introduce a statistical prior based on Schur complement properties [924]. This approach is illustrated on data collected on a saline-filled tank with the ACE 1 system with slices of watermelon and agar to simulate the lungs and heart, respectively; see Figure 7.7. D-bar reconstructions can also be enhanced by using conformal maps [495].

A new way of using the CGO solutions was introduced in [379]. Applying one-dimensional Fourier transform to the polar component of the complex parameter k in the CGO solution reveals

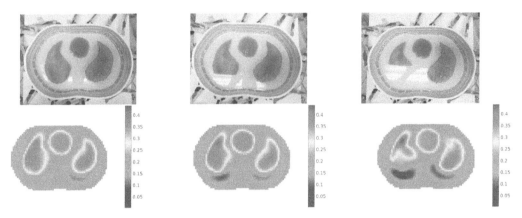

Figure 7.7   Top row: Experimental phantoms, watermelon "lungs" and agar "heart" in saline-filled tank. Bottom row: Absolute images, reconstructions of the experimental phantoms using a statistical prior in the D-bar method. Computations by Talles Santos. (color figure available in eBook)

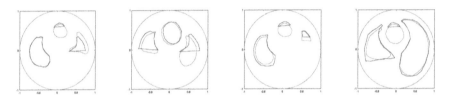

Figure 7.8   Results for multiple simulated injuries using the deep learning method for reconstructing organ boundaries from learned scattering transforms. The actual boundaries are in blue, with the network predicted boundaries are in red. The reader is referred to [179] for details on the method. Computations by Michael Capps. (color figure available in eBook)

information analogous to X-ray projections of the conductivity. This opens up ways to recover partial information about the conductivity, such as parts of interfaces between different tissues.

D-bar methods apply also to modalities other than EIT, such as acoustic tomography[225] and diffuse optical tomography[1022].

Machine learning is a modern technique influencing imaging including EIT. A post-processing method for sharpening D-bar reconstructions with the aid of deep learning was introduced in[415]. A method of detecting organ boundaries using deep learning applied to the scattering transform was introduced in[179]. Example of reconstructed organ boundaries are found in Figure 7.8. The convolutional neural network was trained on 100,000 simulated scattering transforms with a randomly chosen base distribution with perturbed values of conductivity, shape, scaling, organ injury, mollification, and rotation.

Three-dimensional EIT algorithms based on CGO solutions have been discussed and implemented in[108,109,139,211,229,230,231,563].

In conclusion, we hope this chapter will inspire further new advances in the D-bar method. We apologize for any omissions that are inevitable in any survey, and acknowledge the contributions to this field from many.

8 EIT Image Interpretation

Zhanqi Zhao
Department of Biomedical Engineering, Fourth Military Medical
University, Xi'an, China;
Institute of Technical Medicine, Furtwangen University,
Villingen-Schwenningen, Germany

Bin Yang
Department of Biomedical Engineering, Fourth Military Medical
University, Xi'an, China

Lin Yang
Department of Aerospace Medicine, Fourth Military Medical
University, Xi'an, China

CONTENTS

DOI: 10.1201/9780429399886-8

EIT images represent in tissue impedance or changes in this value over time. Since users of EIT systems are typically interested in other (clinical) parameters, there is a need for interpretation of the resulting images. The mechanisms for EIT image interpretation are most advanced for thoracic EIT, but have also seen developed for other applications.

A typical EIT image consists of 32×32 pixels. Depending on the inverse models and reconstruction methods, the number of pixels varies. A spatial interpretation of an isolated EIT image would normally not be useful, since commercial EIT devices can show time-difference of the impedance with respect to a reference value. The sampling frequency of EIT measurement is typically between 20–50 frames per second for thorax and 1 frame per second for brain applications. An EIT measurement includes both temporal and spatial information. Therefore, an interpretation should reflect these aspects at the same time. In this chapter, we describe the basic elements of EIT images, summarize representative functional images and EIT measures and show how to use these tools to interpret clinical applications.

8.1   ELEMENTS OF EIT IMAGES

8.1.1   VALID SIZE OF AN IMAGE

Previously (§6.6) we have discussed different reconstruction methods. The shape and the size of the reconstructed images vary depending on the inverse models and reconstruction methods. Figure 8.1 shows an example of valid pixels in the thorax image depending on the inverse models. Reconstructed images are normally mapped onto a rectangular pixel (or voxel) grid. Since there is no square inverse model for thorax or brain applications, when we see a square reconstructed EIT image, the pixel values at the corners are usually assigned with zeros (or a placeholder such as NaN). To analyze the images, we often focus on certain regions of interest (ROIs) with e.g. pixel that changes its value during the measurement.

8.1.2   COLOUR MAPPING AND COLOUR SCALE

Reconstructed impedance values do not have colours. In order to visualize the results, pseudo-colour with different possible colour mappings have been used by various research groups and manufactures of EIT devices[293]. Some colour maps were used simply because they were default setting of the analysis software or printout version of journals accepted grey scale of figures (e.g. the colour map Jet was the default one in older versions of MATLAB). Colour maps from commercial devices were designed to modify the visual strength of certain conductivity changes (e.g. colour coding from Swisstom and Timpal do not show negative values of changes). Since this chapter

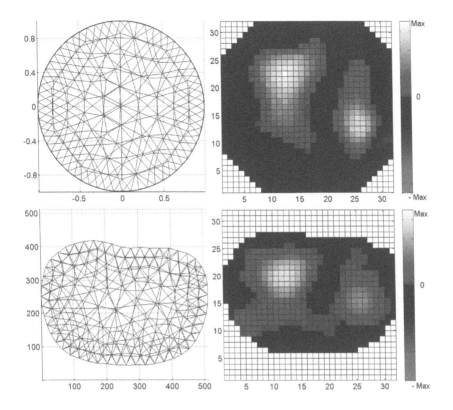

Figure 8.1 Valid size of an image depends on the inverse models and reconstructed methods. Top: reconstruction based on a circle FEM model; bottom: reconstruction based on thoraxs shape extracted from CT of the corresponding patient. (color figure available in eBook)

describes how to process raw EIT impedance values with various functional EIT (fEIT) images and measures, note that pixels may have different meanings even when the images are coded with the same colour map. The colour maps should always be defined and clarified for each use.

When the selection of colour map is determined, a suitable colour scale should be selected in order to visualize the effects of changes (e.g. ventilation, bleeding). Different colour scales may introduce different impressions of the changes (figure 8.2). This impression may be enhanced or attenuated in combination with colour maps. Take the colour map "Jet" for example, blue is assigned automatically to the lowest value in the colour scale. Therefore, the background colour with pixel value of zero may change according to the colour scale selected. The colour coding used by the Dräger Pulmovista 500, on the other hand, assigns black to pixels with values close to zero, which compresses all changes less than a certain level to be represented as black, so that the visual effects of small perturbation are weakened.

8.1.3 SAMPLING FREQUENCY AND MIXTURE OF SIGNALS

Modern EIT devices are able to provide up to 100 frames per second, which is not required for most of the clinical EIT applications. Transient phenomena can be better resolved with increasing sampling rate, however, higher rates may cause a reduction in signal quality and large memory to store the data. For brain applications, e.g. to monitor intracerebral hemorrhage or ischemia, 1 frame per second is adequate to capture the deterioration. For applications of higher dynamics such as

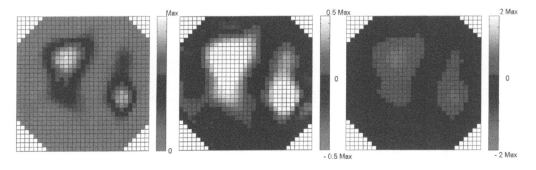

Figure 8.2 Same EIT images as figure 8.1 (top right) presented with different colour scales. (color figure available in eBook)

ventilation and perfusion, a sampling rate of 20–40 frames per second is preferable. With a ventilation mode called high-frequency oscillatory ventilation (HFOV), patients are ventilated between 3 and 15 Hz (180–900 breaths per minute), and the sampling rate should be adjusted accordingly. Otherwise the underlying phenomena (ventilation distribution in this case) cannot be captured correctly. Figure 8.3 shows ventilation distribution during HFOV in a pig. EIT sampling rate were 40, 20, and

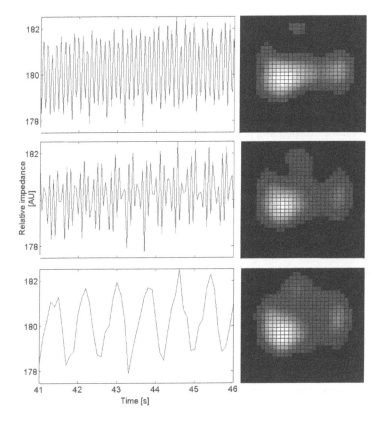

Figure 8.3 Global impedance waveform (left) and ventilation distribution (right) in an anesthetized supine pig during high-frequency oscillatory ventilation. The oscillatory ventilation rate was 9 Hz and the EIT sampling rates were 40 (top), 20 (middle) and 10 (bottom) Hz. AU, arbitrary unit. (color figure available in eBook)

10 frames per second, respectively. According to the Nyquist-Shannon sampling theorem, 20 frames per second would be enough given that the oscillatory ventilation rate was 9 Hz. However, the cyclic minimums and maximums in the global impedance curve were not captured (figure 8.3middle left). With 10 frames per second the data would be aliased and thus misinterpreted (figure 8.3bottom). When several physiological or mechanical signals are captured at the same time, the periodic feature of the signals might not be easy to identify. Figure 8.4 illustrates a mixture of two periodic cardiac-related signals, with the same frequency but phase shift. The whole global signal seems to be irregular (figure 8.4left).

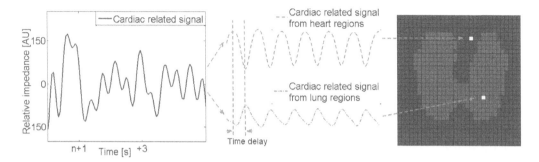

Figure 8.4   An example of cardiac related signal measured on the thorax surface is a mixture of signals from the heart and from the lung regions. (color figure available in eBook)

Signals from other sources in the environment may also interfere with the EIT signal. Figure 8.5 shows an example of EIT measurement with a patient lying on an air mattress. The large signal fluctuations caused by repetitive inflation and deflation of the mattress cushions are observed. The smaller rapid fluctuations reflect the changes in impedance synchronous with HFOV. With high-pass filter, the large disturbances disappeared. Further sources of interference were explored in [303].

Therefore, if certain unknown signals is observed, interpretation must be cautious. It could be caused by low sampling rate, mixture of signals (not necessary signals of interest), noises from the environment or other reasons.

8.1.4   MEANING OF THE PIXELS IN DIFFERENT TYPES OF IMAGES

Raw EIT data usually do not provide directly interpretable results, so that they should be further processed. To visualize the results of data processing, values might be plotted again as an image in the original size. However, the meaning of the image changes from pure impedance values to the advanced indices. Again taking lung EIT for example, in order to quantify tidal ventilation distribution, several functional EIT images were proposed [1213]. Depending on which functional image was used, the pixels in the image may represent impedance change, standard deviation of impedance within certain time, and regression coefficients comparing global and regional impedance-time curves. It is recommended that if the pixels represent parameters other than impedance, it should be explained and use different colour maps, if possible, to avoid confusion. When the images are reconstructed based on frequency-difference [922] or absolute algorithms [402], the pixel values represent totally different information, which require different interpretation.

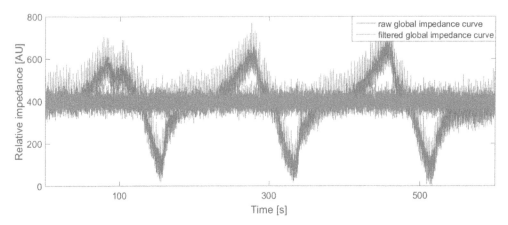

Figure 8.5 Effect of pulsating air suspension mattress on the thoracic EIT examination. Mattress cushions repetitively inflate and deflate within a time interval of about 90 seconds. The patient in a patient with chronic obstructive pulmonary disease was mechanically ventilated using high-frequency oscillatory ventilation (oscillatory rate 6 Hz). Blue, the original EIT waveform and red, the corresponding waveform after high-pass filter (cutoff frequency 2 Hz). This data was kindly provided by Prof. I. Frerichs. (color figure available in eBook)

8.2 FUNCTIONAL IMAGES AND EIT MEASURES

EIT is often used to monitor the impedance changes in ROIs. Due to its ability to capture spatial information with high temporal resolution, EIT scans contain valuable functional information that may influence clinical decisions. Such information is only meaningful when the EIT data are analyzed over a period of time. If the information is subtract and presented visually as an image, it becomes a fEIT image (figure 8.6). If the information is summarized as a single number, we may

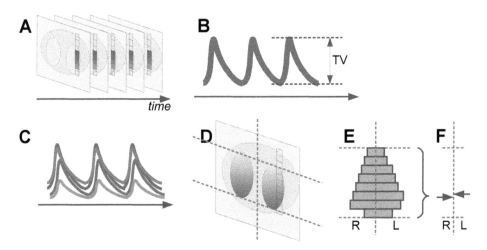

Figure 8.6 Illustration of calculation of functional EIT images and measures (adapted from [293]). (A) EIT image sequence, from which (B) global waveforms calculated, and (C) pixel waveforms are analysed to calculate a tidal variation parameter, from which fEIT images (D) are generated. In (E), a horizontal histogram of breathing in "slices" of the left and right lung is calculated, from which the centre of gravity (F) in the left and right lung are determined. (color figure available in eBook)

call it an EIT measure or index. In a previously published consensus paper on thorax EIT, the fEIT and measures were introduced in separated chapter and grouped according to their target physiological parameters[293]. To include other EIT applications and to avoid a long list of categories, we summarized fEIT and various measures with 4 types based on how the information is derived. Examples of each type are illustrated. At the end of this chapter, a glossary of current fEIT and measures is listed (table 8.2).

8.2.1 SIMPLE DISTRIBUTION OF THE IMPEDANCE CHANGES IN CERTAIN ROIS

To monitor the impedance changes, the simplest way is to compare EIT images at two specified time points. Thus the image, $\hat{x} = R_M(d - d_{ref})$, where R_M is a reconstruction matrix (equation 6.14). The reference measurement, d_{ref} should be chosen at a stable time, such as end-expiration, or can be calculated as an average over a set of points. Unfortunately, clinical environment is far noisier than ideal. To reduce noise level, the impedance changes are usually presented in kind of average over time. Further mathematic algorithms may apply to reduce influences from other signals.

8.2.1.1 Example: Averaging Tidal Variation fEIT

Ventilation related variation during tidal breathing could be assessed by subtracting the end-expiration from the end-inspiration image, which is called tidal image (or subtracting the begin-inspiration from begin-expiration, which makes no differences in case of averaging). Averaging all the tidal images within certain time period increases signal to noise ratio (figure 8.7).

$$V_i = \frac{1}{N} \sum_{n=1}^{N} (\Delta Z_{i,Ins,n} - \Delta Z_{i,Exp,n}) \tag{8.1}$$

where V_i is the pixel i in the fEIT image; N is the number of breaths within analyzing period; and $\Delta Z_{i,Ins,n}$ and $\Delta Z_{i,Exp,n}$ are the pixel values in the raw EIT image at the end-inspiration and end-expiration, respectively.

This fEIT is widely used in thorax EIT applications. By subtracting end-expiration from end-inspiration, every tidal image can be considered to have updated its reference point every breath and the signal to noise ratio is high, especially if the baseline drifts due to sweating, electrodes movement, or posture change. However, reliable auto-detection of end-expiration and end-inspiration is sometimes difficult, especially in patients with irregular breathing (e.g. during spontaneous breathing or in presence of pendelluft, a type of asynchronous alveolar ventilation, which is usually caused by different regional time constants or dynamic pleural pressure variations)

8.2.1.2 Example: Regression fEIT

To calculate this type of fEIT, each pixel value is the regression coefficient of the following linear regression formula:

$$\Delta Z_i(t) = \alpha_i \sum_{m=1}^{M} \Delta Z_m(t) + \beta_i + \varepsilon_i \tag{8.2}$$

where $\Delta Z_i(t)$ denotes time dependent relative impedance value of pixel i of the raw EIT images; α and β are regression coefficients and ε is the fitting error; M is the total number of pixels in the reconstructed EIT images. The result, α_i will be the value plotted in pixel i in the fEIT image.

The assumption of this fEIT calculation, is that the global impedance-time curve is dominated by a desired signal (figure 8.8). Signals not in phase compared to global one will be suppressed.

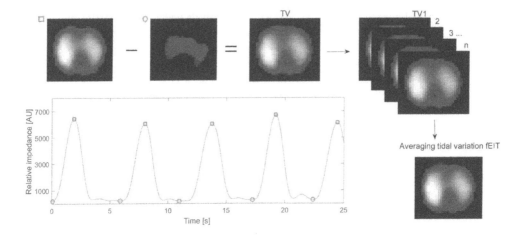

Figure 8.7   Illustration of functional EIT by averaging tidal variations (TV). fEIT, functional EIT; AU, arbitrary unit; green circles, beginning of inspiration; red rectangles, end of inspiration / begin of expiration. (color figure available in eBook)

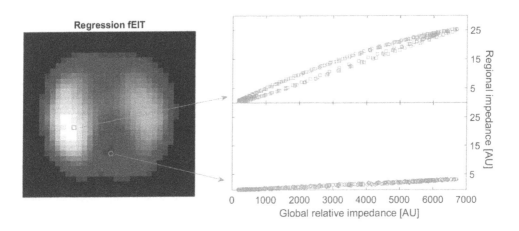

Figure 8.8   Illustration of functional EIT by regression. Signals that not in phase compared to global one will be suppressed (e.g. green circle). fEIT, functional EIT; AU, arbitrary unit. (color figure available in eBook)

8.2.1.3   Example: Cardiac-related fEIT

A sequence of EIT images may often contain impedance information on several underlying physiological or pathological activities simultaneously. If these physiological activities behave at different frequencies, we can extract the EIT signals of interest and filter out the unwanted or noises with frequency filtering. If these activities occur at different frequencies, it is possible to use frequency-filtering to separate the activities.

For thoracic applications, EIT captures ventilation and cardiac-related impedance signals at the same time (see also Chapter 11). As the cardiac-related variation in electrical impedance ($<5\%$) is much smaller than the ventilation-related variation ($>50\%$) and these two signals have different frequencies, frequency filtering can be used to attenuate the ventilation-related signal. Typically, we calculate a Fourier Transform on the global impedance signal to obtain the corresponding power spectrum and determine the expected frequency range for breathing and heart rates. Next, a Fourier Transform is performed again on each waveform of individual pixels in the sequence of raw EIT images. The signal amplitudes at the cardiac frequency are then assigned to individual pixels to construct the cardiac-frequency EIT (figure 8.9). The corresponding fEIT can be calculated using the techniques mentioned for ventilation (e.g. §8.2.1.1 and §8.2.1.2). Since the cardiac-frequency fEIT includes information on lung perfusion as well, it was called "perfusion image" in some publications. However, as pointed out in a previous review[293], there are several physiological reasons why the EIT image filtered at the cardiac frequency may over- or underestimate lung perfusion[131].

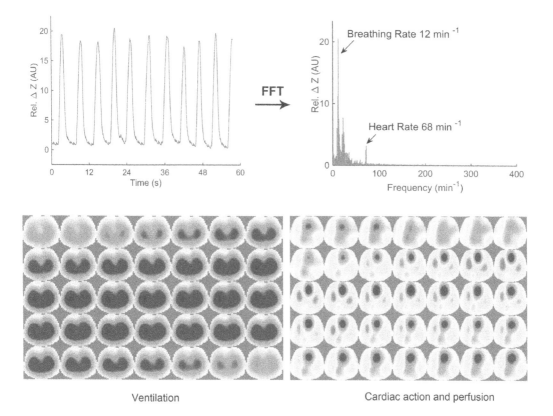

Figure 8.9 Global EIT waveform by summation of individual pixels of EIT images (top left) and the corresponding frequency spectrum (top right). By filtering, the ventilation (bottom left) and cardiac-related EIT images (bottom right) could be shown, respectively. FFT, frequency filtering technique. (color figure available in eBook)

8.2.1.4   Example: Identifying and Tracking of Intracranial Resistivity Changes

Intracranial resistivity changes in humans could be detected and quantified by EIT, which has been validated by a clinical feasibility study of in vivo imaging of twist drill drainage for subdural hematoma[220]. The identification of the impedance change region is determined through the reconstructed values using a predefined threshold $(Z_{peak} - Z_k)/Z_{peak} \leq T$, where Z_{peak} is the peak value of all elements; the threshold parameter T is set to 20% as empirical value. Then, average reconstructed value (ARV) and the size of ROI (sROI) on the EIT images are calculated as follows:

$$ARV = \frac{1}{E} \sum_{k=1}^{E} x_k A_k \tag{8.3}$$

$$sROI = \frac{\sum_{k=1}^{E} A_k}{\sum_{i=1}^{N} A_i} \times 100\% \tag{8.4}$$

where X_k represents the reconstructed value on the kth element in ROI, A_k is the area and E and N represents the total number of elements in ROI and imaging domain, respectively.

Time difference EIT has been used to identify and track intracranial regional impedance changes caused by pathological or physiological processes in clinical trials (figure 8.10), such as EIT monitoring for cerebral hemorrhage and focal cerebral infarction.

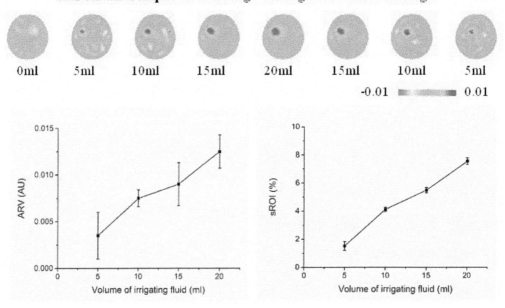

Figure 8.10 EIT images and analysis results of a subdural hematoma patient with twist drill drainage (data from [220]). Top, EIT images and irrigation fluid volume during the twist-drill drainage operation for a patient with subdural hematoma. Bottom left, plot of ARV against the volume of irrigation fluid volume. Bottom right, plot of sROI against the volume of irrigation fluid. ARV, average resistivity value; sROI, size of region of interest; AU, arbitrary unit. (color figure available in eBook)

8.2.2 DIFFERENCES OF IMPEDANCE VARIATION CALCULATED IN SPATIAL CORRELATIONS

These measures are designed to describe variations in the series of EIT images by comparing the spatial correlations among regions. Such characterizations are needed for several features of the image, depending on the physiology of interest. In the application of thorax EIT, the most common measures of this type are used to describe the overall degree of spatial heterogeneity of ventilation. It is important to characterize the level of inhomogeneity in a fEIT image, since pathological or therapeutic processes compromise the uniform distribution of ventilation under normal conditions.

8.2.2.1 Example: Center of Ventilation (CoV)

CoV, sometimes called center of gravity, was first introduced by Frerichs *et al.*[299]. It measures the anteroposterior distribution of ventilation. Therefore, it is often calculated for a fEIT image characterizing tidal variation (e.g. §8.2.1.1 and §8.2.1.2). CoV may be computed by the following equation (figure 8.11).

$$CoV = \frac{\sum(y_i \times Z_i)}{\sum Z_i} \times 100\% \tag{8.5}$$

where y_i is the pixel height and of pixel i scaled so the most ventral row is 0 and the most dorsal row is 100%. As the center of the distribution of ventilation moves dorsally, CoV increases. Variation of CoV may be found in publications e.g. to scale it between -1 and $+1$[1161], or to reverse the direction such that 100% is the ventral[849] or to simplify it as anterior-to-posterior ventilation ratio (i.e. $\sum Z_{ventral}/\sum Z_{dorsal}$), which is less robust compared to CoV[120]. It might also be calculated as $\sum Z_{ventral}/(Z_{ventral}+Z_{dorsal})$, but its specificity of ventilation shift is much smaller than conventional CoV. Thus, distribution within ventral or dorsal regions would not be distinguishable. This calculation would not distinguish whether ventilation is located only in the most ventral row or row 15 towards the center. It is also possible to calculate CoV for the horizontal axis, which is less common in clinical practice.

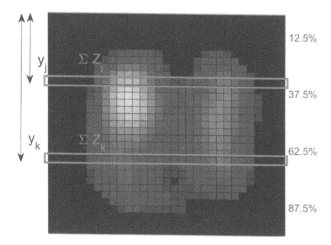

Figure 8.11 Illustration of computation of the center of ventilation (CoV). The image is divided into horizontal regions (of pixel width). In each horizontal region, the pixel impedance $\sum Z_j$ and $\sum Z_k$ are weighted with the corresponding horizontal positions y_j or y_k. The CoV is calculated using the formula provided in the text above. Z, impedance. (color figure available in eBook)

8.2.2.2 Example: The Global Inhomogeneity Index (GI)

The global inhomogeneity (GI) index was introduced by Zhao *et al* to assess the degree of inhomogeneity[1208]. The calculation of GI is based on the difference between each pixel value and the median value of all pixels. These values are normalized by the sum of impedance values within the lung area:

$$GI = \frac{\sum |Z_i - \text{Median}(Z_{lung})|}{\sum Z_i} \tag{8.6}$$

where Z_i denotes impedance value of pixel i in the identified lung regions ($\in$ lung); Z_{lung} are all pixels in the lung regions under observation.

Similarly to CoV, the GI index is often calculated based on a fEIT image of tidal variation (e.g. §8.2.1.1 and §8.2.1.2). As pointed out in the original research, the critical part of GI index calculation is the identification of lung area (figure 8.12). Since in fEIT, only ventilated area can be observed, lung area identified for a single status will mislead the GI value and draw an incorrect conclusion of homogeneous ventilation distribution. Combining lung areas in fEIT at different status improves the accuracy of GI calculation. If different status are not available, interpretation of corresponding findings needs to be cautious (e.g.[107]). CT scans of the patients (if available) could be helpful to identify the real lung area via embedding the shape information into EIT image reconstruction. An updated measure GI_{anat} has recently been proposed which uses subject-specific lung regions for accurate calculation[1182].

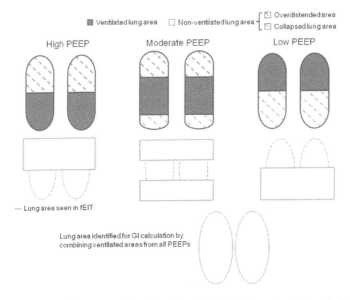

Figure 8.12 Illustration of lung area identification for GI index calculation during a scenario of PEEP titration (positive end-expiratory pressure). At each PEEP level, the lungs are only partially ventilated. Only by combining all lung areas identified at different PEEP levels, the GI calculation could be reliable. fEIT, functional EIT. (color figure available in eBook)

8.2.2.3 Example: Spatial Related Classification of Intracranial Resistivity Changes

For the clinical application of brain EIT, the positional information of intracranial tissue resistivity changes is closely related to the judgment of disease progression and the evaluation of treatment effect. Therefore, the brain images may be divided into six ROIs according to the brain structure in the analysis: left frontal lobe, right frontal lobe, left temporal lobe, right temporal lobe, left occipital lobe, and right occipital lobe (figure 8.13) following[53]. The average impedance changes, which are generally represented as the average reconstructed impedance value, for six ROIs during the course of monitoring, are calculated. Then, evaluation with statistical tests (e.g. paired-t test or repeated measurement data of variance analysis) is performed to identify any significant difference among ROIs[1180].

In a previous clinical experiment[316], EIT was used to monitor in real time changes in brain water content in a clinical model with patients undergoing clinical dehydration treatment. Each of the six regions was classified as either a normal or a lesion lobe (figure 8.13). The average impedance changes for these two individual classes were calculated during the course of treatment. The results of statistical analysis showed that different brain tissues have different dehydration effects, and

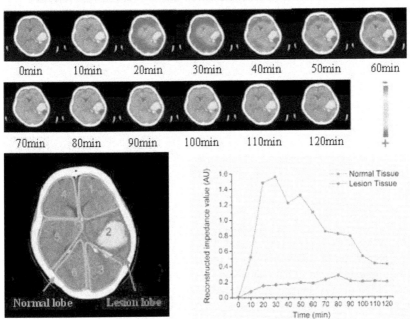

Figure 8.13 The reconstructed impedance change during the clinical dehydration treatment using mannitol. Top, serial images of brain EIT; Bottom left, the definition of six regions of interest according to anatomy; Bottom right, reconstructed impedance values of normal tissue and lesion tissue. Figure based on data from [316]. (color figure available in eBook)

normal brain tissues were more dehydrated than the diseased tissues. This finding agrees with a previous clinical study which has found that in models of ischemic infarction, the reduction in brain water content after mannitol infusion is greater in the normal than in the damaged hemisphere[445].

8.2.3 SUBTRACTING TEMPORAL INFORMATION (TAKING ADVANTAGES OF THE HIGH SAMPLING RATE)

8.2.3.1 Example: Intra-tidal Volume Distribution (ITVD)

ITVD, which is also called intra-tidal gas distribution, was first introduced by Lowhagen *et al*[655]. This measure captures the changes of ventilation distribution during inspiration phase to reflect heterogeneous lung mechanics in various regions. Let Z_n be the EIT frames dividing inspiration into eight iso-volume slices ($n=1, 2, \ldots, 7$). Here, Z_0 and Z_8 represent the EIT image at the start and end of inspiration. Instead of calculating the tidal image, iso-volume images ΔZ_n are built by subtracting Z_{n-1} from Z_n ($n=1, 2, \ldots, 8$). In the original study, the authors divided the lung into four anteroposterior ROIs with equal height. The ITVD is calculated for each ROI and plotted against each other.

$$ITVD_{ROI} = \frac{\Delta Z_{n_{ROI}}}{\Delta Z_n} \tag{8.7}$$

where $\Delta Z_{n_{ROI}}$ is the sum of impedance changes in the ROIs at n, ΔZ_n is the sum of all ROIs ($n=1, 2, \ldots, 8$). A variation of this measure was proposed[1210] to divide inspiration with equal time length instead of iso-volume and to calculate the differential images $\Delta Z'_n$ by subtracting Z_0 from Z_n ($n=1,$

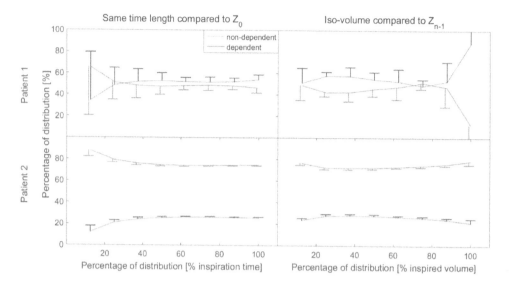

Figure 8.14 Comparison of two variations of intra-tidal ventilation distribution in two patients under support ventilation. Z is an EIT frame at specified time point. Z_0 is the starting frame of inspiration. Left: inspiration time is divided into 8 equal-time periods. Difference images are calculated by subtracting Z_0 from the 8 frames. Ventilation distribution in dependent (red) and non dependent (blue) regions of the corresponding difference images are plotted. Breaths within a minute were analyzed and averaged. Right: inspired volume is divided into 8 iso-volume slices. Difference images are calculated by subtracting Z_{n-1} from Z_n ($n=1, 2, \ldots, 8$). (color figure available in eBook)

$2, \ldots, 8$). Figure 8.14 compares the results of these two types of ITVD in two patients receiving support ventilation. An inspiration is usually shorter than 2 seconds and, with a sampling rate of 20 frames/second, around 5 frames are included between Z_{n-1} and and Z_n, depending on time or volume division. The influence of noise may be high (e.g. last slice in top right subfigure of figure 8.14). Besides, in presence of spontaneous breathing, impedance (volume) change becomes irregular (figure 8.15). Iso-volume slices may not be easily identified automatically (e.g. for inspirations marked red in figure 8.15). In case of recruitment with sudden rapid volume increase, some iso-volume slices may contain only a few EIT frames, which makes it vulnerable to noise (e.g. first slice in top left subfigure of figure 8.14).

8.2.3.2 Example: Regional Ventilation Delay RVD, both fEIT and Index Available

RVD was introduced in two studies[742,1169], and was designed to identify lung regions with late opening, which may be caused by cyclic recruitment/derecrutment. During low-flow maneuver, inspiratory flow is set to 2 l/min or similar. Collapsed regions have late inflation compared to open regions. RVD is defined either as absolute time or percentage to total inspiration time when a regional impedance time curve reaches a defined threshold to tidal variation (10% in[1169] and 40% in[742]; e.g. figure 8.16). A fEIT of RVD corresponds to the pixel values of RVD mapping. A RVD index could refer to the average value of all pixel RVD or standard deviation[742]. We would like to point out that RVD is not validated during conventional control ventilation or support ventilation, where inspiration time is short. The number of sampling points is limited during brief inspiration. Ventilation delay among regions, and could be within 1–2 sampling intervals, making the calculation would be too sensitive to noise (especially during spontaneous breathing).

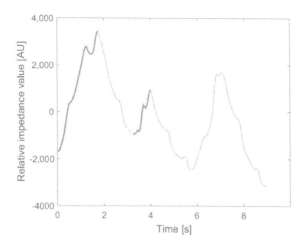

Figure 8.15   Global impedance-time curve of a ventilated patient with spontaneous breaths. Inspirations with non-linear increase of volume are marked in red. AU: arbitrary units. (color figure available in eBook)

8.2.4   NEW UNITS OR DIMENSIONS DERIVING FROM IMPEDANCE

These measures or fEIT parameters have "new units" in the sense that they are not measured in terms of impedance or time, which are directly reconstructed or measured. Instead, new units or dimensions are derived, based on terms and parameters that used in clinical practice. With this type of development, EIT findings can be more easily accepted by the physicians.

8.2.4.1   Example: Regional Compliance as an fEIT and the Application in PEEP Titration

Respiratory system compliance (C_{rs}) is a useful parameter to understand the lung status, and reflects the elasticity of the respiratory system, including the chest wall and lungs. It is traditionally divided into a static and dynamic measurement. Static C_{rs} assesses the volume increase in response to an airways pressure increase when airway flow is small enough to be negligible, while dynamic C_{rs} is measured when airflow is non-negligible. Dynamic C_{rs} may provide different information compared to static one because the mechanical ventilation routine is not interrupted[1001]. Given that the impedance changes are proportional to volume changes, C_{rs} can be assessed by substituting volume with impedance in the calculation (resulting in a different units). Regional distribution of EIT-C_{rs} may provide useful information on the mechanical properties of the lung tissue. Unfortunately, regional pressure changes cannot be obtained. Therefore, a key assumption of the EIT-C_{rs} measurement is that pressure changes are uniform throughout the lung when flow values reached zeros at the end of both inspiration and expiration.

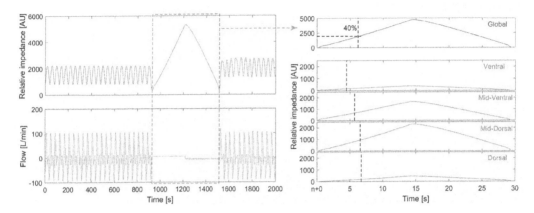

Figure 8.16 Illustration of regional ventilation delay. During low-flow maneuver, time to certain inspiratory volume in various regions of interest is compared. Tidal recruited regions have late opening compared to other regions, which presents as a delay in time. This data was kindly provided by Mr. E. Teschner, with permission from Dräger Medical. (color figure available in eBook)

The assumption of regional C_{rs} is that pressure changes are uniform throughout the lung and use the pressure changes reading from a ventilator when flow values have reached zero at both end-expiration and end-inspiration. This assumption may not be valid with obstructive lung diseases, especially if there were regions of trapped gas or pendelluft. In analysis of an individual breath, pixels in the fEIT image of tidal variation are divided by pressure difference between end-inspiration and end-expiration, which is assumed the same for every pixel. Therefore, the regional fEIT images would be the same as images of ventilation, only in a different scale and a more familiar unit (than impedance) to physicians. If the impedance changes are calibrated to volume changes, the EIT-C_{rs} could also be normalized to volume/pressure units.

A practical contribution of regional C_{rs} is the application in titration of positive end-expiratory pressure (PEEP). During PEEP titration, a decremental PEEP trial is performed where PEEP level is decreased step by step. Typically, the global C_{rs} measured while ventilator first increases and then decreases pressure, showing an ∩ shape. Similarly, one would expect regional C_{rs} to exhibit the same shape but heterogeneous among regions (figure 8.17). Regional compliance was computed for all pixels in the lung regions at each PEEP level. Then, the accumulated collapse and overdistension percentages were estimated based on the decrease of regional compliance curve during decremental PEEP titration, either towards lower or higher PEEP levels (8.8 and 8.9). No air is in collapsed (atelectatic) lung tissue and the corresponding regions do not participate gas exchange during ventilation. Lung volume is reduced accordingly. Overdistension, on the other hand, refers to lung tissue that is overdistended, containing large amount of air that reduces its elasticity. These regions will not participate air exchange during ventilation either.

Assuming $C_{reg,i} = \max(\Delta Z_{i_{allPEEP}}/\Delta P_{aw})$,

$$Col_i = \frac{C_{reg,i} - \Delta Z_{i_{currentPEEP}}/\Delta P_{aw}}{C_{reg,i}} \times 100\% \tag{8.8}$$

If current PEEP is lower than the PEEP level where $C_{reg,i,max}$ is located, then Col_i represents the collapsed percentage ($i \in lung$). If current PEEP is higher than the level of $C_{reg,i,max}$, Col_i can be substituted by another term $Over_i$ and represents the overdistension percentage. The cumulated collapsed or overdistension can be calculated as follows:

$$Col_{cum} = \frac{\sum Col_i \times C_{reg,i}}{\sum C_{reg,i}} \tag{8.9}$$

Similarly, Col_{cum} represents collapsed or be substituted by $Over_{cum}$ and $Over_i$ representing overdistension, depending on the current PEEP and the PEEP where $C_{reg,i,max}$ is located.

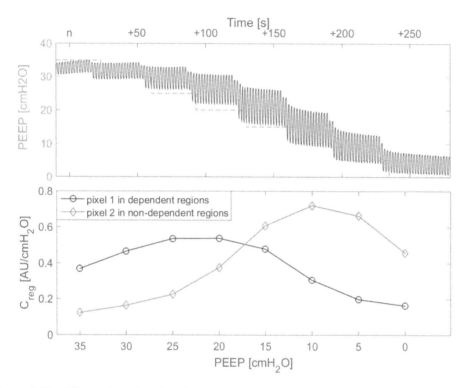

Figure 8.17   Illustration of regional compliance (C_{reg}) changes along with a decremental PEEP trial. In dependent regions, best C_{reg} is obtained at a higher PEEP level (black circles) compared to that in non-dependent regions (red diamonds). (color figure available in eBook)

8.2.4.2   Example: Regional Findings for Pulmonary Function Test

The idea of pulmonary function tests is to examine the flow, volume or other lung function limitations. Impedance changes can be considered as proportional to volume changes in certain ranges[686]. Therefore, flow changes can be approximated with derivative of impedance. Since commercial EIT delivers only relative impedance values, a direct comparison using the EIT-derived lung function parameters (such as $FEV_{1,EIT}$ and FVC_{EIT}) should be avoided. But their ratios could be inter-patient comparable because of the normalization (e.g. $FEV_{1,EIT}$ / FVC_{EIT}). In figure 8.18, an example of forced vital capacity maneuver is presented, which was taken from a healthy volunteer. From the global curve (right bottom) it is noted that the peak flow (approximated with derivative of relative impedance) was late, which indicates a poor subject effort. With traditional spirometry, this would be an unacceptable test and no useful information can be drawn. With regional information, a region with particularly low $FEV_{1,EIT}$ / FVC_{EIT} can be observed in ventral right lung (figure 8.18 left bottom). Unlike normal lung regions, the EIT-derived flow-volume curve (left top subfigure) reveals a smaller expiratory volume compared to inspiratory one, which indicates possible obstructions.

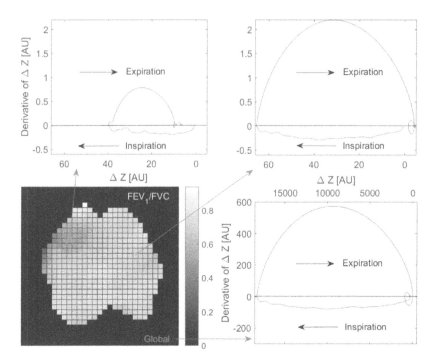

Figure 8.18 Illustration of regional lung function calculated with EIT. Left bottom: regional FEV_1/FVC, ratio between the volume expired within first second of forced expiration and forced vital capacity. Although the maneuver did not meet the ATS/ERS criteria of acceptable test, regional obstruction can still be observed. Top: regional flow-volume curves derived from EIT. Right bottom: global flow-volume curve derived from EIT. (color figure available in eBook)

8.2.4.3 Example: Determining Impedance Measures Using Contrast Agents

As discussed in §8.2.1.3, cardiac-related fEIT images calculated with frequency filtering do not correlate fully to lung perfusion. Therefore, in order to measure regional lung perfusion more accurately, using impedance contrast agents might be an option. Generally, a bolus of hypertonic NaCl solution (5.85% or 20% as reported), as impedance contrast agents, were injected into the vena cava via the central catheter[131,1008]. As saline decreased the blood impedance, it would be visible in the raw EIT images during its propagation from heart to lung (figure 8.19). A monitoring scenario, where small quantities of NaCl are injected every few hours, could represents Na intake levels below recommended consumption guidelines. For example, the 2300 mg recommended daily Na intake of the American Heart Association corresponds to 23 boluses of 10 mL of 3% saline. The main concern of this technique is patient safety, as hypertonic NaCl might provoke osmotic demyelination syndrome. Up to now, the highest concentration we are aware of being used in human subjects is 10%[450].

8.2.4.4 Example: The Linear Correlation Metric Images for Frequency Difference EIT

The impedance spectra between malignant carcinomas and normal tissue are significantly different in freshly excised human breast tissue, which suggested that multi-frequency EIT has the potential to assist mammography in improving detection rate of breast cancer and reducing false-positive rate of standard screening methods[522]. In a clinical study, a linearized image reconstruction algorithm of multi-frequency EIT was proposed for computing the complex admittivity at six frequencies[140].

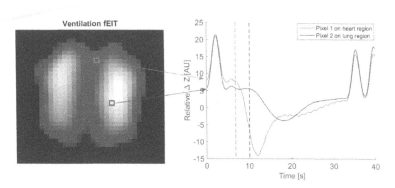

Figure 8.19   Illustration of regional impedance changes after injection of hypertonic saline solution. Left, ventilation fEIT. Right, regional impedance changes. Pixel 1 located the heart region and Pixel 2 the lung region. The region EIT waveforms on Pixel 1 and 2 showed a significant impedance decrease in the heart and lung region due to saline flow from heart to lung, with a noticeable time delay. (color figure available in eBook)

Then an "EIS plot" was created by plotting the imaginary component of the admittivity versus the real component, parameterized by frequency, for each voxel within the breast.

For condensing the information within this EIS plot for each voxel into a single figure of merit, a linear correlation metric (LCM) was introduced, which essentially is a nonlinear transformation that quantifies the resemblance of the EIS plot to a straight line[140]. In order to compute the LCM, the best least-squares fit for a linear relationship was computed between the conductivity σ and the absolute permittivity ε for each point in space for which we compute a reconstruction $\varepsilon_{lin} = a\sigma + b\mathbf{1}$, where σ is a vector composed of the reconstructed conductivities at all frequencies and $\mathbf{1}$ is a column vector of all 1. The LCM was computed at each point in space by applying a nonlinear transformation of the correlation between the reconstructed and linearly predicted permittivities, ε being a vector of the reconstructed permittivities:

$$\mathrm{LCM} = \cfrac{1}{1 - \cfrac{\varepsilon_{lin} \cdot \varepsilon}{\|\varepsilon_{lin}\|_2 \|\varepsilon\|_2}} \qquad (8.10)$$

LCM was displayed in the center layer of the reconstructed image alongside the central slice of the tomosynthesis reconstruction[140].

8.3   CLINICAL APPLICATIONS

8.3.1   USING EXISTING fEIT AND MEASURES

In this section we describe the use of fEIT and measures for clinical applications. To analyze EIT data, we may choose one or more already existing fEIT and measures. In order to do that, we need to understand the assumptions and limitations of each fEIT and measures. Otherwise, the data might be incorrectly interpreted. One good example is EIT measures used for PEEP optimization. Table 8.1 summarizes different EIT measures proposed for PEEP optimization. Most require offline analysis. If we do not understand the assumptions and limitations, wrong conclusion might be drawn (i.e. a suboptimal PEEP is selected) or extra work might be in vain.

For example, ITVD is widely used to assess the ventilation distribution with spontaneous breathing[121,1210] since the distribution changes during inspiration may reflect the diaphragm activities.

However, a so-called "ITVD index" was proposed to summarize the ITVD information by calculating the ratio of sum of 8 ITVD in dependent regions and sum of 8 ITVD in non-dependent regions[120]. Mathematically speaking, however, with all this extra calculation the ITVD index delivers the same results as a simple anterior-to-posterior ventilation ratio, since iso-volume slices were used.

$$\text{ITVD}_{\text{index}} = \frac{\sum \Delta Z_{\text{dependent}}}{\sum \Delta Z_{\text{non,dependent}}} \tag{8.11}$$

where $\Delta Z_{\text{dependent}} = (Z_8 - Z_7) + (Z_7 - Z_6) + (Z_6 - Z_5) + (Z_5 - Z_4) + (Z_4 - Z_3) + (Z_3 - Z_2) + (Z_2 - Z_1) + (Z_1 - Z_0) = Z_8 - Z_0$, and these extra calculation would thus not provide new information in the ITVD-index parameter.

8.3.2 RECOMMENDATIONS FOR DEVELOPMENT OF FEIT IMAGES AND MEASURES

n this section, we present recommendations on how the new fEIT and measures could be developed. When the existing fEIT and measures are not able to summarize the observations, we need to modify or further develop measures to describe the observations. The development must be based on the characters of the underlying observations. Typical way to develop measures can be divided into three steps:

1. Defining the problem (clinical needs).
   We do not develop measures because we can. We may design very beautiful mathematic models or apply complicated mathematic algorithms to EIT data analysis. However, all these would be meaningless if the parameters in the models do not have a clear physiological meaning. The clinical needs are the motivation of the measure development. For example, in early ages where EIT was just developed, clinicians wanted to evaluate ventilation distribution in ICU patients. Ventilation is inhomogeneously distributed in patients under mechanical ventilation, especially ARDS. Of course EIT can capture this because of heterogeneity, but the raw EIT images provide too much information that the clinicians were not able to use without further processing (32×32 pixels, ≈ 20 frames per second).
2. Understanding the problem.
   Researchers need to understand the problem defined in the first step. In the example of ventilation distribution, we have to understand what ventilation is (air in and out), how the lungs behave during mechanical ventilation, what features from the EIT data should be extracted (spatial, temporal), and how to extract (different mathematical methodologies).
3. Generating thoughts and evaluating them with data (simulation, phantom, subjects).
   Many thoughts may come out after the problem is analyzed and understood. However, only those measures should be used, which are evaluated with data and proven to be useful in a clinical settings. To assess ventilation distribution, fEIT and measures from simple ROIs to single indices (e.g. CoV and GI) were proposed. They were evaluated in different clinical settings[305,846,1212]. Evaluation can be done in simulation, phantom, animal or human subjects. Human subjects are the final validation of the measures. No matter how carefully we design the simulation and animal models, there are still chances that measures may fail the final test in a clinical environment. For example, a simple procedure and EIT measure was proposed to quantify pulmonary edema in lung injury[1052]. The design and performance of the animal model were adequate. However, the findings could not be replicated on patients later in a clinical settings[1207], indicated that animal models are still very different from human subjects. The differences between simulation and human subjects would be even bigger.

Here we discuss an example modifying the existing measures to suit our needs. A patient was mechanically ventilated under airway pressure release ventilation (APRV) mode. We observed that

Table 8.1

EIT measures for PEEP optimization.

Name	Calculation method	Optimal	Limitation
C_{reg} trend [213]	Quantify the decrease of current Creg compared to its maximum during PEEP titration.	Lowest decrease overall	1. In case of volume-controlled ventilation, ΔP changes at each PEEP and may not be available for calculation. 2. Calculation of C_{reg} assuming ΔP is the same for all pixels, which may not be valid in certain cases (in presence of pendelluft). 3. In case of limited number of PEEP steps (e.g. only 3–4 steps during titration), C_{reg} cannot be accurately depicted, which lead to inaccurate optimization.
CoV [299]	Height weighted pixel sum divided by pixel sum.	$\approx 50\%$	1. Lung areas are not identical in the upper and lower halves of the image, which means 50% might not be optimal in the case of uniform ventilation. 2. In case of pressure-controlled ventilation, tidal volume changes at each PEEP, so that CoV would not be a desired measure.
EELI trend [259]	Impedance levels at end-expiration	Start to decrease	1. A PEEP step may need to be long to observe an EELI decrease. 2. No good criteria to determine "decrease". Method proposed in literature may lead to different results when the baseline is selected differently.
GI [1208]	Sum of pixel differences from median over the sum of pixels within lung regions	lowest	Results strongly depend on the lung regions identification. In non-recruitable patients, collapsed lung regions cannot be identified which makes GI not suitable.
ITVD index [120]	Ventilation distributions in dependent and non dependent regions during inspiration were divided into 8 iso-volume slices. The index is the sum of ITVnon over the sum of ITVdep	≈ 1	1. The value can be arbitrarily large if the dependent region has a low value. The change of this ITV index may not be proportional to the change in the position of ventilation. 2. Lung areas are not identical in the upper and lower halves of the image.
ΔTV [705]	Differences between TV	decrease	1. Not clear when overdistension and collapse are present at the same time 2. May not be suitable in case of volume controlled, where tidal volume is constant.
RVD [742]	Delay of regional inspiration begin compared to global. Expressed in time or standard deviation.	Lowest PEEP with lowest RVD	Tested in experimental low-flow ventilation settings. During conventional mechanical ventilation where inspiration time is shorter and less sample points, noise level of the results is high and may not be reliable.

C_{reg}, regional dynamic respiratory system compliance; CoV, center of ventilation; ΔP, delta pressure; EELI, end-expiratory lung impedance; GI, global inhomogeneity index; ITV, intra-tidal variation; TV, tidal variation; RVD, regional ventilation delay.

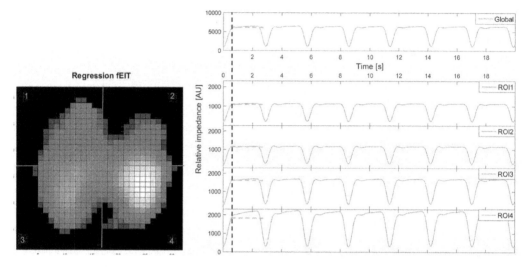

Figure 8.20 EIT measurement on a patient under airway pressure release ventilation. Impedance curves of global and regional variations are plotted against the time. This data was kindly provided by Mr. E. Teschner, with permission from Dräger Medical. (color figure available in eBook)

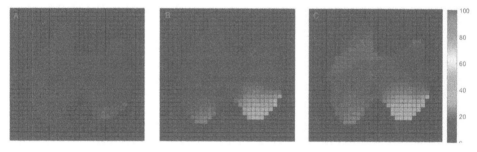

Figure 8.21 Comparison of traditional regional ventilation delay (RVD) and two modifications according to the patient data presented in figure 8.16. (A) RVD threshold 40% to maximum tidal variation. (B) starting point of inspiration reset to skip the rapid increase period. (C) RVD threshold 90%. Pixel value, percentage to full inhalation in time. (color figure available in eBook)

in left dorsal region (figure 8.20ROI4), although the tidal variation was larger than other regions, the opening of this region was delayed. In the impedance-time tracing, ROI4 was still filling up when the other regions had already stopped. However, due to the feature of APRV mode (rapid increase at the beginning of inspiration), the traditional RVD with 10% or 40% threshold would not work well (e.g. figure 8.21A with 40% threshold). In order to capture this observation, we may modify the RVD accordingly. The RVD is not functioning well because of the rapid increase, so the first thought would be to find a new starting point after the rapid increase, and apply RVD from the new starting point (figure 8.21B). This is a modification targeting the character of APRV. Another possible target would be the present observation. ROI4 needs more time to inhale, so the time for inhalation (e.g. to ≈90%) must be longer than other regions (figure 8.21C). Compared to the traditional RVC, both two modifications were able to characterize the observation and more sensitive to changes.

Table 8.2

List of fEIT parameters and measures

Name	Other Name	Usage & applications	Citation
CoV	Center of ventilation, center of gravity	Ventilation distribution	[299]
Creg trend (fEIT)	overdistension / atelectasis	PEEP titration	[213]
EELI trend (fEIT)	End-expiratory lung impedance	Change of end-expiratory lung volume	[259]
GI index	Global inhomogeneity index	Ventilation distribution	[1208]
ITVD	Intra-tidal ventilation distribution, ITV	Ventilation distribution within inspiration	[120]
Lung function fEIT	Named after various lung function parameters	Regional lung function with lung function test	[1092]
Opening & closing pressure (fEIT)	–	Pressure needed for regional recruitment (alveoli opening) and decrecruitment (alveoli collapse / closing).	[844]
Regression fEIT	Linear regression, filling capacity, polynomial fit	Ventilation distribution	[587]
RVD (fEIT)	Regional ventilation delay	Differences in regional inspiration starting time	[742]
SD fEIT	Standard deviation	Ventilation distribution	[403]
Silent spaces fEIT	Poorly ventilated regions	Ventilation distribution	[999]
Time constant (fEIT)	Regional expiration time constant	Regional obstruction	[533]
TV fEIT	Averaging tidal variation	Ventilation distribution	[1213]
Cardiac-related fEIT,	Pulsatility fEIT Cardiosynchronous fEIT, heart beat-related fEIT	Impedance distribution related to cardiac action and perfusion	[304]
Perfusion using contrast agents	None	Perfusion distribution	[131]
ARV & sROI	Average reconstructed value and size of region of interest	Intracranial local impedance changes	[220]
LCM	Linear correlation metric	Detect and localize breast tumors	[140]

8.4   LIST OF FEIT IMAGE AND MEASURES

The fEIT and measures are summarized in the following list with short explanations and references. The presentation is sorted in alphabetic order according to the common used names or abbreviations (See also the online supplements of[293]).

As described, fEIT and measures cover various aspect of information provided by EIT scans. More novel parameters are emerging to cover new application fields of EIT. As discussed in table 8.1, there are still room for improvement for many parameters. We believe that the journey to optimizing various fEIT and measures won't be smooth but promises to give novel insights into the tissue under EIT's investigation.

Section III

Applications

9 EIT for Measurement of Lung Function

Inéz Frerichs
Department of Anaesthesiology and Intensive Care Medicine,
University Medical Centre Schleswig-Holstein, Kiel, Germany

CONTENTS

9.1 INTRODUCTION

9.1.1 BASICS OF LUNG PHYSIOLOGY AND PATHOPHYSIOLOGY

The main physiological role of the lungs is to secure the exchange of respiratory gases, oxygen (O_2) and carbon dioxide (CO_2), between the body and the surrounding atmosphere. Lungs are the essential part of the respiratory system. They consist of air conducting airways, the respiratory tissue, where the transfer of O_2 and CO_2 between air and blood takes place, and the blood conducting vessels.

After the passage through the upper airways, the air enters the trachea, the partially cartilaginous tube starting below the vocal cords. From here the air is further conducted through the branching system of airways. Until the 16th branch of the airways, the so-called terminal bronchioles, no gas exchange between the inhaled gas and the body takes place. The main transport mechanism is represented by convection. Because of the ongoing branching, the overall cross-sectional area of the airways increases and consequently the airflow velocity falls on its way between the trachea and the terminal bronchioles during inspiration. Trachea, main, lobar and segmental bronchi are the locations with the highest resistance to airflow in the healthy subjects. Under pathological conditions,

DOI: 10.1201/9780429399886-9

the airway resistance may additionally increase by constriction of airway muscles, like in patients with asthma, or by swelling and increased secretions, like in bronchitis.

Once the air has passed further into lung periphery beyond the terminal bronchioles it enters the respiratory zone where respiratory gases are transported by diffusion. The respiratory zone consists of alveolar ducts and alveoli. The alveoli are miniature sacs surrounded by a network of the smallest blood vessels called capillaries. The walls of the alveoli and capillaries are extremely thin allowing an easy transport of O_2 from the alveolar gas into the capillary blood and of CO_2 in the opposite direction. Because of the ongoing pulmonary gas exchange, the composition of the exhaled gas differs from the inhaled (atmospheric) air – it contains less O_2 and more CO_2. In healthy subjects, the alveolar-capillary barrier is generally not a limiting factor for gas exchange but its thickening, caused for instance by accumulation of fluid as in lung edema, affects lung mechanics and also worsens the efficiency of respiratory gas transfer. This worsening affects mainly the uptake of O_2, which is the less soluble and diffusible gas of the two respiratory gases.

The transport of blood into the lung respiratory zone between the right heart ventricle and left atrium is called pulmonary perfusion. The circulation of blood in lungs is realized by heart action. The volume of blood entering the gas exchange zone depends on the stroke volume and the frequency of heart beat. It equals about 5 l/min during rest and rises to about 20 l/min during exercise in adults.

The repetitive process of inhalation and exhalation of air into and from the lungs is called pulmonary ventilation. Under physiological conditions, it is accomplished by the contraction of respiratory muscles. The inspiratory muscles (mainly the diaphragm and the external intercostal muscles) are responsible for the inspiration. During quiet breathing, expiration is passive. However, this situation changes during exercise, where the expiratory muscles (mainly abdominal and internal intercostal muscles) are activated to achieve proper exhalation of air. A healthy adult subject inhales about 500 l of air during each inspiration at a rate of about 12–16 breaths per min. Thus, minute ventilation is about 7 l/min. About two thirds of this inhaled volume enters the respiratory lung zone. Minute ventilation may massively increase during exercise to about 150 l/min or more.

Spontaneous ventilation is negative-pressure ventilation, meaning that inspiratory muscles actively increase the dimensions of the chest cavity during inspiration with the lungs passively following this dimensional change. Thereby, the pressure in the lungs falls below atmospheric pressure and air flows into the lungs. If spontaneous breathing is not present (for instance due to brain injury) or suppressed (for instance during anesthesia) then typically positive pressure ventilation is needed to secure adequate air transport into the alveoli. This can be accomplished by mouth-to-mouth resuscitation in a first-aid setting or by means of mechanical ventilators. In both cases, higher than atmospheric pressure is generated at the mouth of the patient or at an artificial airway opening, for instance when a tube has been inserted into the patient's trachea, to propel air into the lungs.

The distribution of inspired air in the lungs is not homogeneous. It is affected by anatomical structure and by gravity, meaning that it is posture-dependent. It further changes with age since it depends on lung tissue maturation and aging. It is also modulated by the type of ventilation – spontaneous breathing activity results in different ventilation distribution patterns than artificial ventilation. And, finally, it is affected by lung pathology. Moreover, since the lungs always contain a certain amount of air on top of which the tidal ventilation takes place, also this basic "resting" aeration of the lung tissue is heterogeneously distributed and affected by the mentioned factors.

9.1.2   LUNG-RELATED APPLICATIONS OF EIT

Since thoracic EIT is capable of detecting physiological and pathological changes of electrical properties of lung tissue at very high scan rates it allows the assessment of regional distribution of lung ventilation and perfusion as well as of changes in regional lung aeration and fluid content. Among all body organs, lungs represent the most frequent target organ for EIT application. As of September 2020, PubMed.gov lists over 900 EIT lung publications.

Lung-related applications of EIT have been dominated by its use focusing on lung ventilation and aeration and their regional changes, where clinical use in patients has already been successfully established. Especially critically ill patients of all age groups (neonatal, pediatric and adult) and undergoing ventilator therapy benefit from EIT examinations and monitoring. In these patients EIT is mainly used to determine the distribution of ventilation, changes in regional aeration and regional respiratory mechanics and their variation over time or with interventions. This information is then applied to guide the ventilator therapy by enabling personalised selection of most adequate ventilator settings and for early identification of adverse events associated with mechanical ventilation. To a lesser extent, chest EIT has been applied in the perioperative field as well, i.e. in patients undergoing surgery and also subjected to artificial ventilation. The indications of use in this latter group are similar to those in critically ill patients.

An increase in EIT use has also been noted in another group of patients, in patients suffering from chronic lung diseases and with preserved spontaneous breathing. This application of EIT has been documented mostly in smaller clinical studies with no routine use yet. The results of these studies imply a potential future role of EIT in the diagnostics, therapy and management of patients with such diseases as cystic fibrosis, chronic obstructive pulmonary disease (COPD) and asthma. The main application field of EIT in these patients has been proposed to be the monitoring of natural disease history and of therapy effects. This conclusion has been drawn based on EIT findings obtained from examining the regional lung function and using a variety of EIT indices. In addition, it has been postulated that EIT might be applied also in lung disease staging and phenotyping but no evidence has been provided with this respect yet.

The majority of both experimental and clinical research studies pertaining to lung-related applications of EIT have been conducted with the aim of establishing EIT as a medical imaging method in routine patient treatment and care. This process is still ongoing. However, it needs to be mentioned that a certain portion of chest EIT studies have been driven by other goals as well. The most relevant ones were the examinations of some still unknown phenomena in lung physiology and pathophysiology that were not (or only with severe limitations) accessible by other existing methods. The full non-invasiveness of EIT and its radiation-free measuring principle were the two decisive factors facilitating its use in this indication, especially in studies performed in both preterm and term neonates. Lung-related research has also been conducted for purposes of general EIT validation and it served goals like improvement of image reconstruction or technology development.

In summary, lung-related EIT research has been very diverse regarding its motivations, examined subject groups, results and clinical relevance. We have decided to group the information presented in the chapters 9 and 10 based on the type of examined ventilation: spontaneous and artificial. By using this approach we can focus on the two major clinical chest EIT applications in spontaneously breathing patients, suffering from chronic lung diseases, and in mechanically ventilated patients. In addition, we can cover relevant physiological studies within the first section dealing with EIT use during spontaneous breathing and address advanced tools for analysis of EIT data in the subsequent section on EIT use during mechanical ventilation.

9.2   EIT EXAMINATIONS DURING SPONTANEOUS QUIET TIDAL BREATHING

EIT examinations carried out during quiet breathing allow the analysis of ventilation distribution in an otherwise unaltered setting without the need of any cooperation of the studied subjects. Interventions, like posture changes, administration of drugs, physical therapy, etc. can additionally be performed, allowing the evaluation of changes in regional ventilation and aeration in response to these events.

9.2.1 ANALYSIS OF EIT DATA ACQUIRED DURING SPONTANEOUS QUIET TIDAL BREATHING

Ventilation distribution during quiet tidal breathing is typically evaluated from the continuously acquired EIT data by calculating regional tidal volumes (V_T) from the pixel EIT waveforms. To calculate regional V_T, the tidal variation of the EIT signal (i.e. the difference between the end-inspiratory peak and the preceding (or the following) end-expiratory trough values) is determined in each pixel. The corresponding values can be plotted as colour-coded functional EIT images either from single breaths to visualise the ventilation distribution in the examined chest section during one inflation (or deflation) or the values from multiple breaths are averaged to show the mean ventilation distribution over a period of time. Examples of the latter functional images acquired over a period of one minute in a healthy spontaneously breathing adult subject are shown in figure 9.1.

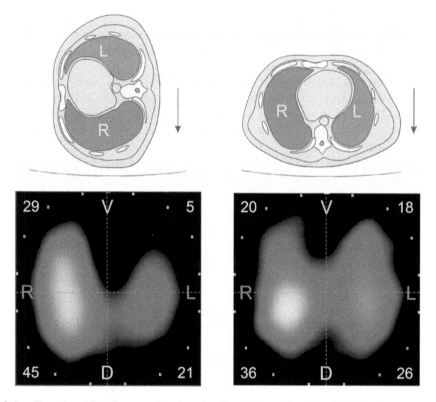

Figure 9.1 Functional EIT images showing the distribution of regional tidal volumes in a healthy adult 32-year old man during quiet breathing in the right lateral and supine body positions. The letters V (ventral), D (dorsal), R (right) and L (left) characterise the image orientation; the numbers in the image corners represent the percentages of ventilation in each image quadrant. Schematic chest drawings at the top show the respective body orientation in relation to the gravity vector (vertical arrows). (color figure available in eBook)

The calculation of tidal impedance variation is the most straightforward and nowadays the most widespread approach to assess ventilation distribution. In the past, the standard deviation of pixel EIT values acquired during tidal breathing was often used to generate functional EIT images showing the distribution of ventilation[300]. Since the scan rate and signal quality of early EIT devices was lower than at present, the identification of tidal end-inspiratory and end-expiratory values was less reliable than with modern devices. The calculation of standard deviation was a pragmatic solution as to how ventilation-related signal variation could be assessed; however, non-ventilation-related signal variation caused by physiological events like heart action and perfusion or noise impacted the information.

Another less-frequent approach to assessing the ventilation distribution is based on the calculation of the slope of pixel EIT values plotted against the sum (or average) of EIT values calculated from of all pixel values in the chest cross-section during tidal ventilation[587,1213]. The individual slope values can be imaged in the respective pixels, giving rise to yet another type of functional image representing regional ventilation. The "pixel vs. global" plots of EIT data can be fitted not only to the linear function but also to the polynomial function of the 2nd order and the curvature of the fitted curve used to characterise the time course of regional filling of lung regions with air during inspiration or emptying during expiration[298].

The characteristics of the most frequent functional images of regional ventilation during tidal breathing and their advantages and limitations have been described in[1213]. Detailed description of all different approaches in quantifying tidal ventilation distribution can be found in the Consensus statement on chest EIT[293]. See also chapter 8.

To characterise the homogeneity of tidal ventilation distribution, different approaches are used that are described in the following text. Their common feature is that they are all based on the calculated pixel values of the mentioned functional EIT measures, mostly of tidal impedance variation.

The values of tidal impedance variation may be summed up in regions-of-interest (ROI) and their proportion on the global ventilation in the chest cross-section determined. The simplest method is to split the image into halves along either the vertical or horizontal axes to describe the ventilation distribution between the right and left or the ventral (near the chest front) and dorsal (near the chest back) lung regions. Figure 9.1 shows another example: the distribution of ventilation in a healthy adult subject is characterised by calculating the percentage of ventilation in image quadrants. Image layers spanning the image in the ventrodorsal direction (from the front to back) represent other frequently used ROIs. An alternative approach is that the ROIs are not defined in the whole image but only within the lung regions.

Another way in which tidal ventilation distribution is described is by using simple one-number measures of ventilation inhomogeneity. The most frequently used ones are the global inhomogeneity index[309,1208] and the coefficient of variation[1090,1091], typically calculated from pixel values of tidal impedance variation. These are robust numerical measures that have been shown to detect ventilation inhomogeneity and assess the degree of airway obstruction in a simulation study[945]. They provide the information on the overall degree of ventilation inhomogeneity without any information how the inhomogeneity is distributed in the chest cross-section.

Another one-number measure of ventilation distribution, the centre of ventilation, offers this directional information[308,986]. The centre of ventilation measure is the projection of the ventilation "mass" on the ventrodorsal or right-to-left axis and its location is quantified in relation to the chest diameter. It was first introduced to describe the ventilation distribution during spontaneous tidal breathing before and after surgery and its difference in comparison to mechanical ventilation[299].

Although most of the EIT studies conducted during tidal breathing focused on the spatial heterogeneity of ventilation and the evaluation was based only on calculating the amplitude of the tidal impedance variation, a few studies also analysed the temporal heterogeneity of ventilation. The occurrence of posture-dependent phase shifts in regional filling with air between the right and left lung regions was identified by EIT in healthy subjects[401] and higher temporal heterogeneity in regional lung filling determined in patients with COPD compared with a healthy cohort[717].

9.2.2   FINDINGS OF EIT STUDIES PERFORMED IN HUMAN SUBJECTS DURING QUIET TIDAL BREATHING

EIT studies reporting findings obtained during quiet tidal breathing have been conducted either in subjects during only this type of ventilation or when the spontaneous tidal breathing was analysed along with the EIT data obtained during other types of breathing activity, like intentional ventilation manoeuvres (reported in §9.3), or artificial ventilation. EIT studies have been carried out in both spontaneously breathing healthy subjects and in patients. The general aims were to obtain information on the still unknown or not well described aspects of lung physiology, to determine the regional changes in tidal lung ventilation in response to various interventions and to characterise the differences in regional lung ventilation between healthy and diseased lungs.

Spontaneously breathing subjects of all ages, including neonates, infants and children, can be examined by EIT during quiet tidal breathing. Unlike EIT that is easy-to-use and not harmful, several established medical techniques cannot be used in otherwise healthy neonates and children for ethical reasons. Therefore, only limited knowledge on regional lung ventilation was available in this age group. Moreover, this knowledge was often based on studies carried out, not in healthy, but in sick patients who were frequently sedated and face masks were used. These effects confounded the results because they modified the natural ventilation distribution pattern. The characteristics of natural breathing could be followed on the regional level only thanks to the use of EIT.

EIT was applied in newborns to describe the spatial pattern of progressive lung aeration development immediately after birth, unravelling an unknown physiological role of crying[1042]. The natural breathing pattern of preterm and term neonates is characterised by frequent sighs that help the neonates preserve their lung volumes. Thanks to EIT, it was determined that the ventilation distribution during sighs differs from how air is distributed in the neonatal lungs during tidal breaths[306,940].

It has been known for a long time that gravity affects the distribution of aeration and ventilation in the lungs. Early physiological studies using radioactive tracer gases confirmed the existence of spatial heterogeneity of air distribution in the lungs in adults. EIT reproduced these findings and additionally showed that this heterogeneity is reduced during weightlessness[298] and increased by hypergravity[130,298].

An easy way how the effect of gravity can be studied without exposing humans to parabolic and space flights and centrifuges is to change the posture of the studied subjects and thereby the orientation of the chest in relation to the gravity vector. In healthy adult spontaneously-breathing subjects, the dependent lung regions are better ventilated than the non-dependent ones. For instance, when a subject is lying on the right side the right lung receives a higher portion of the inspired air than the left one (figure 9.1, left panel). If the subject lies on the back, the dorsal lung regions are preferentially ventilated. Several EIT studies have confirmed the ability of EIT to detect this gravity-dependent heterogeneity of ventilation by examining the subjects in upright (standing, sitting), tilted and lying (supine, prone, left and right lateral) positions[300,870,884,961].

The typical adult ventilation-distribution pattern, favouring the dependent lung regions, reverses with aging due to the continuously changing respiratory system mechanics mainly caused by reduced lung tissue elasticity. EIT has detected these changes in ventilation distribution in the elderly in different postures[297]. The effect of artificially modified chest wall mechanics on ventilation distribution induced by thoracic and abdominal strapping has also been captured by EIT[845] and the posture-dependency of this effect demonstrated.

Numerous   EIT   studies   have   been   carried   out   in   spontaneously   breathing neonates[306,457,490,690,825,883,940,1068] and children[659], examined in different body positions. These studies provided new insights on how posture affects air distribution in immature lungs and how this distribution changes with body growth. The respiratory system mechanics of preterm and term neonates differs from the adults mainly because the alveolisation of the lung tissue is not completed at birth and the chest is highly compliant. Therefore, the spatial ventilation distribution pattern at this early stage of life is not identical with the adult one. It is also characterised by higher variability caused be less regular breathing rate and V_T than in adults. The distribution of air in the lungs may change rapidly from one spontaneous breath to another. In neonates even the head position exerts a small effect on ventilation distribution in the lungs as determined by EIT[457] (see figure 1.3).

At the end of this section a few further studies are summarized, where EIT examinations were conducted in spontaneously breathing patients suffering from various diseases, affecting mainly the lungs and, to a lesser extent, other intrathoracic organs.

A novel approach of using EIT during quiet tidal breathing was proposed in patients suffering from pulmonary or pleural carcinoma during the preoperative assessment of their lung function[961,962]. EIT provided comparable findings to the conventionally used radionuclide scanning. Adult patients suffering from community acquired pneumonia were examined by EIT upon their admission to hospital along with chest radiography and followed during their stay in the course of antibiotic therapy[537]. Good correlation between the ventilation asymmetry determined by EIT on admission and the radiographic findings was found. EIT identified the increasing homogeneity of ventilation in the later course of the disease. The effect of two breathing aids on regional ventilation distribution was examined in spontaneously breathing adult patients with cystic fibrosis[1139]. Continuous positive airway pressure increased the gravity-dependent spatial ventilation inhomogeneity whereas positive expiratory pressure therapy neutralised this effect. Spontaneously breathing children with ventricular septum defect were examined by EIT before and after cardiac surgery[939]. The analysis of regional lung ventilation showed higher ventilation of ventral lung regions after surgery but the main finding was first of all the reduced lung perfusion after the repair of the shunt.

9.3 EIT EXAMINATIONS DURING VENTILATION MANOEUVRES AND PULMONARY FUNCTION TESTING

Various ventilation manoeuvres can be performed by humans by intentional modification of the quiet breathing pattern. These can be achieved by changing the deepness of breaths, the airflow rate, the level of air volume in lungs or breath-hold. The analysis of EIT data obtained under such ventilatory conditions offers the possibility of obtaining deeper insights into local lung function in healthy subjects or to better analyse the regional functional deficits associated with lung diseases and the degree of their improvement with therapy or their deterioration with disease progression.

The ability of humans to perform specific voluntary ventilation manoeuvres is clinically used during conventional pulmonary function testing using spirometry, and the same manoeuvres can also be analysed by EIT. A typical manoeuvre carried out by the subjects during this examination is the forced full expiration manoeuvre. It is preceded by full inspiration to total lung capacity after which the subject is asked to exhale as quickly and as deeply as possible. Figure 9.2 shows the typical lung volume vs. time waveform (left panel) during this manoeuvre and some of the volumetric measures obtained from this examination. At the same time, peak and maximum flow rates at defined time points are determined (figure 9.2, right panel).

EIT examinations can be continuously carried out in subjects performing the described forced full ventilation manoeuvres either alone or in parallel with spirometry. Impedance changes determined by EIT correlate with the volume changes measured by spirometry both during forced expiration manoeuvres and tidal breathing[775]. The advantage of EIT over spirometry is that it is capable of assessing lung function not only on the global but also regional level. Figure 9.3 shows two representative EIT waveforms originating from two separate image pixels acquired in a healthy subject performing the forced ventilation manoeuvre. The volumetric changes during the full expiration manoeuvre were higher than during tidal breaths and resulted in higher impedance changes (figure 9.3,

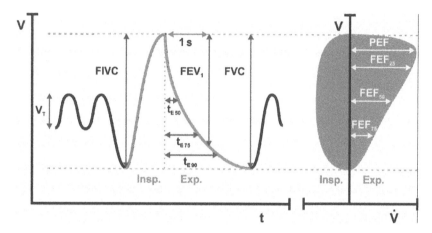

Figure 9.2   Schematic drawing of typical waveforms obtained during conventional pulmonary function testing with some highlighted pulmonary function measures derived from this examination. The volume (V) vs time (t) waveform on the left shows the lung volume changes during quiet tidal breathing followed by the forced full inspiration (Insp.) (red) and expiration (Exp.) manoeuvre (blue). Some of the most frequently derived measures are: tidal volume (V_T), forced inspiratory vital capacity (FIVC), forced expiratory volume in 1 s (FEV_1) and forced expiratory vital capacity (FVC). Expiration times needed to exhale 50% ($t_{E,50}$), 75% ($t_{E,75}$) and 90% ($t_{E,90}$) of FVC are less frequently analysed during conventional spirometric testing. These measures are introduced here due to their relevance for regional pulmonary function testing by EIT mentioned later in this chapter. The diagram on the right shows the volume vs. flow ($\dot{V}$) waveform obtained from the forced full inspiration (red) and expiration (blue) manoeuvre with the following frequently determined measures of peak expiratory flow (PEF) and forced expiratory flows at the times when 25% (FEF_{25}), 50% (FEF_{50}) and 75% (FEF_{75}) of FVC have been exhaled. (color figure available in eBook)

the top two functional images). The regional waveforms exhibited small but clearly discernible differences even in this healthy subject. Such minor differences become far more pronounced in patients suffering from lung diseases that heterogeneously affect regional lung function.

The forced full expiration manoeuvres were the most frequently used voluntary manoeuvres examined in EIT studies[90,296,297,298,309,585,614,752,775,870,1088,1089,1091,1092,1202,1206,1209]. This is natural because this widely accepted standard manoeuvre is routinely used worldwide for diagnostic purposes in patients with lung diseases. A few studies utilized the capacity of EIT to capture the regional lung characteristics during other voluntary forms of non-tidal breathing, like singing[1050], maximum voluntary ventilation[307] and breathing at very low and high volumes[297,942,1139]. EIT examinations were also reported in human subjects during breath-holding[378,746,772]. The primary purpose of carrying out this voluntary manoeuvre was to better visualise the heart beat-related impedance changes in the cardiac and lung regions associated with heart action and pulmonary perfusion.

9.3.1   ANALYSIS OF EIT DATA ACQUIRED DURING VENTILATION MANOEUVRES AND PULMONARY FUNCTION TESTING

Similar to the evaluation of EIT data acquired during quiet tidal breathing (9.2.1) also the data obtained during ventilation manoeuvres are first of all analysed based on the amplitudes of the EIT signal variation in individual image pixels that are associated with the manoeuvre.

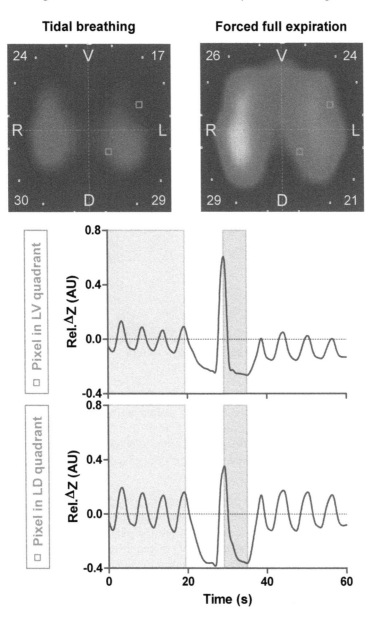

Figure 9.3   Functional EIT images showing the distribution of regional tidal volumes (top left) and forced vital capacities (top right) in a healthy adult 32-year old man during quiet breathing and forced full expiration manoeuvre performed in the sitting position. The respective measurement periods from which these images were generated are highlighted by green rectangles in two representative regional EIT waveforms at the figure bottom. The waveforms show regional variation of the EIT signal representing normalised impedance differences (or relative impedance changes, rel. ΔZ) in arbitrary units (AU) in two image pixels located in the left ventral (LV) and left dorsal (LD) image quadrants. (The exact locations of the pixels are highlighted by small green squares in both images.) The letters V (ventral), D (dorsal), R (right) and L (left) characterize the image orientation; the numbers in the image corners represent the percentages of ventilation in each image quadrant. (color figure available in eBook)

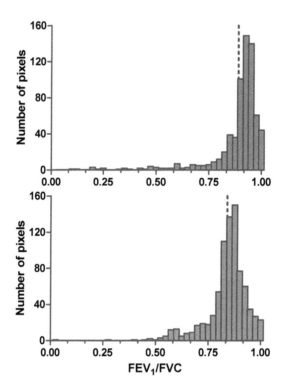

Figure 9.4   Frequency distributions of pixel values of forced expiratory volume in 1 s (FEV$_1$) to forced vital capacity (FVC) ratios obtained from EIT examinations in a 10-year old healthy child (top) and an 11-year old child suffering from cystic fibrosis (bottom). The histogram of the child with cystic fibrosis exhibits less homogeneous distribution of pixel FEV$_1$/FVC values with a shift toward lower values. (The dashed lines show the centre values.) (color figure available in eBook)

For instance, if a subject is examined by EIT during the forced full inspiration and expiration manoeuvre then all the well-known volumetric parameters established in global spirometric lung function testing (figure 9.2) can also be calculated from the pixel EIT waveforms. Some of the most common parameters are calculated as follows: 1) forced inspiratory vital capacity (FIVC) as the amplitude of impedance change between the trough value after full expiration to residual lung volume and the peak value after the subsequent forced full inspiration to total lung capacity, 2) forced expiratory volume in 1 second (FEV$_1$) as the difference between the impedance value at total lung capacity at the time point directly preceding the onset of expiration and the value reached after 1 second of forced exhalation and 3) forced expiratory vital capacity (FVC) as the amplitude of the impedance variation between the peak value after full inspiration to total lung capacity and the trough value after the following forced full expiration to residual lung volume. Forced expiratory volume in 0.5 second (FEV$_{0.5}$) is often used in younger children instead of FEV$_1$, because children can quickly exhale air during this manoeuvre and reach their residual lung volume within less than 1 second. Many other parameters can be calculated from the forced full inspiration and expiration manoeuvre that are not described here because they are less frequently used both in conventional spirometry and in EIT examinations.

Another parameter often used in conventional spirometry is the ratio between FEV$_1$ and FVC (or FEV$_{0.5}$ and FVC in children). This parameter is utilized in staging of disease severity, for instance in patients with COPD. FEV$_1$/FVC (or FEV$_{0.5}$/FVC) can be determined from EIT waveforms in individual image pixels.

To characterise the spatial distribution of volumetric parameters, like FEV_1 or FVC, similar approaches to those applied in the assessment of regional V_T distribution (see 9.2.1) are used. These comprise the summation of the pixel values in various ROIs and calculation of one-number measures of ventilation heterogeneity, like the global inhomogeneity index or coefficient of variation. The parameters calculated as ratios of two amplitude values, like FEV_1/FVC, have frequently been presented as histograms, showing the frequency distribution of the pixel values of the corresponding parameter (figure 9.4).

Several conventional spirometric lung function parameters measure the maximum air flow rates reached by the examined subject at predefined relative expiratory (figure 9.2, right panel) but also inspiratory volumes. Examples of such parameters are the peak flow rate (PEF) or the forced expiratory flows after the exhalation of 25%, 50%, and 75% of FVC (FEF_{25}, FEF_{50}, FEF_{75}). Regional flow-volume loops, similar to the global loop shown in figure 9.2, can be generated by EIT from the forced inspiration-expiration manoeuvre[774,1206] and the mentioned flow values calculated on the regional level.

Finally, time-based EIT parameters can be derived from the ventilation manoeuvres. For instance, regional expiration times needed to exhale predefined percentages of FVC can be calculated from the pixel waveforms recorded during the forced full expiration as explained in the global spirometric waveform in figure 9.2. These parameters have been shown to be very sensitive to temporal inhomogeneities in regional lung emptying. Typically, the pixel expiration times have been presented in form of histograms in EIT articles (figure 9.5). These histograms allow the assessment of delayed emptying (the values are generally shifted to longer expiration times) and high temporal heterogeneity (the values exhibit a broad instead of a narrow distribution peak).

9.3.2 FINDINGS OF EIT STUDIES PERFORMED IN HUMAN SUBJECTS DURING VENTILA-TION MANOEUVRES AND PULMONARY FUNCTION TESTING

Clinical studies, during which human subjects were examined by EIT while performing ventilation manoeuvres, were mostly conducted in patients suffering from chronic lung diseases but occasionally also in healthy subjects. The examinations of healthy subjects often provided reference control data allowing better assessment of the impact of lung disease on regional lung function measures that were determined in patients. Occasionally, studies involving only healthy volunteers were carried out with the aim of describing the differences between tidal and non-tidal forms of ventilation, as well as of the effects of gravity, posture or EIT examination plane[298,584,870]. An interesting finding was obtained in an EIT study performed in an allegedly lung healthy patient population presenting no pulmonary symptoms[1088]. EIT was able to show significantly higher spatial and temporal heterogeneity of regional forced full expiration in past and current smokers compared with non-smokers. The most sensitive EIT parameters were the regional FEV_1/FVC values and the expiration times need to exhale 75% and 90% of FVC. A study performed in healthy school children provided reference EIT data for future studies in pediatric patients[1089]. It also demonstrated that exercise did not significantly affect the spatial and temporal distribution during the forced full expiration.

Patients with a history of lung disease and examined by EIT during the forced ventilation manoeuvre mostly suffered from COPD[604,1091,1092], asthma[309,774] and cystic fibrosis[585,614,752,1089,1206,1209]. These diseases are associated with structural and functional lung tissue changes that increase the spatial and temporal ventilation heterogeneity. EIT examinations were generally very sensitive in identifying this increased heterogeneity of ventilation during the forced ventilation manoeuvre using a variety of volumetric, flow and time based measures described in §9.3.1. Example findings comparing the distribution of regional FEV_1/FVC values in a child with cystic fibrosis with a healthy child of a similar age confirmed the higher heterogeneity and generally

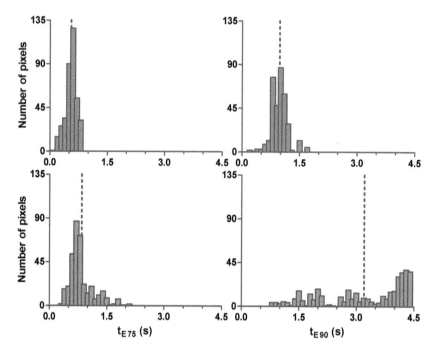

Figure 9.5   Frequency distributions of pixel values of expiration times needed to exhale 75% ($t_{E,75}$) and 90% ($t_{E,90}$) of forced vital capacities (FVC) obtained from EIT examinations in a 10-year old healthy child (top) and an 11-year old child suffering from cystic fibrosis (bottom). The histograms of the child with cystic fibrosis exhibit less homogeneous distributions of pixel $t_{E,75}$ and $t_{E,90}$ values with rightward shifts toward longer expiration times. (The dashed lines show the centre values.) (color figure available in eBook)

lower FEV_1/FVC values in the sick child (figure 9.4). The regional expiration times required to exhale 75% and 90% of regional FVC determined in the same child with cystic fibrosis during forced full expiration (figure 9.5) highlighted how heterogeneous and slow regional lung emptying was in this child when compared with a healthy child.

A few EIT studies have examined the reversibility of airway obstruction in patients with obstructive lung diseases. A bolus of inhaled broncholytic agent was applied in adult and pediatric patients with asthma[309,774] and adult patients with COPD[1092]. The data was obtained before and after the administration. Reduced spatial and temporal heterogeneity of EIT measures could be confirmed in responders.

9.4 SUMMARY

Many years passed since the first publication of a chest EIT image in 1985[159]. This image was obtained during an examination of a healthy spontaneously breathing man who performed a voluntary deep inspiration. It visualised the ventilated right and left lung regions as areas of increased electrical resistivity. Numerous EIT studies have been performed in both healthy subjects and patients with preserved spontaneous breathing of all ages since then. The study participants were examined during spontaneous tidal breathing as well as while performing different voluntary ventilation manoeuvres. Important physiological and pathophysiological knowledge on the spatial and temporal ventilation distribution in the healthy and diseased lungs was obtained. The regional effects of lung maturation and aging or posture could be determined. A multitude of EIT measures, both simple and sophisticated ones, was used to characterise regional lung function. These measures enabled the detection of the deteriorated regional lung function in patients and the clear discrimination from the normal lung function in healthy subjects. This is a promising feature of chest EIT examinations because it highlights the diagnostic and monitoring potential of the method in future clinical use in spontaneously breathing patients suffering from chronic lung diseases like COPD, asthma and cystic fibrosis. The socio-economical and medical impact of these frequent diseases is immense. If EIT can be established in routine care of these patients for their individual and general health-care benefit then this would be a great achievement. Some of the major tasks ahead are 1) large validation studies comparing EIT with other medical methods assessing ventilation inhomogeneity, like multiple-breath inert gas washout or forced oscillation technique, 2) standardization of the EIT examination, data analysis and reporting of findings in these specific patient populations, 3) generation of reference values based on large populations of healthy subjects, 4) long-term follow-up studies confirming the sensitivity and specificity of EIT in detecting lung function deteriorations, 5) development of wearable EIT instruments suitable for remote patient monitoring (www.welmo-project.eu), and 6) integration of EIT findings into existing diagnostic, monitoring and therapeutic pathways.

10 EIT for Monitoring of Ventilation

Tobias Becher
University Medical Centre Schleswig-Holstein, Kiel, Germany

CONTENTS

10.1 INTRODUCTION

Patients suffering from acute respiratory failure are frequently treated with positive pressure ventilation in interdisciplinary intensive care units (ICUs). Due to the severity of their illness and the effect of sedative medications that may be necessary during the course of mechanical ventilation, these critically ill patients are often unable to communicate with their caregivers. Moreover, the lungs of critically ill patients may suffer further damage due to the potentially detrimental effects of ventilator-induced lung injury (VILI,[981] and sometimes patient self-inflicted lung injury (P-SILI,[156]). Therefore, precise monitoring is essential for early detection of changes in the patient's condition as well as for individual adaptation of care. Conventional monitoring of ventilation in non-ventilated ICU patients comprises continuous monitoring of oxygen saturation with pulse oximetry and monitoring of respiratory rate by thoracic impedance measurements through electrocardiography (ECG) electrodes. In mechanically ventilated patients, the above-mentioned conventional monitoring parameters are commonly complemented by monitoring of airway pressure, tidal volume and expired carbon dioxide. These parameters provide important information about the effectiveness and invasiveness of ventilation on a global scale.

The lungs of patients with respiratory failure, however, are frequently characterized by substantial local inhomogeneities of ventilation distribution which are caused by variable distributions of atelectasis (partial or complete), normally aerated lung and overdistended areas. Regional overdistension and cyclic opening and closing of atelectatic lung areas are potentially dangerous phenomena as they may aggravate the pre-existing lung injury. Since all conventional monitoring methods for ventilation rely on global measures of respiratory mechanics and gas exchange, they are not suitable for identifying regional inhomogeneities and for adjusting ventilator parameters accordingly.

In the recent years, chest EIT has received increasing attention as a non-invasive tool for continuous visualization of ventilation distribution and for assessment of ventilation-induced changes in lung volume. EIT provides reliable bedside information on ventilation distribution[301,1169] and its homogeneity[1212]. Moreover, when used in conjunction with specific ventilator maneuvers and information derived from the ventilator, EIT can identify regional overdistension, atelectasis formation and cyclic opening and closing of lung tissue at the bedside[213,742,1217].

In the following, we will discuss different EIT measures that can be used to assess the lungs of critically ill and mechanically ventilated patients at the bedside. Furthermore, we will describe potential strategies for prospective individualization and, ideally, optimization of ventilator settings using these EIT parameters.

10.2   ASSESSMENT OF VENTILATION DISTRIBUTION WITH EIT

Changes in electrical bioimpedance during respiration are closely correlated to changes in pulmonary air content[301]. With current EIT devices, ventilation distribution is commonly assessed based on tidal images displaying the pixel impedance differences between end-expiration and end-inspiration. This information can be displayed graphically as tidal images, which are the basis for some by now well-established EIT parameters describing ventilation distribution in a quantitative way.

By dividing the anteroposterior and the right-to-left chest diameter in their respective middles into one ventral and one dorsal as well as a right and a left half, the ventral, dorsal, right and left-sided percentages of ventilation can be expressed. These measures are relatively inaccurate but provide intuitive information for treating clinicians, which is why they have been used in several clinical studies[693,695] and, among other parameters, have been considered as generally "useful" in a survey on clinical usefulness of different EIT parameters[294]. An example for adjustment of ventilator settings using the dorsal percentage of ventilation is given in figure 10.1.

The center of gravity of ventilation distribution is another way of expressing anteroposterior or right-to-left ventilation distribution. It has a value between 0 and 100% and provides more accurate information on ventilation distribution than the dorsal / ventral or right / left percentages of ventilation but may be less intuitive to clinicians at the bedside[293,299].

A

PEEP = 10 mbar

B

PEEP = 16 mbar

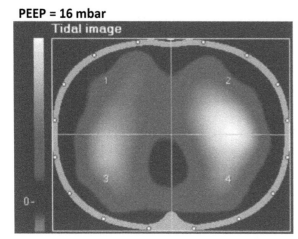

Figure 10.1   Effect of positive end-expiratory pressure (PEEP) on percentages of ventilation in 4 regions of interest (ROI). At a PEEP level of 10 mbar (Panel A), dorsal percentage of ventilation was 43% (12% ROI 3 + 31% ROI 4). After increasing the PEEP level to 16 mbar in 2 consecutive steps, dorsal percentage of ventilation increased to 50% (23% ROI 3 + 27% ROI 4). (color figure available in eBook)

10.3   MEASURES OF VENTILATION INHOMOGENEITY

In addition to the above-mentioned measures of ventilation distribution, tidal images can be used to quantify the homogeneity of ventilation distribution using measures like the global inhomogeneity index (GI index) or the coefficient of variation (CV)[96,1212]. In general, higher values of these parameters denote greater degrees of inhomogeneity in ventilation distribution, as assessed by the tidal image that is the basis for this calculation.

For a meaningful interpretation of inhomogeneity and changes in inhomogeneity using the afore-mentioned indices, it is necessary to define the lung area within the tidal image. Failure to do so will result in very high values of GI index and CV that are not necessarily caused by a high degree of ventilation inhomogeneity but instead by the differences between the ventilated lung pixels and other intra- and extra-thoracic pixels that do not represent the ventilated lung but are still part of the tidal image, although an improved "anatomical" GA metric (GA_{anat}) has recently been proposed[1182].

Various methods for determining the lung area have been proposed. Functional regions of interest (fROIs) define the lung area by analyzing the functional behaviour of pixels during ventilation. For example, an fROI of 10% of maximum tidal impedance change defines all pixels as "lung area" which exhibit a ventilation-related impedance change of more than 10% of the maximum ventilation-related impedance change within the image. This definition automatically excludes all non-ventilated pixel from the lung area, which is desired for extrapulmonary regions but not desired for pixels belonging to the actual lung area that are not ventilated because of atelectasis or severe overdistension. By automatically excluding these non-ventilated lung areas from the analysis, fROIs may lead to an underestimation of actual ventilation inhomogeneity. To overcome this limitation, Zhao *et al* introduced the lung area estimation method, which is based on the assumption that lung dimensions should be relatively symmetric between the right and left lung in the EIT electrode plane[1212]. Obviously, this is only a rough approximation that may lead to an underestimation of the actual lung area, especially in patients with bilateral atelectasis. In the future, individual anatomical regions of interest based on computed tomography (CT) scans may help to detect the actual lung area before EIT data analysis. Until then, measures of ventilation inhomogeneity must be interpreted with extreme caution, bearing in mind the possibility of misinterpretations due to the dependency of GI index and CV on the regions of interest used for these analyses.

10.4   INTRATIDAL VENTILATION INHOMOGENEITY AND ALVEOLAR CYCLING

When calculated from tidal images, GI index and CV describe the homogeneity of ventilation distribution at the time point of end-inspiration. Lung areas that are fully inflated at the time of end-inspiration will contribute to lower values of these indices, even if they are subject to alveolar cycling during inspiration. Therefore, it is important to obtain information on homogeneity of ventilation distribution not only during the comparatively static time point of end-inspiration but also during the course of inspiration, when alveolar cycling takes place. The GI index can be used to assess intratidal inhomogeneity during a slow inflation maneuver[1211]. If the fully inflated lung region (based on EIT data obtained at the end of the slow inflation maneuver) is used as region of interest for calculation of GI index, relatively high values can be observed at the beginning of the maneuver, when a low airway pressure leads to a partial inflation of the lungs. During the maneuver, the increasing airway pressure leads to increased lung inflation. This translates into a progressive decrease of GI index from the initially high values to lower values that can be observed at end-inspiration.

The standard deviation of regional ventilation delay (SD_{RVD}),[742] assessed during a slow inflation maneuver is another measure for intratidal ventilation inhomogeneity. It describes the temporal inhomogeneity of lung filling and is closely correlated to alveolar cycling. Higher values of SD_{RVD} can be found when a large proportion of alveoli are closed at end-expiration and opened at end-inspiration. Of note, the close relationship between SD_{RVD} and alveolar cycling was found only when analyzing a slow-inflation maneuver, during which the lung was slowly inflated with reduced air flow up to an inspiratory tidal volume of 12 ml/kg. During such a maneuver, there is a slow and progressive increase in airway pressure throughout the course of lung inflation. Lung units that are collapsed at end-expiration will typically require a higher airway pressure for opening, which results in delayed inflation during such a maneuver and hence higher values of SD_{RVD}. It is questionable whether SD_{RVD}, when calculated during conventional mechanical ventilation with rapid increases in airway pressure during inspiration will exhibit any correlation with alveolar cycling, because the rapid increases in airway pressure, especially during pressure-controlled ventilation, may lead to rapid opening of previously collapsed lung units during inspiration, masking any heterogeneity in regional opening pressures that could be detected with SD_{RVD}. Therefore, we recommend using a slow inflation maneuver for calculation of SD_{RVD} because the results during ongoing mechanical ventilation may be unreliable for assessment of alveolar cycling. Figure 10.2 presents a patient example of SD_{RVD} before after raising the PEEP setting on the ventilator from 8 to 14 mbar.

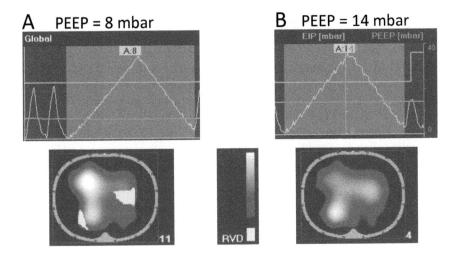

Figure 10.2   Standard deviation of regional ventilation delay (SDRVD) before (A) and after adjusting positive end-expiratory pressure (PEEP) from a level of 8 (A) to 14 (B) mbar. Above: global impedance time curve during a slow inflation maneuver, starting from PEEP to an inspiratory tidal volume of 12 ml per kg predicted body weight. Below: Tidal images obtained from slow inflation maneuvers. Areas with regional ventilation delay above 12% of global inflation time are highlighted. SDRVD was 11% with PEEP 8 mbar and 4% with PEEP 14 mbar. (color figure available in eBook)

10.5   IDENTIFICATION OF OVERDISTENSION AND ALVEOLAR COLLAPSE AT THE BEDSIDE USING REGIONAL COMPLIANCE ESTIMATION

In 2009, Costa and coworkers published a technical note presenting a novel EIT-based method for estimating recruitable alveolar collapse and overdistension at the bedside [213]. Their method is based on calculation of pixel compliances during a stepwise reduction in positive end-expiratory pressure (PEEP). During this maneuver, which is frequently referred to as "decremental PEEP trial", the PEEP level is initially set to the highest value possible (typically in the range of 20 to 24 mbar) and then reduced in small steps of 2 or 3 mbar that are conducted every minute until a very low level of PEEP (typically around 5 mbar) is reached. The maneuver may be conducted in volume-controlled or pressure-controlled mode. To yield valid results, a precise assessment of respiratory system compliance (C_{rs}) must be ensured. This typically requires neuromuscular paralysis or deep

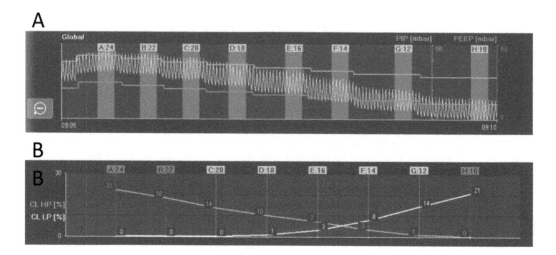

Figure 10.3   Panel A: Time course of global impedance signal during a decremental positive end-expiratory pressure (PEEP) trial, starting with 24 mbar (section A) in steps of 2 mbar until a PEEP level of 10 mbar (section H) was reached. The last five breaths of every PEEP step were averaged and used for analyzing regional compliance changes. Panel B presents the time course of cumulated compliance loss towards higher PEEP levels (CL HP, which is commonly interpreted as relative overdistension) and compliance loss towards lower PEEP levels (CL LP, white, which is commonly interpreted as alveolar collapse). The PEEP level located closest to the intersection of both lines is commonly interpreted as "best compromise" PEEP. (color figure available in eBook)

sedation to eliminate any spontaneous breathing activity that might interfere with a valid assessment of C_{rs}. For every PEEP level investigated, the maneuver yields one value for relative overdistension and one value for relative alveolar collapse. Typically, the "best" PEEP is considered to be the PEEP level where both alveolar overdistension and collapse are minimized. Figure 10.3 presents the results of a decremental PEEP trial analyzed according to this approach.

Since its publication, this approach has become one of the most widely used methods for bedside adjustment of positive end-expiratory pressure (PEEP) with EIT. It was successfully applied in patients with respiratory failure[455], on extracorporeal membrane oxygenation[288] and suffering from COVID-19[954,1069]. In a study including 40 intraoperative patients ventilated during elective abdominal surgery, adjustment of PEEP with the "Costa-Approach" led to improved intraoperative oxygenation and reduced postoperative atelectasis[815].

The ability of the "Costa-Approach" to identify the PEEP level associated with the "best compromise" between alveolar overdistension and collapse makes it appealing for clinical use, whenever finding the best PEEP level for an individual patient is challenging. Despite these advantages, the required decremental PEEP trial is a time-consuming ventilation maneuver that requires adjusting PEEP to a level that is presumably too high for most patients at the beginning, and then reducing it to a level that may be too low for most patients at the end of the maneuver. It is only after performing such a ventilation maneuver that the "best" PEEP for an individual patient can be identified and, subsequently, used for further therapy. The dynamics of critical illness and mechanical ventilation, however, may require frequent adjustment of PEEP on an hourly or, at the very least, daily basis. In clinical practice, when patients are already ventilated with a previously selected PEEP level, the clinical question is frequently whether it is safe to reduce PEEP by, e.g., 2–3 mbar or whether it should instead be maintained or even increased by a similar figure. Performing a complete decremental

PEEP trial every time that an adjustment of PEEP is required is not feasible and may be potentially dangerous.

This frequently encountered clinical challenge may be addressed by performing a brief variation in PEEP or tidal volume (V_T) and by analyzing its effects on regional C_{rs}[95,1217]. If a reduction in V_T leads to a decrease in regional C_{rs}, this may be interpreted as being caused by an end-expiratory collapse and delayed opening of the corresponding lung areas. In this case, an elevation of PEEP to overcome this "alveolar cycling" would be advisable. If, on the other hand, a brief reduction in V_T leads to an increase in regional C_{rs}, this may be interpreted as relative overdistension of the corresponding lung area with the previously applied V_T. The clinical consequence, in this scenario, would be to decrease PEEP or V_T to reduce this overdistension. A patient example of a temporal decrease in V_T to identify alveolar overdistension and alveolar cycling is presented in figure 10.4.

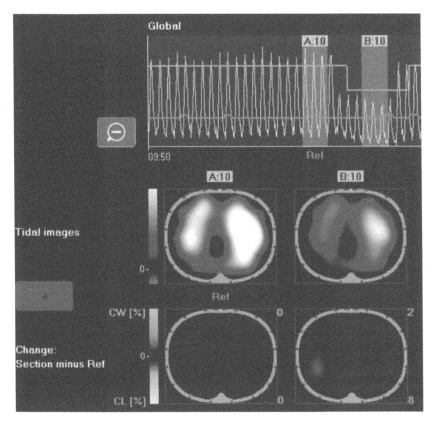

Figure 10.4   Patient example of a temporal decrease in tidal volume (V_T) to identify alveolar overdistension and alveolar cycling. While positive end-expiratory pressure (PEEP) was kept constantly at 10 mbar, V_T was decreased from originally 6 ml per kg predicted body weight during the reference measurement (Ref, section A) to 3.5 ml per kg during the maneuver (section B). This maneuver revealed 2% relative compliance win (CW), that could be interpreted as relative overdistension, and 8% relative compliance loss (CL), that could be interpreted as alveolar cycling. This generates the hypothesis that the patient should be ventilated with a higher PEEP level to avoid alveolar cycling. This was consistent with the results of ventilation distribution obtained in the same patient (cf. figure 10.1). As a consequence, PEEP was increased to counteract alveolar cycling and V_T was decreased to reduce overdistension. (color figure available in eBook)

Especially in patients with acute respiratory distress syndrome, end-expiratory collapse and overdistension occur at the same time but in different lung regions[97]. This clinically challenging situation can be addressed by increasing PEEP to counteract end-expiratory collapse and by simultaneously decreasing V_T to reduce end-inspiratory overdistension.

10.6 IDENTIFICATION OF POORLY VENTILATED LUNG AREAS

While EIT allows reliable identification of ventilated lung areas by means of the ventilation-associated impedance changes that can be detected in the ventilated lung, identification of non-ventilated or poorly ventilated lung areas may be more challenging. In principle, lung areas may become non-ventilated or poorly ventilated due to complete alveolar collapse at low airway pressure or due to severe overdistension caused by excessive airway pressure. In both cases, these areas exhibit no detectable impedance change making them difficult to distinguish from extrapulmonary thoracic tissue. Thus, a pixel exhibiting little or no tidal impedance change in the functional EIT image can be atelectatic, severly overdistended or may be located outside the anatomical lung region.

Distinguishing these situations requires prior knowledge of the presumed lung area within the EIT image. This information can be derived from a CT image of the patient investigated or, more pragmatically, by obtaining the lung contours from a previously established database of three-dimensional thoracic models created from computed tomography scans of different patients, taking into account the patient's demographic characteristics[1064]. If the presumed lung contours are known, any pixel within the lung contours that exhibits no or very little tidal impedance variation can be classified as "low tidal variation region"[293] or "Silent Space"[1064]. Silent Spaces in the gravitationally non-dependent lung regions may be interpreted as being caused by overdistension, whereas Silent Spaces in the gravitationally dependent lung regions may be interpreted as being caused by atelectasis. In a study including 14 patients with acute respiratory failure, Spadaro and colleagues demonstrated a correlation between changes in dependent Silent Spaces and lung recruitment measured using the pressure-volume-loop technique[999]. Non-dependent Silent Spaces, however, may be less sensitive for detection of overdistension with higher PEEP. An example of the influence of higher PEEP on Silent Spaces is given in figure 10.5.

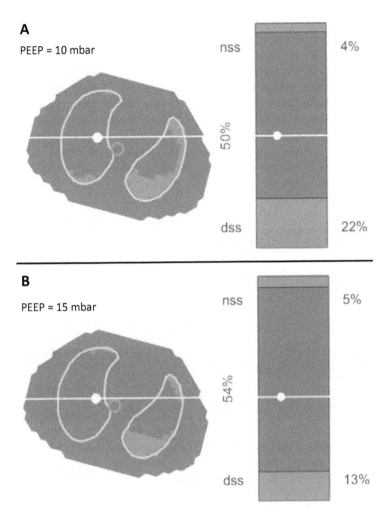

Figure 10.5   Non-dependent SilentSpaces (nss), dependent SilentSpaces (dss) and center of ventilation (CoV) before and after adjusting positive end-expiratory pressure (PEEP) from 10 to 15 mbar. Silent spaces are highlighted. White dot = current CoV (50% for PEEP 10, 54% for PEEP 15); blue circle = "ideal" center of ventilation (55%); white line = "ventilation horizon", dividing nss and dss along the gravitational axis. (color figure available in eBook)

10.7   END-EXPIRATORY LUNG IMPEDANCE CHANGES FOR QUANTIFICATION OF LUNG RECRUITMENT AND DERECRUITMENT

Images of end-expiratory lung impedance change display the regional impedance differences between different breaths comparing the time points of end-expiration. As tidal impedance differences are closely correlated to regional tidal volume, it has been postulated that changes in end-expiratory lung impedance between different breaths may be used to quantify regional changes in end-expiratory lung volume. A study comparing changes in end-expiratory lung volume as determined by the Helium dilution method to changes in end-expiratory lung impedance found an acceptable agreement between both methods[696]. In a feasibility study, the time course of end-expiratory lung impedance following a brief recruitment maneuver was utilized to determine whether the applied PEEP level was sufficient to maintain alveolar recruitment[259]. Despite these promising results, it must be highlighted that end-expiratory lung impedance may be influenced by other factors than changes in end-expiratory lung volume. Pulsation therapy with inflatable mattresses, which is a rather common intervention in today's intensive care units, can cause substantial changes in end-expiratory impedance levels that cannot be explained by changes in lung volume[303]. The same applies to patient movement and changes in torso and arm position[1090]. Even intravenous fluid therapy, which is applied to mechanically ventilated patients on a regular basis, may cause substantial changes in end-expiratory lung impedance[93,985]. To avoid serious misinterpretations, such possible sources of interference must always be considered when assessing changes in end-expiratory lung volume by observing the time course of end-expiratory lung impedance. Figure 10.6 presents two examples of the time course of end-expiratory lung impedance after a decrease in PEEP (panel A) and a recruitment maneuver (panel B).

A

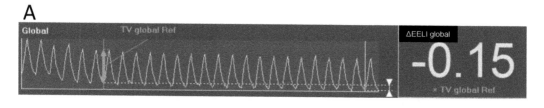

B

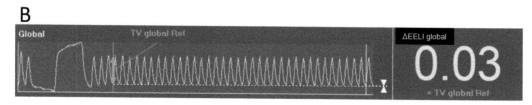

Figure 10.6   Time course of global end-expiratory lung impedance (EELI) following a decrease in positive end-expiratory pressure (PEEP, panel A) and a recruitment maneuver (panel B). ΔEELI global = change in end-expiratory lung impedance relative to the global tidal impedance variation (TV) at time point "Ref" (highlighted by blue arrow). While panel A presents an example for progressively decreasing EELI that could be interpreted as ongoing loss of end-expiratory lung volume (i.e., derecruitment), panel B presents and example of stable EELI that could be interpreted as stable end-expiratory lung volume (i.e., no ongoing recruitment or derecruitment). (color figure available in eBook)

10.8 EXPIRATORY TIME CONSTANTS FOR MONITORING AIRFLOW LIMITATION

Some respiratory diseases like Asthma and chronic obstructive pulmonary disease (COPD) are characterized by delayed lung emptying caused by expiratory airflow obstruction. By approximation, the time course of lung emptying during exhalation can be described with an exponential function:

$$V(t) = V_0 \times e^{-t/\tau} + V_{exp}, \tag{10.1}$$

where

- $V(t)$ = remaining lung volume at any time point during exhalation,
- V_0 = lung volume at the beginning of exhalation,
- t = time elapsed from the onset of exhalation,
- V_{exp} = lung volume remaining at end-exhalation and
- τ = expiratory time constant.

The time constant (τ) of this exponential function is proportional to the time required for complete exhalation. By approximation, τ describes the time required for exhaling $1 - 1/e \approx 63\%$ of tidal volume, with 3τ leading to almost complete exhalation of at least 95% of tidal volume. It can be obtained by multiplying C_{rs} with expiratory air flow resistance (R_e). Increased R_e is a distinctive feature of obstructive lung diseases like Asthma and COPD, leading to high values of τ and delayed lung emptying.

In tracheally intubated mechanically ventilated patients with COPD, the global value of τ can be obtained from the ratio of exhaled tidal volume to expiratory air flow[653]. This approach requires accurate measurements of air flow and volume, that can easily be obtained in tracheally intubated patients but are usually not feasible in patients undergoing non-invasive ventilation or unassisted spontaneous breathing. Furthermore, COPD and Asthma may exhibit regional differences in time constant, that cannot be assessed with a global measurement approach. EIT allows regional assessment of time constants, that should technically be feasible in patients undergoing non-invasive ventilation or unassisted spontaneous breathing. For this purpose, a curve characterized by the exponential function $V(t) = V_0 \times e^{-t/\tau} + V_{exp}$ is fit to the expiratory part of the regional impedance-time curve starting at 75% of tidal volume. Patients with COPD exhibit high levels of regional τ that can be modified by application of PEEP and, presumably, other therapeutic measures like inhalation of bronchodilators[533].

10.9 COMPARISON OF DIFFERENT APPROACHES FOR OPTIMIZING MECHANICAL VENTILATION WITH EIT

In this section, the various approaches that have been proposed for optimization of mechanical ventilation with EIT are compared. The advantages and disadvantages of each are indicated and a recommendation given.

10.9.1 GRAVITY-DEPENDENT VENTILATION DISTRIBUTION

Practical approach:
- Assess ventilation distribution along the gravitational axis by analyzing ventral and dorsal fraction of ventilation (depending on patient position) or geometrical center of ventilation at the bedside (cf. figure 10.110.5).
- Aim for "normal" values (dorsal fraction of ventilation 50-60%; center of ventilation 55%).
- If ventilation of dependent lung areas (e.g., dorsal fraction of ventilation in supine position, ventral fraction of ventilation in prone position) is lower than ventilation of non-dependent lung areas, increase PEEP.
- If ventilation of dependent lung areas is above normal, decrease PEEP or V_T.

Advantages:
- Easily applicable at the bedside, straightforward.
- Feasible during assisted mechanical ventilation.

Disadvantages:
- Normal values of ventral / dorsal fraction of ventilation and center of ventilation have not been established in a large cohort and may depend on patient demographics (height, weight, sex).
- A shift in ventilation distribution from non-dependent to dependent lung areas may be due to recruitment of dependent lung or overdistension of non-dependent; this cannot be differentiated with this approach.
- No clinical outcome data available.

Recommendation: Ventilation distribution along the gravitational axis may be used to generate a first hypothesis on whether PEEP should be increased or decreased. More accurate assessments may be necessary to confirm this hypothesis.

10.9.2   GLOBAL INHOMOGENEITY INDEX AND COEFFICIENT OF VARIATION

Practical approach:

- Calculate GI index or CV at the bedside, increase or decrease PEEP until both parameters are minimized.

Advantages:

- Easy to understand at first glance.
- May be feasible during assisted mechanical ventilation.

Disadvantages:

- Both parameters are highly dependent on lung area investigated; if used with functional regions of interest results may be misleading; other approaches like calculation of GI index and CV from anatomical lung areas have not been validated.
- No clinical outcome data available.

Recommendation: At present, do not use GI index and CV for optimization of mechanical ventilation in a clinical setting.

10.9.3   INTRATIDAL VENTILATION INHOMOGENEITY

Practical approach:

- Perform slow inflation maneuvers ("low-flow pressure-volume loop") at different PEEP levels, starting each maneuver with the respective PEEP value and inflating the lung at an air flow of ≤ 12 l/min up to an inspired volume of 12 ml/kg.
- Calculate SD_{RVD} from slow inflation maneuver performed at each PEEP level; select PEEP level with lowest value of SD_{RVD} (cf. figure 10.2).

Advantages:

- Helps to identify the PEEP level corresponding to minimized alveolar cycling.

Disadvantages:

- Provides no information on regional overdistension.
- Slow inflation maneuvers must be performed correctly and EIT analysis must be restricted to these maneuvers to yield valid results.
- Not feasible during assisted mechanical ventilation or in the presence of spontaneous breathing efforts.
- No clinical outcome data available.

Recommendation: May be used to compare different PEEP levels in terms of alveolar cycling. If used, it should be complemented by measures of overdistension.

10.9.4   QUANTIFICATION OF ALVEOLAR OVERDISTENSION AND COLLAPSE DURING A DECREMENTAL PEEP TRIAL ("COSTA-APPROACH")

Practical approach:

- Perform decremental PEEP trial starting at the highest clinically acceptable PEEP level (e.g., 24 mbar in ARDS patients or 20 mbar in non-ARDS patients). Observe hemodynamic stability, adjust vasopressors / fluid therapy as necessary.

- Reduce PEEP by 2 or 3 mbar every minute, until the lowest clinically acceptable PEEP level is reached (usually around 5 mbar).
- Analyze changes in pixel compliance according to the "Costa Approach", interpreting compliance loss towards higher PEEP levels as overdistension and compliance loss towards lower PEEP levels as collapse.
- Select PEEP level with "best compromise" between overdistension and collapse (cf. figure 10.3).

Advantages:

- Allows simultaneous detection of alveolar overdistension and collapse for every PEEP level.
- Yields "best compromise" PEEP level.
- Has been successfully applied in multiple clinical studies.

Disadvantages:

- Requires PEEP trial starting at very high levels (that may be associated overdistension while applied) and continuing to very low PEEP levels (that may lead to alveolar collapse and atelectasis formation).
- Should not be repeated too often because of possible side-effects.
- Not feasible during assisted mechanical ventilation or in the presence of spontaneous breathing efforts.
- Limited clinical outcome data.

Recommendation: Method of choice for initial PEEP setting with EIT in patients with early ARDS on controlled mechanical ventilation. After an initial PEEP level is found with this method, other methods for continuous monitoring of ventilation distribution should be preferred. Repeated decremental PEEP trials should be avoided.

10.9.5   ASSESSMENT OF CHANGES IN REGIONAL COMPLIANCE WITH DIFFERENT V_T OR PEEP LEVEL

Practical approach:

- Perform variation in V_T, for example by halving inspiratory pressure difference during pressure-controlled ventilation or set V_T during. volume-controlled ventilation for a few breaths only.
- Assess changes in regional C_{rs} with lower V_T.
- Decreases in regional C_{rs} with lower V_T are interpreted as tidal recruitment and PEEP should be elevated by 2-3 mbar, then maneuver should be repeated.
- Increases in regional C_{rs} with lower V_T are interpreted as overdistension and V_T should be reduced to a lower value, if possible. If no reduction in V_T is possible, consider PEEP reduction (cf. figure 10.4).

Advantages:

- May allow simultaneous detection of alveolar overdistension and alveolar cycling for every PEEP level.
- Individual adjustment of PEEP and V_T.
- No need for decremental PEEP trial.
- Can be used during ongoing mechanical ventilation to decide whether PEEP / V_T should be increased or decreased.

Disadvantages:

- Not feasible during assisted mechanical ventilation or in the presence of spontaneous breathing efforts.
- Clinical experience still limited.
- No clinical outcome data available.

Recommendation: Perform initial PEEP adjustment according to "Costa Approach", then perform variation in V_T whenever a new assessment is required. Do not use in patients with spontaneous breathing efforts!

10.9.6 POORLY VENTILATED LUNG AREAS ("SILENT SPACES")

Practical approach:

- Assess dependent and non-dependent Silent Spaces at clinically selected PEEP.
- If dependent > non-dependent Silent Spaces, increase PEEP.
- If non-dependent > dependent Silent Spaces, decrease PEEP.
- Assess changes in SilentSpaces after PEEP-step (cf. figure 10.5).

Advantages:

- Straightforward.
- Can be used in patients with spontaneous breathing efforts.

Disadvantages:

- Relies heavily on presumed "lung contour"; may be inaccurate if a patient's anatomy is different from the selected model.
- Non-dependent Silent Spaces may underestimate the actual degree of overdistension.
- No clinical outcome data available.

Recommendation: May be helpful for assessing changes in lung recruitment. Should be complemented by other measures for excluding overdistension during PEEP titration.

10.9.7 ANALYZING CHANGES IN END-EXPIRATORY LUNG IMPEDANCE

Practical approach:

- Perform recruitment maneuver (could be sustained inflation maneuver or "staircase" recruitment maneuver with stepwise PEEP increments).
- Assess time course of end-expiratory lung impedance after recruitment maneuver while ventilating the patient with constant PEEP.
- If progressive decrease in end-expiratory lung impedance with constant PEEP is observed, this is interpreted as alveolar derecruitment and a higher PEEP level should be selected to stabilize end-expiratory lung volume. Repeat analysis at higher PEEP level. (cf. figure 10.6A)
- If progressive increase in end-expiratory lung impedance is observed, this is interpreted as ongoing lung recruitment. The set PEEP level should be maintained.
- If no increase or decrease in end-expiratory lung impedance is observed, this is interpreted as stable end-expiratory lung volume. If sufficient lung recruitment has been achieved prior to this analysis, the set PEEP level should be maintained (cf. figure 10.6B).

Advantages:

- Easy to perform and relatively straightforward.
- Changes in end-expiratory lung impedance with constant PEEP are sensitive parameters for detection of recruitment and derecruitment.

Disadvantages:

- No assessment of overdistension.
- Highly susceptible to artifacts and misinterpretation caused by patient movement, fluid therapy, pulsating mattress etc.
- No clinical outcome data available.

Recommendation: Under very controlled conditions (no patient movement, no intravenous fluid administration, no pulsating mattress, no other sources of interference, constant PEEP level), changes in end-expiratory lung impedance may be a sensitive marker for lung recruitment and derecruitment. Should be analyzed with caution and must always be complemented by other analyses (e.g., regional compliance changes) for excluding overdistension.

10.9.8 REGIONAL EXPIRATORY TIME CONSTANTS

Practical approach:

- In patients with severe expiratory airflow obstruction, assess regional values of τ before and after therapeutic measures like adjustment of PEEP or inhalation of bronchodilators.
- If therapeutic measure is associated with a decrease in average regional τ, this can be interpreted as potentially beneficial (less severe airflow obstruction).
- If therapeutic measure is associated with no decrease or even an increase in average regional τ, this is interpreted as not beneficial (e.g., increased expiratory airflow obstruction).

Advantages:

- Allows assessment of expiratory airflow obstruction in patients with COPD and Asthma.
- Requires no specific ventilatory mode or maneuver.
- Theoretically feasible in assisted mechanical ventilation or unassisted spontaneous breathing.

Disadvantages:

- A decrease in regional τ may not only be caused by a decrease in R_e, but also by decrease in regional C_{rs} (which might be caused by alveolar overdistension).
- For subjects undergoing assisted mechanical ventilation or unassisted spontaneous breathing, this approach has not yet been validated.
- No outcome studies.

Recommendation: Regional τ in patients with obstructive lung diseases offers valuable bedside information on the emptying characteristics of the patient's lungs. However, the results should only be interpreted by trained specialists with a deep understanding of respiratory physiology.

10.10 CONCLUSION

Ever since its availability for clinical use, the interest in chest EIT has been growing rapidly. Current devices allow EIT examinations to be carried out for prolonged periods of time and present a variety of potentially useful results at the bedside. Still, the correct application of EIT and the interpretation of its results remain challenging and time consuming. Factors like the high dependency of EIT results on the position of the electrode plane[538] and the numerous sources of interference in the electromagnetically "noisy" environment of an intensive care unit[303] complicate the use of EIT in clinical practice.

The most important challenge to a wider application of EIT in clinical practice, however, is the correct interpretation of EIT measures and their potential clinical consequences. In this chapter, we intended to provide background information on some of the most widely used EIT parameters as well as recommendations on their use in clinical practice. Nevertheless, prospective optimization of ventilator settings with EIT should only be performed by experienced practitioners with both expertise in mechanical ventilation and a sufficient understanding of EIT technology and its limitations.

Future improvements in EIT hardware and bedside evaluation software should focus on faster and easier applicability in clinical practice and increase the robustness of EIT results with respect to confounding factors like patient movement and fluid administration. Furthermore, the current dependency of results on the EIT electrode plane should be reduced, perhaps by implementing three-dimensional EIT that might correct for this confounder by automatically analyzing multiple image planes simultaneously. Decision support tools might facilitate interpreting the additional information provided by EIT given the limited amount of time available for decision-making in modern intensive care medicine. These and other developments may foster the clinical application of EIT by users with variable degrees of training and expertise.

11 EIT Monitoring of Hemodynamics

Lisa Krukewitt, Fabian Müller-Graf, Daniel A. Reuter
Stephan H. Böhm
Department of Anesthesiology and Intensive Care Medicine, Rostock
University Medical Center, Germany

Huaiwu He
Department of Critical Care Medicine, Peking Union Medical College
Hospital, Peking Union Medical College, Chinese Academy of Medical
Sciences, Beijing, China

CONTENTS

DOI: 10.1201/9780429399886-11

11.1   INTRODUCTION

The dynamics of blood flow within the circulatory system are of key clinical interest. The relevant mechanical parameters are pressures, flows and volumes, and, traditionally, require invasive or inconvenient measurements procedures. Due to its high temporal resolution and its ability to detect impedance changes not only due to ventilation but also to blood volume changes, EIT offers the opportunity for non-invasive real time measurements of various hemodynamic parameters. The reader is also recommended reviews of this area[615,778].

The function of the hemodynamic system is to deliver blood flow to tissues, and to produce these flows, the heart must raise the blood *pressure* in the arteries. The blood flow leaving the heart is called *cardiac output*, CO, and the flow in tissues is called *perfusion*. At steady state CO=HR×SV, or the *heart rate* times the blood ejected per beat, the *stroke volume*. The most common hemodynamic parameter is the heart rate, HR, largely because the ease with which it can be measured. Many techniques and technologies reliably measure HR, and we don't further discuss its measure in this chapter. Next, arterial pressure is very commonly measured. It is relatively easy to measure, and gives useful information on cardiac effort, but does not necessarily correlate with the delivery of blood and oxygen to the organs. However, this nutrition by flow is crucial for survival of tissue. For this purpose organ perfusion or cardiac output should be assessed. Clinically, the parameters to determine how to improve the individual situation of a patient are *functional volume status parameters*, which, for example, let the physician know how an organism would react to fluid administration.

Assessment of the hemodynamic status is fundamental for patient-centered decision making. EIT is able to measure blood flows (Box 11.12) and can be used for this purpose, due to its thoracic localization, relatively high frame rate. EIT is currently most often used in the perioperative and intensive care settings, where circulatory disorders are most frequent. See also §8.2.1.3 for a discussion of interpretation of cardiac-related EIT signals.

All of these three classes of hemodynamic parameters – pressure, flow, and functional volume status – have been addressed in EIT-research and show promise for clinical use. In the remainder of this chapter, we discuss by their classic measurement. Then we will present the origins of cardiosynchronous impedance changes before discussing how interfering signals in EIT can be addressed. Afterwards, we will focus on research performed in the EIT-community to adapt the classical hemodynamic parameters to the measurement through impedance change.

We start with the two parameters used to assess functional volume status by EIT – stroke volume variation[670,989] and extravascular lung water[1052], before then covering pressure measurements, i.e. aortic blood pressure[148,378,988] and pulmonary artery pressure[147,841,842]. Lastly, we will discuss how EIT can be used to measure parameters of blood flow such as pulmonary perfusion[131,268,269,304,778,982] and stroke volume[670,828,1100,1101].

11.2 CLASSICAL METHODS AND KEY PARAMETERS OF HEMODYNAMIC MEASUREMENTS

In the following section we will discuss the basics of intra-vascular pressure, flow parameters, and functional volume parameters.

11.2.1 INTRA-VASCULAR PRESSURE MEASUREMENT

In general, the circulatory system (Figure 11.1) is divided into the systemic circulation and the pulmonary circulation, which are separated by the two chambers of the heart. This results in four vascular compartments which are all separated on one side by a capillary bed and a cardiac valve on the other (systemic arterial, pulmonary arterial, systemic venous, pulmonary venous). In each of these compartments the blood pressure is roughly constant.

11.2.1.1 Systemic Arterial Pressure

Historically and physiologically, the arterial pressure, especially the systemic one, is important. In 1896, the pressure cuff method was invented by Riva Rocci and used a sphygmomanometer for indirect measurement of blood pressure by compressing the tissue of a limb. Cuff-based methods are still, with several optimizations, standard for assessing systemic pressure. This method is robust but is uncomfortable for repetitive use and can only be repeated every minute or so. Another common technique for assessing systemic arterial pressure is puncturing an arterial vessel and measuring the intra-arterial signal with a pressure transducer. This technique allows continuous assessment of intra-arterial pressure variations up to a frequency of $\sim 100\,\text{Hz}$ but is invasive with the risk of bleeding or vessel occlusion by thromboemolism. Therefore, it can only be performed in a perioperative or intensive care setting.

The systemic arterial pressure is generated by the cardiac output of blood into the aorta and is the perfusion pressure for the capillary bed. This perfusion is often a more interesting parameter to a physician (see section 11.2.2). The systemic arterial pressure can be calculated by the EIT-measured pulse arrival time[989] (see §11.5.1).

11.2.1.2 Pulmonary Arterial Pressure

The pulmonary artery pressure (PAP) is the force against which the right chamber of the heart has to work. The pressure is normally around one third of the systemic arterial pressure. Pulmonary artery hypertension occurs due to lung disease (e.g. chronic obstructive pulmonary disease, COPD), heart disease, and other conditions such as high-altitude pulmonary edema. To measure the PAP indirectly, sonographically tricuspid regurgitation can be used to calculate systolic pulmonary artery pressure. For direct assessment, a catheter has to be placed through the right heart chamber into the pulmonary artery, a Pulmonary Artery Catheter (PAC) also known as Swan-Ganz Catheter. The PAC is invasive with potential damage to all passed structures, and it's use is discouraged. PAP is thus clinically interesting but hard to measure. EIT-based algorithms have been developed[148] (see §11.5.1).

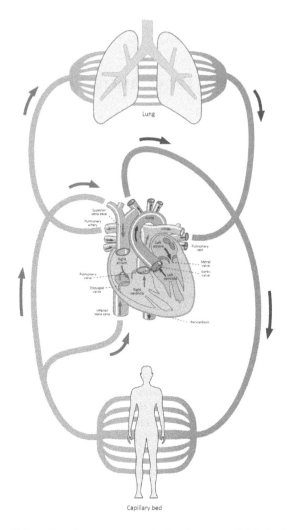

Figure 11.1   Circulatory system. (color figure available in eBook)

11.2.1.3 Systemic Venous Pressure

For direct assessment of the systemic venous pressure (central venous pressure), a catheter must be placed in a central vein and has to be connected to a pressure transducer. By echocardiographically measured inferior vena cava filling systemic venous pressure could be calculated. The systemic venous pressure is much lower than the systemic arterial pressure. To our knowledge no EIT-based technique has been developed to asses systemic venous pressure.

11.2.1.4 Pulmonary Venous Pressure

With the pulmonary artery catheter, pulmonary venous pressure can be assessed. A small balloon at the tip can occlude a pulmonary artery, so that all blood between the balloon and the pulmonary vein shows the same pressure. With this technique, the filling pressure of the left heart chamber can be assessed. This is an important parameter in left heart failure. We are also not aware of EIT-based techniques to asses pulmonary venous pressure.

11.2.2 FLOW PARAMETERS

The cardiac output (CO) is the main flow parameter and calculated as the heart rate (HR) multiplied by the stroke volume, SV, the blood volume that is ejected by one heart beat from the left chamber into the aorta. For each organ an individual organ perfusion/flow can be determined. Pulmonary perfusion can be separately calculated from the cardiac output, but absent left-right shunt, the average systemic and pulmonary perfusion is equal. Perfusion measurement is a "holy grail" of hemodynamic measurement which can be assessed by several techniques, all with their own advantages and disadvantages. Readers are referred to the review by Vincent et al[1084].

The most frequent method and gold-standard to assess the perfusion is the thermodilution technique. A small cold isotonic saline bolus is injected into a central vein. The bolus travels through the right atrium, the right ventricle, passes the lung, the left atrium and ventricle, through the aorta to a thermistor which is placed an arterial vessel in the arm or leg. Thermodilution is also measured by the Swan-Ganz Catheter where a thermistor is placed at the tip of the catheter which is in a pulmonary artery. The thermistor senses a drop in intra-arterial temperature induced by the cold saline bolus. The area under the curve inversely correlates with cardiac output and is calculated by the Stewart-Hamilton-equation (11.1)[875].

$$CO = \frac{V_1}{t} = \frac{V_0 \left(T_B - T_0 \right) K_1}{\int_t \Delta T_B dt} \tag{11.1}$$

where CO is the cardiac output, V_0 the injected volume, T_B the temperature of blood at baseline, T_0 the temperature of injected volume, and K_1 a heat capacity factor.

11.2.3 VOLUME STATUS PARAMETERS

To our knowledge, so far two functional parameters for volume status have been adopted in the use in EIT, which we describe in this section. The heart-lung interaction by Solà et al[989] finds its corresponding clinical parameter in the *Stroke Volume Variation* and lung water ratio$_{EIT}$ by Trepte et al[1052] corresponds to the clinical parameter of the *extravascular Lung Water*. The stroke volume variation reflects low intravascular fluid and increased extravascular Lung Water describes one kind of fluid overload. Many other volume status parameters are used in clinical routine. For a synopsis readers are referred to the review of Pinsky[829] and Jozwiak et al[526].

11.2.3.1 Stroke Volume Variation

Stroke volume variation (SVV) is the beat-to-beat variations in stroke volume. Such variations occur with the breathing cycle as breathing pressures change the thoracic pressure and thus systolic pressures and stroke volume. In healthy conditions, SVV should be low, but it can increase, especially if blood volume is too low. Thus, the variability in CO is a useful indication of whether the patient has sufficient blood volume. SVV assesses the variation of the beat-to-beat stroke volume induced by every respiratory cycle to the mean value and is calculated by equation 11.2[874].

$$SVV = (SV_{max} - SV_{min})/SV_{mean} \tag{11.2}$$

During inspiration, intrathoracic pressure decreases with the respiratory cycle and ventricular diastolic filling is increased. As described in the Frank Starling Mechanism, this could lead – depending on the diastolic fiber distension – to a higher contraction force of the myocardium and thereby increased stroke volume[1002]. A value of SVV below 10–12% is considered as optimal (euvolemia), higher values (hypovolemia) should be treated with fluid administration if possible (figure 11.2).

11.2.3.2 Extravascular Lung Water

Instead of deficient intravascular volume, the opposite is harmful as well. One side effect of a fluid overload is a pulmonary edema, an accumulation of fluid (extra vascular lung water) in the tissue and air spaces of the lung. Pulmonary edema is commonly associated with heart failure. Traditionally, extra alveolar lung water ($EVLW$) is calculated by equation 11.3, where $ITBV$ is the intrathoracic blood volume, and $ITTV$ is the intrathoracic thermal volume as defined in equation 11.4, with CO being the cardiac output and MTt the mean transit time. Increased values are considered as hypervolemia (figure 11.2), and could be treated for example by increased diuresis if possible.

$$EVLW = ITTV - ITBV \tag{11.3}$$

$$ITTV = CO \cdot MTt \tag{11.4}$$

The $EVLW$ is traditionally indexed to the ideal patient body weight and should range between 3.0–7.0 ml/kg [919,920].

Measurements of the hemodynamic status can help guide individualized therapy for each patient. Both sides of euvolemia – the moment of optimal fluid load – hypervolemia and hypovolemia should be avoided (figure 11.2, [545]).

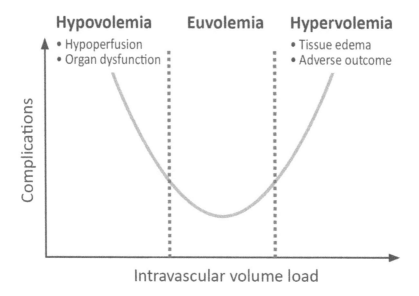

Figure 11.2   In Euvolemia the cardiac function is ideal and rate of complications is low. Hypovolemia is associated with organ dysfunction like kidney failure and cognitive impairment. Hypervolemia causes tissue or lung edema and is associated with inflammatory reactions. Hemodynamic monitoring allows diagnostic of hypovolemia and hypervolemia, follow up their changes to treatment and thereby improving patient outcome (adapted from [545]). (color figure available in eBook)

11.3 ORIGINS OF CARDIOSYNCHRONOUS SIGNALS IN EIT MEASUREMENTS

> **Box 11.12: Does EIT Measure "perfusion" or "pulsatility"?**
>
> There are two ways to measure blood flow: using conductivity-contrasting injections or analyzing heart-frequency changes in the EIT signals. No doubt has been expressed that conductivity contrasts reflect flow (perfusion). Unfortunately, bolus injections are invasive and not suitable for continuous monitoring. The other approach is much more convenient. Typically, EIT signals are digitally filtered at the heart rate. The results are available continuously and non-invasively. However, the EIT community has debated the extent to which these signals represent perfusion[293].
>
> - not all blood flow leads to EIT signal variations. For example, consider blood flowing uniformly and continuously through a capillary bed. Without changes over time, no EIT signals variations will be seen.
> - not all EIT signal variations originate in flow. For example, the movement of the heart in the chest accounts for a large fraction of the signal, which will vary with the patient, posture, and electrode placement[843].
>
> Several studies[131,870,1008] have compared these approaches, and suggest that filtering-based approaches are a fairly good measure of perfusion.
>
> Clearly for "marketing" of EIT, it's best to say it measures *perfusion*. Many prefer to use this terminology. For others, filtering-based approaches are clearly indirect indices. They recommend terms like *pulsatility*. The terminology we recommend is *perfusion-related*, which is scientifically accurate and easy to understand for non EIT specialists.

The fact that EIT can measure impedance changes related to blood flow is the basis of hemodynamic measurements using EIT. Two approaches have been used, injecting conductivity contrasts into the blood, or filtering the cardio-synchronous component of the impedance signal. Since blood is more conductive than tissue, EIT is sensitive to changes in blood concentration. When measuring the pressures, flows, and volumes of interest, it is often assumed that the observed heart-beat related changes in impedance are caused by perfusion. The exact cause of these changes however is still subject to some debate – how exactly are perfusion-related impedance changes caused and which other factors contribute to cardio-synchronous impedance changes? Box 11.12 discusses the debate about the correct terminology.

Analysis of of blood flow, heart movement, and ventilation, and also how to separate signals of cardiac- and ventilatory frequency will be discussed in §11.4, and also in §8.2.1.3.

The physical mechanisms causing cardiosynchronous signals have been summarized in[30], and illustrated in figure 11.3:

- Cardiosynchronous mechanical deformations: heart motion and the resultant movements of other structures.
- Blood volume changes: within the heart, a vessel or an organ.
- Red blood cell reorientation due to pulsatile blood flow[333].
- Reorientation of anisotropic structures such as the heart muscle.

The influence of mechanical deformation and volume changes is widely agreed on and has been subject of much research (see 11.4); the influence of red blood cell orientation seems less well verified and we have not been able to reproduce these effects.

ard EIT technique to measure perfusion is with the injection of a bolus of a sting saline[131,301,656]. Brown *et al* first showed in 1992 that intravenous injecsaline bolus can be used for EIT-based imaging of blood flow[158]. Bolus-based ample been used to detect regions of interesti[167,1032,1159], (see also §11.4.2.1

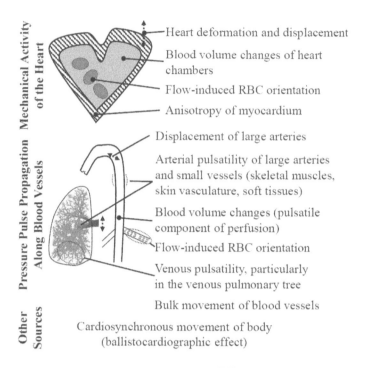

Figure 11.3   Origins of cardiosynchronous signals.[30] (color figure available in eBook)

and §11.5.3.2) or to measure pulmonary perfusion[131] (see also §11.5.2.1). These techniques have seen wide usage in animal experiments, and recently, bolus-based methods have been used in humans to detect pulmonary embolism (10 ml, 10% sodium chloride)[452], and lung perfusion 10 ml 5% sodium chloride[697]. No side-effects of bolus injection in humans have been reported; however, adverse effects cannot be ruled out and some groups have observed temporary reduced blood pressure when using saline boli in animal experiments. An additional concern is the possibility of hypernatrimia if repeated bolus is given (see §8.2.4.3). One additional concern with saline bolus injections is transient global reduction in the EIT thoracic signal due to saline absorption in lung tissue.

11.4   INTERFERING SIGNALS IN HEMODYNAMIC EIT MEASUREMENTS

During thoracic EIT measurements for hemodynamic monitoring, several phenomena simultaneously cause impedance changes: The signal of largest relevance for hemodynamic EIT are impedance variations caused by blood volume changes. This signal, however, is overlaid with the much larger ventilation-related impedance change, especially in and around EIT pixels in the lung region. Furthermore, heart movement during a cardiac cycle also causes impedance changes. Possible causes for cardio-synchronous impedance changes have been discussed in[304] and[30] (see also 11.3).

This § describes the different causes of thoracic impedance change as well as some approaches to limit their influence on hemodynamic measurements an to separate the different signals that are commonly present in an EIT measurement.

11.4.1 BLOOD FLOW AND BLOOD VOLUME CHANGES

The resistivity of blood of about $150\,\Omega\text{cm}$ is well-differentiated from tissues such as muscle $(450\,\Omega\text{cm})$, fat $(2000\,\Omega\text{cm})$ and lung tissue $(1325\,\Omega\text{cm})$[1115]. Therefore, volume changes of blood vessels and heart chambers will cause impedance changes that can be measured by EIT. See also §3.2 for a discussion of tissue electrical properties.

In 1997, Vonk Noordegraaf *et al*[1101] measured calf blood flow by EIT based on the volume changes induced by blood flow that in turn caused impedance changes. Cardiac-related volume changes occur in several structures: the heart chambers themselves change significantly in volume, e.g. the left atrium shows a change of 65% of its cross-sectional area from end diastole to end systole[1115]. The pressure pulse that travels though the vascular system leads to distension of the blood vessels; for example the cross sectional area of the descending aorta changes by 22% during a cardiac cycle[1115]. Besides this distension, the orientation of red blood cells in the arteries changes during different blood flow stages, which also leads to a cyclic change in blood electrical conductivity during a cardiac cycle[727].

Instead of monitoring the small volume changes that blood flow causes, an impedance change can be induced by using a contrast agent of different resistivity such as hypertonic saline. At about $60\,\Omega\text{cm}$[656], the resistivity of even 0.9% saline is much lower than that of blood and other tissue and decreases with the saline's concentration. Therefore, the conductivity of the blood, and hence the contribution of perfusion to the EIT image, can be greatly increased by the intravenous injection of a bolus of hypertonic saline[131,301,656]. There is some evidence that contrast-agent imaging by EIT can view perfusion of organs, for example[21] imaged cerebral perfusion in a rat.

It is interesting to observe that many possible contrasts have not yet been used in medical EIT. Low concentration saline (with sugar solutions) have rarely been attempted, but should give an inverse signal. Additionally, isotonic saline has a higher conductivity than blood and could be a contrast (in larger doses). Finally, thermal contrasts function well in EIT, and cold saline is less conductive. It is worth noting that hypotonic solutions can be dangerous due to the osmotic effects in lung tissue.

The cardiac-related impedance changes discussed here are much smaller than the changes caused by ventilation and separating these signals is challenging, as will be discussed in the next section. Furthermore, due to the partial volume effect, blood volume and resistivity changes in one intrathoracic structure will also be visible in surrounding EIT pixels, e.g. heart volume changes will also affect the signal in EIT pixels representing lung tissue.

11.4.2 VENTILATION

The presence of ventilation- and cardiac-related signals in EIT data and the resulting need to separate them is one of the main challenges when using EIT for hemodynamic applications. Generally, a bioimpedance signal ΔZ is comprised of a cardiac (ΔZ_C) and a respiratory (ΔZ_R) component and a stochastic disturbance (*noise*) as given in (11.5)[583].

$$\Delta Z \approx \Delta Z_C + \Delta Z_R + noise \tag{11.5}$$

The cardiac signal may be one order of magnitude smaller than the respiratory signal component and may be modulated by the respiration[583]. Several approaches to separating the cardiac and ventilatory component of the signal will be discussed here: by dividing the EIT image into regions of interest (ROIs) such as heart and lung regions, areas where one of the signals is more dominant can be assessed separately[279]. Averaging techniques amplify the cardiac-related signal and attenuate the ventilatory component and noise by summing over several cardiac cycles. Other decomposition methods rely on frequency filtering or more elaborate algorithms. Lastly, during short periods of apnea the influence of breathing can be eliminated entirely.

A typical spectrum of the EIT signal in the thorax will contain a cardiac signal and a lower frequency ventilation signal with several harmonics that may overlap with the cardiac signal as shown in figure 11.4. Especially during mechanical ventilation where variations of the breathing rate are absent these higher harmonics are particularly noticeable[656]. Frequency-based approaches of separating signals perform worse if the spectra overlap or the ventilatory frequency varies such as during spontaneous breathing or certain modes of mechanical ventilation.

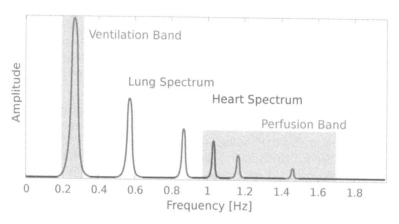

Figure 11.4   A simplified spectrum of an EIT-measurement containing ventilation and cardiac-related impedance changes. The typical frequencies of ventilation and perfusion are marked in grey. The lung spectrum shows several harmonics that may overlap the heart spectrum. (color figure available in eBook)

11.4.2.1   Regions of Interest

Since the ventilatory signal is strongest in the lung area, evaluating regions representing the different cardiovascular structures separately limits the influence of the ventilatory signal. Several methods for assigning EIT pixels to different ROIs have been proposed that are based on information from EIT measurements such as frequencies[278] or phase differences[842]. A bolus of hypertonic saline can also be used to identify ROIs in EIT images[987].

Possibilities of defining heart and lung ROIs from EIT image series include assigning EIT pixels to ROIs based on their maximal amplitude[96,378,846] or on their energy in the EIT image at estimated cardiac and respiratory frequencies[278]. Other approaches are based on phase differences of the pixels compared to an external timing parameter[842] or compared to the global EIT signal after frequency filtering[304].

Due to the low spatial resolution of EIT, there is an overlap of the different signal sources even in the separate ROIs; for example, Braun et al[148] studied the signals present in the aortic region and found pulmonary and cardiac influence in all pixels. Similarly, the ventilation-related signal is present even in heart pixels[549]. Therefore, even when the EIT signal in only one ROI is studied, additional filtering may be necessary.

11.4.2.2   ECG-Gating

The influence of ventilation-related impedance changes on the data can be reduced by recording ECG-gated EIT images and averaging the data over multiple cardiac cycles[1100]. Averaging over at least 100 cardiac cycles N_c has been suggested to attenuate the respiratory component[262]. Furthermore, synchronous averaging reduces the random noise in a signal as given in equation 11.6[661], meaning that when averaging over 100 cardiac cycles, the signal-to-noise ratio (SNR) will be increased by a factor of 10. Averaging over N_c cardiac cycles will amplify the amplitude of the cardiac component by a factor of $\sqrt{N_c}$[615], and thus amplify its power by a factor of N_c.

$$\frac{SNR_{averaged}}{SNR_{non-averaged}} = \sqrt{N_c} \tag{11.6}$$

The main drawback of this method is the long data collection time needed to achieve the necessary number of cardiac cycles. This delay may also prevent real-time analysis[615]. Furthermore, this method does not allow for separation of cardiac- and ventilation-related signal if the heart rate is an exact multiple of the respiratory rate[656]. Moreover, signal averaging methods can only separate signals successfully if they are independent. Therefore, a main drawback of this method is that due to heart-lung interactions, ventilation and heart rate are not usually independent. In healthy subjects respiration affects the circulation mostly through changes in pleural pressure but heart lung interactions are complex and can be affected by disease or ventilation[339].

11.4.2.3 Decomposition of Signals

Without the use of additional information such as ECG data, cardiac- and ventilation-related EIT signals can be separated using frequency filtering or more complex separation algorithms.

The ventilation-related and cardiac-related impedance changes are well separated by frequency with the normal rate of ventilation ranging from 0.2 to 0.23 Hz and the rate of perfusion ranging from 1.08 to 1.25 Hz[1200]. Filters can be applied to either the voltage measurements or the reconstructed image data.

One approach to frequency filtering of EIT images is to use a fast Fourier transform of the global EIT signal to determine the frequency bands related to breathing and cardiac-related changes and subsequently bandpass filter the regional pixel time courses to determine the components of the EIT signal[304]. High-pass and low-pass filtering as shown in figure 11.5 is also possible, although the influence of higher harmonics of the ventilation-related signal may be lager in this case. The data collection period required for frequency filtering is much shorter than that needed for averaging methods, with the frequency resolution improving with increased length of data collection.

However, even in the case of band pass filtering, the cardiac band may contain harmonics of the ventilation band, making the separation of the two sources by frequency filtering impossible[615].

Kerrouche et al use a singular value decomposition on an EIT time series and find that the first principal component shows a respiratory time series while in the second principal time series respiratory- as well as cardio-synchronous changes were seen[549].

Similarly, Deibele et al[228] suggest a method for the decomposition of ventilation- and cardiac-related signals in EIT images based on principal component analysis (PCA) that, unlike the simple frequency filtering described above, can separate the signals even in the presence of overlapping harmonics without any additional information sources[228]. In contrast to the approach by Kerrouche et al[549] mentioned above, this method is able to extract not only the ventilatory but also a cardiac-related time series. By PCA, they identify template functions that can be used for time domain filtering. The first principal component score, i.e. the one with highest variance, is used as an approximation of the ventilation component[228]. Since the EIT signal is composed as given in equation 11.5, the cardiac signal can be approximated by subtracting the ventilation signal from the input signal. In several steps, the algorithm then improves the approximation for cardiac- and ventilation-related signal by finite impulse response band pass filtering and fitting the signals to principal component scores[228].

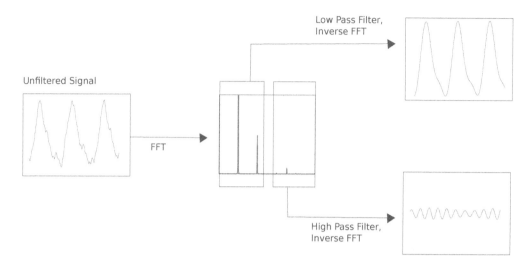

Figure 11.5 Separation of ventilation- and perfusion-related signals by high-pass and low-pass filtering, adapted from[1200]. (color figure available in eBook)

Unlike averaging methods such as ECG gating, the algorithm used by[228] preserves heart rate variability, allows for fast processing and does not need information other than EIT data. Despite these advantages, methods based on ECG gating and frequency filtering seem to be more commonly used in current EIT-based hemodynamic monitoring than PCA-based approaches.

Besides PCA, independent component analysis (ICA) has been used to separate cardiac and ventilatory signals. Recently, Jang *et al* used PCA to extract a a shape-reference waveform for the ventilatory signal, and a combination of PCA and ICA to extract a shape-reference waveform of cardiac blood flow[510]. In contrast to the method developed by Deibele *et al*, their approach does not use any frequency filtering in the process of extracting shape-reference waveforms. They applied this method successfully to simultaneously measure tidal volume and stroke volume[510].

11.4.2.4 Apnea Measurements

One simple and efficient way to eliminate the influence of breathing, is recording EIT data during apnea[268,269]. However, slow changes in lung volume still occur during periods of apnea: the removal of oxygen from the lungs during this period leads to a drop in intrathoracic pressure if the airway is occluded[979]. The obvious drawback of apnea measurements is that these do not allow for continuous monitoring as only short measurements are possible. Furthermore, it has not yet been clarified if perfusion remains the same during episodes of apnea as during ventilation[615].

11.4.3 HEART MOVEMENT

Especially when trying to image impedance changes within the heart region, heart movement plays a significant role. When measuring volume changes within the heart, only changes within the electrode plane will appear in the EIT images, making the measurements critically dependent on electrode placement[262,1099,1100]. When placing the electrodes in a transverse plane, ventricles and atria will overlap in the EIT image and the ventricles will move through the EIT plane during a cardiac cycle due to the heart's contraction along the long axis of the left ventricle[1099].

Vonk Noordegraaf *et al*[1099] were able to improve the isolation of the ventricles by using an oblique electrode placement instead of a transverse one[1099]. They showed that the impedance change in ventricles and atria over time is inversely proportional to their area as determined by MRI (see figure 11.6)[1099]. Furthermore, they found a higher reproducibility of the analysis of the region of interest of the ventricular region during periods of exercise – where heart movement is increased – when using an oblique EIT plane compared to a transverse plane[1099].

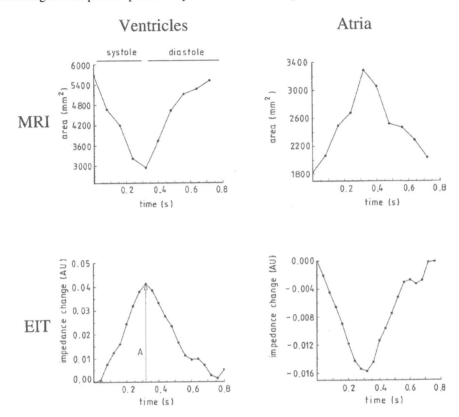

Figure 11.6   Changes of the cross sectional area of ventricles and atria as determined by MRI and the average pixel values in regions of interest representing ventricles and atria are inversely proportional. The regions of interest were defined during systole when using an oblique plane for EIT measurements[1099].

While the isolation of the ventricles can be improved by adjusting the electrode plane, tissue displacements within the plane will still influence the measurements. Proença *et al* investigated impedance change in the ventricular region in a simulation using 32 electrodes in an oblique plane. They found that approximately 56% of the impedance change in the ventricular region was due to myocardial deformation, and thus more than to ventricular volume changes[843]. This is explained by the in-plane displacement of tissues with different impedances, for example during ventricular ejection, the myocardium is displaced by adipose or lung tissue while blood is displaced by my-ocardium[843].

11.5   EIT MEASUREMENTS FOR HEMODYNAMICS

When using EIT to monitor central hemodynamics, the relevant parameters are usually calculated from the impedance signals in either the heart region[828,1100] or central blood vessels such as the aorta[255,670,989,1159]. Perfusion measurements can also assess impedance changes caused by volume changes in smaller blood vessels, e.g. in the lungs.

EIT-based hemodynamic measurements can be assigned to three categories: measurements of pulse transit times, which have been used to determine intra-arterial pressures, flow measurements that relate the impedance change in a region of interest to parameters such as pulmonary perfusion or cardiac output, and volume status measurements that quantify heart-lung-interaction or extravascular lung water. These approaches are described in the following.

11.5.1 INTRA-ARTERIAL PRESSURE MEASUREMENT BY PULSE TRANSIT TIME

Pulse wave velocity (PWV), the velocity with which pressure pulses propagate through arteries, depends on two parameters: arterial stiffness – which is assumed to be constant over short periods of time – and arterial blood pressure[842,987]. This means that for short measurement periods, PWV can be used to assess arterial blood pressures. The PWV of a pressure pulse can be assessed by measuring the pulse transit time (PTT) between two points in the arterial tree. This can be determined from the difference in pulse arrival time (PAT) at these points or the difference between an external timing parameter and one PAT[20]. Higher pressure will lead to faster propagation of pressure pulses and thus shorter pulse transit times.

EIT can be used to measure pulse arrival times: as a pressure pulse travels through the arteries, the arterial walls distend, leading to an increased local blood volume which in turn induces a small increase in electrical conductivity[842]. This allows for PATs to be measured by EIT even without knowing the characteristics of the respective vessel. Local pressure and conductivity changes behave synchronously and will have identical PTTs[842].

Several approaches to determining PWV exist. These can then be used to assess pressures in arteries such as the aorta or pulmonary artery. So far, no studies on EIT-based measurements of central venous pressure measurements exist.

11.5.1.1 Determining Pulse Arrival Times with EIT

As mentioned above, pressure and resistivity signals are synchronous. Thus, it seems reasonable to use similar approaches for determining PATs from EIT signals as are used in traditional PAT measurements from pressure signals. PATs can be determined from a single point of the waveform such as its minimum, the maximum of its first or second derivative, or the point where a tangent to the systolic upstroke intersects a horizontal line through the curve's minimum[197]. However, the results of these methods may become erratic in the presence of extensive noise[197] which may hinder their applicability to EIT data.

Solà et al[988] proposed a parametric model for estimating PATs that uses not just one characteristic point of the pressure pulse but rather takes into account the whole wavefront as determined by a set of parameters Ω. Their parametric model of a pressure pulse waveform p_Ω is given in equation 11.7 where the amplitude of the wavefront is given by A, and the position of its inflection point and its slope are given by μ and σ respectively[988]. Here, the PAT is determined by the position of the inflection point μ.

$$p_\Omega(t) = A \tanh\left(\frac{t-\mu}{\sigma}\right) + const \qquad (11.7)$$

The PAT is then determined by minimizing the quadratic error of the measured pulse waveform compared to the parametric model p_Ω, as shown in figure 11.7. This hyperbolic tangent parametric estimator has been found to outperform state of the art techniques when applied to noisy photoplethysmography data[988].

When pulse arrival times in two regions of interest – such as the left ventricle and the descending aorta or the right ventricle and the pulmonary artery – are known, the PTT can be calculated as the

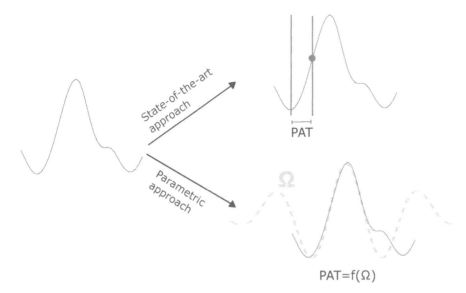

PAT

Ω

PAT=f(Ω)

Figure 11.7   Parametric estimation of PTT as proposed by Solà *et al*: Instead of using just one point in the photoplethysmographic time curve, this method takes into account the entire waveform (adapted from [988]). (color figure available in eBook)

difference between these times [987] as given in equation 11.8. In place of one of these times, an external trigger signal such as the opening of the aortic or pulmonary valve [842,987] or the R-wave peak of the ECG [841] may be used .

$$PTT = PAT(ROI_1) - PAT(ROI_2) \tag{11.8}$$

The PAT is not constant throughout the lung. Using 3D-EIT, Braun *et al* found pulmonary pulse arrival in healthy humans to have little cranio-caudal variation but significant lateral variation within the thorax, with the shortest pulse arrival times in the middle of the thorax [150].

11.5.1.2   Aortic Blood Pressure

The ability of EIT to continuously and non-invasively monitor central blood pressure was first shown by Solà *et al* in 2012 [987]. They defined the PTT as the duration from the opening of the aortic valve as determined by arterial line to the arrival of the pressure pulse in the region of the descending aorta in the EIT plane and observed a strong correlation between this PTT and the central blood pressure measured by arterial line in an anesthetized pig [987]. They used the above discussed parametric detection algorithm [988] to create a PAT-based image and detect an aortic ROI [987]. The PTT for the impedance pulses in the aortic ROI showed a very strong negative correlation with invasively measured central blood pressure, suggesting that assessing arterial pulsatility by means of EIT is capable of non-invasively monitoring central blood pressure [987]. So far, this cannot be used to measure aortic blood pressure without calibration – which requires invasive measurements – but could be used for trending.

In simulations using finite element models, Braun *et al* [148] found best performance for belt placements at the height of the heart (between the 9th and 10th thoracic vertebra) or higher which would allow for blood pressure estimations with an error below 1.4 mmHg. They also found a dependency on the reconstruction algorithm, with higher accuracy for images reconstructed with Gauss-Newton than for those reconstructed with the GREIT algorithm [148].

11.5.1.3 Pulmonary Artery Pressure

Proença et al[842] showed that in simulations, pulmonary artery pressure (PAP) could be monitored reliably by EIT. They neglected impedance changes due to respiratory activity; for in-vivo measurements, methods such as ECG-gating were applied in later studies to separate the signals (see also §11.4.2.3). They analyzed pulsatile changes in lung impedance to extract parameters related to the pulmonary pulse wave velocity from which the PAP can be determined. To select a pulmonary ROI, they used the phase shift of the first cardiac harmonic as determined by Fourier analysis. They then calculated the PTT using the intersecting tangent method with the opening time of the pulmonary valve as $t = 0$. In the simulations, correlation coefficients of $r > 0.99$ were found when comparing PTTs from pressure waveforms to EIT-derived PTTs.

Furthermore, an inverse exponential relation between pulmonary artery pressure and PTT exists, which suggests that PTT-based approaches will work best for monitoring pulmonary hypertension in its early stages where small changes in pulmonary pressure induce large changes in PTT[842].

In healthy volunteers, Proença et al[841] compared PAP calculated from EIT to values from Doppler echocardiography during hypoxia induced variations of PAP and found strong correlation scores. Here, the PTT was approximated by the PAT using the R-peak of a simultaneously recorded ECG as the time of the pulse onset. They showed that the sensitivity of EIT was sufficient to track changes in PAP of less than 10 mmHg as they found standard errors of less than 2 mmHg when translating PAT values to PAP through a least-square linear fit. This shows the feasibility of EIT for tracking trends of PAP. For the measurement of absolute PAP values, however, a calibration would be necessary, since different relations between PAT and PAP as measured by reference method exist for each subject.

Using the PAT-based method of measuring of pulmonary artery pressure described in[842] in neonatal lambs, Braun et al[151] found good correlation with measurements with a Swan-Ganz catheter. This was the first study using the gold standard for PAP measurement, i.e. a pulmonary artery catheter, as a reference.

Overall, the feasibility of EIT-derived PAP measurements has been shown. Studies using the gold standard for PAP measurement, i.e. a pulmonary artery catheter, as a reference in humans have not yet been done. Furthermore, it remains to be seen if one-time calibration would allow for calculating absolute EIT-derived PAP values over a longer time frame since so far only its application for tracking PAP changes has been successful.

11.5.2 FLOW PARAMETERS

Since the electrical resistivity of blood is well separated from that of other tissues, the resistivity of most tissues changes significantly with blood perfusion[262]. Perfusion leads to impedance changes as discussed in §11.4.1.

11.5.2.1 Pulmonary Perfusion

For measuring pulmonary perfusion by EIT, pulsatility-[268,269,304,778,982] and bolus-based[131,301] methods were described. Pulsatility-based approaches are hindered by cardiac-related impedance changes that are due not to perfusion but to impedance changes due to heart motion. Their influence can be reduced by improved belt placement and selection of a ROI that excludes the heart, but cannot be removed completely[778]. See Box 11.12 for a discussion of terminology.

EIT detects the distension of pulmonary vessels during perfusion. Since the small pulmonary vessels are the most distensible and contain the largest amount of blood in the lungs, they contribute the most to the pulsatility[269]. One of the earliest studies on EIT-based pulmonary perfusion measurements showed significant differences in the systolic impedance change in the lung region between patients with idiopathic pulmonary arterial hypertension and healthy controls[982].

For pulsatility-based analysis of pulmonary perfusion, the cardiac-related signal is isolated as described in §11.4.2 (e.g. by frequency filtering or ECG gating) and the pulsatile changes of the EIT signal in the lung region during systole ΔZ are analyzed. Thus, the focus of these studies has been the decomposition of signals[131] and the definition of lung regions of interest[304,982].

An alternative to pulsatility based measurements are bolus-based measurements (see also 11.4.1). Due to the higher conductivity of the saline bolus compared to blood, the injection causes an impedance change which depends on the two conductivities and the volume fraction of the bolus[158]. Frerichs *et al*[301] used saline bolus injections to generate local time-impedance dilution curves and showed that these could be used to image lung perfusion.

To assess pulmonary perfusion, Borges *et al*[131] analyzed the pixel-wise time impedance curves after injection of 20% NaCl into the right atrium as a contrast agent during apnea. Due to the difficulty of determining a transit time, they opted for calculating the blood flow not as relative volume per mean transit time but rather based on a maximum slope method as given in equation 11.9[131].

$$\text{bloodflow}_{\text{pixel}} \propto \left[\frac{dm(t)}{dt} \right]_{\text{max}} \tag{11.9}$$

Here, $m(t)$ represents the mass of contrast in the respective pixel. Based on this, they were able to calculate the relative perfusion of lung regions as the proportion of the maximum slope of the lung perfusion related EIT time signal relative to the sum of slopes of all other pixels[131].

When comparing the results, this method showed better agreement with lung perfusion measured by single-photon-emission computerized tomography than pulsatility based EIT perfusion analysis, especially in the presence of atelectasis[131].

11.5.2.2   Regional $\dot{V}/Q$ Matching

Ventilation ($\dot{V}$) and perfusion (Q are two functionally linked steps in the oxygen transport chain. In healthy conditions, $\dot{V}$ and Q match both at a global level and regionally within the lungs. Mutiple mechanisms exist to enforce this regional match, the most important of which is hypoxic vasoconstriction which reduces blood flow to poorly ventilated lung regions. Information of regional $\dot{V}/Q$ matching is an invaluable diagnostic and therapeutic tool for critically ill patients at the bedside. One of the exiting promises of EIT is its ability to measure $\dot{V}$, Q and their match.

Superficially, creating $\dot{V}/Q$ EIT images appears simple: simply divide the pixel values of the two functional images. However the challenge is that low Q leads to large Q^{-1} and large spikes in the image. Computational strategies manage these images are required.

One recent approach caculates regional $\dot{V}/Q$ as follows[450,451]: starting with the values of perfusion image P_i and ventilation image $\dot{V}_i$ for each pixel i, a threshold is chosen such that region k is ventilated if

$$\dot{V}_k > 20\% \cdot max(\dot{V}_k) \tag{11.10}$$

and a region g is perfused if

$$P_g > 20\% \cdot max(P_g) \tag{11.11}$$

Three regions were identified based on ventilation/perfusion patterns: regions that were only ventilated (R_V), regions that were only perfused (R_P) and regions both ventilated and perfused (R_{V+P}). The following $\dot{V}/Q$ matching from EIT-derived parameters were calculated:

$$\text{DeadSpace}_\% = R_V/(R_V + R_P + R_{(V+P)}) \cdot 100\% \tag{11.12}$$

$$\text{Shunt}_\% = R_P/(R_V + R_P + R_{(V+P)}) \cdot 100\% \tag{11.13}$$

$$\text{VQMatch}_\% = R_{V+P}/(R_V + R_P + R_{(V+P)}) \cdot 100\% \tag{11.14}$$

Clinical studies have shown bolus based EIT method has the potential to quantitatively assess regional V/Q match pattern at various pathophysiologic conditions (such as PE, ARDS, response to PEEP change etc.)[450,451,697,1040]. He *et al* found the patients with pulmonary embolism (PE) had significantly higher DeadSpace$_\%$ and lower VQmatch$_\%$ (calculated by EIT) than patients without PE in a prospective observational study[451]. A cutoff value of 30.37 for DeadSpace$_\%$ resulted in a sensitivity of 90.9% and a specificity of 98.6% for the PE diagnosis[451]. Mauri *et al* using the similar EIT method to show a larger prevalence of ventilated nonperfused lung units (dead space) in comparison to perfused nonventilated units (shunt) in the COVID-19 patient with ARDS[697]. Moreover, a real-time monitoring of regional $\dot{V}/Q$ matching could be generated by the pulsatility method. Another approach is to calculate instead the difference $D_{\dot{V}-Q} = \dot{V}_{norm} - Q_{norm}$, which does not suffer from division by zero. Nomalized values $\dot{V}_{norm}$, Q_{norm} can be calculated by dividing by the average value. In this case normalized values of zero represent uniformly ventilated and perfused regions.

Regional $\dot{V}/Q$ matching assessment from the combined measurement of ventilation and perfusion by EIT would lead to a clinically useful combination of key physiological variable, and be helpful to characterize the etiology of respiratory failure and guide the individual mechanical ventilation setting.

11.5.2.3 Cardiac Output and Stroke Volume

Cardiac output (CO) is defined as the product of stroke volume (SV) and heart rate (HR) as given in equation 11.15 and discussed in §11.2.2.

$$CO_{l/min} = SV_{l/beat} \cdot HR_{beats/min} \tag{11.15}$$

While a good correlation of HR determined from variations of cardiac impedance with reference methods has been shown[828], determining stroke volume from EIT is more challenging. Several approaches to EIT-based SV-measurements exist. They calculate SV from a part of the EIT image such as the cardiac ROI[828], ventricular ROI[1100], or aortic ROI[670].

Vonk Noordegraaf in 1996 *et al*[1099] suggested that EIT might be used to assess stroke volume, when they found a mean impedance increase of 34(13)% in the ventricular region during exercise in a group of ten healthy male volunteers while expecting a stroke volume increase of 40% in the same period. In 2000, Vonk Noordegraaf *et al*[1100] calculated EIT-derived stroke volume using ECG-gated EIT measurements from 25 patients and found a strong correlation between these measurements and stroke volume measured by thermodilution.

By means of regression analysis, Vonk Noordegraaf determine the stroke volume SV_{EIT} from the EIT signal in a ROI containing the ventricular region from the signal's amplitude and length in time[1100]. Despite the correlation found in this study, the applicability of the derived equation may be limited since the signal amplitude will commonly depend not only on stroke volume but also electrode placement and patient physiology.

Another approach is used by Pikkemaat *et al*[828]. They calculate a cardiac impedance signal from the sum of all EIT pixels in the cardiac region and from this derive a parameter Z_{SV} that is calculated during each cardiac cycle as given in equation 11.16.

$$Z_{SV} = Z_{Card}(dia) - Z_{Card}(sys) \qquad (11.16)$$

Here, $Z_{Card}(dia)$ is the impedance in the heart region during diastole and $Z_{Card}(sys)$ during systole. Pikkemaat *et al* then calculate a linear regression between this parameter and SV determined by transpulmonary thermodilution to determine an EIT derived stroke volume in ml[828]. They observe that changes in PEEP level may lead to a scaling of the SV determined by EIT, possibly due to changes of lung volume and heart position[828].

Later studies also focused on deriving changes in stroke volume from measurements in the region representing the aorta in the EIT image[670,1051]. These measurements of stroke volume variation will be discussed in the next section.

A more recent work calculating stroke volume from the cardiac-related impedance change in the heart region (as given in 11.16) was presented by Jang *et al*[510]. After separating cardiac and ventilatory impedance signals using a PCA- and ICA-based approach (see section 11.4.2.3), they were able to simultaneously measure stroke volume and tidal volume in six pigs.

In contrast to the work focused on the impedance changed in the heart region, in 2020, Braun *et al* found it impossible to use this signal for measuring stroke volume changes, yet found acceptable trending performance for cardiosynchronous impedance changes in the lung region[152].

For future studies, machine learning based approaches have shown some promise. Murphy *et al*[756] showed that a machine learning regression model can extract left ventricular volume even in the presence of poor electrode contact and validated this approach on a simulated thorax and in a tank measurement.

While many authors have described promising results for CO measurement, there are some good reasons to think that it will be difficult to obtain accurate measures between patients. First the thoracic signal is only partially determined by blood flow; it has a very large component of heart movement[843]. The relative contributions will be affected by many factors: the posture of the subject, the exact electrode postion, the exact position of the heart in the thorax, and the amount of tissue between the heart and the body surface electrodes. Calibrating for each factor is possible in principle, but requires information on the anatomy and advanced techniques. However, it is likely that these factors remain mostly constant over time, and thus trends in CO should be more reliable[152].

11.5.3   VOLUME STATUS

As discussed in 11.2.3, the main goal of volume status management is avoiding hypovolemia – as indicated by increased heart-lung-interaction – and hypervolemia, which leads to an excess of extravascular lung water. Both sides of this spectrum can be assessed by EIT-derived parameters which will be discussed in the following.

11.5.3.1   Stroke Volume Variation and Heart-Lung-Interaction

One method for measuring EIT-derived stroke volume variation SVV_{EIT} was proposed by Solà *et al*[989]. They patented a method to calculate EIT-based heart-lung interaction (*HLI*) from which SVV_{EIT} can be obtained by a pre-trained linear transformation based on fitting with reference SVV[989]. Their approach is based on the EIT time signal in a ROI in the descending aorta from which they calculate the frequency values using a Fourier periodogram technique as shown in figure 11.8[989]. The frequency with maximum power density in the periodogram can be either the

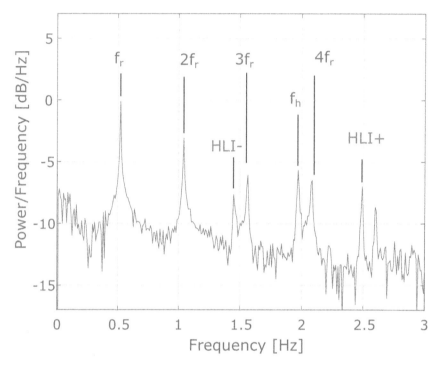

Figure 11.8   A Fourier periodogram obtained from a time series of one EIT pixel can be used to determine the frequencies of cardiac- and respiratory related activity f_h and f_r and the frequencies of heart-lung interaction $HLI+$ and $HLI-$. These can be used to calculate a heart-lung-interaction factor as described in [989]. Also marked are the harmonics of the respiratory frequency. (color figure available in eBook)

frequency of heart activity f_h, in which case the frequency with a maximum power density lower than this frequency will be the frequency of respiratory activity f_r; or the frequency with maximum power density is f_r – as in figure 11.8 – in which case f_h will be the frequency with maximum power density at frequencies higher than f_r (excluding harmonics of f_r) [989].

Then, a frequency value of heart lung interaction f_{HLI+} or f_{HLI+} can be calculated as in equation 11.17 [989]; these frequencies are marked as HLI+ and HLI- respectively in figure 11.8.

$$f_{HLI+} = f_h + f_r \qquad (11.17)$$

The energy at the frequencies of the heart activity E_h and at that of the heart-lung interaction frequency E_{HLI} can be calculated from the periodogram and from this HLI can be obtained with equation 11.18 [989].

$$HLI = 2 \cdot \frac{E_{HLI}}{E_h} \qquad (11.18)$$

To calculate SVV_{EIT} from this heart-lung interaction factor, a linear transformation obtained by fitting HLI against reference SVV values is necessary [989].

A similar approach is calculating SVV_{EIT} by applying a pre-trained linear transformation directly to the energy at intermodulation frequency $f_h - f_l$ [670]. With this approach, Maisch *et al* found significant correlation of SVV_{EIT} and SVV measured by aortic ultrasonic flow probe as well as

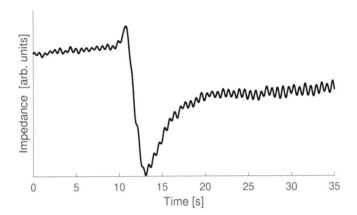

Figure 11.9   Typical change in impedance in a pixel of the EIT image representing the aorta during hypertonic saline bolus injection during apnea.

arterial pulse contour analysis, but limits of agreement were wider than commonly accepted[670]. Later investigations found reasonable correlation between SVV_{EIT} and SVV measured by pulse contour analysis in healthy porcine lungs, but wider limits of agreement after acute lung injury[1051].

One aspect that makes measuring HLI by EIT challenging is the weakness of the respective impedance changes. While the cardiac-related impedance changes only account for less than 10% of the total EIT signal (with most of the impedance change being due to ventilation), the impedance changes induced by HLI in turn only account for about 10% of the total hemodynamic signal[1051]. Therefore, current algorithms for EIT-based measurement of SVV require a low signal-to-noise ratio[1051].

When using an approach based on the EIT signal in the region of the descending aorta, one of the main challenges is the detection of aortic pixels. In recent years, detection of the aortic region of interest has been discussed in multiple studies[255,670,989,1159] and will be described below.

11.5.3.2   Detection of Aortic ROI

One way to detect the aorta and other structures of interest is based on calculating the pulse arrival time for all pixels in an EIT image[670,989,1051]. During a cardiac cycle, a pressure pulse travels through the cardiovascular system resulting in a corresponding impedance pulse. To each pixel in the EIT image, a time value of pulse arrival can be assigned. Since the propagation velocity in the aorta is high, pixels representing the descending aorta can be identified as those pixels anatomically located behind the pulmonary arteries that have the shortest pulse arrival times and can be identified and clustered automatically[670,988,1051].

Another possibility of aortic detection in the time and space domain is the use of a hypertonic saline bolus that is injected into the aortic arch[167,1032,1159]. When passing through the aorta, the bolus causes an increase in conductance as shown in figure 11.9. This change in conductance becomes clearly visible in the EIT image in pixels near the aorta as can be seen in figure 11.10. Thuerk et al calculated a prominence image from local maxima and neighbouring minima and define the aortic pixel as the maximum of this image[1032].

When comparing the position of the aorta as detected by EIT to its position in a CT image in the same plane as the EIT belt, an offset can be observed[1032]. This displacement of the aorta is dependent on image reconstruction which also affects the spatial extension of the detected aorta[1034].

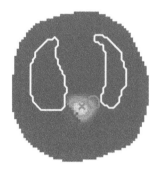

Figure 11.10   EIT image of a domestic pig during injection of hypertonic saline into the aortic arch. Contours of lungs (white) and aorta (red) as detected by CT are overlayed; the detected aortic pixel is marked by a red asterix. (color figure available in eBook)

Trepte *et al*[1051] used pooled functional EIT images of 30 domestic pigs to define a standard aortic location, but acknowledged that this location may not be perfectly suited for all animals.

11.5.3.3   Extravascular Lung Water

Healthy lung tissue consists of 5% fluid and 80% air[157]. The differences in conductivity of fluid, lung tissue and air allow for monitoring of pulmonary edema by EIT.

Different methods for EIT-based measurement of extravascular lung water have been proposed. The earliest approach by Newell *et al*[771] and others[17] showed a correlation between lung fluid volumes and the EIT measurements. In 2000, Noble *et al*[783] showed a relationship betwen the change in total lung impedance after diuretic induced fluid loss and a correlation between the decrease in impedance and urine output.

In a study with eleven patients suffering from pleural effusion, Arad *et al*[54] found a high correlation between lung resistivity measured with an eight-electrode system and and the volume of fluid removed.

In 2016, Trepte *et al*[1052] quantified extravascular lung water in pigs based on changes in EIT images during lateral body rotation and found that their results showed significant correlation with lung water as measured by postmortem gravimetric analysis. Assuming that lung water shifts downwards towards the dependent parts of the lung and thus redistributes with changes in position, they calculate an imbalance coefficient *IM* in three different positions as given in equation 11.19 that describes the imbalance between left (TV_L) and right (TV_R) tidal volume[1052]. The lung water ratio$_{EIT}$ quantifies the change in this imbalance coefficient with the rotation angle[1052]. In 2019, this work was followed up in humans by Zhao *et al*[1207] who could not reproduce its results in all subjects. The authors suggested that the fluid induced by lavage is inherently more mobile due to posture change than the fluid from clinical edema.

$$IM = \frac{TV_L - TV_R}{TV_L + TV_R} \qquad\qquad (11.19)$$

11.6   SUMMARY AND OUTLOOK

The use of EIT for hemodynamics is very promising, especially for use in a clinical setting where EIT may already be in use for visualizing ventilation parameters. So far, the results of EIT-based measurements for pressure, flow, and volume status monitoring have confirmed the applicability of EIT for these measurements. However, more studies will be necessary before these measurements could be validated in clinical use or may be part of clinical practice. So far, to the best of our knowledge, none of these technologies are in clinical use.

The inherent link between blood pressure and pulse wave velocity means that the ability to reliably measure pulse wave velocity with EIT will allow EIT-derived blood pressure measurements. However, all studies so far needed an external measurement parameter such as ECG or even invasive measurements to determine the pulse arrival time. While aortic pressure measurements have been successfully performed in animal experiments and pulmonary artery pressure measurements have shown good results in simulations and a small number of volunteers, none of these measurements have been compared to the respective gold standard method in humans. Furthermore, without individual calibration, these measurements can so far only be used for tracking pressure changes. All of these calculations are based on quantification of cardiac-related pulsatility in the EIT data, which have been shown to have various origins besides perfusion.

Despite the challenges described above, perfusion measurements seem to be the EIT-based hemodynamic measurements that are the most likely to become widely used in the near future. Pulsatility-based measurements require no additional information, though ECG-gating may be used. They can be done at the same time as ventilation measurements and have been used to determine ventilation-perfusion ratios and from this to diagnose pulmonary embolism or assess response to PEEP. Although bolus-based measurements have also been used, these methods are more invasive. In contrast, pressure measurements remain more challenging since they require not only the separation of cardiac- and ventilatory signals but also aim to measure pulse-transit-times that are of the same order of magnitude as the sampling rate of most EIT devices, meaning the precision of the results may be limited.

While studies on measuring cardiac output with EIT have been done for more than 20 years and many different approaches have been suggested, a consensus on the most suitable measurement setup and data analysis has not yet been reached. The same is true for measurements of heart-lung-interaction.

There is dramatic promise for reliable EIT-based hemodynamic measurements, since they would be continuous and non-invasive. To advance EIT-based hemodynamic measurements, we identify the following requirements:

- Precision and accuracy of EIT monitoring for various pathophysiologic conditions needs to be validated in big sample clinical studies.
- Consensus on the EIT-based hemodynamic parameters, analysis, cutoff values and potential application principle needs to be reached.
- Validate whether EIT-based hemodynamic measurements can diagnose some significant pathophysiologic change and help the management of hemodynamics in critically ill patients. Promising areas are the assessment of fluid response, the identification of severely low cardiac output, and the identification of severe regional lung perfusion defects.

In the long run, hemodynamic EIT measurements could become widely used in the perioperative setting wherever EIT-based monitoring is already in use. The fact that a wide array of studies have shown its applicability when used with existing hardware makes this technology very promising. Whether the methods can one day replace invasive measurements or if they will remain an additional approach to estimate hemodynamic parameters or track their changes remains to be seen. Ideally, a future EIT device could combine the current ventilatory monitoring with a wide array of hemodynamic parameters and thus provide a comprehensive real-time monitoring of patients.

12 EIT Imaging of Brain and Nerves

David Holder
Departments of Clinical Neurophysiology and Medical Physics,
University College London and University College London Hospitals,
London, UK.

CONTENTS

DOI: 10.1201/9780429399886-12

12.1   INTRODUCTION

In the neurosciences, two broad areas may be defined in which non-invasive imaging methods could provide useful information – imaging of variations or abnormalities in structure, and imaging of normal or abnormal functional activity.

The ease of diagnosis of structural abnormalities in neurology has been transformed since the development of X-ray computed tomography and magnetic resonance imaging (MRI). Both now are capable of imaging structural abnormalities in the brain with an accuracy of less than one millimetre. For the great majority of diagnostic requirements, the advantages of accurate spatial resolution outweigh the expense and inconvenience of these methods. The advantages of Electrical Impedance Tomography (EIT) are that it is relatively inexpensive, safe, non-invasive and portable. Set against this is a relatively poor spatial resolution. In currently available devices, this is about 10% of the electrode array diameter. Unless there is a breakthrough in image reconstruction, its spatial resolution will be limited by the physical process that current spreads out throughout the whole subject, so that the inverse problem is less well defined than in X-ray CT or MRI. It therefore seems most unlikely that EIT will be able to compete directly with these techniques for high resolution structural imaging in the foreseeable future. However, its advantages may still enable it to be indispensable for monitoring structural changes at the bedside, in casualty departments, or in remote locations where large scanners are too expensive or impractical, such as in acute stroke.

On the other hand, there is a great need for improved methods of imaging functional activity in the nervous system. At present, a great deal is known about behaviour, localization of activity in the brain and cellular neurophysiology, but there is limited understanding of how information is processed in neuroanatomical pathways. The challenge is that such activity is widely distributed and occurs with a timescale of the order of milliseconds. No system yet exists which could measure such activity non-invasively and with a high temporal resolution. One avenue of approach is to image changes in blood flow and metabolic activity events which are related to nervous activity. These are caused by the accumulation of the effects of many action potentials or depolarizations. They are therefore easier to image, being large, but can only give an indirect guide to nervous activity. Such changes may already be imaged by Positron Emission Tomography or functional MRI. The temporal resolution of these techniques is seconds or tens of seconds because this is the timescale over which these changes in the brain occur. Measurement of nervous activity with a much greater temporal resolution of tens of milliseconds has been possible for decades with Electroencephalography (EEG) and, more recently, by Magnetoencephalography (MEG), but these do not provide unique solutions and source imaging is of doubtful accuracy, especially for deep or distributed sources.

If neuroimaging with EIT is successful, then it could be used in several clinical areas to which other methods of functional brain imaging are unsuited. These could be in adults and infants receiving intensive care, and the long term imaging of epilepsy on telemetry units, where prolonged periods of monitoring are required in order to localise seizure activity in the pre-operative assessment for epilepsy surgery. EIT may also be suited to provide images of brain impedance changes brought about by cell swelling in cerebral energy failure, in such pathological conditions as stroke, ischemia, head injury, hypoxia or hypoglycemia. It could also be used to monitor shifts in cerebral fluid balance, such as in the use of mannitol to decrease cerebral edema. It also has the unique potential to provide a means of imaging the tiny fast impedance changes due to opening of ion channels during neuronal depolarization. This would provide a means of imaging neuronal activity along neuroanatomical pathways with a temporal resolution of milliseconds, which would constitute a revolutionary development in neuroscience technology.

The development of EIT for imaging brain function commenced in the 1980's. An impedance scanning system for detecting brain tumours was designed and tested but was not followed up with a practical EIT device[100]. Shortly after, I[475] proposed EIT as a novel means for imaging the fast impedance changes known to occur during neuronal activity in the brain. Pilot animal studies were then performed in which simultaneous scalp and intracranial impedance measurements were made of the brain of anaesthetised rats during cerebral ischemia[477]. The conclusion was that measurements of brain impedance could be made, non-invasively, by scalp electrodes, although these changes were attenuated by the skull. This study indicated the practicality that EIT could be used to image impedance changes in the human brain. At that time, the only available EIT system was the Sheffield Mark 1 EIT system[161], which was limited in that current could only be applied through adjacent electrodes and in 2D with a single ring of electrodes. This system was unlikely to be able to image impedance changes in the brain from scalp electrodes, as most of the applied current would be shunted through the scalp. As the EIT technology was not at the stage to inject current with more widely spaced electrodes, the Sheffield Mark 1 was used, and experiments were designed to eliminate the effect of the skull. In these, the effect of the skull was excluded by using a ring of electrodes placed on the exposed cerebral cortex of anaesthetised rats or rabbits. The first EIT study of brain activity was in artificially induced stroke[478], followed by EIT imaging during cortical spreading depression[126], physiologically evoked responses[483] and during electrically induced seizures[863]. The impedance changes varied between a 2–5% decrease during somatosensory or visual stimulation, a 10% increase during seizures or up to 100% during stroke, due mainly to cell swelling and blood volume changes. Taking the evidence that functional activity changed brain impedance in the rabbit by 2–5%, and that from rats that the skull attenuated peak impedance changes by a factor of 10, it seemed plausible that scalp impedance changes of 0.2–0.5% might be detected non-invasively during functional activity in humans.

These initial studies were undertaken in our group at University College London. In the intervening two decades, we have pursued the goal of imaging nervous activity with EIT. Sadly, the original hope of a non-invasive EIT system with scalp electrodes able to image localised cerebral fast neural or fMRI-like slow activity has not yet been possible. However, we have been able to produce EIT systems able to image fast neural activity in brain and peripheral nerve but with intracranial or perineural cuff electrodes respectively. These are complete and ready for use by collaborating groups in experimental and clinical neuroscience. We were also unable to produce EIT images in acute stroke in the brain with scalp electrodes, but there is now considerable interest in these applications from several groups. It may still be possible to realise non-invasive stroke imaging with advances in machine learning algorithms.

When the first edition of this book was published in 2005, our group had published the great majority of papers. Fifteen years later, it is gratifying that several groups are active in this field. A leading group in this respect has been at the 4th Military Medical University in Xian, China. They have worked on both technical development and physiological and clinical studies. Their principal focus has been in time difference measures with a 2D ring of scalp electrodes in conditions such as intracranial hemorrhage and fluid shifts after head injury. At the time of writing, brain and nerve EIT is still not yet in routine hospital use. However, both the UCL and 4MMU groups are developing commercial collaborations with the aim of making systems available for end-users in science and medicine. By the time of the next edition of this book, neural EIT may well have found its way into widespread experimental and hospital use.

In this chapter, I initially review the physiological basis for expecting impedance changes during these conditions. I then review the development and testing of hardware and reconstruction algorithms specifically for imaging brain and nerve function. Finally, I review animal and human studies in the development of EIT for imaging brain function in the areas of EIT of normal brain function, epilepsy, other disorders such as cerebral edema and stroke, and peripheral nerve.

12.2   PHYSIOLOGICAL BASIS OF EIT OF BRAIN FUNCTION

12.2.1   BIOIMPEDANCE OF BRAIN AND NERVE AND CHANGES DURING ACTIVITY OR PATHOLOGICAL CONDITIONS

The bioimpedance of tissues in the head is relevant in two main ways. EIT of the brain poses an especially difficult, but not insuperable, problem because the brain is encased by a conductive covering, the cerebro-spinal fluid, two layers with high resistivities, the pia mater and skull, and then the scalp, which has a moderate resistivity. Secondly, there are time related changes in impedance in the brain itself, which provide the opportunity for imaging with EIT. Absolute imaging with EIT is not yet possible, (see §6.7), so structural imaging of pathology such as brain tumours or infection is not considered here. These time related changes fall into three main categories: i. Changes over periods of hours or days related to fluid balance or stroke. ii) Changes over tens of seconds, due to cell swelling and blood flow, which are relatively large, of the orders of ten to one hundred per cent. iii) Those due to the opening of ion channels during neuronal activity, which occur over milliseconds, and are much smaller: $\sim$0.1 – 1% recorded locally in the brain or nerve and $\sim$2–3 orders of magnitude smaller if recorded on the scalp.

These are reviewed in reasonable detail in this section, as their magnitude is critical to the design of experiments to ascertain the utility of EIT in imaging brain function. A knowledge of the basic anatomy and histology of the brain has been assumed. (In this section, published impedivity and resistivity values have been converted to Ω·m).

12.2.1.1   Impedance of Resting Brain

Within the brain, applied current will be distributed through several anatomical or physiological compartments. The cerebral blood volume fraction has been estimated at 3–10%[1071]; blood has a low resistivity of about 1.25 Ω·m at 50 kHz[823]. The extracellular space has been estimated by dye dilution techniques in rats as 12–18% of the brain volume[239]. Its resistivity can be estimated from measurements of the ion concentration of the extracellular space cat sensorimotor cortex which is similar to 0.9% saline at 2.0 S/m[335].

Neurones and glial cells comprise the remaining 80% of the volume of the brain. Their contribution has been analyzed in rabbit cerebral cortex[859]. It was He calculated that the path of a low frequency current in the brain would be predominantly through the large volume, low resistivity, glial cells, conductive extracellular fluid space and blood volume. This is because, although the blood and extracellular space have a lower resistivity than glial cells, they have less conductive volume, and the bulk of the current flow would be through the glial cells. Glial cells are conductive because they are permeable to potassium and chloride ions[660], unlike neurones which have a highly insulating membrane which is only permeable to ions during depolarization with the action potential or during cell energy failure. As a result, only a small amount of current will conduct through the intra-cellular space of neurons at rest. A little conduction does occur through neurones at low frequencies because some of the long processes which enable transmission of nervous impulses – axons and dendrites – may be aligned with the direction of current flow. Compared to the transverse case, the surface area of an individual neuronal process is much greater if the current flows along it, so the resistance is lower and more current enters the intracellular space.

On the macroscopic scale, the brain mainly comprises grey matter, which is made up of neuronal cells and their immediate branching processes, and white matter, which comprises tracts of long nerve fibres which connect different regions of the brain. Nerve fibres in the mammalian brain are largely surrounded by an insulating myelin sheath, and so are anisotropic. There was anisotropy of about 10:1 in the impedance of cerebral white matter in cats over 20 Hz to 20 kHz[779] – for example, 0.11 S/m for the longitudinal fibres compared to 1.25 S/m for the transverse ones at 20 Hz. Grey matter is largely isotropic as nerves and their processes run randomly. However, Ranck[859] noted that there is lamination in the cortex, so this is only true at distances greater than 200 μm. In rabbit cerebral cortex in vivo, at 5 Hz, the conductivity was 0.31±0.04 S/m (mean±S.D.), falling to 0.43±0.06 S/m at 0.5 kHz. When the shunting effect of the blood vessels was taken into account, the conductivity values fell to 0.28 S/m for 5 Hz and 0.39 S/m at 0.5 kHz. Latikka[606] recorded the impedance of white and grey matter in situ using a needle electrode in human subjects undergoing brain surgery for deep brain tumours. Koessler *et al* also recorded in human epileptic subjects using intracranial depth electrodes and observed resistivities of 0.36 or 0.85 Ω·cm for grey and white matter respectively[566]. In summary, brain grey matter resistivity at frequencies below 100 kHz is about 3 Ω·m in vivo, and white matter, depending on orientation, is about 50% higher (table 12.1).

Table 12.1
Conductivity of cerebral white and grey matter in vivo. All measurements were made at body temperature (37–38°C) in vivo.

Ref	σ_{Cortex} (S/m) ± SD	$\sigma_{White Matter}$ (S/m) ± SD	Frequency	Method
[310]	0.44±0.02	0.29	1 kHz	4 electrodes
[779]	1.18–0.125	20 Hz–20 kHz	20 Hz–20 kHz	Point electrode and remote electrode
[1070]	0.48±0.01 0.46 with correction for blood conductivity	Specific impedance of white matter/cortex = 4.6±0.2	1kHz	Grey and white matter combined 2 electrodes used
[859]	0.28-0.39		5 Hz–5 kHz	Point electrodes on cortex.
[606]	0.28	0.26	50 kHz	Monopolar needle electrode
[566]	0.26	0.17	50 kHz	Intracranial depth electrodes

12.2.1.2   Anoxic Depolarization and Cerebral Ischemia

When cerebral grey matter outruns its energy supply, a characteristic sequence of events takes place. This is termed "anoxic depolarization"[170], because it occurs during pure hypoxia, but the term has been extended to include the similar events which occur in ischemia, spreading depression or epilepsy. These events have been mostly studied in the cerebral cortex, but also occur in other areas of grey matter in the brain. When measured in the cerebral cortex, the characteristic event is that spontaneous electrical activity ceases and a sustained negative shift of tens of millivolts is recorded with an electrode on the cortical surface. These events are accompanied by a substantial movement of ions and water, as ionic homeostasis fails. Water follows sodium and chloride into cells, so that the extracellular space shrinks by about 50%[434]. At frequencies up to 100 kHz, the great majority of current applied to the brain passes through the extracellular fluid. This component of current is resistive and so is measured by EIT systems, which usually measure the in-phase component of the impedance. During anoxic depolarization, the impedance of grey matter in brain therefore increases, because the extracellular space shrinks. Changes in temperature, the impedance of neuronal membranes, and blood volume may also contribute, but the effect due to cell swelling is greatly predominant (figure 12.1). Changes of this type occur to differing degrees in the pathological

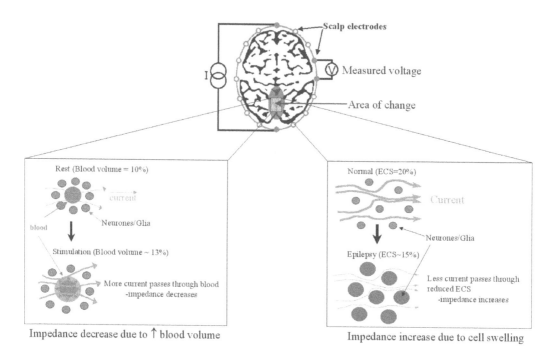

Figure 12.1   Mechanisms of time dependent slow impedance changes in the brain. Left figure: Impedance decrease due to increased blood volume. During physiological activity, a signal is sent to the blood vessels which increases blood flow and blood volume to that cortical area. As blood has a lower resistivity than the surrounding brain (1.5 Ω·m and 3.5 Ω·m, respectively), the increase in the lower resistivity volume of blood will allow more current to flow through that area of tissue and decrease the bulk impedance of that volume of cortex. Right figure: Impedance increase due to cell swelling. Cells expand during cell swelling (bottom). At rest, the size of the conductive extra-cellular space (ECS) is about 20% of the brain volume. During epilepsy, moderate cell swelling occurs as water and ions enter the glial cells and the neurones, and the volume of the low resistivity ECS is reduced. This increases the bulk impedance of that volume of cortex. Larger changes of cell swelling and impedance occur during ischaemia and spreading depression. (color figure available in eBook)

conditions of stroke (or cerebral ischemia), spreading depression, and epilepsy. In each case, the cells run out of energy needed to maintain the balance of water and solutes between the intra- and extra-cellular spaces. In stroke, this is because blockage of arteries leads to an insufficiency of blood; in spreading depression or epilepsy, it is because intense neuronal activity exceeds the capacity of the blood to provide energy supplies.

Large impedance increases of about 20–100% occur during cerebral ischemia in species such as the rat[477], cat[692] and monkey[327]. Spreading depression is a phenomenon which can be elicited in the grey matter of experimental animals by applying potassium chloride solution or mechanical trauma. Intense activity in depolarized cells occurs, so that potassium and excitatory amino-acids pass into the extracellular space. These then excite neighbouring cells by diffusion. In this way a concentric "ripple" of activity moves out from the site of initial disturbance like a ripple in a pond. It moves at about 3 mm per min, and has been postulated to be the cause of the migraine aura in humans[184]. Impedance increases of about 40% occur in various species[170]. During epilepsy induced in experimental animals, reversible cortical impedance increases of 5–20% have been observed during measurement at 1 kHz with a two electrode system in the rabbit or cat[1072]. The changes had a duration similar to the period of epileptic EEG activity and were due to anoxic depolarization-like processes, as a negative DC shift occurred. Similar changes have been observed in cat hippocampus, amygdala, and cerebral cortex[254,963]. Impedance increases of about 3% have been recorded in humans during seizures[480].

These increases during cerebral ischemia may be reversible if it is transient and reversed within a few minutes[478,692]. If it is established, then irreversible changes take place by about 3 hours. The impedance spectra of normal and ischemic brain and clotted blood vary significantly (figure 12.2). Although the absolute spectra vary widely, it is the relative changes over frequency that appear to be most relevant for EIT. This is because EIT at present cannot recover absolute images or spectra. Imaging of the relative spectra over frequency may be possible. In this regard, the largest differences occur below 250 Hz. Healthy in vivo rat brain had a decrease of 15% between 0–250 Hz while ischemic brain decreased by 7%, and clotted blood by 9%. Thereafter blood was largely invariant up to 2 MHz[245].

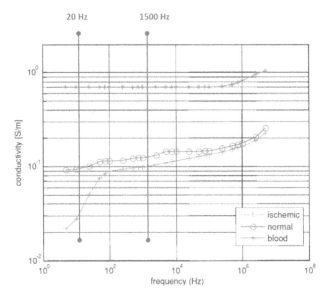

Figure 12.2   Conductivity spectra of ischaemic and normal brain and blood. (color figure available in eBook)

12.2.1.3   Slow Impedance Changes During Functional Activity

Impedance has been shown to change in the brain during physiological stimulation, but by a much smaller amount. Adey *et al.*[12] measured impedance at 1 kHz using chronically implanted electrodes in the limbic system of the cat. They observed impedance decreases of about 2% which lasted for several seconds during physiological stimuli, such as presentation of milk or exposure of a female to a male. Aladjalova[35] observed similar impedance changes in cerebral cortex after direct electrical stimulation.

The cause of such changes has not, to my knowledge, been directly investigated. The most likely explanation is that blood volume and flow alter. Changes in blood volume will alter tissue impedance, either by replacing a fluid of different resistivity (such as CSF), or by changing the cross-sectional area available to current flow. Changes in blood flow can also alter impedance, because erythrocyte alignment alters[215]. It is well established that blood flow and volume increase in the brain during functional activity. For example, in cat visual cortex during visual stimulation, changes in volume, recorded by reflected light at 570 nm, occurred almost immediately after stimulus onset and preceded change in flow recorded by Laser Doppler flowmetry by 2 seconds; both changes peaked at 5–6 seconds after stimulus onset and decayed to baseline within 6s of stimulus cessation[680]. The blood volume therefore increased prior to changes in blood flow, probably because of venous pooling in advance of arterial dilation. In rats, contrast MRI was used to give high resolution maps of changes of cerebral blood volume during forepaw and hindpaw stimulation[805]: a stimulus lasting 5 minutes increased blood volume 3–6 s after the onset of stimulation, which returned to baseline 13–51 s after stimulus cessation. In humans, similar changes of regional cerebral blood flow during visual stimulation have been observed with PET and functional MRI. The time course of the blood flow response from fMRI studies is similar to that measured in animals: blood flow increases 1–2 s after stimulus presentation, rises to a peak at 5–7 s and then decays to baseline blood flow within 6-10 s of stimulus cessation.

12.2.1.4   Functional Activity with the Time Course of the Action Potential

In both the possible applications described above, similar changes can at present be imaged by other, existing, methods; the advantages of EIT would be of a practical nature. There, is, however, a third possible application of EIT in neuroscience, in which it would have a unique advantage in being able to image nervous activity with a temporal resolution of milliseconds. The application would be based on the well-known change in impedance of neural membranes which occurs on depolarization as ion channels open. In the squid giant axon, impedance falls forty-fold[203] when measured directly across the axon. There should therefore be an impedance change across populations of cells in nervous tissue during activity. The effect could be due to action potentials in white matter, or to summated effects of synaptic activity in grey matter, which is the origin of the EEG.

At the frequencies of measurement with EIT, most current passes in the highly conductive extracellular space. The amplitude of the impedance changes across tissue is therefore likely to be small. Klivington and Galambos[560] measured impedance changes during physiologically evoked activity in the auditory cortex of anaesthetized cats at 10 kHz. A maximum decrease of about 0.005% was observed, which had a similar time course to the evoked cortical response. Similar changes were measured in visual cortex during visual evoked responses[559] and less reproducible impedance decreases of up to 0.02% were observed in subcortical nuclei during auditory or visual evoked responses in unanaesthetized cats[1082]. Freygang and Landau[310] observed a maximum decrease in impedance of 3.1% measured with square wave pulses 0.3 – 0.7 ms long during the evoked cortical response in the cat.

Over the past decade, our group at UCL has recorded impedance changes during neuronal depolarization during activity in brain and nerve. In brain, there is a predominant negative impedance change of a peak of -0.13% at 1475 Hz, falling off rapidly at lower or higher frequencies (-0.04 at 1025 Hz or -0.05% at 9975 Hz[270]. The peak signal-to-noise ratio was 30 at 1475 Hz. Changes were slightly larger when recorded in the thalamus, -0.4% at 1475 Hz[270], and during epileptic spikes in rat cerebral cortex, -0.36%[429]. During the compound action potential in peripheral nerve, changes were similar at -0.9% at DC[345], -0.25% at 125 Hz decreasing to -0.06% at 825 Hz[790] in unmyelinated crab walking leg nerves and -0.2% at 6kHz in myelinated rabbit sciatic nerve[56].

The biophysical basis for these changes is as follows. At rest, applied current will not enter neuronal cells or their processes, because the cell membrane is lipid and so highly resistive. When ion channels open during depolarization and activity, the applied current can enter the intracellular space, which contains additional conductive fluid, so the overall bulk tissue impedance decreases. This effect is likely to be largest at DC. The change decreases at higher frequencies. This is because applied current starts to cross the highly capacitive neuronal membranes at frequencies above $\sim$2 kHz, so there is a smaller relative change as ion channels open during neuronal depolarization. This has been modelled in detail in our group but, so far, only for such changes in peripheral nerve. For simulated unmyelinated human C fibres, the resistance change decreased from -1.3% at DC to

−0.13% at 2 kHz[1030] and was ∼ 45% at DC reducing to −11.4% at 8 kHz for myelinated nerve. The largest signal therefore occurs near DC applied current. However, in this frequency band up to c. 500 Hz, impedance changes are obscured by potentials produced endogenously from nervous tissue – the resting EEG and evoked potentials or epileptic spike wave activity in the brain, or the compound action potential in nerve. There is therefore a "sweet spot" for impedance recording or EIT during activity, which lies above the low frequency range with endogenous potential overlap, but below higher frequencies where the signal decreases and is obscured by noise. For the cerebral cortex during normal and epileptic activity, this was determined to be at c. 1.5 kHz[270].

12.2.2   OTHER MECHANISMS OF IMPEDANCE CHANGE: TEMPERATURE AND CSF MOVEMENTS

There are two additional factors which may influence the impedance of brain but for which there is little experimental information. During increased neuronal and, therefore, metabolic activity, an increased generation of heat may occur which would increase brain temperature. Decreased brain temperature increases brain resistivity by approximately 2–3% per °C[619,1071]. Cortical temperature changes of up to 1°C during functional activity, with an average 0.2C decrease in temperature after 1–2 minutes of visual stimulation, have been detected with MRI and fMRI, in humans[1178]. Such cortical temperature changes could produce changes in impedance, which could be detected by EIT, but these would occur over minutes, rather than changes over seconds expected by blood volume change.

The thickness of the cerebro-spinal fluid which overlies the activated cortex is another possible cause of apparent impedance change with recording with scalp electrodes: an expansion of local cerebral blood volume, such as during epileptic seizures, might shift small amounts of CSF overlying adjacent superficial cortex to areas of lower volume[1096]. Changes of CSF pressure, monitored by indwelling intracranial pressure sensors, have been recorded during seizures in 7 subjects[317,718].

12.2.3   EFFECT OF COVERINGS OF THE BRAIN WHEN RECORDING EIT WITH SCALP ELECTRODES

The principal problem in imaging with scalp electrodes is the relatively high impedance of the skull. Bone is anisotropic. Rush and Driscoll[906] found the effective resistivity of skull, when soaked with a conducting fluid, to be 80 times that of the fluid. A different study suggested that the contrast between skull resistivity and that of brain was much less than this value, which has been widely used in EEG inverse source modelling calculations[795]. At 100 Hz-10 kHz, using a 4 electrode measurement method, they found that in vitro the average resistivity was 65 Ω·m at 37°C. They also made in vivo estimates of resistivity in two subjects by fitting a model of the head to impedance recordings made with scalp electrodes. The resistivities calculated gave a lower ratio of 1:15:1 for brain: skull: scalp, with values of 4.9 Ω·m for brain and scalp and 76,Ω·m for skull giving a ratio of 15.7. A detailed and technically sound study was undertaken in human skull samples immediately after surgical excision but at the higher frequencies of 1-4 MHz. Resistivities were relatively constant from 1 Hz - 10 kHz. At 1 kHz, the resistivities varied from 57 to 265 Ω·m depending on one of six bone types identified in the skull[1023]. There is therefore some disagreement in the literature over the correct value for the skull resistivity at EIT frequencies which are typically below 50 kHz. A reasonable approximation is to use values of 80–210 Ω·m[1204] which is approximately 50 times the resistivity of brain.

The resistivity of scalp has not been accurately measured, to my knowledge, but is probably similar to that of mammalian skeletal muscle, which has been reported to be between 2.5 and 10 Ω·m measured at 100 Hz–100 kHz in various species[335]. Cerebrospinal fluid has a low resistivity of 0.55 Ω·m at body temperature.

It is therefore clear that the resistivity of the skull is substantially higher than that of brain and scalp (Tables 12.1 and 12.2). Current applied for impedance measurement to the scalp will therefore tend to flow through the scalp and not pass through the skull into the brain. The relative size of these values determines how much current flows into the brain compartment. This has been investigated by applying current to a skull inside a saline filled tank[906]. Closely spaced current injection electrodes produced negligible current penetration within the skull, but when electrodes were widely spaced across the skull in polar positions, 45% of the applied current entered the skull cavity. The current that does traverse the skull will tend to shunt through the highly conductive cerebro-spinal fluid. The effect of all this will be to decrease the signal-to-noise ratio, in the sense that the signal will be sensitive to local changes in the scalp, and relatively insensitive to events in the brain. One of the challenges in attempting brain EIT has been to try and maximize the current flowing into the brain itself.

Table 12.2
Resistivity of Brain Tissues

Tissue	Resistivity ($\Omega \cdot$m)	Ref	Freq. (kHz)
White matter	3.4	[310]	1
Grey matter	2.3	[310]	1
Blood	1.3	[823]	50
CSF	0.7	[91]	0.01–10
Skull	150	[1204]	1

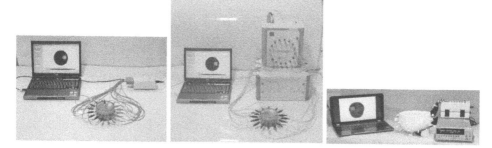

Figure 12.3   Examples of EIT systems developed at UCL. Left: UCH Mk2.5 (single channel record and current drive, multiplexed to 16 or 32 electrodes, switches serially, $\tilde{3}$ images per second[702]; Middle: KHU Mark 1 (16 parallel record single drive, $\tilde{1}$ image per second[792]; Right: Current "ScouseTom" UCL EIT system based on Keithley current source, 32–128 research EEG Brainproducts record, and custom multiplexer[74]. (color figure available in eBook)

12.3   EIT SYSTEMS DEVELOPED FOR BRAIN IMAGING

12.3.1   HARDWARE

12.3.1.1   Relevant Instrumentation Principles

The principle of EIT is to record multiple transfer impedances using a ring or arrays of skin or tissue electrodes. The principal instrumentation variables are as follows:

1. Each transfer impedance may be made with 2 or 4 electrodes. Typically, this is achieved using a high precision constant current source with a high output impedance applied to two electrodes. The resulting voltages may be recorded from the same (2 electrode method) or two different

electrodes (4 electrode method). The 2 electrode method is simpler but suffers from the inclusion in the transfer impedance of the skin-electrode impedances which are additional unknown variables. All successful clinical data series have been collected with the 4 electrode method which avoids this – a constant current can be delivered to the subject across variable skin-electrode impedances, and voltages may be accurately measured as the skin-electrode impedance is small compared to the input impedance of voltage recording amplifiers if of high quality with a high input impedance. Measurements may be made serially with current injection through one pair of electrodes and recording at one other pair. Many measurements may be rapidly collected by multiplexing different electrode combinations with a single impedance recording circuit. More commonly, current is injected through a single pair of electrodes but recorded simultaneously at all other available electrodes using parallel voltage recording amplifiers; this reduces the time needed to collect a full data set. Voltage measurements from current carrying electrodes include the electrode impedance and so are usually discarded. A full data set for image reconstruction usually comprises the entire set of linearly independent transfer impedances e.g. 104 for a ring of 16 electrodes. It is also theoretically possible to acquire transfer impedance with injection of spatial current patterns across many electrodes simultaneously. This provides a better penetration of current into the subject but has not been used in practice for successfully implemented clinical systems. The reason for this is unclear. It may be because errors in current application between the many channels cause more error than the theoretical advantages.

2. The applied current is limited by medical safety standards to about 10% of the threshold for skin sensation according to IEC 60601. This is typically 5 mA at 50 kHz for lung imaging or 100A below 1 kHz for applications in the nervous system. Although all possible combinations of electrode paths may be obtained in theory from a minimum linearly independent data set as above, in practice differing protocols for electrode addressing are used according to the application, in order to maximise signal to noise. For lung imaging in the chest, current is typically applied to adjacent electrodes as it can pass relatively easily into the thorax between the ribs. In the head, the skull acts as a high impedance barrier so that the best SNR is obtained with widely spaced or diametrically opposed current injection electrodes. For a single ring of 16 electrodes, as in chest EIT, all linearly independent measurements are generally made. At the typically used recording frequency of 50 kHz, an entire data set of 104 measurements may be made in N 100msec and so give image frame rates of 10 frames/sec or more. Where larger numbers of electrodes are used, as in imaging brain function, it may be necessary to choose a subset of all possible independent combinations with the highest sensitivity. The choice of such electrode addressing combinations may be hard coded in simpler EIT systems such as for lung imaging; other systems usually have a more flexible design in which software chosen multiplexers may flexibly address different electrode combinations according to the application.

3. The frequency of the applied current depends on the physiological application. A significant source of error in impedance measurement is the skin-electrode interface impedance which may interact with non-idealities in electronic instrumentation to produce errors related to stray capacitance. In general, this is lower as frequency increases. On the other hand, errors from stray capacitance increase with frequency and become significant above ∼100 kHz. For imaging lung ventilation, the impedance contrast is provided by the entry of highly resistive air into conductive lung tissue. This is relatively independent of frequency. Thus, most clinical EIT lung systems apply current at ∼50 Hz, as a reasonable "sweet spot" with lower skin impedance and also errors due to stray capacitance. For EIT in the nervous system, the physiological basis for the impedance change differs, and frequencies as low as 100 Hz may be needed.

4. Time and frequency division multiplexing (TDM and FDM). Most EIT applications, such as imaging lung ventilation, may be made by imaging differences over time at a single frequency. As this may be accomplished rapidly at frequencies used of ∼50 kHz, the approach is usually simplified to "time division multiplexing" in which serial impedance measurements from differing

electrode combinations are made rapidly over time. In some applications, especially those using lower frequencies, or those with a low signal to noise ratio, it may be desirable to record differing transfer impedances simultaneously. This may be accomplished by recording simultaneously from multiple current/recording combinations at different frequencies. This is termed "frequency division multiplexing", a well known approach in telecommunication. This requires the provision of multiple highly specified current sources sufficient to inject simultaneously on many channels – typically 32 for the applications in the nervous system. In contrast, a time division multiplexing system only usually requires a single current source. Most applications are based on changes in impedance over time and so only need a single injected frequency (or closely spaced set of frequencies measuring the same physiological change for FDM). However, there are some clinical applications where it is necessary to record over a spectrum of multiple frequencies. This is where there is a need for a single EIT image in time, as in breast cancer or stroke imaging. Unfortunately, it is not possible to produce absolute images at a single frequency and time point with EIT. This is because the inverse solution to reconstruct images is ill-posed and underdetermined; small errors between the real recording situation and FEM models used for image reconstruction yield large errors. For time difference imaging, this is obviated by making images of a data set normalized to reference data recorded earlier; the geometric errors largely cancel and images may be produced accurately. The alternative approach is to make images of transfer impedances made across multiple frequencies at a single time point. The principle is that geometric errors will be similar at the different frequencies and still cancel, but the spectra corresponding to different tissues will differ and yield images with significant contrast between the different tissues. This is termed a "multifrequency" approach. Single frequency measurement is generally accomplished using a digital waveform generator and digital-to-analog converter which produces a constant current sine wave. Multifrequency EIT may utilize similar hardware in which a single such current source produces a composite calculated multifrequency waveform. It may also more accurately be produced by multiple current sources each producing a pure sine wave which are combined as analog signals. The resulting voltages are then deconvolved digitally after recording.

5. A/D bit resolution. The resulting analog voltages are usually recorded with two stage instrumentation amplifiers with a buffer amplifier as close to the electrodes as possible, to minimize stray capacitance. They are then digitized. The accuracy of this depends on the physiological variable. For lung imaging, the impedance contrast during ventilation is large at $\sim$30%, and 12-bit A/D usually suffices. For applications in the nervous system, the SNR may be far smaller. Typically, a standing recorded voltage is $\sim$10 mV and changes of $\sim$1 or 0.1 µV need to be resolved after demodulation. It is usually necessary to employ 24-bit A/D converters to provide sufficient overhead for electrode combinations with lower standing voltages. It may be possible to obtain sufficient accuracy with a resolution of 18 or 19 bits in practice, or even less if methods such as adaptive programmable gain are employed to ensure that each electrode combination yields a standing voltage near full scale deflection.

Below, several different EIT system configurations are described and vary according to the physiological application. To summarize, for each, these variables are relevant:

- Applied current and frequency
- Single or multiple current sources
- Single or multi-frequency current injection
- Number of current application and recording channels
- Need for adaptive multiplexing or fixed recording
- A/D bit resolution
- All systems described below for brain and nerve imaging use a 4 electrode method driving constant current through a single pair of electrodes at a time

12.3.1.2   EIT Systems Developed in the UCL Group

12.3.1.2.1   Systems for imaging "slow" time difference changes during evoked physiological activity, epilepsy, spreading depression or stroke.

EIT systems for imaging the slow time difference changes have less exacting technical requirements. The changes are reasonably large – usually $\sim 1 - 10\%$, and change relatively little over frequency. Thus, time division multiplexing (TDM) works well. Higher frequencies of up to 100 kHz still yield a good SNR so image data sets may still be acquired rapidly – up to 10 times per second even with TDM. As signals change over seconds or less, a narrow bandwidth of ~ 1 Hz may be used in voltage recording, which reduces noise. As such, it has generally been possible to acquire good signals with A/D converters with 12 or 16 bits.

The first EIT recordings of brain function were made with the Sheffield Mark 1 system[161]. This employed 16 electrodes in a ring; 5 mA current at 50 kHz was applied and voltage was recorded serially through adjacent pairs of electrodes; the algorithm employed the assumption that the problem was two-dimensional and that the imaged subject initially had a uniform resistivity. It therefore was a time division multiplexing system and had a 12-bit A/D converter. This was used in specialized circumstances, where the experimental preparation was designed to match the limitations of the system. In anaesthetized rats or rabbits, the entire upper surface of the skull and brain coverings (the dura mater) were removed, and a ring of sixteen spring mounted electrodes were placed on the exposed upper brain surface. As most of the activity occurred in a layer of cerebral cortex about 3 mm thick, and the upper surface of the brain in these species is almost planar, this was a good approximation to a two-dimensional uniform problem. It is not possible to record fast neural ion channel opening changes at 50 kHz[476] so images were of "slow" changes only. Images were successfully obtained during stroke[477], epilepsy[863], spreading depression[126] and evoked activity[483].

Imaging in this way was helpful in demonstrating proof of concept, but, clearly, for use for imaging fast neural activity or slow or stroke multifrequency changes with scalp electrodes in humans during neurological conditions, the requirements were greater. In our group at UCL, we therefore set out to develop a new hardware design which permitted the following:

1. Software selectable electrode driving, so that different electrode protocols could easily be produced in an experimental setting. The particular idea was to use diametrically opposed electrodes for current injection, as this would enable more current to pass through the brain.
2. The ability to image at low frequencies – about 200 Hz, as the theoretical considerations above indicated that changes in cell swelling during stroke or epilepsy would be larger at low frequencies, as more current would pass in the changing extracellular space.
3. A system should be suitable for recording in ambulatory patients. For example, we wished to record in patients with epilepsy being monitored on a ward over days until they had several seizures documented. This could be achieved by changing the physical configuration of the EIT system so a small headbox could be worn on the subject, with a long lead of 10m or so, which passed back to a base station and PC.

Our first system permitted the first two of these. It was based on a Hewlett-Packard HP4284 impedance analyzer. This was adapted to make four electrode impedance recordings through a multiplexer able to address any combination of 32 electrodes. The HP impedance analyzer was highly accurate but slow, as it utilized a balancing bridge procedure. As a result, a single image, comprising about 300 serial recordings from different electrode combinations, took about 25 seconds. It was used to make the first series of EIT recordings with scalp electrodes in humans and neonates during physiologically evoked responses[1038]. The next system, termed the "UCLH Mark 1a or 1b", was similar, but was purpose built and based on a single impedance measuring circuit similar to the Sheffield Mark 1 system. It differed from the Sheffield Mark 1 in that it could record at much lower frequencies, electrodes could be addressed flexibly from software, and it was suitable for ambulatory recording. Recording could be performed at one of 18 frequencies from 77 to 225 kHz; up to 64 electrodes could be addressed (16 in the Mark 1a and 64 in the Mark 1b). It comprised a headbox about the size of a paperback book into which the electrode leads were inserted, which could be worn in a waistcoat by the subject; this connected to the base station by a lead 10 m long. It had a 12 bit A/D converter and was primarily intended to produce images of the larger "slow" changes over time. In order to improve recording accuracy limited by the relatively coarse A/D converter, a setup routine interactively set programmable gain on each electrode combination so that the recorded standing voltage was close to the full scale deflection[1188]. It produced acceptable images down to 200 Hz in saline filled tanks[1036]. It was used in recordings in human subjects during epilepsy and epileptic seizures[265].

12.3.1.2.2   Multifrequency systems able to image at one point in time for stroke.

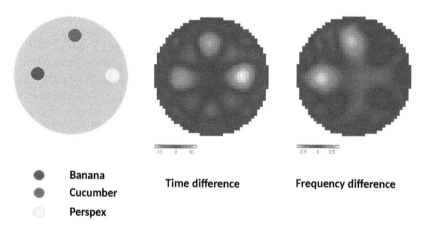

Figure 12.4   EIT images acquired with the UCLH Mark 2 EITS system. Banana, cucumber and Perspex were placed in 0.2% saline in a cylindrical tank with 16 electrodes in a single ring. Time difference imaging was performed at 640,kHz. The frequency difference image was collected at 640,kHz and referenced to 8,kHz[1187]. (color figure available in eBook)

Although these systems were capable of applying currents of different frequencies, they were not optimized for multifrequency measurement and were used for time difference imaging. The next generation device – termed the "UCLH Mark 2 or 2.5" was designed with the aim of imaging stroke, where time difference imaging was not practicable – a single image needed to be acquired in a novel subject who already has brain pathology (figure 12.3). This was achieved by making difference images across frequency. The design was again time division multiplexing but with a complex multifrequency current waveform from 2 kHz to 1.6 MHz injected serially through chosen pairs of 64 electrodes[702]. Acceptable and reproducible images of multifrequency objects such as a banana in a saline filled tank could still be obtained[1187] (figure 12.4). It was employed for studies in human subjects with stroke or tumours[893,894]. A similar time division multiplexed multifrequency EIT system was then developed in collaboration between the UCL group and that at the Kyung-Hee University in South Korea[792]. This addressed 32 electrodes over a similar frequency range and used fixed wiring to address electrodes to reduce errors from stray capacitance and series resistance in multiplexers (figure 12.3). This reduced the previous capability of flexible addressing of any electrode pair but it was possible to arrive at a close performance using ad hoc electrode addressing protocols[264]. Such an approach was used to collect the first dataset in acute stroke in human subjects[370].

Frequency division multiplexing offers the advantage in principle of more rapid data collection. Simultaneous channel acquisition could also be enhanced by additional phase [247] or Code Division Multiplexing [1058]. In practice, for frequencies above ~5 kHz, TDM is possible with frame rates of more than 10 per second, so it has not been necessary to use it. However, FDM could offer improved data collection speeds if low frequencies below 100 Hz are needed for stroke or other imaging, and the theoretical advantage of decreased noise if different channels are collected contemporaneously. It was successfully used to produce images in saline filled tanks but has not yet been implemented for physiological or human recording [75,247].

12.3.1.2.3 *EIT systems for imaging fast neural activity.*

The above systems worked in saline filled tanks and the Sheffield mark 1 studies demonstrated proof of principle for imaging slow changes with intracranial electrodes. The later systems were able to collect reproducible data in human subjects with scalp electrodes. However, sadly, it was not possible to produce accurate EIT images of fast or slow activity, nor stroke, in human subjects with scalp electrodes. Our group then changed direction to concentrate on imaging fast neural activity over milliseconds but using intracranial electrodes. This avoided the signal reduction produced by the skull if scalp electrodes were used. However, this placed exacting requirements on the instrumentation as small voltage changes of a few µV needed to be discriminated from standing voltages of tens of mV. The designs were again time division multiplexing but with parallel recording from up to 128 electrodes used in electrode mats placed on the exposed brain surface in anesthetized rats. 24-bit A/D converters were used to give the required bit resolution. Different designs used custom or commercial current sources which usually applied ~50 µA at ~1.5 kHz, a custom multiplexer, and highly specified commercial 24 bit A/D converter able to sample at 50kHz on each channel. Its most recent implementation was termed the "UCL ScouseTom" system after its designer Tom Dowrick (figure 12.3). He grew up in Liverpool, where the English dialect made famous by the Beatles is termed "Scouse" after a stew of this name popular with sailors and dock workers at the Liverpool docks [74]. This was successfully used to image fast neural changes in the brain and nerve presented below.

Other groups have also been interested in EIT of the head. An active group at the 4th Military Medical University in Xian China have developed a system optimized for imaging brain function and pathology with a single ring of scalp electrodes, similar in principle and performance to the UCL Mk 1a system. This too operated at a wide frequency range of 1-190 kHz with a single frequency with TDM, 16 bit A/D converter and electronic multiplexer addressing 16 electrodes [966]. Another design proposed for brain imaging utilized a capacitively coupled approach [517]. This has the advantage that electrodes were contactless but this approach requires higher frequencies of 200 kHz – 15 MHz. It therefore might be suitable for imaging slow changes or stroke, but not fast neural activity. Other groups have undertaken preliminary studies using TDM single frequency systems [76,861].

12.3.2 RECONSTRUCTION ALGORITHMS FOR EIT OF BRAIN FUNCTION

12.3.2.1 Reconstruction Algorithms for Time Difference EIT Imaging Based on a Linear Assumption

In parallel with the historical development of hardware, there have been developments in reconstruction algorithms for the especially difficult case of imaging brain function within the head. The majority of effort has again been within our group at University College London, but there have been contributions from other groups too who, like us, have been intrigued by the special problem which the high resistivity of the skull poses.

When we first attempted EIT of brain function, the only available hardware was the Sheffield Mark 1 EIT system, which employed a 2-D filtered back projection reconstruction algorithm. This proved to be remarkably effective in imaging brain function, for the limited circumstances in which a ring of 16 electrodes was placed on the exposed superior surface of the brain of anaesthetized rats or rabbits (see §12.3.1). This system employed a back-projection approach. More recent experimental and human studies in the Xian group have employed a similar 2D approximation approach with a single ring of scalp electrodes.

Since then, the standard approach in mainstream EIT of lung function has been a time difference approach with a linear approximation and use of inversion of a sensitivity matrix and a geometric or numerical anatomical Finite Element Model. As the thorax is roughly cylindrical, most lung reconstruction algorithms approximate the thorax to a cylinder and approximate reconstruction to a 2D slice or use a single ring of electrodes with modelling of out-of-plane current. Imaging of time difference fast and slow changes in the brain and head has been based on this approach. However, this has the additional requirements that substantial errors are introduced if the head is approximated to a slice or if current is injected through adjacent electrodes[78]. We have therefore adopted approaches with widely spaced current injection, to maximise current entering deeply into the brain, and a 3D tomographic image production based on highly accurate anatomically realistic head or brain models.

Since current is not confined to 2-dimensions in the three dimensional head, it is more appropriate to employ 3-D models. Our group at UCL took a first step towards 3-D imaging with an algorithm in which the forward solution employed a model of the head as a uniform homogeneous sphere. This was used to generate a sensitivity matrix and images were produced by matrix inversion employing truncated singular value decomposition (tSVD). This algorithm is clearly based on an oversimplification, but we adopted it as we did not at the time have the ability to implement more realistic models. The approach was used in the first series of human recordings with scalp electrodes, during physiologically evoked responses[1038]. Unfortunately the resulting images were not sufficiently similar to fMRI or PET during similar activation, but it was not clear which of several factors were responsible.

Following development in which the head was modelled as concentric spheres[633], we progressed to use of an anatomically realistic Finite Element Model (FEM) mesh of the head. These may be produced by segmentation of surfaces apparent on MRI or CT of the head. The advantages of using a realistic head model were first demonstrated in computer simulation, saline filled tanks with a real skull, and human studies[78]. Since then, we have used realistic FEM meshes which are now ~10M elements for the human head including scalp[513,1097], 5M for the rat brain[57], and 2.6M for peripheral nerve[869].

Image reconstruction for time difference imaging in the head with scalp electrodes has been achieved using a standard method borrowed from lung EIT. This is to use inversion of a Jacobian matrix which provides estimates of the sensitivity of different electrode combinations to changes in conductivity of image voxels in the head. It employs the assumption that boundary voltage changes are linearly related to internal voxel changes in conductivity. This is strictly incorrect as the underlying physics clearly indicates that the boundary changes start to fall off relatively as conductivity changes inside the head. However, empirically, this appears to be a reasonable approximation for internal resistance changes of up to about 20%[482]. Imaging is achieved by inversion of the Jacobian matrix. As there are far fewer measurements than unknown voxels, a mathematical procure termed regularization is employed to achieve image reconstruction. Although attempts have been made to determine the best assumptions for this, our experience is that these need to undertaken empirically for any situation in anatomically realistic saline filled tanks or in vivo studies. Using it with an optimization method termed "Zero order Tikhonov" has been settled on as the best compromise in our group[515]. It yields reconstructed perturbations which are relatively smooth but has the advantage of requiring the fewest assumptions about the nature of impedance variations within the subject. Calculation of the Jacobian sensitivity matrix may be time consuming as it requires an iterative solution with a large FEM mesh. However, an efficient parallel procedure[514] allows computation of the Jacobian matrix to be achieved in a few minutes with modern powerful workstations[514,515]. Image reconstruction by matrix inversion may be undertaken with much smaller meshes – typically about 300 k elements, without appreciable loss of image quality. This is a single matrix inversion and may be accomplished in a few seconds.

There has been interest from several groups as to the best method for time difference EIT reconstruction in the head. A method with Total Variation regularization with the Split-Bregman approach has been proposed as producing the best images in simulation and tank studies with a 3D[1215] or 16 single electrode ring[621] approach. It is our experience that the image quality is acceptable for most practical use in the head with 16-32 scalp electrodes, at least in anatomically realistic saline filled tanks (figure 12.5). We recommend equal distribution of electrodes over the scalp; use of a single 2D ring is likely to yield off-plane errors. However, the Split-Bregman approach is computationally demanding and was evaluated with relatively small meshes. For the larger meshes of 5–10 M FEM elements currently employed in our group, we recommend an approach with up to 128 electrodes, a fine mesh as above, and time difference inversion using Zeroth order Tikhonov regularization[57,58].

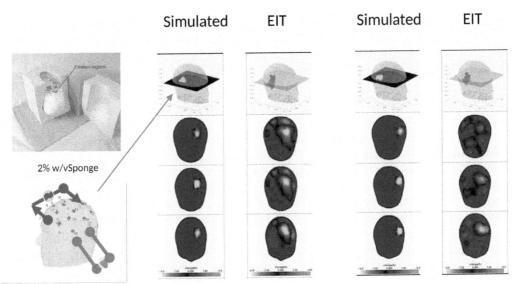

Figure 12.5   Image quality with time difference EIT in a saline filled head shaped tank. Upper left – Tank with real human skull and 31 Ag/AgCl silver electrodes placed in a 10–20 distribution over the surface with scalp simulated by a layer of saline 5 mm thick; lower left diagram indicating electrode placement for one electrode combination of 258 used with recording at 4 kHz. Right: reconstruction of simulated and tank recoded images in the plane of an inserted sponge (middle row) and 1 cm above and below (upper and lower rows)[263]. Impedance changes are shown as arbitrary units. (color figure available in eBook)

12.3.2.2   Non-linear Reconstruction Algorithms for EIT of Brain Function

All the above methods employ an assumption of a linear relationship between changes in conductivity in a subject and the resulting change in voltage on the boundary. This approximation appears valid in saline filled tanks up to changes in resistivity of 20%[482]. In time difference imaging, in which there are relatively small changes in a small region in the subject, this assumption appears to yield acceptable images. However, there are potential applications in biomedical EIT where time difference imaging is not practically possible. A leading one is in acute stroke. This requires a single image in time. Absolute imaging has never successfully been implemented in clinical data series, presumably because small errors between the real-life recording arrangement and numerical model are too great. An alternative approach is multifrequency imaging of the brain. The underlying principle is that the spectra of blood, normal or ischemic brain vary widely. The resistivity of blood is flat with respect to frequency up to about 2 MHz, and it differs from brain by a factor of 2 or 3. The difference between normal and ischemic brain is greatest at low frequencies below 100 Hz. In principle, EIT imaging with recovery of the spectrum of tissue impedances might be able to yield images which represent these different tissues.

However, this is a challenging imaging problem. Although these changes are large in the brain itself, modelling indicates that they translate into differences of only several per cent when recorded with scalp electrodes, and similar differences are present through confounding factors such as instrumentation errors or variation in the natural impedance of tissues in the head. Mathematical approaches based on Bayesian priors have been developed in our group[678,679] and have been shown to produce acceptable images when used in tanks filled with vegetable materials intended to simulate the impedance spectra of tissues in the head[679]. However, because the impedance differences are greater than 20%, the reconstruction process cannot be approximated to a linear solution. Multiple iterations of solutions are needed to arrive at the best fit solution, which may take several hours even for a reduced density mesh. Other approaches have included comparison of side-to-side differences[699] or the use of neural network and machine learning approaches[177]. While these are supported by modelling studies their accuracy in clinical data series is not yet established.

12.3.3 DEVELOPMENT OF TANKS FOR TESTING OF EIT SYSTEMS

In order to test hardware and reconstruction algorithm improvements, we have developed a series of saline filled tanks. Before starting this work, there were some published methods for this approach. Griffiths developed the "Cardiff" phantom which comprised a circular array of resistances and capacitances[381], which has been widely used for calibration of hardware in our and other laboratories. Several groups have employed saline filled tanks in which highly conductive metal or resistive perspex objects are suspended e.g.[978], but this poses a large impedance contrast which does not fully examine the ability of the system to image the lower contrasts of 0.1–10% which are usually seen in in vivo applications in neuroscience applications. Other groups have produced lesser contrasts by using agar test objects[910] or semipermeable tubing containing fluid[1031] which contain a salt concentration different to the bathing solution. Unfortunately, these different saline concentrations will diffuse, leading to uncertain boundaries of the test objects.

This may be overcome by the use of a test object such as a gel or sponge immersed in saline, in which the impedance contrast is produced by insulating material itself; the bathing fluid permeates the pores of the test object, so it is stable over time. Resistance increases of 10–200% were produced using polyacrylamide gels[482]. For testing multifrequency systems, it is desirable to utilize test objects which comprise capacitance as well as resistance. Unfortunately, it appears that only biological materials contain the high capacitance needed to simulate human tissues. Cucumber in potassium chloride solution and packed red cells appeared suitable and were stable over several hours. Impedance contrasts of 5–20% in both resistance and reactance could be produced in the packed red cell solution by immersing polyurethane sponges of differing densities[481].

The above tanks were all cylindrical. We wished to develop tanks for testing our 3-D reconstruction algorithms. Initially, we constructed a tank which contained a real human skull in a surround of watertight latex, filled with 0.2% saline. A gap 5 mm wide surrounded the skull, which simulated the scalp[634,1037]. This contained the barrier of the skull and realistic geometry, but not a layer to

Figure 12.6 Tanks intended to simulate the human head. Left: Human skull covered with skin of the giant zucchini[1036]; middle and right: simulated neonatal skull created with a 3D printer and containing tank. Multiple perforations were made in PLA (polylactic acid) to simulate the resistivity of human skull in vivo according to the measurements of[1023] at 1 kHz[73]. (color figure available in eBook)

simulate the impedance properties of skin. The most realistic tank simulated electrical properties of skin with the use of the skin of the marrow, or giant zucchini. This was plastered over a human skull and a layer of slow setting dental alginate, to simulate the scalp. The interior cavity was watertight, and the brain was simulated by 0.2% saline (figure 12.6). Acceptable images could be acquired with this and a reconstruction algorithm in which the head was modelled with realistic geometry[803,1036]. More recently with the advent of 3D printers, it has been possible to construct similar tanks with an anatomically realistic skull simulated by an impermeable plastic with multiple holes to admit current[74] or ABS (acrylonitrile butadiene styrene)/carbon black composites[1204]. These 3D printer tanks have the advantage that the skull and scalp surface is anatomically accurate, but the cranial contents and scalp are simulated by a uniform saline solution and there is no layer simulating skin. To my knowledge, only the use of the giant marrow skin has simulated the challenge posed for in vivo recording by the impedance barrier of the skin.

12.4 EIT OF SLOW IMPEDANCE CHANGES IN THE BRAIN RELATED TO CHANGES IN BLOOD VOLUME AND CELL SWELLING

12.4.1 DURING PHYSIOLOGICAL EVOKED ACTIVITY

There are good grounds for expecting that EIT could produce images of increases in blood flow and volume, and related changes, which occur when part of the brain is physiologically active. These changes have been the basis of functional MRI and PET studies, and have been reviewed in §12.2.1.3. If successful, EIT could provide a low-cost portable system, which would produce similar images to fMRI and be widely used in cognitive neuroscience in healthy and neurological or psychiatric subjects.

The local changes in the brain are small – a few per cent – and occur over seconds or tens of seconds following the onset of activity. As the mechanism of impedance difference is probably changes in resistivity due to a changed proportion of blood to brain, these may be imaged at any suitable frequency which can distinguish these. The ideal frequency has not been studied. Most measurements have been undertaken at similar frequencies of about 50 kHz to those used in lung EIT, as this represents a reasonable balance between electrode impedance, which decreases with frequency of applied current, and stray capacitance, which increases. They may be imaged with a time difference approach. As the physiological changes occur over seconds, any EIT system similar to the Sheffield mark 1, which produces several images per second, will have sufficient time resolution without any special adaptation.

The first EIT images during evoked physiological activity were collected with a Sheffield Mark 1 system using a ring of 16 spring mounted electrodes placed on the exposed superficial surface of the brain of anaesthetized rabbits (figure 12.7). In 8 rabbits, evoked responses were produced by stimulation with flashing lights or forepaw stimulation, lasting 2.5 or 3 minutes respectively, and EIT images were collected by averaging over consecutive 15 second periods. Reproducible impedance decreases of 2.7±0.8% (visual) and 4.5±0.9% (somatosensory) (mean±SE) were consistently observed in the appropriate region of the brain. An unexpected finding was that, in addition to the expected impedance decreases, there were adjacent smaller increases. The explanation was unclear – it could have been a "ringing" artefact due to the Sheffield back-projection reconstruction algorithm, or due to steal of blood volume from neighbouring regions. These initial studies used intracranial electrodes and established that there were reliable impedance changes which could be successfully imaged. There were resistance decreases of about 2% consistent with increased blood volume in active brain regions.

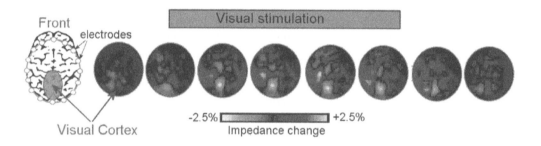

Figure 12.7   EIT images of rabbit cortex during visual stimulation. Images displayed were collected every 30 s. An impedance decrease may be seen over the posterior visual cortex which persists for about 30 seconds after cessation of stimulation[483]. (color figure available in eBook)

The natural next step was to attempt to image these changes with scalp electrodes. Unfortunately, this has not been successful. Reproducible scalp impedance changes have been recorded during visual and somatosensory evoked activity in humans using an early and accurate but slow Hewlett-Packard impedance analyser based EIT system. Significant reproducible impedance changes of about 0.5% appeared to arise from within the skull, as there was no significant changes in local scalp impedance recorded simultaneously[1038] (figure 12.8). Unfortunately, it has not been possible to produce reliable reconstructed images of this slow evoked activity with scalp electrodes in humans, even using more advanced reconstruction algorithms with anatomically accurate finite element meshes of the head[8].

At present, therefore, no method is available to enable the attractive goal of an inexpensive portable EIT system able to deliver fMRI-like images with an array of scalp electrodes in experimental animals or humans. The reason for this is not entirely clear. It most probably is that that intracranial changes of a few per cent fall below the noise level of about 0.5% present during physiological scalp recording[265].

A related potential technique is to use injection of a bolus of saline as a contrast-enhancing agent. This concept was first suggested in 1998 for imaging venous calf flow[158]. It has been suggested if could be adapted for use in imaging local increased blood flow during cerebral activity in a similar fashion to the BOLD response in fMRI[21] but it is not yet clear if there will be sufficient signal.

12.4.2   EIT OF SLOW CHANGES DURING EPILEPTIC SEIZURES

Because EIT systems can produce several images a second, and are portable and safe, they could be suited to image blood flow and cell swelling known to occur during epileptic activity with a high time resolution. These have the theoretical advantage that there is intense activity so that increases of impedance occur due to cell swelling which are about 5-10 times larger than those that occur during physiological activity (§12.2.1.3). EIT could be employed to localize the part of the brain that produces epileptic seizures, so that resective surgery can be performed. At present, about 80% of patients with epilepsy can be satisfactorily treated with medication. Of the remainder, some can be cured or improved by surgery. In order to perform this, it is essential that the correct source of epilepsy in the brain is localised. This is usually performed with a prior stay in hospital of several days. EEG and video are monitored continuously, so that it is recorded when a number of seizures occur. The EEG is usually performed with scalp electrodes but, if the seizure onset zone is unclear, it may be performed with subdural mat or depth electrodes, inserted at operation. Together with psychometry and neuroimaging studies, the onset zone is usually localised, and a decision as to whether to embark on surgery is undertaken.

EIT could be run concurrently with scalp EEG during this pre-surgical EEG telemetry. EIT images would be recorded about once a second over a period of days while the patient was observed on the ward. When a seizure occurred, the EIT images would be retrospectively analysed to see if changes in impedance occur at the same time as EEG activity. Imaging of this nature, with a temporal resolution or seconds, is not presently possible by any other method. In principle, the same information could be obtained if a subject had a seizure when in an fMRI scanner, but this is not practicable. Recent advances in neuroimaging have lessened the need for invasive recordings with depth or subdural mat electrodes, but these still need to be performed in patients in whom pre-surgical findings are not congruent. While subdural electrodes carry a low risk, depth electrodes which penetrate into the cerebral substance carry a significant although low morbidity and mortality[753]. In addition, the success rate of surgery is only about 70%. The cause of this is not entirely clear, but it may be partly because intracranial electrodes can only sample at a limited number of sites. In principle EIT could be undertaken on telemetry and produce images of slow or fast neural activity during seizures or interictal epileptic activity. The ideal would be if this were possible with scalp electrodes. However, there might also be advantages over existing methods even if it were only possible with subdural or depth electrodes as EIT would enable information could be obtained from all sites in the sensitive volume of brain. A modelling study has been performed, using anatomically realistic FEM meshes from 3 subjects with intracranial electrodes undergoing evaluation for epilepsy surgery. This simulated EIT undertaken with depth and 32 scalp electrodes. It concluded that EIT offered greater coverage and sensitivity compared to the current method of visual inspection of EEG recorded from depth electrodes during seizures[1158].

12.4.2.1   Proof of Concept in Animal Studies

The first EIT studies in epilepsy were, as for evoked responses, collected with a Sheffield Mark 1 system using a ring of 16 spring mounted electrodes placed on the exposed superficial surface of the brain of anaesthetized rabbits. Epileptic seizures were induced by focal electrical stimulation and were either localised or spread to involve the entire brain[863]. Reproducible predominant impedance increases of $7.1\pm0.8\%$ (localised) and $5.5\pm0.8\%$ (generalised) were present in EIT images in 9 animals at the sites where the epilepsy was initiated. As in the previous animal study in evoked responses, there were smaller adjacent impedance decreases apparent in the images. In this study, two probes were placed on the brain near the site of seizure onset and about 10 mm away, to try and elucidate the physiological mechanisms and establish if the impedance increases and decreases were physiological or due to reconstruction algorithm artefact. Local impedance measured at both sites

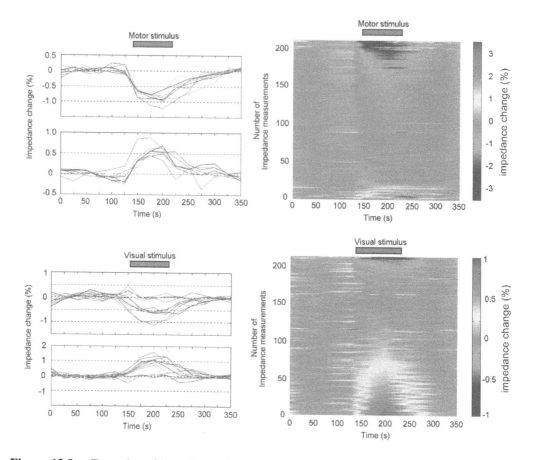

Figure 12.8 Examples of impedance changes in the raw impedance data. Impedance changes from single channel 4-electrode impedance recordings, during motor (top row, 8 repetitions) or visual stimulation (bottom row, n=12). On the left, data from a single electrode combination is shown; all repetitions are superimposed. Reproducible impedance changes are seen at selected electrode combinations with the same time course as the stimulation paradigms. The y-axis indicates the percentage change from baseline impedance. Impedance measurements were made every 25 s; the lines between these measurements are drawn for clarity. Both impedance increases and decreases were observed. On the right are shown all 258 electrode combinations for the same subjects, displayed as a sorted waterfall graph. The 8–12 runs for each electrode combination were averaged together. The averages were sorted due to the size of the impedance change during stimulation and stacked on the vertical axis. Measurements with baseline noise greater than the impedance changes are excluded from these plots so that these changes are not obscured. Significant stimulus related impedance increases and decreases are seen in approximately 25% of electrode measurements in these subjects[1038]. (color figure available in eBook)

was always an increase. Extracellular potassium, temperature, DC potential, and Laser-Doppler flowmetry were all consistent with the expected mechanism of cell swelling as the explanation for the increased impedance. The probable increase was about 10%, but was offset slightly by a concurrent decrease of a few per cent due to increased temperature and blood volume. The decreases appeared to be due noise or reconstruction artefact.

Recently, we have been able to image similar slow changes in the cerebral cortex or hippocampus in the anaesthetised rat using an epicortical electrode mat with 54 electrodes during electrically induced seizures. It was possible to reconstruct slow changes with an accuracy of <200 μm (figure 12.9[432]). The ability to localize activity in the hippocampus using epicortical electrodes alone supports this ability of EIT to offer true 3D tomographic imaging. Another study simulated the human Telemetry situation. EIT images with intracranial depth electrodes during a benzylpenicillin model of epilepsy in 5 pigs. Images were reconstructed without the need for averaging with the use of a frequency division multiplexing system injecting current simultaneously on 32 electrode pairs at frequencies spaced 50 Hz apart in the range 8.5-10kHz. It was possible to image a slow change

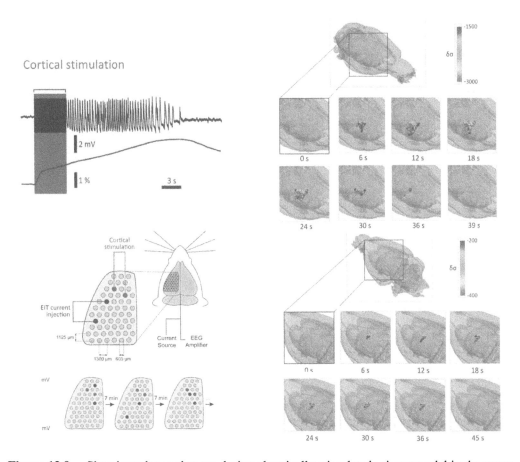

Figure 12.9 Slow impedance changes during electrically stimulated seizure model in the anaesthetised rat. Upper left : Electrical stimulation with an electrode on the surface of the exposed cerebral cortex induces a focal seizure lasting about 20 sec. An impedance increase of 5% over tens of seconds occurs as a result. Right : EIT images every 6: neocortex (upper), hippocampus (lower, 3 mm deep to cortical surface). Scale: t-score of impedance change. Accurate to seizure onset site <200 μm[432]. (color figure available in eBook)

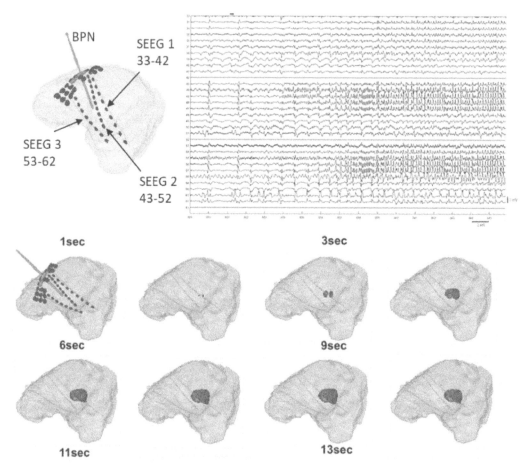

Figure 12.10   EIT images of seizure induced by benzylpenicillin (BPN) in anaesthetised pig recorded with 3 depth electrodes with 29 electrode contacts. Top: EEG shows seizure onset over 15 s. Bottom: EIT images of the reconstructed seizure onset and propagation. The regional spread among the adjacent contacts seen on EIT images correlated with EEG findings. The impedance change started 12 s after the EEG onset[1157]. (color figure available in eBook)

related to each seizure which developed over the course of the seizure and lasted about 40 s. The EIT onset occurred with 7.5±3.2mm of the EEG onset site (37 seizures in 5 pigs) (figure 12.10[1157]).

12.4.2.2   Human Studies

In relation to this, some single channel studies were performed in humans. The previous literature (§12.2.1.3) demonstrated impedance increases during seizures in animals. The proportion of glial cells in humans is greater, so the theoretical possibility existed that impedance changes might not occur in humans during seizures. Impedance was recorded in 5 subjects during telemetry, at 50 kHz, using four contacts on subdural mats over the temporal cortex or fine wire electrodes in deep temporal structures such as the amygdala[480]. In two patients with superficial parietal foci and recording with subdural mats, reproducible impedance increases of 4.5±0.3% and 2.4±0.3% were observed. In a third patient with a superficial temporal focus, consistent impedance increases of 3.6±0.2%, p<0.05) were observed with both temporal subdural and amygdala depth electrodes. The changes commenced within 20 s of the onset of ictal EEG activity and lasted for 1–2 min. These results indicated that substantial impedance changes do occur in the cerebral cortex of at least some human subjects.

Encouraged by these preliminary findings, we undertook a study in human subjects with epileptic seizures, using the UCLH Mark 1b system at 38 kHz, scalp electrodes and the linear reconstruction algorithm with an idealized realistic model of the head[265]. EIT images were recorded continuously 3 times per second in 9 subjects with temporal lobe epilepsy receiving continuous EEG monitoring as in-patients on a Telemetry ward. Several seizures were recorded in each subject, and the EIT changes were correlated with the EEG and other investigations to localize the site of onset. Unfortunately, as with scalp EIT evoked responses, it was not possible to produce reliable EIT images during seizures; this was attributed to the baseline noise using scalp electrodes in ambulant subjects.

In summary, slow EIT has been shown to localize the seizure onset zone reliably in animal studies but only with depth electrodes. Unfortunately, it has not yet been possible to produce reliable images with scalp electrodes. Complete studies with intracranial electrodes have not yet been undertaken in humans. However, the above animal data and a modelling study suggest that EIT could offer a powerful additional imaging modality to intracranial EEG in patients with intracranial electrodes. Unfortunately, unless there are major technical advances, it seems unlikely that EIT could offer a non-invasive imaging system with just scalp electrodes for the foreseeable future.

12.5   EIT IN CEREBRAL PATHOLOGY OVER HOURS OR DAYS

The section above refers to time difference changes over seconds or minutes. Cerebral EIT with scalp electrodes also has the potential to offer a novel method for diagnosis and monitoring of longer lasting cerebral pathology. In principle, EIT might be able to yield images similar to structural MRI or CT, but with a low-cost portable device at the bedside or at point of contact in the community where expensive imaging scanners are not available.

12.5.1   TIME DIFFERENCE CEREBRAL EIT OVER HOURS OR DAYS

In principle, EIT with scalp electrodes could provide a bedside monitor for the management of intracranial pathology such as intracranial hemorrhage, stroke or cerebral edema after head injury. This could be especially valuable in an unconscious patient where level of consciousness and neurological signs are no available to signal a deterioration and need for intervention.

Unfortunately, the ability of EIT to deliver clinically useful time difference data is limited by physiological drift in scalp impedance recording. In general, time difference EIT during lung ventilation or slow changes as above over seconds or tens of seconds is successful because drifts in boundary impedance measurements are negligible over these short periods and may be further obviated by averaging. In contrast, the drifts become limiting if imaging over hours. For use in clinical neurology, the most likely scenario is that EIT would be left to run over days and detect slow changes in pathology due to cerebral edema. It is possible it could also be used to detect more rapid pathologies such as a sudden intracranial hemorrhage or infarction. However, drifts in scalp EIT recording of a few per cent over minutes appeared to be the limiting factor in our unsuccessful attempt to image epileptic seizures with scalp electrodes[265]. We have investigated this objectively with measurement of drift in healthy volunteers over 6 hours. Using the most stable Ag/AgCL gelled electrodes with recording at 2 kHz, there were drifts of $\sim0.5\%$ per hour in boundary impedances which varied between different four-electrode combinations. These appeared to be due to natural variations in scalp impedance itself as drift attributable to the hardware was 8 times smaller on a resistor network. The drift fell naturally into longer term drift over hours, 1–10 min, and <10 s. The longer term drift was the largest at 1.3% mean per hour. Although these appear to be small, they become significant in relation to boundary changes which result from intracranial pathology. The

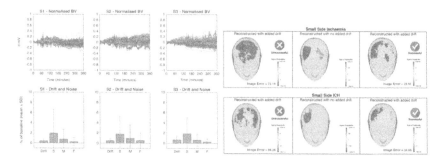

Figure 12.11   Scalp impedance drifts and simulated effects on reconstructed image quality in stroke and intracranial haemorrhage (ICH)[369]. Left: Scalp impedance in healthy volunteers over 6 hours varied by 0̃.5% per hour on average. Right: Examples of EIT images reconstructed with drift over 1 hour for a simulated lateral 10 ml infarct or haemorrhage. Only 6,4,33 or 11% of pathologies reconstructed successfully for 25/10ml, stroke/ICH respectively. (color figure available in eBook)

contrast between ischemic, normal brain relatively large at low frequencies of 20 Hz – about double, and both differ by a factor of 3 times or more to blood at all frequencies up to 1 MHz. However, partial volume effects and the presence of the CSF and skull result in boundary changes of only a few per cent on the scalp[487]. As a result, the drifts are of a similar magnitude to the expected pathologies, at least for localized discrete pathologies such as stroke or hemorrhage. This was evaluated by simulating 10 or 25 ml stroke or hemorrhage with drifts recorded from human volunteers. Even for the optimistic case of EIT imaging at one hour, only a small proportion of simulated pathologies could be successfully reconstructed (figure 12.11,[369]).

Proof of principle has been accomplished in time difference images. Initial studies confirmed that over time, it was possible to record significant scalp impedance changes of 15-60% and also EIT image changes with a ring of intracranial electrodes at 50 kHz during global cerebral ischemia[477,478]. Several studies from the Xian group have further demonstrated proof of principle. It was possible to identify significant changes during intracranial hemorrhage in piglets[221] and during unilateral stroke in the rabbit[1180] or rat[178], all with scalp electrodes. These studies employed the 4MMU approach which has a single ring of 8 electrodes with injected current at 50 kHz. In contrast, time difference EIT during stroke or ICH in the rat with either 40 spring-mounted scalp electrodes was not successful[246]. Clear impedance changes were detected but it was not possible successfully to reconstruct them into reproducible and biophysically plausible images. The first successful human study employed the Xian group time difference approach with a ring of 16 electrodes. it was possible to detect significant changes during infusion of saline over 4 minutes into the subdural space[220]. The same group has also recorded EIT with 16 scalp electrodes in human patients in whom cerebral water balance was altered by mannitol infusion. There were significant correlations with intracranial pressure[316,622,1179].

12.5.2 MULTIFREQUENCY EIT IN ACUTE STROKE

Stroke is a leading cause of death and long-term disability in the UK and is associated with high costs. Treatment with thrombolytic (clot-dissolving) drugs is effective for ischemic stroke due to occlusion of arteries but needs to be undertaken within three hours from the onset of symptoms. A brain scan is required prior to treatment onset to differentiate between ischemic and hemorrhagic strokes; thrombolytic drugs cannot be used where there is a hemorrhage as they may extend it. In practice, it is difficult to obtain rapid scans because of the difficulty of obtaining access to a scanner and rapid reporting. There is therefore a need for a neuroimaging system which could be utilized in casualty departments or health centres, which is inexpensive, rapid and safe. EIT could be ideal for this purpose. It could be used with an array of elasticated scalp electrodes, which may be easily applied by a technician or nurse in a few minutes. Interpretation could be performed by a trained technician or by a radiologist using remote reporting over a network or the internet. It could also be useful for research studies in which new treatments for stroke needed to be assessed over days as a stroke evolved. Unfortunately, time difference imaging could not be performed as the clinical need is for a single image on presentation. This could be achieved by absolute imaging, but this has not yet been shown to be practicable for clinical studies. The possibility, however, exists for achieving this by multifrequency imaging in which difference images are produced by referencing one frequency against another. The main principle will be that the impedance spectrum of blood in the range of 1 kHz – 1MHz will be different from brain and recently ischemic brain.

This poses a challenge for image reconstruction. Human studies in our group have confirmed that significant differences in impedance spectra may be observed in patients with stroke or similar pathology[370,894]. In principle, it might be possible to reconstruct such changes using the frequency difference – instead of making images of the change in two EIT data sets over time, this is attempted with a linear assumption based reconstruction algorithm applied to a data set in which data at one frequency is normalized to another lower frequency. This gave acceptable images in studies with tanks filled with chopped vegetable suspensions to simulate the complex impedance of the normal and ischemic brain[803]. A new mathematical approach was developed, in which the correct non-linear approach was used to attempt to reconstruct entire impedance spectra of the underlying component tissues. This too was successful in liquid filled tanks. However, unfortunately, it took hours or even days to produce images on powerful workstations and so was not yet ready for practical use[678,679]. In addition, these measurements place exacting requirements on the instrumentation. Non-idealities in electronic components and stray capacitance may lead to variability in recorded impedances with differing loads inevitable in in vivo recording. We were able to reduce these to <0.2% for transfer impedances of 5–70 Ω encountered in human scalp recording[701,702]. Nevertheless, such variability is not greatly less than expected signal differences of a few per cent over frequency and could contribute to poor image quality unless minimized. We have collected the first human multifrequency data set in 23 patients and 10 human volunteers[369,370]. It is possible to discern differences in impedance spectra consistent with the underlying biophysical principles. However, to our knowledge, there is not yet a suitable robust and timely reconstruction algorithm able to yield accurate images from such data. As presented in §12.3.2.2 there is interest in the use of machine learning methods to achieve this goal[177].

At the present time, therefore, there have been encouraging proof of principle studies which encourage the potential use of EIT with scalp electrodes in clinical neurology. However, it is not possible at the moment for acute stroke. EIT images for discrete cerebral pathology are limited and variable; it is unclear if the method could provide a new approach that is clinically reliable. It seems unlikely that instrumentation will advance materially to enable accurate image production – the bottleneck in producing good images is more probably variability in subject geometry compared to the geometric models used for image production. Thus it may be the use of machine learning methods, if coupled to training in large data sets, may offer a path to successful development in the future.

12.6   EIT OF NEURONAL DEPOLARIZATION

The novel applications presented above all make use of the low cost and portability of EIT, but similar images can already be obtained with fMRI or PET. However, EIT could in principle be used to image neuronal activity over milliseconds (§12.2.1.4) The proposed application would be to record EIT images from electrode arrays in 3D. These might be on or in the brain or nerves, in experimental animals or human surgical subjects, or, ultimately, around the scalp. Data could be gathered after a

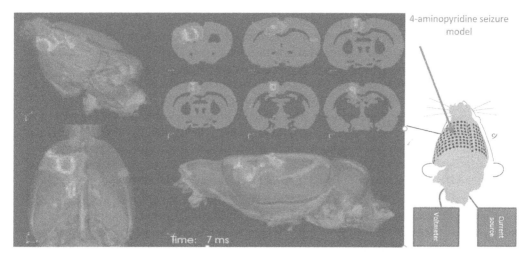

Figure 12.12   EIT images of epileptic spikes in the anaesthetised rat using 120 epicortical electrodes[57]. (color figure available in eBook)

repeated stimulus, in the same way as somatosensory or visual evoked responses, or other activity such as epileptic seizures. An EIT image would subsequently be reconstructed for each millisecond or so of the recording window. In this way, it would be possible to determine the waveform of activity in any selected pathway in 3D. This is not currently possible by any existing method, and, if possible, this would be a substantial advance. Unfortunately, it poses a formidable technical challenge. The linear sensitivity matrix based reconstruction algorithms developed for EIT of the brain (§12.3.2) could be employed as they stand, as the small changes are suitable for linear reconstruction approaches. However, the signal is small – less than 1% – so obtaining sufficient signal-to-noise for imaging deep changes is a challenge. Thus it has not yet been possible to produce satisfactory images of brain function with scalp electrodes, but a notable advance has been development of a method to produce reproducible and accurate fast neural images in the brain and peripheral nerve.

Fast neural EIT may be compared with EEG inverse source modelling, as both methods record boundary voltages. However, EIT has the advantages that the inverse solution is in principle unique, unlike inverse source modelling, and many more independent measurements are made from the same number of electrodes e.g. 4096 compared to 64 from a 64 channel array.

12.6.1   FAST NEURAL IN THE BRAIN DURING EVOKED PHYSIOLOGICAL ACTIVITY AND EPILEPTIC SEIZURES

The first attempts to produce fast neural EIT in brain were undertaken with scalp electrodes to inject current and electrodes or magnetic field detection with MEG. This was performed with a 2 Hz square wave, as theory indicated this would yield the largest signal[345,346,347,348]. This confirmed that a signal existed which matched biophysical modelling. Unfortunately, the impedance change was obscured by the endogenous EEG or MEG activity of the brain so that prolonged averaging was needed which led to unacceptably long averaging times. Images were then successfully achieved with determination that the greatest signal-to noise was at 1.7 kHz. The method developed utilized a custom made epicortical electrode array with 30-60 electrodes over each hemisphere, a detailed adaptively refined 5M Finite Element Model mesh of the rat brain, Tikhonov regularization, and display of reconstructed impedance changes as a statistical t-score (figure 12.12,[57]). Image sets would be produced by averaging to a repeated trigger – either a somatosensory evoked response or spike wave complex during epileptic seizures. This would be undertaken over 15–30 minutes but the final reconstructed image would comprise an image every 1–2 ms of fast neural impedance changes due to neuronal depolarization over the averaged epoch of 50-100 msec. This was accomplished for somatosensory evoked potentials and shown to have an accuracy of <200 m compared to local field potentials and intrinsic optical imaging[58]. It was possible to track the trajectory of the volley of evoked activity. This entered at 1 mm deep in the cortex, corresponding to Layer IV, the layer at which the largest changes are expected on physiological grounds. Similar images were obtained during epileptic activity by triggering from spike wave complexes both in cerebral neocortex[429] and deeper in the hippocampus still using epicortical electrode mats on the brain surface[431]. Methodical characterization indicated the peak signal-to-noise was measured at 1475 or 1355 Hz for evoked responses or spike wave complexes respectively[270,429]. The current used of 50 A was shown not to damage the cerebral cortex or significantly alter neuronal function, and a modelling study indicated a field of coverage to the outer third of the brain with epicortical electrode mats alone[271,430].

12.6.2   FAST NEURAL EIT IN PERIPHERAL NERVE

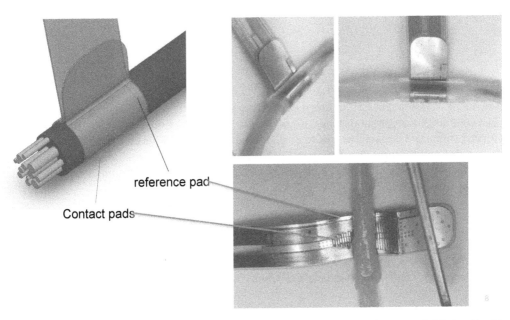

reference pad

Contact pads

Figure 12.13   Silicone rubber nerve cuff with 14 electrode pads. (color figure available in eBook)

Fast neural EIT can be used with a nerve cuff and 14 longitudinal electrodes in a ring to image compound action potential activity in multiple fascicles in peripheral nerve in real time. This is achieved by imaging the impedance decreases of 0.1–1% known to occur during the compound action potential. The same electrode cuff may then be used to stimulate or block the identified fascicles by injection of intelligent current patterns. This arose from interest in the new medical technique of "Electroceuticals" -- treatment of disease by electrical stimulation of autonomic nerves. This may be limited by the inadvertent stimulation of undesired organs when stimulation takes place in complex nerves such as the vagus nerve in the neck. EIT offers a means to identify activity in fascicle activated by an organ of interest, and then a means to selectively activate or block that function.

A method has been developed which employs the time difference fast neural UCL ScouseTom EIT system and a cylindrical soft rubber nerve cuff with laser cut stainless steel foil electrode pads wrapped around the rat sciatic nerve (figure 12.14). Imaging of large myelinated fibres is

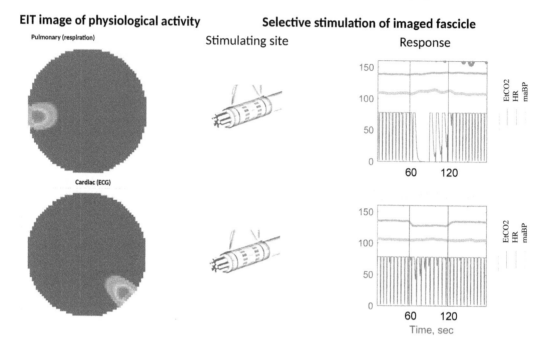

Figure 12.14   Fast neural EIT in sheep cervical vagus nerve. Left : EIT images triggered from the ECG (cardiac) or respiration (lung). Right: Validation by selective stimulation. The nerve was electrically stimulated through two electrode pads at same radial position on the nerve, 4 mm apart, all 14 possible pairs in turn. Stimulation at 10 O'clock caused apnoea (diminished respiration) but ECG unaffected; stimulation at 4 O'clock decreased heart rate but not respiration[869]. (color figure available in eBook)

best undertaken at 6 kHz. The method has been validated against microCT and neural tracers for the three fascicles in rat sciatic nerve, with an accuracy of $<200\,\text{m}$[868]. It has also been employed in sheep and pig cervical vagus nerve where it has been possible to identify fascicles relating to recurrent laryngeal, cardiac and pulmonary function (figure 12.14,[869]).

12.7   CONCLUSION

In the 1st edition in 2005, this chapter contained some proof of principle slow change EIT images and preliminary ideas for applications in stroke and imaging fast neural activity. In the intervening 15 years, studies in the nervous system have prospered and this area could justly be considered to lie second to those in the thorax. Unlike the imaging of lung ventilation, brain and nerve applications have not yet entered commercial production or hospital use. However, methods are now developed and validated for fast neural EIT in brain and nerve, and monitoring cerebral edema, which are ready for production and use in experimental and clinical neuroscience. It seems not implausible that use of fast or slow EIT and cerebral monitoring could be in clinical use in epilepsy, cognitive neuroscience or electroceuticals by the time of the next edition.

13 EIT for Imaging of Cancer

Ryan Halter
Thayer School of Engineering, Dartmouth College, Hannover, NH,
USA

CONTENTS

DOI: 10.1201/9780429399886-13

This chapter provides an overview of clinical applications of Electrical Impedance Tomography with a specific emphasis on cancer imaging applications. The rationale for using EIT for cancer imaging in terms of the reported contrast mechanisms between benign and malignant tissues is introduced. This is followed by a summary of EIT technologies developed for breast, prostate, and other anatomic sites (lung, cervix); the focus of these summaries is on the anatomy- and application-specific hardware and reconstruction algorithms used and the clinical data collected and analyzed with these systems. Much of the section on breast imaging has been adopted from the 1st edition of this book with several updates made to reflect recent research and clinical studies undertaken in this space [484].

Earlier chapters of this book describe in general the underlying tissue characteristics that impact bioimpedance signatures (chapter 3), the hardware required for recording impedance (chapter 4), and algorithms used to reconstruct electrical property images of tissue (chapter 6). As we look at specific applications it is important to emphasize that a single EIT system and reconstruction algorithm are not typically ideal for a broad range of applications. Instead of this "one size fits all" approach it is common to optimize hardware and reconstruction algorithms for the intended application.

When we think about optimizing hardware and reconstruction algorithms for specific applications we need to answer several questions about the application prior system design. Questions to answer include:

Is this a static or dynamic application? Cancer imaging with EIT has primarily focused on static applications (i.e. the tumour does not change dynamically on the order of ms to sec to min). This static imaging typically relies on so-called "absolute imaging" approaches in which a reference frame is not available for time-difference EIT (§6.7). Because of the ill-posedness and ill-conditioned nature of the EIT problem, absolute imaging has much stricter requirements on data acquisition system accuracy and channel-to-channel errors than time-difference EIT in which absolute accuracy and channel-to-channel variations are mitigated through the difference operation. Several groups have been exploring frequency-difference imaging [640,677,957,1170] to help overcome some of the challenges in static absolute imaging, but these also are limited by requirements on spectral accuracy. There has been some initial work on imaging hemodynamics in cancer using time-difference imaging in breast cancer patients [408], but this is still under development. Answering this question helps to define the hardware accuracy, channel-to-channel variations, and precision requirements of the hardware and the type of reconstruction algorithm one might select.

What are the characteristics of contrast mechanism? Because the impedance of tissue varies so much between tissue types and pathologies it is critical to define the exact contrast that one intends to image and the characteristics of that contrast. For example, if you are interested in imaging breast cancer in an older population the contrast of most interest is likely the difference in impedance between adipose tissue and malignant breast lesions, while if you are targeting a younger population of patients with denser, more glandular rich tissue, you might be more interested in the contrast between a malignant lesion and glandular mammary tissue. Once the tissue types or pathologies of interest are defined, it is important to then decide over which band of frequencies is the contrast maximum. In some cases, you might have a single frequency that exhibits a large contrast while in other cases it might be that some spectral parameter (e.g. one derived from a Cole model) provides the most contrast. In the single frequency case, the hardware required for the application is much simplified as compared to the spectral case in which a multi-frequency data acquisition system is required. Answering this question helps to define the bandwidth of the required hardware.

What is the clinical use case? Both the anatomy of interest and the specific way in which the EIT technology will be used and acquired will drive much of the system design. In particular, the electrode array deployed, the proximity of the hardware to the electrode array, and required speed of the reconstruction algorithm are all folded into answering this question.

13.1 EIS IN CANCER

13.1.1 CANCER BIOLOGY AND IMPEDANCE CONTRAST MECHANISM

Tissues, like any material, can let currents flow with more or less ease and, given an applied potential, hold more or less electric charge (see also §3.1.2). Conductivity is a material's ability to allow current flow: as it increases, greater current is established for a given potential difference. Permittivity is a measure of a material's ability to hold charge, with greater permittivity corresponding to greater amounts of charge stored, for a given potential difference.

Conductivity (σ) and permittivity (ε) are distributed properties of a tissue. They affect the flow of alternating currents and this effect can be measured. For a given geometry, the impedance (Z) or its inverse, admittance ($Y = 1/Z$), relates directly to the distributions of σ and ε in the material. Admittance can be expressed as the complex value $Y = G + i\omega C$, with G the conductance relating to the tissue's conductivity (σ) and C the capacitance relating to the tissue's permittivity (ε). Distributed complex properties have been defined which correspond to impedance and admittance; they are impedivity and admittivity, respectively. Admittivity can be expressed in terms of conductivity and permittivity: $\sigma^* = \sigma + i\omega\varepsilon_0\varepsilon_r$, where i is the imaginary unity, ω the angular frequency, ε_0 the dielectric constant of vacuum and ε_r the relative permittivity of the tissue. Impedivity can be expressed as the inverse of admittivity.

Early experimenters reported data obtained from excised tissues which indicated significant differences between cancerous and non-cancerous tissues. Since not all the data were collected under comparable conditions and although there are some reasons to believe that even freshly excised tissues will have their properties altered in the process, these data nevertheless should be interpreted as indicating that a measurable difference in electrical properties does exist in cancerous tissue compared with their surroundings.

13.1.2 ANATOMY-SPECIFIC IMPEDANCE CONTRAST

Because EIT systems for breast and prostate cancer have been the most prominently developed clinical applications, details regarding the electrical properties of these tissues are highlighted below. The literature is rich with studies comparing the electrical properties of malignant and benign normal tissue in other tissues. Pathiraja *et al* provides a systematic review of the literature in this space and points to relevant studies in breast, prostate, bladder, hepatic, lung, tongue, skin, gastric, renal, thyroid, colorectal, and esophagial tissues[809].

13.1.2.1 Breast

Research on the use of impedance measurements for breast cancer screening has been ongoing for some time, with some of the earliest data on breast tissue impedance published in 1926[314]. In this study Fricke found a significant difference in the capacitances of their excised samples, with benign tumours having lower capacitances than cancerous tumours. Review articles have been published which present good overviews of the field with the article by Zou and Guo providing a modern overview[1221]. We would like here to present a brief summary of the existing experimental data which provides the rationale for using electrical impedance measurements for breast cancer detection.

Jossinet conducted two studies, in 1985[523] and in 1996[521], both of which reflect measurable differences. In the earlier study, it is reported that the magnitude of impedivity is smaller for cancerous tumours than for surrounding tissue by a factor of approximately 5 at 1kHz. In the later study, the magnitude of impedivity of cancerous tumours is compared with several other classes of tissues. It is found that carcinoma has lower impedivity (magnitude) than subcutaneous fat and connective tissue, but is greater than fibroadenoma. However, at higher frequencies cancer tissue has the greatest reactive (capacitive) response of all the tissues tested. Furthermore, no significant differences have been observed between the impedivity of the normal or benign tissue types.

Several other studies have been published which generally confirm these results[176,185,456,668,1005], although not all cover the same frequency range. One study found no significant differences in the conductivity or permittivity of benign and malignant tumours[185]; however, the data were recorded at a very high frequency (3.2 GHz), at which different phenomena may be taking place in tissues.

Many more studies can be found that have published data on ex vivo breast impedance. Most of the results reviewed here seem to concur that cancer tumours have lower impedance than normal surrounding tissues.

Many fewer studies have published data based on in vivo invasive measurements. One of the few groups to publish such data, Morimoto *et al*[728], used a specially designed probe inserted in breast tumours on anaesthetized patients, and measured impedances between the needle tips and an abdominal patch electrode, using a three-lead technique. Measurement data from these studies was presented in the form of equivalent lumped components R_e, R_i and C_m, forming a network in which R_e is in parallel with a series combination of R_i and C_m. This way of presenting the data makes it difficult to compare with other studies. In this study R_e and R_i were found to be higher in tumours, while C_m decreased in tumours, compared with normal tissues. Although this study showed that significant differences in the electrical responses of the different types of tissue could be used for differentiation, it is largely in disagreement with other data regarding the direction of the changes, presenting an increase in impedance instead of a drop for cancerous tumours.

One study has compared ex vivo point-based impedance measurements to EIT derived in vivo measurements of benign and malignant breast[414]. Direct intra-patient correlations were explored between in vivo EIT images and resected tissue specimen measurements and it was found that the ex vivo conductivity and permittivity were both elevated with respect to in vivo measurements. Similar to the majority of other explorations, this study suggested that tumour conductivities were higher than the background benign tissue.

A few groups have performed non-invasive two points impedance measurements on breasts with and without tumours[793,976]. The reports based on these experiments indicate again a drop in resistance and an increase in capacitance[976] for cancerous tumours, or at least that differentiation is possible[793].

13.1.2.2 Prostate

The electrical impedance difference between benign and malignant prostate has been suggested as a potential contrast for prostate cancer imaging with EIT. The architectural differences between benign epithelia-lined glandular lumens loosely spread within a bed of stroma and that of rampant epithelia growth, decreased luminal spaces and loss of stromal density in malignant prostate gives rise to contrast in both conductivity and permittivity.

Lee *et al* were the first to report on resistance (inverse of conductance) differences between malignant and benign prostate[609]. In their ex vivo study, they used a four-needle probe and showed prostate cancer to have a 10% to 15% higher resistance than benign prostate at 100 kHz and 1 MHz, respectively. Salomon *et al* also explored a higher frequency range of 0.3 and 80 MHz and showed cancer to have an average 10% higher resistance than benign tissue with their "custom designed probe"[921]. Halter *et al* completed a series of ex vivo impedance measurement studies between 100 Hz and 100 kHz and similarly showed prostate cancer to have an approximately 10% lower conductivity than benign prostate[404,407,410,411,412,412,414]. These studies also showed permittivity to be significantly higher in malignant prostate, with the largest contrast exhibited at the higher frequencies. Spectral Cole-based parameters provided enhanced contrasts as high as 4:1 for the characteristic frequency (1.5:1 for high frequency conductivity, 2:1 for the conductivity increment) suggesting that multifrequency impedance spectroscopy may provide better clinical benefits than single frequency data acquisition.

Important in the diagnostic work-up for prostate cancer is identifying the grade of the disease present. This is specified as the Gleason grade which is a histological assessment of the architectural structure of the cancer with grades spanning from 1 to 5 with grade 1 representing low-grade disease in the form of well-differentiated glands with well-defined lumens and grade 5 representing high-grade disease in the form of a fluid sheets of grouped epithelial cells and absence of luminal spaces. Halter *et al* conducted a sub-analysis of some of their early work in this space and showed significant differences between cancer of different Gleason grades[413]. Prostate cancer treatment is typically based on the cancer grade identified, however, differentiating Gleason grade using currently-deployed clinical tools is challenging. The promise of Gleason grade differentiation with impedance represents a potentially impactful pathway to clinical adoption.

13.2 BREAST EIT

13.2.1 INTRODUCTION

Approximately one woman in eight will develop breast cancer over a lifetime in the US[971]. The prognostic for women diagnosed with the disease is greatly influenced by the stage at which it is discovered. Long term survival is greatly improved for women found with small tumours in the early stages of development. Periodic mammograms for women over 40 or 50 years of age constitute the principal tool used in screening for breast cancer and can be credited with saving many lives. However, mammography has not reached the level of perfection desirable for a mass screening tool.

Exposure to X-rays, although minimal in mammograms, is one objection that is raised, particularly for women who are advised to have more frequent exams and to start at an earlier age, due to a family proclivity. It is thought that the cumulative X-ray exposure, beyond a reasonable lifetime quota, may itself become a health risk.

More immediately of concern for women who undergo the examination is the significant discomfort caused by the need to squeeze the breasts to a thickness of a few centimetres against a detector plate. The procedure is thought to discourage some women from submitting to regular examinations.

From a public health point of view, the greatest objections to X-ray mammography is its imprecision as a diagnostic tool. Studies estimate that a woman with a tumour may remain undiagnosed following a mammogram (false negative) 10–25% of the time[546,732,899]. This means a sensitivity of up to 90%. Conversely, women who undergo periodic examinations will have a high probability of an abnormal finding; nearly a 50% chance after 10 visits according to one study[257]. Such findings typically call for biopsies to be performed, which in 80% of cases reveals benign abnormalities[938]. Improvements with digital mammography have been realized in recent studies, however the sensitivity and specificity are still limited to 87.8% and 90.5%, respectively[1000]. The diagnostic effectiveness of mammography diminishes as it is applied to women with denser breast tissue. This generally corresponds to younger women who undergo the procedure because of a family history and who are usually at a higher risk. The high rate of false positive findings (lack of specificity) represents a great cost to the health care system, and women undergoing the process could be spared the distress it causes if a better diagnostic tool were available.

The idea of measuring the impedance of tissues is not new, but until methods were devised to measure non-invasively the impedance of internal structures, it was only of interest to researchers. Computers make it possible now, using advanced algorithms, to reconstruct the electric properties of internal tissues from non-invasive surface measurements. Electrical impedance tomography, in its various forms, has been applied to several areas in medical diagnosis and monitoring, including the measurement of breast tissue impedance. Preliminary data strongly indicate that cancerous tissues have electrical properties that are significantly different from their normal surroundings. This has spurred a wave of activity, which it is hoped will result in improved screening for breast cancer.

13.2.2 OTHER METHODS IN USE FOR BREAST CANCER DETECTION

Standard practice has established X-ray mammography as the primary and most used method of breast cancer screening. Breast self-examination has been advocated as a possible alternative, but its effectiveness compared with X-ray mammography is very limited. The size of tumours that are detectable by palpation is typically much larger than that of abnormalities that are detectable by mammography.

Given the less than perfect performance of mammography as a screening tool, several procedures are in use to specify the nature of abnormalities that are detected in normal periodic examinations. These diagnostic modalities are used mostly as follow-up on the results of mammography and clinical guidelines do not generally recommend their used as primary screening tools[106]. In this category are ultrasound and MRI.

Ultrasound alone does not compete with mammography. The image quality, although greatly improved in the past few years, is not comparable with X-ray mammography. However, because it is much more flexible and interactive, with the user able to scan the desired area repeatedly and from different angles, it is often used to inspect more closely suspicious masses or cysts. It is typically used to distinguish between tumour types for diagnostic purposes, and also for the placement of biopsy needles. Automatic whole breast ultrasound systems have recently been designed to image the entire breast using a fixed ultrasonic transducer that revolves around the breast and captures a full 3D image. In combination with mammography, this multi-modal approach has been reported to improve the accuracy of breast cancer detection and confidence in callbacks for biopsy especially in women with dense breasts[543].

MRI mammography is usually used to verify a diagnosis, and rarely as the primary screening tool. The cost of MRI, particularly when compared with inexpensive X-ray mammography, will preclude it for the foreseeable future from becoming the standard in breast cancer screening. However, it has being investigated in Britain[835] for screening younger high-risk women who generally have denser breasts, which proved to be more difficult to screen with X-ray mammography. Similarly, a randomized control trial in the Netherlands reported value in using supplemental MRI in women with extremely dense breast[79]. In addition, centres have appeared which have dedicated MRIs for breast examinations, although not exclusively for cancer screening. It is offered in one centre, for example, for diagnosis of breast implant rupture, cosmetic surgery planning, staging of breast cancer and treatment planning, post-surgery and post-radiation follow-up, dense breast tissue evaluation, and monitoring of high-risk patients with a non-radiation alternative.

While electrical impedance is still in the research stage, its early clinical effectiveness has been evaluated in a number of studies, though complete clinical translation remains to be demonstrated. There are other technologies also being investigated in view of applying them to breast cancer screening. Many groups worldwide are investigating light attenuation and scattering, particularly in the near infrared (NIR) region, as a method of detection[287,834,1056]. Microwave imaging (MWI) is also being investigated in several groups worldwide for breast imaging[273,720]. In a fashion very reminiscent of EIT, it seeks to image conductivity and permittivity of irradiated tissues by reconstructing a tomographic section of a breast, although at a much higher frequency range ($\sim$300 MHz–3 GHz). MRI-based elastography (MRE) is another technique that is being explored for breast cancer screening[972,1073]. It relies on MRI's ability to detect very slight motion ($\sim$100 μm). A periodic motion is imparted by a mechanical shaker to one side of the breast and the resulting displacement field inside the breast is captured by the MRI. Computational techniques can recover the tissue's shear modulus (which corresponds more or less to "hardness") from the motion data. It is thought that hardness may be a reliable indicator of a malignant tumour. It is well established that cancerous tumours are felt as hard nodules when reaching a certain size.

13.2.3 DIFFERENT APPROACHES TO BREAST EIT

Different approaches have emerged for imaging internal tissue impedances using non-invasive techniques. Two categories present themselves based on the arrangements of the electrodes: tomographic systems and planar or mapping systems. The tomographic type systems led to the adoption of the term electrical impedance tomography or EIT. We use the term impedance imaging to encompass all methodologies.

13.2.4 IMPEDANCE MAPPING

Impedance mapping systems are simpler in two respects. The electrode arrangement is planar, usually an $n \times n$ square array of electrodes, which is used to press the breast against the chest wall. In this arrangement, breast tissue constitutes a relatively shallow layer between the array and the rib cage. A current is applied sequentially between each electrode in the array and a distal electrode, usually held in the patient's hand. In the simplest version of this type of system, the impedance sensed at each electrode in the array is represented as a shade of grey in an image, in the position of the electrode. The planar array is easier to construct than circular arrangements, which require being adjustable for different breast sizes, and the reconstruction is usually simplistic, although algorithms have been developed to compute the impedance map at different planes away from the electrodes. Two main versions of this type of system are in existence, one developed in Israel and previously marketed by Siemens[64], another designed by the research group in Yaroslavl in Russia[194].

13.2.5 TOMOGRAPHIC IMAGING

In a tomographic system, the electrodes are arranged so as to surround the region of the body to be imaged, in our case the breast. The electrodes, usually arranged in a circular array, define a plane of intersection for which the spatial distribution of electrical properties is sought. Multiple planes can also be used simultaneously, in which case the region of interest is the enclosed volume rather than a plane. In both cases predefined patterns of currents or voltages are applied and the corresponding voltages or currents are measured. The recorded data are used to reconstruct the desired properties. Tomographic systems further distinguish themselves by the reconstruction methods for which they were optimized[83,444]. Early stage development of a two-plane tomographic system with radiolucent electrodes and coupled to a mammography system has been developed by RPI though has not been used in any large clinical studies to date[532].

13.2.6 LIMITATIONS OF IMPEDANCE MEASUREMENTS

Two-dimensional tomographic systems usually base their reconstruction method on the assumption that current flow is restricted to the imaging plane. This assumption holds approximately for shallow phantoms, but it is clearly not realistic in the case of breasts or other body regions. The effect of ignoring current flow through the out-of-plane volume results in lost accuracy in the reconstructed images. Full 3D solutions represent an advantage in this respect for both planar and 3D data, in spite of their added complexity.

The sensitivity of impedance imaging systems to variations in tissue properties decreases with distance from the nearest electrode. In a circular array configuration, this means that the central portion of the imaged plane has the least sensitivity. In the case of a planar array, sensitivity decreases as the distance from the electrode plane increases.

In addition to uneven sensitivity, impedance imaging techniques suffer from a poor spatial resolution, compared with other imaging technologies. Physicians used to seeing a great deal of detail (sub mm) with X-ray mammography, for example, will be disappointed by the typical $\sim$5 mm spatial resolution of impedance imaging systems. With tomographic systems, the spatial resolution is prescribed by the physical arrangement of the electrodes, their number and the number of different excitation patterns that are used. In a system with 16 electrodes, for example, it is possible to apply 15 optimal patterns forming a complete orthogonal set (i.e. additional patterns would not theoretically add any information). Each pattern corresponds to 16 measurements and so we have 16×15 measurements or 240 independent measurements. For a 10 cm cross-section, if we divide it evenly we have 240 patches of roughly $0.33\,\text{cm}^2$ or square patches 5.7 mm on a side. This is a very simplified estimation of the spatial resolution of tomographic impedance imaging; in reality the resolution is best on the periphery and worst at the centre, as has been shown experimentally[548].

In planar impedance imaging systems, the spatial resolution is more or less equivalent to the electrode density of the array. This will deteriorate with distance away from the contact plane. Planar array with 16×16 electrodes have been presented, which measure 12 cm on a side[194], which corresponds to a spatial resolution at the contact plane of 8 mm.

Adding electrodes may be a way to increase spatial resolution, at least in the case of planar arrays. With tomographic systems, the addition of electrodes on the periphery of the imaged cross-section improves spatial resolution at the periphery, but only slightly in the central region.

13.2.7 ADVANTAGES OF IMPEDANCE AS A SCREENING TOOL

At this time it does not appear likely that impedance imaging will unseat mammography as the primary method of screening for breast cancer. Its poor spatial resolution, compared with X-ray, represents a barrier to its being adopted, even if its sensitivity and specificity were to improve. However, given the current performance of X-ray mammography, it is quite conceivable that impedance will be adopted as a second step in the standard examination, when an abnormality is discovered. EIT systems could be designed to be relatively inexpensive to purchase (<$10 000) and very inexpensive to use. Examinations could be very rapid (<10 min), and very safe. They do not involve X-ray exposure and could be repeated as often as needed.

13.2.8   CLINICAL RESULTS SUMMARIES

A few groups to date have presented clinical results of breast cancer screening. Most of the clinical results published to date have been based on planar array instruments such as the T-Scan (previously marketed by Siemens as the TS2000), which received FDA approval for use as an adjunct to mammography TS2000[272]; this system has been used by several groups worldwide in clinical trials.

The only clinical experiments we are aware of, using the tomographic approach based on circular arrays, have been conducted at Dartmouth starting in 2000[796]. Preliminary and clinical trial findings from this work will be discussed here.

13.2.9   PLANAR GEOMETRY SYSTEMS

13.2.9.1   Piperno 1990[830]

This is a large-scale study based on 6000 patients using the 'Mammoscan' device, an early version of the planar array system marketed under the 'TS2000' name. Although this is not the first such study, it is the largest and most significant evaluation of impedance imaging. Of this patient group, 745 underwent biopsies to verify a suspicious finding in mammography. Every patient in the group submitted to mammography, palpation, trans-illumination, thermography and ultrasound exams. This study set out to compare all these modalities against each other. The first finding was that there were nine cases in which EIS was the only modality to flag as highly suspicious exams that all other modalities did not detect and which were confirmed by histopathology. The presentation of the results in that publication makes it impossible to compute the usual statistics regarding the rates of true positive (etc.) or the sensitivity and specificity of each modality. In their tabulated results, it is shown that the Mammoscan was correct (i.e. true positive + true negative) in 454/745 cases and incorrect (i.e. false positive + false negative) in 119/745 cases. The remaining cases were labelled as 'borderline cases' with no further indication of outcomes. For X-ray mammography the results are 395/745 correct findings, and 154/745 incorrect findings. The number of correct findings is greatest for EIS, compared with all the other exam types, and the number of incorrect findings is the lowest for EIS as well. This early study was interpreted as very encouraging for EIS at the time of its publication.

13.2.9.2   Malich 2000[675]

This study, based on 52 patients with 58 suspicious mammogram findings, was conducted using the TS2000 commercial electrical impedance scanning system marketed by Siemens. The system consists of a planar arrangement of 256 electrodes in a 16×16 arrangement. Patients were examined with the EIS system in low- and high-resolution mode, both breasts in every case. Patients also had breast MRI mammograms and biopsies or surgical removal of the abnormalities. In this study, the high-resolution scanning mode correctly identified 27/29 malignant lesions (93.1% true positive), and also correctly classified 19/29 (65.5% true negative) benign lesions (10/29 benign lesions resulted in false positive rate of 34.5%). The report indicates a negative predictive value of 90.5% and a positive predictive value of 73% for the TS2000 in its high-resolution mode. In standard-resolution mode the results were 22/29 (75.9%) TP and TN, giving the TS2000 a sensitivity of 75.9% and a specificity of 72.4%. Skin imperfections (lesions, scratches, moles etc.) and air bubbles resulting from placement constitute a reported practical limitation to the effectiveness of the TS2000.

13.2.9.3   Cherepenin 2001 [194]

The system used here is very similar to the TS2000, consisting of a planar array of 256 electrodes 12 cm on a side. The image reconstruction is slightly different in that impedances are computed at different planes away from the electrode array in order to reconstruct a 3D map of the volume facing the array. Slices of the volume between the electrode array and the chest wall were computed every 8 mm and for up to 6 cm depth. Twenty-one patients with tumours in sizes ranging from 1.5 to 5 cm in one breast were examined. Both breasts were imaged, with the contralateral breast used as a normal reference. Imaging was performed twice, with the patients standing and reclining, resulting in 84 data sets. The data sets were divided into five groups, including (1) 42 normal breasts and (2) 42 malignant tumours. Group (2) was subdivided, based on whether the tumours were visible as white spots, into groups (3) – 16 studies without focal abnormalities – and (4) – 26 studies with visible abnormalities. Group (5) contained 13 studies, selected from group (4) for their high conductivity peaks. Tumours were correctly identified in 14 out of a total of 21 cases (67%), as evidenced by clearly visible white spots on the reconstructed images. Four more were identified as anomalies due to the inhomogeneous aspect of the images, which brings the TP rate to 85.7%. The analysis was repeated using more sophisticated statistical methods instead of visual inspection. With this approach, groups (3) and (4) (malignant tumours-both types) could be identified in 19 of 21 cases (90.5%), on the basis of significant statistical differences in the property distributions. Although this study shows that malignant tumours can be identified when compared with normal breast examinations, it does not tackle the more important question of whether EIT can be used to discriminate between malignant and other types of abnormalities, which is where mammography comes short.

13.2.9.4   Malich 2001a [674]

This study concentrated on determining the incremental benefit of using EIS in addition to mammography. In this study, 210 women were examined who presented 240 suspicious findings in mammograms or ultrasounds. All lesions were verified by histological inspection of removed tissue. The results were that 86 of 103 malignant lesions and 91 of 137 benign lesions were correctly classified by the impedance examinations (87.8% sensitivity, 66.4% specificity). Predictive values were also presented, with NPV and PPV of 84.3 and 65.2% respectively. A sensitivity of 85.5% is reported for all cases and, for invasive cancers alone, a sensitivity of 91.7%. Ductal carcinoma in situ (DCIS) resulted in a much poorer sensitivity (57.1%, n = 14). Combining impedance mapping to mammography and ultrasound increased the sensitivity of the exams from 86.4 to 95.1%, while the accuracy decreased from 82.3 to 75.7%.

13.2.9.5   Malich 2001b [673]

In this study, the authors set out to determine whether impedance mapping duplicates or augments the clinical results obtained with ultrasound and MRI, as an adjuvant to mammography. One hundred patients were examined with ultrasound, EIS and MRI, following ambiguous abnormal findings in their mammograms. In all, 100 abnormalities were studied. Ultrasound and MRI findings were categorized by experienced radiologists using the LOS (level of suspicion) scale, with LOS values corresponding to: 1 normal, 2 most likely benign lesion, 3 probably benign, 4 probably malignant,

5 most likely malignant. Ultrasound findings with LOS of 2 or 3 were categorized as non-malignant findings, while LOS 4 and 5 were categorized as malignant findings. EIS images were categorized as indicative of a malignant finding if a bright spot was visible, and could not be discarded as artefact due to poor contact or the presence of the nipple. Sixty-four such lesions were identified on impedance maps and were categorized as positive findings. EIS showed a sensitivity of 81% and a specificity of 63%, ultrasound had a sensitivity of 77% and a specificity of 89%, and MRI had a sensitivity of 98% and a specificity of 81% on this group of patients. These findings correspond to the individual modalities- individual performances. Statistical analysis further showed that EIS adds clinical information to ultrasound, but that MRI and EIS are mostly similar in the information they contribute to the diagnosis.

13.2.9.6   Cherepin 2002[196]

This study leverages the same hardware used in 2001[194], with modification to the electrode array: the planar array consisting of 256 electrodes is now arranged in a circular pattern which increases the "utilization factor" for the electrodes. In the previous square arrangement, electrodes in the corners tended to not make contact with the patient. Furthermore, the array is somewhat smaller, increasing the electrode density, and as a result the spatial resolution achievable with the device. This study did not seek to evaluate the performance of impedance mapping in breast cancer screening; rather it sets out to establish baseline measurements for women in several categories. Fifty-seven women were examined in ages ranging from 18 to 61. The patients were selected to fit in five groups: (1) 12 women (18 to 45 years) in the first menstrual phase (1 to 10 days); (2) 12 women (18 to 45 years) in the second menstrual phase (16 to 28 days); (3) 14 women (18 to 39 years) during their pregnancy (37 to 40 weeks); (4) 14 women (18 to 39 years) during lactation (three to five days post labour); and (5) five postmenopausal women (47 to 61 years, one year post menopause).

The findings in this study, although not directed at breast cancer, are still interesting. Differences in the appearance of the impedance images were noted which were consistent within groups. However, a systematic analysis of the data is based on the average conductivity value obtained in the second plane away from the array (1.2 cm depth). Significant differences were few between the groups. Of the five groups presented here, all consisting of healthy women (i.e. no abnormalities expected), only group 5 presented a statistically different and consistent difference in conductivity. Particularly of interest is the fact that hormonal changes during the menstrual cycle may not affect noticeably impedance measurements. This has been a consideration in the clinical application of impedance imaging. Should periodic hormonal fluctuations affect it, then this should be taken into account in scheduling examinations.

13.2.9.7   Glickman 2002[351]

In this study, using data collected with the TS2000 impedance imaging system, the authors implemented an automatic algorithm to identify bright spots which correspond to conductivity increases and generally to malignant tumours. They also refined the algorithm to discriminate more reliably between malignant and benign lesions. Their algorithm is based on two main predictors, the phase at 5kHz and the crossover frequency (where the imaginary part of admittance peaks). A learning process was used to adjust the identification thresholds which were trained on data from 461 examinations, with 83 malignant and 378 benign cases. The designation of every case was based on biopsy results. With this methodology, they applied their algorithm to a separate group of 240 examinations (87 malignant, 153 benign). Under these conditions they reported a sensitivity of 84% and a specificity of 52% in properly identifying malignant and benign impedance images.

13.2.9.8 Martin 2002 [688]

In this study, 74 patients were examined using impedance imaging as well as mammography, and a systematic comparison between the two imaging modalities as well as the histopathology findings was undertaken. Impedance imaging was conducted using the TS2000 on patients from several centres. Of this patient group, 77% were classified as having mammograms suspicious for malignancies. In their findings, histopathological diagnosis and EIS positivity were positively correlated and EIS showed a true positive rate of 92%. In addition, all the cases diagnosed as in situ carcinoma, based on histopathology, were positive in EIS. In cases of ductal carcinoma or as ductal carcinoma plus in situ carcinoma, 92% were positively identified by EIS and all the cases of lobular carcinoma also had positive EIS diagnoses. Only three of 50 cases of malignant disease (6%) had negative EIS diagnoses. The false positive rate of EIS was 17%, while for this group of patients the false positive rate for mammography was 17.5%.

13.2.9.9 Stojadinovic 2005 [1007]

This prospective multi-center study used the T-Scan 2000ED to explore the feasibility for early detection of breast cancer, especially in younger women. Women presenting to the clinical for routine screening or those referred for breast biopsy were eligible for inclusion. All women were given at minimum a clinical breast exam prior to impedance scanning. The T-Scan 2000ED is technically similar to the earlier TS2000 model, but relies on a different scoring algorithm for cancer detection that was being tested as part of this study. Instead of providing a localized "bright spot" in the impedance map like the TS200, this new algorithm simply reports cancer classification as a single green (negative) or red (positive) indicator bar. Of the 1,103 women include in the study 556 were <40 years of age, 450 between the ages of 40 and 49, and 73 >50 years of age. Within the group screened with the T-Scan 2000ED, 29 cancers were identified with 19 of these being non-palpable. Sensitivity and specificity for cancer detection in the younger (under age 40) population was 50% and 90%, respectively; however, the overall sample size of malignancies was noted to be low in this group. The smallest tumour detected with EIS was ~7 mm in diameter and was non-palpable. The authors suggest that EIS-based detection of any non-palpable cancer exceeds the limits of clinical breast exam in which only later-stage palpable masses can be detected.

13.2.9.10 Stojadinovic 2006 [1006]

This study evaluated the efficacy of using the T-Scan 2000ED to identify young women at risk for breast cancer. The prospective, observational, two-arm, multicenter trial specifically enrolled women between the ages of 30 and 45. The much denser breast tissue in this population of women limit the effectiveness of other conventional screening modalities (Clinical breast exam (CBE) and mammography), and this study aimed to explore if EIS could fill this clinical need. Due to the low number of expected cancer cases in this population, the study was divided into a Specificity Arm (1,361 women) and a Sensitivity Arm (189 women); the sensitivity arm included women in this age group scheduled for breast biopsy due to other positive screen findings. A false positive rate of only 4.9%, corresponding to a specificity of 95.1% was reported. Within the sensitivity arm, 26.5% of 189 women had positive cancer findings. The average EIS-based sensitivity for the group was 38%, and the specificity within this smaller group was 80.7% specificity was significantly lower than that found in the Specificity Arm group. It is also noted that the sensitivity is higher in older women (42% vs 29% in age 40–45 and 30–39, respectively) and in smaller tumours (44%, 50% and 19% for tumours <11 mm, 11–20 mm, and >20 mm, respectively).

13.2.9.11 Trokhanova 2008 [1055]

A modified version of the electrical impedance mammography (EIM) system used in the Cherepenin *et al* studies was used in this study of dual-frequency EIM. While this study did not specifically explore the use of EIM for cancer detection, it did evaluate the utility of breast EIM for diagnosing general mastopathy. Data was collected from the 256-electrode system at 10 kHz and 50 kHz and images were generated and analyzed at depths of 0.4, 1.1, 1.8, 2.5, 3.2, and 4.6 cm relative to the electrode plane. Of the 166 women enrolled, 92 had no breast pathology while the remaining 74 had diagnosed mastopathy that was further subdivided into those with (33) and without (41) a cystic component. A series of example images that highlight EIM imaging features distinctive to patient characteristics are provided and showcase the potential features that can be extracted from EIM. While clinical parameters are not explicitly provided, this study reported that the electrical conductivity of normal breast tissue was significantly higher than in mastopathy for both 10,kHz and 50 kHz. This relationship held independent of the menstrual cycle phase during which the impedance data was acquired.

13.2.9.12 Raneta 2013 [862]

A single institution study of the EIM system used in the Cherepin *et al* studies explored the efficacy of this technology to help with differential diagnosis in pathological findings noted on mammographic and sonographic studies. A total of 870 women with suspected pathological breast lesions were imaged with the EIT system and differential diagnosis was evaluated with EIT as a standalone modality or in combination with mammography or sonography. EIT radiographs were interpreted qualitatively and regions of elevated conductivity were noted as suspicious. Lesions were categorized as fibrocystic mastitis, cystic, fibroadenoma, hyperplasia, lipoma, and carcinoma. When compared with mammography (specificity = 79.5%, sensitivity = 87.8%), EIT had an average specificity of 76.6% and sensitivity to carcinoma of 86% and when combined with mammography the specificity dropped to 72.8% and sensitivity increased to 94.5%. When compared with sonography (specificity = 90.2%, sensitivity = 86.7%), EIT had an average specificity of 89.2% and sensitivity of 66.7% that became 86.4% and 93.3%, respectively, when combine with sonography. The authors suggest that using EIT as an adjunct to these more conventional modalities has the potential to improve sensitivity, which may be especially important in a younger screening population of women.

13.2.10 CIRCULAR GEOMETRY SYSTEMS

13.2.10.1 Osterman 2000 [796]

This is an early non-blinded report based on Dartmouth's first-generation EIT system [444]. Examinations of 13 participants were conducted in order to investigate the feasibility of delivering breast examinations on a routine basis. A 16-electrode circular array was used with 10 signal frequencies from 10 kHz to 1 MHz. Both breasts were imaged on all patients except for one who had had a mastectomy (25 data sets). A custom examination table was used, on which the patients lie prone with the breast to be imaged pendant in the electrode array that is located below the table. Measurement data were used to reconstruct absolute images of permittivity and conductivity. In all cases electrode artefacts were evident on the periphery (near the surface) of the images. Several findings were reported. Permittivity images were generally more informative than conductivity images. Specifically, normal breasts appear to have consistent permittivity and conductivity images across subjects (figure 13.1). When abnormalities were present and detectable in the images, their location on the images corresponded with expectation (figure 13.2).

Twelve of the examined breasts were mammographically normal. The remainder included the following pathologies: three invasive breast carcinoma; one benign mass; six cases of fibrocystic

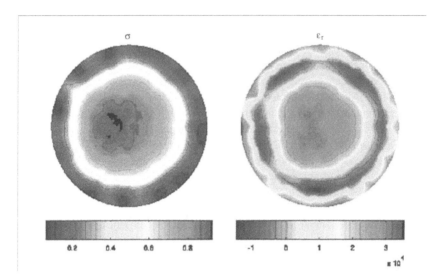

Figure 13.1   Reconstructed conductivity (left) and permittivity (right) image of a normal subject at 125 kHz using Dartmouth generation 1 EIT system.

disease including four cysts; and two cases of fibrocystic change without a discrete cyst. In addition, there were three patients who had had lumpectomies and radiation on one of their breasts. Of the four known tumour cases (3 Cancer, 1 benign), all were confirmed as having abnormalities in the appropriate breast in the EIT images. In one case the heterogeneous appearance of the images was considered to be a false positive for that patient who had no known pathology.

Using the coefficient of variation (standard deviation/mean) as an objective measure of heterogeneity in the central region of the image (60% radius to eliminate electrode artefact zone), abnormal designations were confirmed in 10 out 14 cases (true positive) and wrongly assigned in three out of 11 cases (false positive). Similarly, normal designations were given correctly in eight out of 11 cases (true negative), while misattributed in four out of 14 cases (false negative).

While this preliminary report had few participants and was not conducted in a blinded fashion, it constituted the first data presenting tomographic reconstructions of absolute electrical properties in a comparative normal and abnormal study. Because of the small size of the study, a meaningful comparison of the results between malignancies and all other cases was not possible, which reduces the value of this study. Its findings, however are supporting the notion that spectrographic absolute tomographic images of dielectric properties of breast tissue could be used for breast cancer diagnosis.

13.2.10.2   Halter 2004, 2008c[405,406]

The work presented here describes an EIT system that has multi-frequency broadband capabilities suitable for use in a clinical setting (figure 13.3). fast acquisition rate to minimize exam time and includes patient safety considerations. Also, because of 3D artefacts present in 2D imaging systems it incorporates 3D measurement capabilities. Its range of frequency is 10 kHz–10 MHz, an order of magnitude higher than its predecessor. The design of a second generation of electronics, based on a digital signal processing (DSP) architecture, centred around a 66 MHz ADSP-21065L SHARC processor and a reconfigurable field programmable gate array (FPGA) operating at frequencies up to 80 MHz. These two devices are reprogrammable, giving the design an unprecedented level of flexibility both in terms of the algorithms it can execute and the configuration of the digital circuitry.

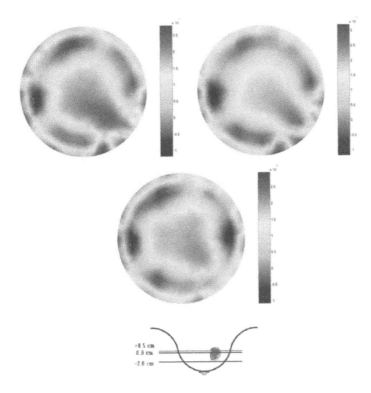

Figure 13.2 (Top) 125 kHz permittivity images in three different places. The left images is 0.5 cm above the lesion, the right one passes through it, and the bottom one is 2 cm below it. A 3.5 cm tumour is present at 4 o'clock. (Bottom) Diagram of where the lesion is located relative to the three viewing planes. From[547].

The signal-level performance of the system shows very significant improvements over previous implementations in accuracy, bandwidth and speed. For example, signal-to-noise ratio (SNR) is better than 80 dB at high frequency, compared with 60–70 dB previously.

With the development of a high-frequency design based on wedge-shaped circuit boards in close proximity to the electrodes, the team realized the breast interface shown in figure 13.3. which consists of four levels of 16 electrodes, where the electronics are integrated with the electrode-positioning system. The radial translation stages utilize electrode holding rods arranged in a sliding pattern under stepper motor control. This results in a very compact unit consisting of 64 channels, with leads from the electrodes to the electronic cards not exceeding 5 in (12.5 cm). We integrated the complete EIS system with a stereotactic biopsy table by fitting it into a sliding assembly that resides on a custom cart designed to dock against the biopsy table. The system engages tracks mounted under the table and is locked in position during an EIS exam. The biopsy table is still fully functional for X-ray exposures for lesion localization and surface fiducial marking prior to an EIS exam.

13.2.10.3 Soni 2004[997]

This preliminary study enrolled 42 participants (18 normal, 24 abnormal) that were imaged with the single plane EIT system used initially in Osterman *et al*. Absolute images were reconstructed in 2D using the dual mesh approach with the conductivity and permittivity estimated on a 353-node coarse mesh. Abnormal cases were defined as any patient with an ACR rating > 1, where the ACR rating

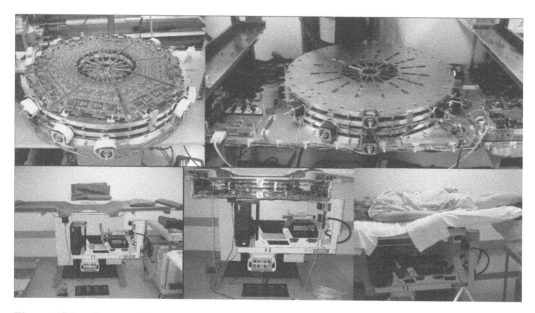

Figure 13.3 Dartmouth Breast EIT system attached to a stereotactic biopsy table. The system consists of multiple 4-channel "wedge" boards (top left) arranged in four layers of 16 electrodes (top right. The unit fits below the exam table and above the x-ray tube (bottom left and center left). The patient is prone on the table during exams (bottom right)[406]. (color figure available in eBook)

is defined by the BI-RADS American College of Radiology (ACR) indexing system. In addition, each breast was classified as fatty, scattered, heterogeneously dense or extremely dense based on its radiodensity. Average conductivities within annular regions of interest specified in the center, middle third, and outer third of each image were used to compare the cohorts. Normal subjects were observed to have higher conductivity and permittivity than abnormal subjects for all of the ROIs compared. While this study did report on specific clinical metrics, it highlighted some of the initial artifacts present in tomographic EIT images and provided some foundational expectations for the expected differences in normal and abnormal electrical properties imaged with EIT.

13.2.10.4   Poplack 2004[839]

This study of alternative imaging modalities for breast imaging included EIT, microwave imaging spectroscopy (MIS), and near infrared (NIR) tomography. The focus of this initial report is on the expected variation in electrical properties of breast tissue in women with negative clinical findings. The single-plane Dartmouth EIT system was used to record data from 23 women with negative mammographic findings. EIT data was collected at 10 distinct frequencies ranging from 10 kHz to 950 kHz using their voltage-drive, parallel-acquisition system. Average conductivity and relative permittivity values were found to be 0.46±0.1 S/m and 12,313±5,067 and 0.47±0.13 S/m and 5,034±1,816 for 125 kHz and 950 kHz, respectively. Significant positive correlations were also noted between relative permittivity and body mass index.

13.2.10.5   Poplack 2007[840]

One hundred and fifty women were enrolled in a study of NIR, MIS, and EIT at the Dartmouth Hitchcock Medical Center (Lebanon, NH, USA). The women were categorized as have normal (36 in the EIT cohort) and abnormal (62 in the EIT cohort) mammographic findings. Multiplane 3D

EIT data was acquired with the Halter 2004[405] system over the frequency range of 10 kHz to 10 MHz. Images were reconstructed in 3D using patient specific meshes that were defined based on the opening diameters of each plane's electrode array. Single plane cross-sectional images were extracted from each volumetric reconstruction for analysis. A region of interest was defined based on mammographic or MR findings in the ipsilateral breast. A symmetric ROI was defined within the contralateral breast as well for comparison purposes. Within the cohort of abnormal mammograms, 16 cancers were identified with the rest of the cases including fibrocystic disease, fibroadenoma, and other benign abnormalities. A maximum area under the curve of 0.78 is reported based on the conductivity ratio between ipsilateral and contralateral breasts.

13.2.10.6   Halter 2008c[406]

As part of developing the hardware used in the Poplack 2007 study[840], 97 women with no negative clinical findings were imaged to evaluate the impact of breast density on the imaged electrical properties of breast tissue. Impedance data was recorded from multiple planes of 16 electrodes over the frequency range of 10 kHz to 10 MHz. Patient specific 3D meshes were generated based on breast diameters (figure 13.4). The women included in this study were categorized as fatty, scattered, heterogeneously dense or extremely dense based on its radiodensity. In general, the average conductivity within the breast was observed to increase with breast density (i.e. fatty breasts had the lowest conductivity while extremely dense breasts had the highest conductivity). These differences became significant at higher frequencies ($\gtrsim$250 kHz).

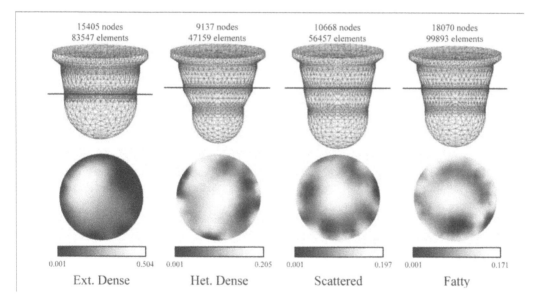

Figure 13.4   Patient specific volumetric meshes generated for four women with no known pathology, but with different breast tissue composition. 3D reconstructions were computed for EIT data recorded at 545 kHz. Cross-sectional images display conductivity at the plane coincident with the second level of electrodes marked by the line intersecting the mesh. Colorbar has units of S/m[406].

13.2.10.7   Halter 2015[408]

Absolute EIT (aEIT), as is conventionally required in cancer imaging, is more technically challenging than dynamic imaging applications; absolute imaging cannot take advantage of the systemic errors that are removed during the time-differencing operation used in dynamic EIT. Halter *et al* attempted to use dynamic EIT to image cardiodynamic changes in the breast in women with and without breast cancer. The rational for this approach is that as tumours form and grow there is a local vascularization process that takes place to provide sufficient nutrients to the growing tumour. The hypothesis was that as blood pumps through the vasculature of the breast, the relationship between conductivity changes and blood pulsation may be different local to tumours. This approach synchronized EIT data acquisition with pulse-oximetry in order to correlate dynamic changes in the breast conductivity with the cardiovascular cycle. Like other dynamic imaging applications in EIT, a time series of EIT data was acquired at a rate of 17 fps over a period of time that included multiple heart beats. The EIT data is averaged over that time and used as a reference frame from which all other frames are subtracted and reconstructed. The nodal-based conductivities within each frame define a time series that is synchronized with the pulse-ox-based cardiac signature. The conductivity time series at each node is correlated with the pulse-ox signal and several spectral and temporal parameters are extracted from the correlation. These parameters are computed at each node in the reconstruction and can be projected as an image. Nineteen women were imaged using this framework; of these women, 10 had cancer and 9 did not. Images of the temporal and spectral parameters were generated for all women and breasts imaged. An 81% specificity and 77% sensitivity were reported for the optimal correlative parameter. The primary advantage of this approach is the use of dynamic EIT to image the vasculature variations associated with tumour growth. Additional studies are needed to further evaluate this approach for cancer detection.

13.2.10.8   Discussion of the Clinical Trials

The first observation one makes regarding clinical results using impedance imaging is that it almost exclusively consists of experiments with planar array devices. Work on the development of such devices seems to have started around 1979[993], and this may explain the predominance of these types of device. In addition, planar devices are generally simpler and do not require a complex procedure to reconstruct impedance maps-in some cases no reconstruction at all is used, the impedances sensed by each electrode simply being displayed in the correct arrangement. When reconstruction computations are used they consist of reconstructing impedance maps at different depths and can be performed very rapidly, allowing real-time updating of the display.

The data discussed here based on a tomographic impedance device are still somewhat preliminary and do not constitute a large-scale clinical trial of the sizes conducted for planar device. Despite the limited trial sizes with tomographic systems, the early results suggest that these approaches could have sufficient clinical efficacy to augment current screening and diagnostic techniques. More clinical trials are needed to further evaluate.

It is worth remarking that imaging is incidental in reaching the goal of using impedance in breast cancer screening. In the Glickman study[351], the group tried with some success to automatically classify impedance maps into different diagnostic categories. Likewise, Stojadinovic used the T-Scan 2000ED to provide a direct cancer classification using a single green (negative) or red (positive) indicator bar. If such an approach is taken, the displaying of images becomes secondary in importance, and may only be of value to assist the operator in performing the examination.

As another example, Demidenko and colleagues[233], explored the use of raw data obtained from a tomographic (circular array) EIT device to compute directly the relationship between the applied currents and the measured voltages-the impedance matrix. This step is much simpler than the computations required to reconstruct a tomographic image, yet in principle the impedance matrix contains the same information. By analysing directly the impedance matrix, using advanced statistical methods, it is possible to distinguish between different imaging domains in phantom experiments, and it is expected that eventually sophisticated algorithms will be able to classify patient examination data reliably as well, based directly on the measurement data.

Kim *et al*[553] reports on spectral data collected from breast using the two-plane mammographic EIT system developed by RPI. Instead of producing images, this analysis looks directly at the spectral shape of the impedances recorded between sets of electrodes and proposes using spectral shape to identify cancer (as opposed to generating images). Given the growing development of machine learning based classification in screening, diagnostic, and general medical imaging applications it stands to reason that imageless analysis of EIT data may represent an easier translational pathway to the clinic. The data rich features inherently recorded during multi-electrode and multi-frequency EIT data acquisition could easily be integrated with these imageless and machine learning classification approaches, and therefore forgo the image reconstruction problem entirely and the limitations associated with image artifacts and moderate resolution typically reported. Exploration of these data analytics methods are worth pursuing.

Of the clinical results presented here, the general conclusion is that notable differences existing between the impedance of malignant and non-malignant tissues (including normal and benign lesions) can be reliably flagged by impedance imaging systems. Most planar scanning systems presented do not use absolute impedance parameters; rather they rely on the relative impedance as sensed across the array, which show as white spots on the display when imaging is used. The reason given for using this approach is that there is a large variation in absolute property values between patients, which is difficult to account for, while the relative local differences are a more consistent indicator of abnormality and easier to identify. In all the results reported here, the ability of impedance imaging to differentiate between malignant and other tissues was confirmed. What is lacking in these reports is an evaluation of how small a tumour impedance imaging can detect.

In addition to the confirmed malignant-to-benign differentiation, the relatively high sensitivity and specificity reported for planar EIT in the several moderately sized clinicals trials summarized here suggests that this technology could meet a clinical need, especially in a younger population of women with denser breast tissue. Given these findings, the question remains, why have these devices not been clinically adopted despite the T-Scan 2000 being FDA cleared in 1999 and marketed by Siemens Medical Systems in the early 2000s. Several potential reasons for this limited clinical translation are worth noting:

- Breast EIT has been primarily suggested as an adjunct to mammography to help decrease the number of negative biopsies performed (i.e. following a mammography-based screen-positive finding), which is why specificity has been a significant metric of interest in these studies. Despite the need for better identifying a benign breast lesion, convincing a clinician that there is no need to perform a biopsy is a challenging proposition as histological review of the tissue will give a near-definitive answer regarding the malignancy of a lesion (as long as the biopsy accurately targets the lesion).
- MRI has become more readily available since the early 2000s and does a good job at specifying lesion types (with better resolution than EIT). Likewise, digital mammography and tomosynthesis have improved significantly since the 90s and early 2000s when EIT systems were being developed. These improvements in already accepted clinical imaging modalities upped the bar for what was required of EIT. The average SN and SP of the studies summarized above for EIT is $\sim$71.2$\pm$18.3 and 76.2$\pm$14.2 does not compete with the average SN and SP reported for digital mammography (86.9% and 88.9%, respectively[613]. Further, the moderate variability in SN and SP reported between studies arises from the different patient cohorts included and lack of consistency in probe types, data, and classification algorithms used.
- Breast lesion classification algorithms for the T-Scan devices were kept mostly proprietary and not published making it challenging to repeat studies and verify results. It would be useful to publish classification algorithms so that others could repeat and verify some of the findings reported in these studies. Finally, it will be important to enlist independent clinical collaborators to evaluate the efficacy of EIT in randomized control trials.

- Introducing EIT as an additional stand-alone technology that competes with X-ray mammography, digital tomosynthesis, MRI, CT, and ultrasound may be a challenging market to enter without a significant value-add. Coupling EIT to existing technologies may represent the best approach for translating EIT to the clinic. Devices such as EIT-coupled ultrasound[759] or EIT-coupled mammography[532] may be more readily accepted by clinicians. The study by Roneta et al[862] suggests improved cancer sensitivity when stand-alone EIT was combined with stand-alone ultrasound or mammography. Direct integration with these technologies to enable precise co-registration between imaging spaces may lead to further improvements and represent a more streamlined translational pathway.

Despite these challenges, there is significant evidence that the malignant-to-benign impedance contrast is sufficiently large to make breast lesions detectable with impedance imaging. It seems to be accepted by most researchers whose work is mentioned here that impedance imaging, if it proves itself clinically, will ultimately be used as an adjunct to :-ray mammography, joining ultrasound and MRM as second-tier examinations, or as a tool for use in a specific cohort of women; e.g. younger women with familial history of breast cancer. Given that there is a significant inter-subject variability in breast tissue impedance signatures, use as a non-invasive, non-ionizing, treatment monitoring technology that images impedance differences over time in a single individual may represent a new clinical scenario worth exploring. It is hoped that the data summarized here along with the limitations highlighted suggest 1) that impedance imaging still holds promise for breast cancer applications and 2) that new approaches designed to mitigate the historical limitations, such as leveraging multi-modal imaging or deployment in specific patient cohorts, might help to translate breast EIT.

13.3   PROSTATE EIT

13.3.1   INTRODUCTION

The prostate is a glandular male reproductive organ situated between the bladder and penis and anterior to the rectum. It contributes to seminal fluid production and encloses the prostatic urethra through which urine and semen pass. The apex (distal end near the penis) and base (proximal end near the bladder) of the prostate each harbor a muscular sphincter that controls fluid flow and helps to maintain continence. A rich supply of blood vessels and nerves enters the lateral pedicles of the prostate through bilateral neurovascular bundles. The nerves within the bundles serve to control continence and erectile function and if disturbed can lead to incontinence and erectile dysfunction.

As a glandular organ the prostate is susceptible primarily to adenocarcinoma-type cancers, although other types including sarcomas and neuroendocrine tumours can also occasionally be found. Prostate cancer (PCa) is the second most diagnosed form of cancer and the second leading cause of cancer-related death in men[971]. Approximately 20% of all diagnosed cancers in the US are expected to be within the prostate and over 30k men die annually as a result of the disease. Prostate cancer detection has matured since prostate specific antigen (PSA)-testing began and as a result, the number of prostate cancers diagnosed in the US has increased from 96.1 to 177.8 per 100,000 men since 1974 with a larger portion of these men having localized disease[51]. Of particular note with prostate cancer is that autopsy studies have shown that up to 50% of men will die with prostate cancer, though not all dying of prostate cancer[506]. This represents a significant clinical challenge in terms of determining how men with prostate cancer need to be managed.

Curative management for prostate cancer includes chemotherapy, radiotherapy, hormonal therapy, and surgical therapy. Because some prostate cancers are more indolent than others, watchful waiting and active surveillance are other less-invasive managements strategies prominently used in prostate cancer care. Recommended prostate cancer screening programs include annual blood testing to track the presence and evolution of prostatic specific antigen (PSA) concentration in the serum and digital rectal examination. For screen-positive cases further imaging and biopsy studies are recommended to stage the disease. The traditional next step has been to perform an ultrasound-guided biopsy procedure in which transrectal ultrasound imaging is used to localize the prostate and several (8–30) needle biopsy cores are extracted for histological evaluation. Unfortunately, ultrasound alone has limited sensitivity and specificity to detecting cancer and is instead simply used to guide the surgeon to sample different regions of the prostate[588]. More recently, cross-sectional imaging with MRI has been growing in prominence over the last decade; the recent PIRADS v2 standards have defined a multiparametric MRI imaging protocol that has promise in defining cancer stage and identifying potential prostate lesions that can be targeted for MR-fused ultrasound-guided biopsy[1136]. Despite these improvements MRI protocols still exhibit challenges including 1) only moderate interreader agreement[748], 2) limited performance in assessment of tumours with Gleason Grades of 4+3 or higher and a volume of 0.5 mL or less[1074], 3) inherent subjectivity in assessing diffusion weighted imaging (DWI)[900], and 4) limited description of other prostatic abnormalities (e.g. prostatitis, peripheral zone benign prostatic hyperplasia (BPH), etc.)[900].

13.3.2   TRANSRECTAL EIT FOR CANCER DETECTION AND BIOPSY GUIDANCE

Clinical Use Case: Transrectal ultrasound (TRUS) has limited soft-tissue contrast making cancer detection and lesion identification challenging. Despite these challenges, TRUS is almost universally used for prostate biopsy guidance since it is capable of delineating the prostate from the surrounding anatomy. Hence, it is routinely used to guide systematic biopsy sampling of the prostate; in a typical biopsy protocol, the prostate is subdivided into twelve volumes (spanning anterior to posterior and laterally right to left) and from each volume a needle biopsy core is systematically extracted. Transrectal Electrical Impedance Tomography (TREIT) has been developed to couple to standard TRUS probes as a multi-modal imaging system with the potential to improve cancer detection. Two types of probes have been explored for this purpose: side-fired and end-fired.

Side-fired TREIT approach (figure 13.5): Wan *et al* developed a multi-modal TRUS/TREIT system by coupling a custom flexible electrode array with a side-fired TRUS probe[1111]. A TargetScan Prostate Biopsy and Treatment Guidance System (Envisioneering Medical Technologies, St. Louis MO, USA) includes a side-fired US probe that is inserted into the rectum and fixed in place while a series of cross-sectional ultrasound images are acquired of the prostate. The flexible electrode array was printed on a polyimide substrate and adhere directly to the US probe. The array consisted of 30 gold-plated electrodes (2 mm×9 mm) arranged in a rectangular pattern that surrounded the US transducer's field-of-view so as not to impede the transmitted and received US pulses. A 31st distal electrode was placed on the patient's abdomen to enable current patterns that passed through the prostate. Impedance data was acquired using a National Instruments data acquisition system (NI USB 6259, National Instruments, Austin, TX) and a custom front-end that supported two 32-channel multiplexors for arbitrarily pairing voltage drive/sink electrodes (converted to current through the transimpedance of the tissue) and 32 single-ended voltage measurements. The system operates over a frequency range of 400 Hz to 102.4 kHz, has an average SNR of 81 dB, and requires approximately 40 seconds to record a complete data set at a single frequency.

The electrode configuration required an open-domain absolute image reconstruction algorithm be used since the electrodes were not able to surround the entire anatomy of interest as is common in application such as thoracic imaging[135]. The benefit of having EIT coupled to US is that the prostate geometry can be segmented from the US images and used to generate a patient specific FEM forward mesh. Because static/absolute imaging poses additional challenges, these patient specific meshes helped to mitigate errors associated with mesh-to-anatomy mismatch. A dual-mesh approach was used to reconstruct only voxels within the prostate.

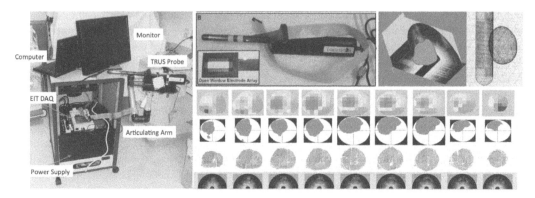

Figure 13.5   Side-fired multi-modal TRUS + TREIT imaging system (left). Open window electrode array (inset) and sonolucent electrode array integrated with side-fired TRUS probe (top middle). Prostate surface segmented from TRUS image and embedded into custom FEM mesh for EIT reconstruction (top right). Example image panel extracted from a patient with large upper quadrant prostate cancer (bottom right). Images represent slices through the prostate from base to apex (left to right) with the rows representing the conductivity, stylized map of cancer location, microscopic view of H&E stained prostate sections, and the TRUS images (top to bottom rows)[1110]. (color figure available in eBook)

This system was used to image the prostate of 50 men undergoing radical prostatectomy. Imaging was performed immediately prior to surgery and the resected prostate was used to provide ground truth localization of cancer within the prostate[1110]. Ultrasound segmentation and mesh generation were used for each case and the TREIT images were registered to the prostate micrographs using a thin plate spline warping procedure to account for shrinkage of the prostate between in vivo conditions and microscopy slide preparation. The average conductivity in the cancer regions was found to be significantly different than benign tissues at all frequencies with cancer conductivity being greater than benign conductivity at 0.4, 3.2 and 25.5 kHz and less than benign conductivity at 102.4 kHz. In a very preliminary study, Wan *et al*[1109] also discussed the possibility of incorporating impedance measurements recorded from electrode-tipped biopsy needles as intraprostatic measurements to enhance EIT sensitivity farther from the TRUS-located electrode array. This approach has the potential to transform a fully open-domain reconstruction problem into one with partial internal boundary measurements

End-fired TREIT approach (figure 13.6): The side-fired approach initially explored for TREIT provided for a fixed electrode array with a well-defined location for the US images; however, end-fired US probes are predominantly used clinically. Murphy *et al* have designed a TREIT system that couples a sonolucent electrode array to an end-fired TRUS probe[761]. This system uses a grid of twenty gold-plated electrodes (size 1.5 mm×2.0 mm) configured in a 4×5 array and printed on a 1 mil polyimide substrate that is secured with a pressure sensitive adhesive over the active elements of a Phillips C9-EC end-fired TRUS probe. The electrode array covers a total area of approximately 20 mm×20 mm which makes imaging structures far from the electrode array (i.e. >~12 mm challenging. To overcome this limitation, they have developed a custom biopsy needle instrumented with 4 or 8 electrodes located at the distal tip of the needle that is capable of recording intraprostatic impedance measurements either between the electrodes on the needle tip or between the needle electrodes and the TRUS-located electrodes. In practice, the TRUS probe and biopsy needle are electromagnetically tracked in space so that precise locations of these instruments are known. These tracked coordinates are used for so-called fused-data TREIT; specifically, the moving electrodes are incorporated into a high-density forward mesh that is mapped to a coarse mesh that encompasses the prostate.

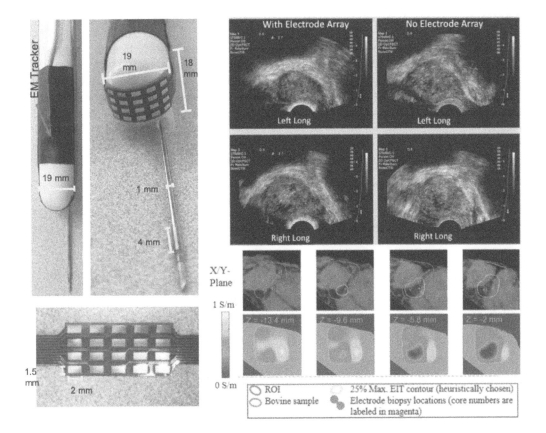

Figure 13.6   End-fired multi-modal TRUS + TREIT imaging system (left). Sonolucent electrode array adhered over US transducer. EM tracking coil embedded with TRUS probe for spatial tracking. Electrode instrumented biopsy needle shown provides intraprostatic impedance measurements. TRUS images acquired of in vivo prostate with and with out the electrode array (right top) demonstrating minimal electrode array artifacts. Ex vivo muscle imaging using end-fired probe and biopsy needle-based EIT data acquisition. MRI shows ground truth (top row) and conductivity images (bottom row) exhibit correct low conductivity associated with adipose tissue and higher conductivity in regions of muscle[755]. (color figure available in eBook)

Simulations and phantom experiments demonstrate that the additional use of the biopsy needle electrodes coupled with the end-fired configuration enable the system to be sensitive to 91.4% of a typical prostate as compared to 1.8% when just using end-fired electrodes[760].

This system was used in a yet-published ex vivo study to evaluate the efficacy of end-fired TREIT in detecting prostate cancer. TREIT-based electrical properties were recorded from 22 ex vivo prostates removed during robot-assisted radical prostatectomy using a simulated biopsy procedure. Specifically, the prostate was secured in an anatomically mimicking environment and a 12-core biopsy was performed using US-guidance. Electrical impedance measurements were recording during each tissue core extracting and fused-EIT was used to reconstruct all images. Histology of the resected prostate served as ground truth. Sensitivity and specificity for cancer detection were reported to be 0.85 and 0.78, respectively, with an area under the curve of 0.9.

13.3.3   SURGICAL MARGIN ASSESSMENT USING ENDOSCOPIC ELECTRODE ARRAYS

Clinical Use Case: Radical prostatectomy (RP) aims to remove the prostate from the body with the objective of 1) eradicating all cancer cells, 2) maintaining urinary continence, and 3) preserving erectile function. Unfortunately, these are conflicting objectives pitting cancer control (Objective 1) against improved quality of life (Objectives 2 & 3). In order to minimize risk of PCa-related mortality and recurrence, surgeons would like to resect wide margins around the prostate to ensure negative surgical margins (NSMs); however, wide resections cause damage to peri-prostatic tissues that decrease post-operative urinary and erectile function. Small, endoscopic-like electrical impedance tomography probes have been in development for use in intraoperatively assessing surgical margin status.

Endoscopic EIT Probes: A number of different endoscopic EIT probes shown in figure 13.7 have been described for use in interrogating surgical margins. These surgical margin assessment (SMA) probes typically include a set of gold-plated electrodes ranging in number from 8 to 21 that

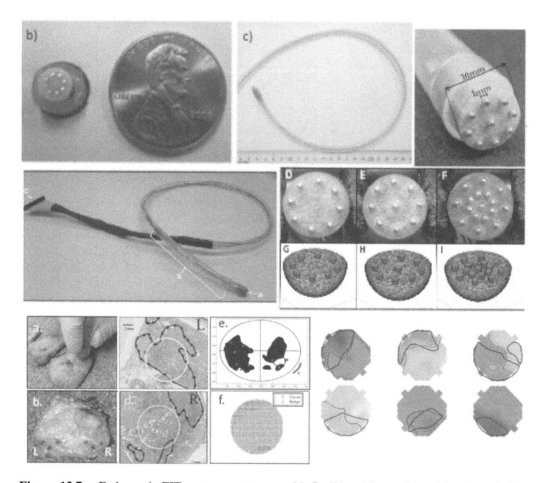

Figure 13.7   Endoscopic EIT system prototypes with flexible tubing and capable of use during laparoscopic surgical procedures (top panels). Ex vivo sampling of resected prostate tissue (bottom left) – probing procedure, histological assessment, and cancer region localization. Example conductivity images of ex vivo prostate samples; cancer verified through microscopic evaluation denoted by black lines and corresponding conductivity threshold denoted by red lines demonstrate good correspondence to cancer regions [667,758]. (color figure available in eBook)

are embedded into a rigid housing and interfaced through a flexible tubing/cable system to an EIT data acquisition system. Mahara *et al* compare probes with 8, 9, and 17 one mm diameter gold-plated electrodes embedded into a garolite housing [Mahara 2015]. The electrodes are interfaced through ~ m long cables to a 32-channel EIT data acquisition system with the entire electrode housing and cable assembly sheathed in an 11 mm diameter silicone tube; the 11 mm diameter enables the probe to be inserted into a 12 mm diameter auxiliary laparoscopic abdominal port used as part of standard of care during robot-assisted laparoscopic RP procedures. These probes have been used in several ex vivo studies and one preliminary in vivo study (figure 13.8).

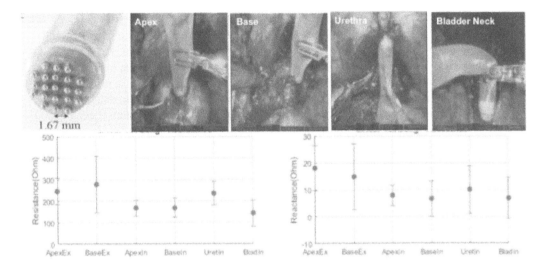

Figure 13.8   Twenty-one electrode endoscopic EIT probe (top left) being deployed during robot-assisted laparoscopic radical prostatectomy (top right). Surgeon probing the apex and base of the resected prostate and different regions of the pelvic space corresponding to the probed prostate surfaces (urethra and bladder neck, respectively). Average impedance acquired from select tetrapolar patterns with 'Ex' and 'In' denoting ex vivo and in vivo measurements, respectively (bottom). Ex vivo measurements were made from the resected prostates apex and base approximately 30 minutes following surgery to correspond with the regions probed intraoperatively. As expected the ex vivo impedances are greater that the in vivo impedances.[409] (color figure available in eBook)

Kahn *et al* used a 9-electrode version of the probe to record data from 14 ex-vivo prostates[551]. They developed a composite impedance metric (CIM) which combines similar electrode patterns into set of features. Significant differences were reported between benign and malignant prostate using the CIM and when classified using a Support Vector Machine (SVM), a predictive accuracy of 90.8 % was reported. Murphy *et al* developed a custom open-domain image reconstruction algorithm that used a non-uniform coarse mesh to account for the drop in sensitivity far from the probe face[757]. They demonstrated localization errors of <0.5 mm and sensitivity to anatomic structures <2 mm in width. Murphy *et al* went on to use a 17-electrode SMA probe to record data from 19 ex vivo human prostates[758]. They explored the data using 3D EIT with the custom open-domain image reconstruction algorithm, a combined EIS metric similar to Khan's CIM, and a single parameter EIT reconstruction. They explored single feature and multi-feature machine learning based classification. EIS and EIT showed good predictive performance with a 0.76 and 0.80 area-under-the-curve (AUC), respectively, when considering discrete frequencies only. An SVM classifier improved the AUCs of EIS and EIT to 0.84 and 0.85, respectively.

13.4   OTHER CANCER EIT

Several other groups around the world are exploring other cancer imaging applications of EIT including in lung and gynecological cancers. While limited in vivo studies have been conducted, hardware and reconstruction algorithms specific to these anatomic locations are being developed.

For instance, while lung imaging in the context of ventilation monitoring has been the most promising clinical application of EIT to date, the EIT group at Tianjin University has explored the potential for using EIT to classify lung tissues as benign or malignant[1012]. They have shown significant electrical property difference between normal and malignant lung tissue and are specifically interested in targeting early detection of lung tumours given the much better outcomes for these patients. One of the primary challenges with imaging lung cancer is the loss in sensitivity far from the boundary electrodes placed around the thorax. In breast and prostate application, the tissue being interrogated is near to the electrode array and the volume is relatively smaller than in the lung. One approach being explored to overcome this limitation is through incorporation of a priori information into the inversion; specifically, X-ray computed tomography (XCT) is used to define a general conductivity distribution that is embedded within the penalty term of a Tikhonov regularization framework in the approach developed here.

Cherepenin *et al* have developed a small form-factor EIT probe suited for gynecological applications[193]. Specifically, the device is a 48-electrode planar array embedded in a 30 mm diameter housing that can be introduced through the vagina for cervical imaging with the aim of detecting and diagnosing cervical intraepithelial neoplasia (CIN). Custom electronics were designed to fit within the housing of the probe to enable active electrodes and minimize signal loss and noise associated with long leads. A 3D weighted backprojection algorithm similar to the approach developed for this group's planar breast imaging probe is used. The system has been used in a cohort of 170 women with no disease (80), inflammation changes (50), low-grade CIN I (20), and mid- and high-grade CIN II&III (12)[1054]. Conductivity for the CIN II&III tissues were significantly higher than the conductivity for all other tissues suggesting ($\sim$1 S/m vs. $\sim$0.85 S/m at 50 kHz).

13.5   PERSPECTIVE ON CANCER IMAGING WITH EIT

While breast cancer imaging and detection is the most mature and widely studied of the EIT-based cancer imaging applications, the significant benign-to-malignant impedance contrast in other cancer types supports further research and development in these other cancer applications as well. There have been significant advances in both data acquisition systems and reconstruction algorithms that improve the accuracy and precision of raw impedances recorded and the resolution and suppression of artifacts of reconstructed images. Given the challenges exhibited in translating breast imaging systems to the clinic (i.e. the TS2000 and T-Scan 2000ED are no longer actively being marketed by Siemens), it may make sense to focus develop EIT technologies with a focus on translation.

In many cancer imaging applications, systems are being designed to explicitly interface with target anatomy or to meet specific clinical use-cases. For example, coupling EIT to existing technologies such as ultrasound has the potential to streamline the process of clinical acceptance. Further, one can consider using EIT as an adjunct as opposed to the primary imaging modality as has been proposed for breast imaging. This could help to overcome the central limitation of EIT, which has been resolution. Using EIT as a functional imaging modality with high-contrast and coupling with a higher-resolution low-contrast imaging modality has the potential to make EIT clinically relevant. As an example, by coupling EIT with ultrasound, one has the option of "turning on EIT, much the way that Doppler imaging is "turned on" to investigate blood flow.

Finally, considering use of imageless EIT in some applications (such as diagnosis) may help to overcome the concerns with image resolution. Big data and machine and deep learning approaches to classification are being explored within the medical device and imaging communities and have started to be applied in EIT applications.

Given the large impedance contrast exhibited between benign and malignant tissues continuously demonstrated in multiple tissues over the past several decades, further development of EIT-based cancer imaging and characterization is recommended.

14 Other Clinical Applications of EIT

Ryan Halter
Medical Physics and Bioengineering, University College London, UK

David Holder
Medical Physics and Bioengineering, University College London, UK

CONTENTS

The principal clinical applications for biomedical EIT are imaging of heart and lung function in the thorax, soft-tissue cancers and brain function. These are all covered by individual chapters elsewhere in this volume. There are several other possible applications, some of these of historical interest – they were started in the first flush of enthusiasm when the Sheffield Mark 1 system became available in the mid 1980s, but then active research was discontinued because of inherent technical problems, or because other areas within EIT appeared more promising. Recently there has been a flurry of new activity in exploring new and revisiting older clinical applications that are being made possible with the availability of smaller form-factor IC-based EIT systems, cost-effective research-based EIT systems that have entered the market, and increased computational power.

While these other clinical applications have not yet been fully optimized nor vetted in clinical trials, both the historical applications and emerging applications are reviewed in this chapter to provide insight into what might be possible with EIT.

14.1 TUMOUR ABLATION MONITORING

Malignant tumours may be treated by artificially increasing temperature by radiofrequency, microwave or laser ablation approaches or by decreasing temperature using cryoablation. It is essential to monitor tissue temperature during these procedures so that normal tissue is not overheated or overcooled, while malignant tissues are heated to the desired temperature of about $50°C$ or cooled to temperatures of about $-40°C$ for a few minutes. At present, this is achieved by inserting thermocouples into the tumour. This is practicable for superficial tumours, but difficult for deep ones. MRI-based thermometry has been proposed as an alternative method for monitoring ablation of deeper tumours, but this approaches requires specialized MR-compatible ablation probes and a high-cost interventional MR. There is therefore a need for an accurate non-invasive thermometry method that can easily be applied and work well for both superficial and deep tumours. In principle, EIT might be suitable for this, because there is a nearly linear monotonic relation between temperature and impedance change in simple aqueous solutions—the impedance of ionic solutions varies inversely with temperature by about 2% per °C[383]. EIT therefore presents a possible non-invasive means of imaging temperature within a subject.

Unfortunately, the relationship between resistivity and temperature is complex. Using a laser probe to heat ground calf liver in a cylindrical tank, Möller *et al*[724] compared changes within the EIT image with temperature determined by thermocouples. The tissue was heated to between 35°C and 60°C using an oscillation inducing thermoregulatory feedback system. There was a qualitative correlation between changes in the EIT image and temperature, but a substantial impedance drift of uncertain origin occurred. A similar study was performed in a tank filled with conducting agar, into which small pieces of foam had been inserted in order to simulate inhomogeneous tissue. Heating was performed with radiofrequency coils[208]. A linear relation was observed between EIT image changes and temperature, but the slopes varied with position in the phantom.

Temperature calibration experiments have also been performed in vivo. In three volunteers, 200 ml of conducting solutions at various temperatures were repeatedly introduced into the stomach, whilst EIT data was acquired from electrodes around the abdomen and conductivity images reconstructed[208]. Acid production was suppressed by cimetidine. It was found necessary to compensate for baseline drifts in the images. After compensation, a linear relationship between the temperature of the infused fluid and region of interest integral was observed, although the slopes varied between subjects.

Unfortunately, reliable clinical use for hyperthermia monitoring requires a high degree of both spatial and contrast resolution. Single images in the thigh[383] and over the shoulder blade[207] of human subjects, with the Sheffield mark I system, during warming showed substantial artefacts, and it was also demonstrated in normal volunteers, without warming, that baseline variability would produce impedance changes which were equivalent to temperature changes of several degrees. Pilot clinical measurements made in the mid-1990s with planar arrays at 12.5 kHz showed encouraging average results, but estimates of some of the tissue temperatures were erroneous by 9°C[741,811].

There has been an increased interest in incorporating a priori information from other diagnostic imaging modalities to improve the EIT reconstruction outputs in hyperthermia monitoring applications. One such approach uses a coupled electromagnetic (Gauss's Law) and heat transfer (Pennes' bioheat equation) model to describe the coupled biophysical processes in action during EIT-monitoring of RF-ablation[277]. In this approach, MR images are used to construct an anatomically accurate mesh with internal organ segmentation in order to specify organ-specific electrical and thermal properties within in the mesh. The RF probe is used in this case as a current drive with a single skin-located return electrode. This approach has been extended by adding a multiplicity of external return electrodes making this a more feasible configuration for EIT reconstruction[175]. While these approaches may be promising, their implementation has been limited to numerical exploration to date.

More recently, advanced EIT data acquisition systems have been used to capture temperature and tissue property changes of ex vivo bovine liver undergoing radiofrequency (RF) ablation[1142]. EIT images were acquired at 10 kHz and 100 kHz and both time- and frequency-difference images were evaluated. RF application and EIT data acquisition were duty-cycled (18 s RF / 2 s EIT) to ensure no RF interference in the EIT data and temperature was recorded via thermocouples during EIT data acquisition. Both time and frequency difference images showed significant conductivity changes associated with ablation regions as compared to unablated regions. It is suggested that while EIT can accurately detect ablation regions, the accuracy and resolution of the reconstruction is limited making clinical utility and translation a challenge.

In the case of EIT-based cryomonitoring, feasibility experiments were initially explored using ex vivo chicken breast and liver tissues[799]. Tissue samples subjected to cryoablation were positioned directly on a set of linearly spaced electrodes and a developing ice front was detectable to approximately half the spacing between the sense electrodes. Positioning electrodes local to the ablation site in clinical applications are an obvious challenge, though co-located impedance sensing probes with the cryodelivery probe is a potential solution to this. In one such implementation, a cryoprobe instrumented with 16 electrodes (two rows of 8 electrodes) was submerged into a saline-filled tank with 16 peripherally positioned electrodes (i.e. representing skin electrodes adhered around the abdomen of a patient)[443]. Ice balls were formed in the tank while impedance was acquired from probe and peripheral electrode combinations. A dual mesh reconstruction algorithm with both peripheral and internal electrodes was used to demonstrate the feasibility of imaging ice ball formation. Significant improvements in imaging the internal ice ball were observed when both probe and peripheral electrodes were used as opposed to peripheral electrodes only.

Additional numerical simulations that explore the use of secondary imaging modalities to serve as a priori information for guiding EIT reconstruction in cryoablation have also been suggested. Specifically, use of ultrasound to define a specific region of interest over which the electrical properties are estimated has been proposed as a way to decrease the ill-posedness of the reconstruction algorithm[251]. As an extension of this work, a level-set method that uses data from multiple narrowband ultrasound transducers to guide boundary shape estimation of the ice ball has been proposed to improve ice front monitoring[990].

In addition to monitoring during ice ball formation, it has been suggested that acquiring a pre-ablation and post-thaw EIT images (i.e. after the ablated tissue has completely returned to normal body temperature) may serve to better assess the actual volume of tissue that was effectively ablated. In this case, difference impedance images would serve to map the treated tissue based on cellular morphology post-ablation rather than acting as a thermometry system to map the temperature distribution as is the case in monitoring during ablation.

In all of these ablation monitoring applications, accurate temperature estimation requires not only accurate imaging, but also an assumed linear (or at least known) relationship between temperature and conductivity. This latter appears to change in a hysteretic fashion during tissue heating. Given this uncertainty in calibration a priori, and the baseline variability in vivo, it seems that EIT may struggle to be an accurate technique unless substantial improvements in system performance continue to be made[119,811]. Improvements including coupled electromagnetic-heat conduction models and use of a priori imaging to better constrain the reconstruction algorithm may help in this effort. Likewise, incorporating both probe-located electrodes and skin-based surface electrodes is likely to provide better sensitivity local to the ablation site. In vivo animal model measurements in which blood perfusion is present are critically needed to better evaluate the potential EIT-based ablation monitoring in a more realistic clinical setting.

14.2 SYSTEM-ON-CHIP AND CELL/TISSUE IMAGING

With the advancements in micromanufacturing and system-on-chip technologies, there has been an increased interest in exploring the use of EIT for in vitro cell imaging. Cell studies typically require destructive staining processes to image and evaluate cellular response; EIT represents a non-invasive, non-destructive modality that can provide real-time feedback on cellular response to different stimuli (i.e. cell growth studies, cancer treatment studies, etc.). Specific, potential applications of cellular imaging include high throughput screening of blood samples to detect circulating tumor cells, pharmaceutical studies of drug efficacy, cell growth studies and non-invasive cell aggregation and health monitoring during tissue manufacturing.

Chen *et al*[188] developed a CMOS-based 96×96 microelectrode array with each square electrode having a 25 μm edge length (approximately the size of a single cell). A single counter electrode was positioned above the array and individual electrode impedances were recorded from each electrode using a benchtop LCR-meter. While tomography was not performed, impedance maps were generated to show cell density and to perform cell counting with each electrode serving as a pixel.

In another example, a more traditional circular EIT configuration was implemented on a printed circuit board (PCB) using standard photolithography techniques[1185]. A 10 mm cylindrical tank adhered around 16 microelectrodes printed on the PCB served as a vessel for imaging breast cancer cell spheroids and high-density cell pellets. While the electrodes were only present on the surface of the PCB, 12-layer 3D conductivity images were successfully reconstructed. This system has been further used for hydrogel, scaffold, and tissue imaging for use in tissue manufacturing applications[1171,1172].

These are nice early examples of what can be achieved with cellular EIT imaging. A good review of electrical impedance methods in general for cell characterization can be found in Schwarz *et al*[946].

14.3   WEARABLES

The increased availability of ASIC-based EIT systems (e.g.[394,610,635,864,865,1018,1053,1171] and other off-the-shelf IC-based EIS sensing solutions (e.g. AD5933 (Analog Devices), AFE4300 (Texas Instruments)) make developing small form-factor, low-powered EIT systems possible. While the clinical applications for which these small form-factor systems will have utility still needs to be determined, early devices have demonstrated proof-of-concept. The below summary includes only systems that are fully wireless and wearable in the sense that a patient would not be tethered to an off-person device.

Hand gesture recognition has been envisioned as once such application. Tomo is a wrist-wearable, eight electrode EIT system designed for detecting various hand gestures[1205]. The idea is that conductivity images of the muscle activation and muscle, tendon, and bone movement in the wrist will be unique for different hand gestures. This wearable design leverages the AD5933 and a pair of 8-to-1 multiplexors for electrode selection. Difference images are computed between measurements recorded at a relaxed state and those recorded for a specific gesture. Accuracies as high as 97% were reported for whole hand gestures when trained on a single individual.

Jiang *et al*[516] extended this work to include two rows of eight electrodes positioned along the forearm. A custom EIT ASIC and analog switch matrix was used in this case. They were able to achieve a gesture classification accuracy of 97.9% when a single band of 8 electrodes is used and an accuracy of 99.5% when two rows of 8 electrodes are used.

Another recent device developed by Yao *et al*[1186] used the Red Pitaya platform, a voltage controlled current source and set of multiplexors to serve as a wrist wearable device. They introduced conductive cloth electrodes with sponge filling to help improve constant electrode contact. This team found a gesture recognition accuracy of 95% in their study.

Finally, Wu *et al*[1175] introduced an ASIC-based forearm wearable EIT system for gesture detection. However, in addition to gesture detection alone they propose using this as a human-machine interface that enables control of a hand prosthesis through use of EIT images acquired of the forearm. They found gesture classification accuracies of 94.4–98.5% depending on the number of gestures they were classifying.

It is important to emphasize that all data collected in these gesture classification studies was acquired in a controlled laboratory setting from small numbers of users (2–10 study participants); in a real-world scenario electrode movement, electrode-skin interface conditions, and general motion artifact are expected to decrease overall accuracy. Also, because all systems used machine learning based classification, it is not specifically clear what the value of tomographic image reconstruction is in this scenario.

Thoracic imaging with wearable EIT systems have also been developed, however the majority of these still require tethering to an off-person power supply or data acquisition system. In one example of an untethered system, an MSP430 (Texas Instrument) microcontroller coupled to a voltage-to-current converter, an amplifier and demodulation circuit, and a multiplexing front end were interfaced to 16 electrodes[493]. An on-board Bluetooth module was used to transmit data wireless to a host laptop for image reconstruction and analysis. Low resolution imaging of lung inhalation and exhalation was demonstrated, though the actual clinical application for continuous ambulatory thoracic imaging was not discussed.

14.4   INTRA-PELVIC VENOUS CONGESTION

Pooling and congestion of blood in the pelvis is a poorly understood phenomenon which is thought to be the cause of pelvic discomfort in women. Thomas *et al* (1991) investigated the possible use of EIT in its diagnosis, on the basis that abnormal pooling would produce impedance changes. EIT images were collected with a ring of electrodes around the pelvis, as the subject was placed in horizontal and vertical positions using a tilt table. The rationale was that this should produce fluid shifts in the pelvis. A central area of impedance change was observed in both normals and subjects, with pelvic congestion diagnosed by venography. A significant difference in the ratio of the areas anterior and posterior to the coronal midline and greater than 10% of the peak impedance change was observed. No difference in mean amplitude of impedance changes was observed between the two groups. Venography is an invasive procedure, so EIT would provide a welcome alternative. However, there is no direct evidence concerning the origin of these changes, although it has been shown that they are at least plausible by comparison with EIT images made in tanks with saline-filled tubing[1031]. This is an intriguing and potentially valuable application, but larger prospective studies will be needed before its use can be established.

14.5   OTHER POTENTIAL APPLICATIONS

Using a 16 electrode system operating at 10 kHz and an algorithm similar to that of the Sheffield system,[589,590] it was responsible to produce EIT images in long bones. Areas of increased resistivity could be identified in the normal subject and 16 weeks after fracture, whilst a similar region showed lower resistivity in another subject, four weeks after fracture (Ritchie *et al*, 1989). It remains to be determined if such results could be used effectively to monitor fracture healing. However, fractures can at present be assessed with great accuracy by X-ray. EIT might offer an advantage if repeated measurement was needed for follow-up, but it is unlikely that it could offer appropriate spatial resolution.

Other proposed applications have included EIT imaging of limb plethysmography[1101], apnea monitoring[612,1163], intra-abdominal bleeding or fluid pooling[913,970,1060], neuromuscular disease[759], vesicoureteral reflux[248] but no direct evidence is yet available to assess the likely clinical accuracy of these possibilities.

15 Veterinary Applications of EIT

Martina Mosing
School of Veterinary Medicine, College of Science, Health, Engineering
and Education, Murdoch University, Perth, Western Australia

Yves Moens
Division of Anaesthesiology and Perioperative Intensive Care,
University of Veterinary Medicine Vienna, Austria

CONTENTS

15.1 INTRODUCTION TO THORACIC EIT IN VETERINARY APPLICATIONS

The scientific development of thoracic EIT in humans relied on translational research performed on experimental animals, especially swine. However, in the veterinary field anesthetists were the first ones to recognize EIT's potential for use in clinical research in other species such as horses, which are too large for conventional lung imaging techniques. Active research in this field began just over a decade ago. Today, research and clinical utilization of EIT in veterinary medicine is still in its infancy, and veterinary-specific literature is scant, in comparison to human medicine.

15.1.1 CREATION OF FINITE ELEMENT MODELS FOR ANIMALS

Anatomically-correct reconstruction of electrical impedance tomography (EIT) images in different species necessitates the use of species-specific finite element models (figure 15.1). The routine method to calculate an average FE model is by using helical computed tomography (CT) scans

DOI: 10.1201/9780429399886-15

during inspiratory hold. From the CT images obtained at the electrode plane of the EIT belt, the heart, lungs and thoracic contours are segmented. Segmented image files from several animals are then used to create an average contour and finally the corresponding FE model for that species. These FE models are then used to guide placement of the electrodes and create the final mesh for image reconstruction[1106].

Species-specific FE models have been created at the height of the 6th intercostal space at half width of the thorax for beagle dogs and calves respectively, as well as for pigs and ponies[1107].

In larger species that do not fit in CT scanners, alternative methods are required to create FE models. One method is to construct models from photographs taken of thoracic sections of anatomically dissected specimens. This obtains the essential topographic anatomy, including the dimensions of the lungs, heart and body-wall contour necessary to create the FE model. This has been performed in a horse and a rhinoceros[145]. Another method used in live horses is to use gypsum to fabricate casts of the thoracic contours. These shapes are subsequently digitized and averaged including assumed organ location based on published anatomical studies[47]. These techniques are illustrated in figure 15.1.

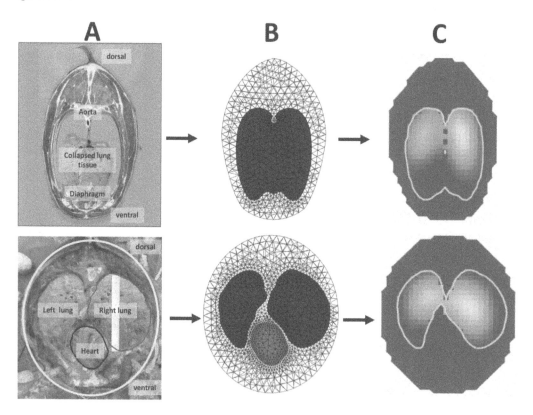

Figure 15.1 Examples of the creation of an FE model using anatomical dissected specimens (A). The FE model (B) is then used for anatomically correct image reconstruction of the EIT raw data (C). Horse (top); rhinoceros (bottom). (color figure available in eBook)

15.2 TRANSLATIONAL RESEARCH IN ANIMALS

Similar to other new medical technologies, new EIT algorithms that show promise are validated using animal models before they are released as clinical applications in human medicine.

15.2.1 PIG

The pig is the most commonly used species for translational studies in EIT research as they have a similar thoracic circumference for EIT belts to humans (compared to rabbits and rodents), their topographic anatomy is relatively similar to humans (compared to ruminants such as sheep), they grow relatively quickly (compared to dogs) and there are relatively lower ethical concerns (compared to primates). As of this writing, more than 175 papers have been published on EIT in different pig models when using Pub Med search using following search terms: (electrical impedance tomography) AND (pig OR swine OR porcine). The majority of reports using pigs evaluate changes in EIT algorithms in acute respiratory distress syndrome (ARDS) after induced lung injury. When interpreting EIT ventilation data obtained from pig studies, the anatomical and physiological differences between the human and porcine lung have to be considered; pleura and the non-respiratory bronchioles show grossly the same anatomical features, but the interlobular and segmental connective tissue, and the alveolarized respiratory bronchioles, are different between the two species[813]. One important difference is that in humans, the interlobular connective tissue only partially surrounds the lung lobules whereas in pigs this tissue forms complete interlobular septa. This extensive interlobular connective tissue prevents collateral ventilation between adjacent lobules in pigs[40]. This difference must be acknowledged when extrapolating EIT data after specific ventilatory interventions from pigs to humans as it is collateral ventilation which partially prevents the development of lung collapse after distal airway obstruction[888]. This makes the diseased human lungs less susceptible to atelectasis compared to pig lungs. These physiologic and anatomical differences between species might result in unknown effects on pulmonary function in response to different external insults, which needs to be taken into consideration when using EIT as a functional lung imaging technique.

Pig models are also often used in cardiovascular studies in translational research although pronounced differences in the anatomy of pulmonary vessels, muscular layers of arteries and location of pulmonary veins exist between pigs and humans. The latter should be considered when results from EIT studies using pig models are used for EIT interpretation in humans[541]. In particular, the pronounced hypoxic pulmonary vasoconstrictor response present in pigs[888] requires caution when the results of EIT lung perfusion studies are extrapolated to humans.

15.2.2 DOG

From an anatomical point of view dogs are the opposite of pigs. They have an excellent collateral ventilation and minimal hypoxic pulmonary vasoconstriction[813]. This at least makes dog lungs functionally more comparable with human lungs[1166]. This also makes dog models more reliable in translational research to monitor specific ventilatory interventions using EIT. Dogs are not used extensively anymore in translational research; however, partly because they are more difficult to breed and house for research purposes compared to pigs, and partly due to ethical concerns. Some early EIT papers, which describe the use of dog models and a 16 electrode EIT system, showed that EIT can be used to estimate changes in tidal volume, lung inflation at the end of expiration and liquid content in the lung and acute pulmonary edema[17,31,771]. Ambrisko and co-workers used a dog model to evaluate the influence of anesthesia on the distribution of ventilation[48]. They were able to show that anesthesia, but not recumbency, affected the centre of ventilation and inhomogeneity factor in these dogs suggesting that this may also apply in humans.

15.2.3 LAMB AND SHEEP

Sheep are straightforward to house, breed and handle and are therefore often used in translational research. Lambs have become an established animal model for preterm lung research using EIT.

Sheep usually have only one or two offspring in comparison to pigs or dogs, where litter sizes are normally larger. The construction of EIT belts for preterm lambs is less technically challenging as their body size is comparable to the one of a human neonate. Ethical concerns are also lower in lambs than in preterm non-human primates. Tingay and colleagues used EIT monitoring to compare the effects of different ventilation strategies used to inflate the lungs of premature newborn lambs[1041,1043,1044,1045]. EIT was able to diagnose the occurrence of a spontaneous pneumothorax for the first time[716], and even predict its occurance[714].

The use and interpretation of thoracic EIT in adult sheep is challenging as the rumen occupies a large volume on the left side of the body. The rumen is one part of the multicompartmental stomach in ruminants and represents a gas filled viscus adjacent to the diaphragm, partially covered by the ribcage. This can interfere with the ventilation signal when the cranial part of the rumen moves into the EIT plane. The latter is even more likely in recumbent anesthetized sheep as the natural escape mechanism for accumulating gas (eructation) is disabled, causing ruminal tympany. Therefore, EIT data in sheep and other ruminants have to be interpreted carefully. A sheep model has been used to investigate EIT as a monitoring tool to detect pulmonary embolism[777]. EIT images of lung perfusion were created by injecting different concentrations of saline solution into the right atrium.

15.2.4   HORSE

Given what is known about equine lung physiology and anatomy, horses are potentially the best lung model for translational research. In both the human and the horse, the separation of lung lobules is incomplete and collateral ventilation and hypoxic pulmonary vasoconstriction are comparable[813]. Furthermore, after induction of anesthesia and recumbency, horses show reproducible small airway collapse, atelectasis and increases in venous admixture within a short time[789]. Therefore, different ventilation strategies can be studied to re-inflate these atelectatic lung regions without the confounding variables of inflammation and bronchoconstriction which play a role in all other induced lung-injury animal models. Following up on all these arguments in favour of horses as a lung model, they are not often used in translational research due to their size, which makes their handling difficult and housing very expensive.

One EIT paper evaluated the differences of 3D image reconstruction in humans and horses[391]. In both species, the data obtained from a 2-plane belt (2×16 electrodes) instead of the traditional one-plane 32 electrode belt showed anatomically plausible images. Image reconstruction of the horse data revealed a gas-filled organ ventral to the caudal lungs, corresponding to the anatomical position of the large dorsal colon in horses.

15.3   CLINICAL RESEARCH IN ANIMALS

In this section, we review the use of EIT clinical research where the primary goal is to improve diagnosis and treatment of the animal, rather than as a translational vehicle for human disease.

15.3.1   HORSES

Most clinical research involving EIT in animals has been performed in horses. This research was guided by the need to understand and treat ventilation/perfusion mismatching and the resultant venous admixture which occurs during anesthesia in this species[789]. Until the implementation of EIT into equine anesthesia research, evaluation of the pathophysiology of ventilation, and possible treatment, was difficult as routine diagnostic imaging is very challenging (radiography, scintigraphy) or impossible (CT or MRI) due to the size of the animal. The use of EIT in horses has been made possible by the development of large custom-made rubber or neoprene belts with metal electrodes attached at equal distances.

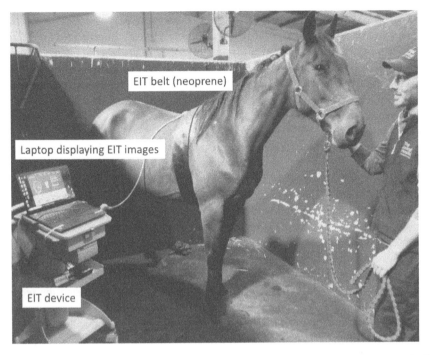

Figure 15.2   Set-up for an EIT measurement in a standing horse. The custom made EIT belt, which has 32 electrodes equally distanced mounted on neoprene, is connected to the EIT device via a cable and EIT images are displayed on a laptop computer. (color figure available in eBook)

1. **Distribution of ventilation**

The distribution of ventilation has been described using EIT in standing unsedated horses[738,944]. Figure 15.2 shows collection of EIT data in a standing horse. The effect of pregnancy on the distribution of ventilation was studied in Shetland ponies before and after parturition[944]. The results confirmed that the increasing size of the uterus influences the distribution of ventilation by shifting ventilation towards dorsal lung areas and that this distribution abruptly normalizes after birth.

A feasibility study to evaluate the distribution of ventilation by evaluation of the fEIT parameters, left-to-right ratio, centre of ventilation and global inhomogeneity index was performed in standing horses[47]. Functional EIT images were created using standard deviations (SD) of pixel signals and correlation coefficients (R) of each pixel signal compared with a reference respiratory signal. Different interventions (sighs, CO_2-rebreathing and sedation) coupled to plethysmographic spirometry were used to confirm feasibility of EIT to detect changes in ventilation. An interesting finding was that an inverse respiratory signal was found in the most ventral dependent region of the EIT image. This is thought to be due to the cyclic appearance of a gas-filled abdominal organ (most likely the dorsal large colon) into the EIT plane at the 4-5th intercostal space. This signal should not be interpreted as being of lung and needs to be distinguished from heart origin when detailed analysis of the thoracic EIT signal in the horse is required[740].

The breathing pattern of horses and the regional distribution and dynamics of ventilation were also evaluated prior to and following recovery from general anesthesia in dorsal (supine) recumbency[738]. To determine regional time delays within the lungs, the inflation period of seven regions of interest (ROIs) evenly distributed along the dorso-ventral axis of the lungs was calculated. The regional filling time was defined as the time at which $\Delta Z(t)$ of each region reached 50% of its maximum impedance change during inspiration, normalized by the global inspiratory time. The regional inflation period was defined as the time period during which $\Delta Z(t)$ of each lung region remained above 50% of its maximum inspiratory impedance change, normalized by the global breath length. After recovery of anesthesia in dorsal recumbency a respiratory pattern with inspiratory breath-holding was observed for six hours after standing. During these episodes, EIT showed that ventral lung areas were emptying while the dorsal areas were still filling, suggesting redistribution of air from ventral into dorsal lung regions. This is considered an auto-recruitment mechanism of lung tissue in areas which were dependent, and likely atelectatic, during anesthesia.

Several studies have specifically addressed the distribution of ventilation in horses during the anesthetic period [46,72,733,735,738,739]. In mechanically ventilated ponies in right lateral recumbency, the centre of ventilation (CoV) was in the non-dependent left lung [72]. A shift of CoV from the dependent to the less perfused non-dependent areas of the lungs was also shown after initialisation of volume-controlled ventilation (CMV) in dorsally recumbent anesthetized horses [735]. The initiation of CMV was accompanied by an increase in silent spaces and a decrease in ventilation in the dependent lung areas suggesting additional atelectasis development. This may be one of the reasons why a lack of improvement in oxygenation is frequently seen when switching from spontaneous to controlled ventilation in anesthetized horses. EIT has also been used to show the effect of specific ventilation interventions; one case report evaluated the recruitment of collapsed lung areas in the dependent lung under EIT guidance. The changes in ventilation distribution were measured in real time [721]. Regional lung compliance was calculated by simultaneous EIT and airway pressure measurements. This case report also discussed preliminary perfusion-related information from the EIT signal. Changes in a ventilation/perfusion mismatch algorithm based on the EIT signal was compared to changes in blood gas analysis and showed promise for EIT-based evaluation of ventilatory interventions for V/Q mismatch.

In anesthetized horses, as in humans, recruitment manoeuvres or continuous positive airway pressure (CPAP) can be used to re-inflate collapsed alveoli or to maintain alveolar patency, respectively – a process that can be demonstrated and confirmed by EIT [46,733]. As described in the human literature, the simultaneous recording of EIT signal and airway pressure allows the evaluation of regional lung compliance. A positive relationship between increases in dependent lung compliance following recruitment manoeuvres and blood oxygenation has been found [46]. In the CPAP study, the CoV shifted to dependent parts of the lungs indicating a redistribution of ventilation towards those dependent lung regions, thereby improving ventilation-perfusion matching. Dependent silent spaces were significantly smaller when CPAP was applied compared to anesthesia without CPAP, and non-dependent silent spaces did not increase demonstrating that CPAP leads to a decrease in atelectasis without causing overdistension [733].

2. Tidal volume

The routine measurement of tidal volume in anesthetized horses is not commonly performed due to a lack of practical and affordable equipment for tidal volumes of up to 20 litres per breath. In some situations, spirometric information would be highly desirable because capnography results can falsely suggest adequate ventilation despite the presence of pronounced ventilation/perfusion mismatch [734]. In one study in mechanically ventilated anesthetized horses, a very close relationship between measured tidal volume and impedance change was found for a wide range of tidal volumes (4–10 L per breath). This relationship was further improved when total impedance change in the entire image was calculated compared to impedance changes within predefined lung ROIs especially at high tidal volumes. The authors concluded that during breaths with high tidal volumes the lungs become overinflated and extend over the ROI borders leading to a loss of correct "volume" information when the FE model is used [739]. In a clinical study the linear relationship between the total impedance change and measured tidal volume was confirmed. Based on the individual line of best fit the tidal volume was estimated by impedance measurement and was within 20% of the tidal volume recorded using spirometry (Author's unpublished data).

3. **Airflow**

Equine asthma, also known as heaves, is a chronic performance-limiting disease which affects horses of all ages and breeds. The disease is characterised by bronchoconstriction, mucus production and smooth muscle remodelling, however the clinical signs are nonspecific and can be subtle, which makes diagnostic evaluation challenging, particularly when examining horses in the field. This is also true for monitoring the effectiveness of treatment in severe cases suffering from respiratory distress.

Global airflow signals in the lungs can be generated by calculating a smoothed first derivative of the EIT volume signal. This has been used to demonstrate global and regional ventilation characteristics related to changes in airway diameter in people[774].

Changes in global and regional peak airflow calculated from the EIT signals have been compared with a validated method (flowmetric plethysmography) in standing horses before and after histamine-induced bronchoconstriction[951]. The expiratory peak flow calculated with EIT changed in accordance with the flowmetric plethysmography variable, particularly in the ventral lung regions.

In another study in horses, peak flows and flow-volume curves were generated from the EIT signal before and after inhalation of nebulized histamine. When signs of airflow obstruction appeared, bronchodilation was obtained with aerosolised salbutamol[950]. This study showed that EIT can verify histamine-induced airflow changes and also subsequent reversal of these changes by using a bronchodilator. EIT therefore has potential as an additional method for pulmonary function testing and as a non-invasive monitoring tool to guide therapy in horses with equine asthma.

The same concept of monitoring airflow with EIT was used to evaluate airflow changes in healthy and asthmatic horses before and after exercise when respiratory rates reached pre-exercise levels[464]. Global and regional EIT gas flow changes were seen after exercise in horses suffering from equine asthma. More pronounced flow changes were noted in the ventral dependent areas of the lung, which is in accordance to the observed changes in the ventral lung areas after histamine-induced bronchoconstriction.

4. **Heart Rate**

Cardiovascular monitoring is essential in equine patients as the anesthezia-related mortality is 1:100 horses even with basic anesthezia monitoring used. Measurement of heart rate is a standard parameter in monitoring anesthetized patients. Therefore, every anesthetic monitor is expected to be able to measure heart rate. A recent study showed an excellent agreement between EIT and standard anesthetic monitor heart rate measurements when evaluating impedance changes of 4 pixels within the anatomical heart region[855]. The horses were positioned in dorsal recumbency during normotension, hypotension and hypertension.

5. **Conclusion**

To conclude, EIT has the potential to qualify as a new diagnostic tool in equine medicine to evaluate lung pathology where routine diagnostic tools used in other companion animals fail due to the size of the patient. This explains why most clinical research on EIT in animals has focused on the equine species. The scientific results to date have increased our knowledge of dynamic changes in lung function occurring during anesthesia or accompanying equine asthma. The EIT technology is only a step away from providing a novel practical way at the stall side to monitor lung disease especially equine asthma and follow up treatment success. Including recent research EIT has the capability to be used as a multiparameter monitor in anesthetized horses allowing the veterinarian to follow tidal volume changes and heart rate.

15.3.2 DOG

Dogs are companion animals in close daily relationship with their owners. This highlights the need for advanced medicine from complex surgeries to high level life support and intensive care. Under these circumstances, where veterinarians are often facing severe lung pathology, a respiratory monitor like the EIT would support these patients the same way as it does aid human patients. Up to the date of writing, there are relatively few clinical research studies assessing EIT measurements in dogs. This can be primarily related to the fact that good image quality is more difficult to achieve. Dogs have longer fur than horses and therefore need to be clipped when flat EIT electrodes are used. Furthermore, their skin seems to have a higher resistance to the current (possibly due to the lack of sweat glands), making it difficult to collect good EIT raw data. Dog skin and subcutis is highly mobile over the muscles which makes it difficult to have a stable belt position and heart-related impedance changes are very prominent in the thoracic plane which provides maximum lung area (personal experience). The optimal EIT belt position for clinical use has been determined based on radiographic and CT images using "fake" EIT belts [890], and shown in figure 15.3. The sternum was used as an external landmark as this is palpable even in obese dogs.

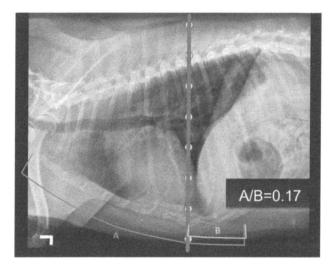

Figure 15.3   Optimal belt position for dogs based on sternum length. The length of the sternum is measured (A) and multiplied by 0.17 (B). The belt (blue line) is placed by this length in cm cranial to the caudal end of the xyphoid. (Courtesy: A. Rocchi) (color figure available in eBook)

Two studies have been conducted in anaesthetised dogs evaluating the distribution of ventilation using a 16 electrode EIT system (Pulmovista 500, Dräger, Germany). One study evaluated the effects of different PEEP levels in dogs in sternal (prone) position [353], while the other study looked at the change in the distribution of ventilation before and after a recruitment manoeuvre using two different tidal volumes [49]. In both studies, EIT was able to verify a change in the CoV and ventilation in four regions of interest ordered from dependent to non-dependent. Results were comparable with changes observed in CT images and other measured functional ventilation variables, confirming the validity of EIT to monitor distribution of ventilation in anaesthetised dogs.

Two other studies used EIT to monitor ventilation in sedated dogs. In both studies, a 32 electrode EIT belt was used (BBVet, Sentec AG). In one study the effect of the application of three CPAP interfaces on end-expiratory lung impedance (EELI) was evaluated as a surrogate for functional residual capacity, which is expected to increase under CPAP ventilation. Changes in thoracic impedance change (ΔZ) were used as estimates of changes in tidal volume [706]. Whereas all devices induced a significant increase in EELI when CPAP was applied, no difference between the interfaces was found in EIT variables.

Another study used EIT to monitor breathing pattern and changes in tidal volume following the application of two sedatives (dexmedetomidine and medetomidine). Results show that both drugs decrease respiratory rate and increase tidal volume (ΔZ_{TV}) suggesting that minute ventilation remains essentially unchanged [833].

15.3.3 RHINOCEROS

Rhinoceroses are anaesthetised for different purposes among which are relocation procedures to bring them to safer areas away from poachers. This "life-saving" procedure invariably has a profound negative impact on gas exchange characterised by severe hypoxemia and hypercapnia [848]. The pathophysiology of these severe complications is still incompletely understood as monitoring due to the size of the animal and to the fact that they are immobilised in the field is difficult. Therefore, EIT measurements were considered on several occasions and successfully performed making use of battery power, a custom made extra-long EIT belt and sufficient electrical current generation at the electrode/skin interfaces. Rhinoceroses are the largest mammals where EIT signal collection has been performed and this proved the use of EIT as a versatile in the field monitoring and research device (figure 15.4D).

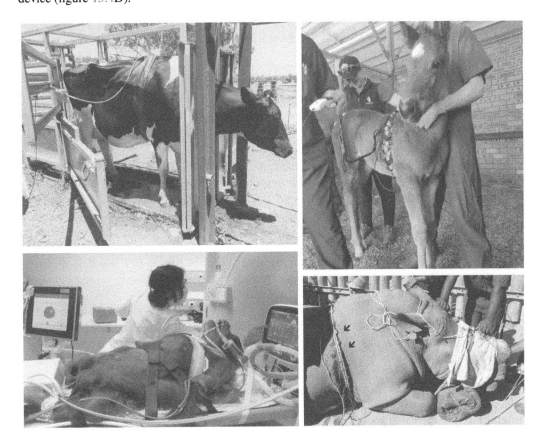

Figure 15.4   Use of EIT in different species: A) Measurment of respiratory disease process in a cow; B) EIT measurement in an anaesthetised orangutan using a human belt; C) Foal with EIT belt for a study evaluating EIT values in healthy foals; D) immobilised rhinoceros in lateral recumbency with EIT belt in place. (color figure available in eBook)

In a study in immobilized laterally recumbent rhinoceroses pronounced shifts in ventilation and perfusion towards the non-dependent lung were demonstrated. The EIT images including perfusion mapping after injection of hypertonic saline showed that the dependent lung did not collapse but stayed aerated while only minimally perfused. The decrease in perfusion is very likely due to hypoxic pulmonary vasoconstriction. Breath holding in some rhinoceroses shifted ventilation towards the dependent lung[740].

The gas exchange is worse in lateral compared to sternal recumbency. EIT has been successfully used to compare distribution of ventilation in anaesthetised rhinoceroses in lateral and sternal recumbency[737]. A large difference in ventilation between the two lungs was observed. As in the first study the non-dependent lung contributed 80% of the total ΔZ in lateral recumbency, while the dependent lung was only minimally ventilated. In sternal recumbency the difference in ventilation between the two lungs was only 10%. These findings show that the impairment in gas exchange in anaesthetised lateral recumbent rhinos is likely due to marked ventilation-perfusion mismatch, rather than an increase in pulmonary dead space, as previously suggested. It also highlights the potential of EIT to elucidate lung pathophysiology non-invasively in animals that normally cannot be monitored due to the fact that they are living in the wild or are too big for all other known monitoring devices.

15.4   CLINICAL APPLICATIONS IN ANIMALS

To date, EIT has not been used extensively as a routine monitoring tool to manage clinical situations. One of the reasons might be that the displayed variables do not yet represent the information the clinician needs to change case management. Reports on the use of EIT in clinical cases are mostly anecdotal. However, some case reports and case studies can be found in the veterinary literature.

1. **Horse**

    A case study with six horses endorses the potential usefulness of EIT as a diagnostic imaging technique in large animal species[463]. Two healthy horses, one with exercise-induced pulmonary hemorrhage (EIPH), two with pleuropneumonia and one with cardiogenic pulmonary edema (CPE) were included. EIT was recorded in these standing non-sedated horses over 10 breath cycles and the distribution of ventilation was evaluated (CoV and silent spaces). In the horses with lung disease, a shift of CoV towards the non-dependent lung fields/segments and an increase in dependent silent spaces was demonstrated compared to healthy horses. EIT was thus able to detect changes in the distribution of ventilation as well as the presence of poorly ventilated lung units in horses with naturally occurring pulmonary pathology.

    In a separate study EIT data from horses suffering from left sided cardiac volume over load disease and therefore different degrees of pulmonary edema were collected and compared to a healthy cohort of horses (unpublished data). A distinct dorsal shift of the ventilation with loss of ventilated lung area in the ventral parts of the lungs was observed. The EIT also revealed that the left lung is more effected by the volume overload disease. Specific EIT variables describing the distribution of ventilation were able to distinct between healthy, subclinical and clinical cases of cardiac volume overload.

2. **Orangutan**

    A case report describing the clinical use of EIT in an orangutan with recurrent airway disease undergoing diagnostic evaluation demonstrated how EIT can help at the bedside to detect ventilatory issues quickly and allow goal-directed treatment[736]. A belt designed for humans (Sentec, Switzerland) was placed directly after induction of anaesthesia to continuously monitor ventilation during planned transport between different locations within the hospital (radiology, CT imaging, and bronchoscopy room, figure 15.4C). Immediately after endotracheal intubation one-lung intubation and ventilation became apparent on the EIT. The endotracheal tube was withdrawn until both lungs were visibly ventilated on the EIT images. Nevertheless, one lung continued to show reduced regional stretch and large silent space which was subsequently verified as a large pathologic process (abscess) in the lung on CT. After a diagnostic bronchoscopy procedure, EIT revealed increased silent spaces now in the peripheral lung area, which were indicative of excess residual lavage fluid within the lung tissue. Subsequent weaning from the ventilator was guided by improvements in lung function as suggested by the EIT image. In this case EIT played a key role in safe anaesthesia monitoring and non-invasive diagnostics.

15.5 FUTURE OF EIT IN VETERINARY APPLICATIONS

As EIT becomes established as a useful non-invasive and affordable method for dynamic general and regional lung function examination in the veterinary field, this new technology is expected to be increasingly used for a broader range of indications, conditions and animal species over the coming years.

The authors of this book chapter have collected data from cows, sheep, pigs, dogs, calves, foals, rhinoceroses, an orangutan and even birds evaluating anaesthesia- and ventilatory-related variables (figure 15.4). New applications are currently being investigated. Preliminary data on the evaluation of heart rate in standing animals based on thoracic EIT measurements look very promising. Also evaluation of lung disease over time in cattle which has a big economical impact in the agricultural industry is on its way. The equine racing industry is more and more interested in early diagnosis of lung diseases effecting athletic horses like exercise induced pulmonary hemorrhage, but also common lung diseases in foals. EIT is the most promising new technology that can fulfil those requirements. The option of heart rate measurement in addition to the ventilation signal would further improve the clinical usefulness of EIT in veterinary applications.

EIT has a bright future in veterinary medicine as a non-invasive respiratory monitoring tool that can be used at the stable-side and in the field with minimal time effort and impact on the animals health and welfare.

Section IV

Related Technologies

16 Magnetic Induction Tomography

Stuart Watson
Dept. of Medical Physics, Salford Royal NHS Foundation Trust, Stott
Lane, Salford, M6 8HD

Huw Griffiths
College of Medicine, Swansea University, Singleton Park, Swansea
SA2 8PP

CONTENTS

16.1   INTRODUCTION

The development of tomographic techniques for imaging the low-frequency (<2MHz), passive elec-tromagnetic properties of materials non-invasively has been an active area of research since the 1980s. Most of this interest has been in the areas of medical imaging, where cross-sectional images

DOI: 10.1201/9780429399886-16

of the human body are sought[138,479], and industrial imaging for the visualisation and control of processes in vessels and pipelines[1154]. Electrical imaging has also been used in environmental monitoring for tracking the migration of pollutants underground[222] and in archaeology for imaging submerged remains[784].

The oldest of the electrical imaging techniques is electrical impedance tomography (EIT) which normally involves attaching an array of surface electrodes around the region to be imaged. Currents are injected and electric potentials measured via the electrodes, resulting in a set of four-electrode measurements of transimpedance from which a cross-section of electrical conductivity and permittivity can be computed. In some EIT systems, sinusoidal patterns of current are injected involving all electrodes at once as this has been shown theoretically to provide optimal measurement sensitivity. EIT is sometimes referred to as electrical resistance tomography (ERT) in applications where the permittivity is negligible.

Another technique, electrical capacitance tomography (ECT), is very similar to EIT in that it also uses an array of electrodes and applies an electric field to the material. It differs only in the way the measurements are made; instead of a measurement of transimpedance involving four electrodes at a time, capacitance is measured between different pairs of electrodes. ECT is designed for materials of low permittivity and negligible conductivity imaged through an insulating boundary.

The most recent and least developed technique is magnetic induction tomography (MIT), the first reports of which appeared in 1992–3. MIT applies a magnetic field from an excitation coil to induce eddy currents in the material and the magnetic field from these is then detected by sensing coils. This, in effect, measures the changes in mutual inductance between the coils. Direct contact with the material is not required. The technique has been variously named 'mutual inductance tomography' (also MIT), 'electromagnetic tomography' (EMT), 'electromagnetic inductance tomography' (EMIT) and 'eddy current tomography'. MIT is sensitive to all three passive electromagnetic properties: conductivity, permittivity and permeability.

A number of hybrid systems have been reported involving either magnetic excitation with coils and measurement of surface potentials with electrodes[291,336,568,1218] or current injection via electrodes and sensing of the external magnetic field with coils[1048]. Improvements have been claimed by injecting current and measuring both surface potentials and external magnetic field[618] or by both injecting and inducing current and measuring surface potentials[847]. The terminology has become very confusing, as these methods have been named 'magnetic impedance tomography', 'electromagnetic impedance tomography' (EMIT again) or 'magnetic EIT', but not being true MIT they will not be discussed further here.

Adding still more to the confusion, the passive (and altogether different) technique for locating the equivalent electrical sources in the brain from the recorded electroencephalogram activity, has been named 'low resolution brain electromagnetic tomography' (LORETA). Indeed, the term 'electromagnetic tomography' could equally well be applied to the familiar X-ray CT or to optical or microwave tomography, all of which employ electromagnetic waves. It is desirable, therefore, that the term 'magnetic induction tomography' be universally adopted for the class of techniques in which eddy currents are induced and the external magnetic field sensed (whether by coils or by other types of magnetic-field sensors).

16.2 THE MIT SIGNAL

There are three contributions to the signal detected by the sensing coil. The first is directly induced by the field from the excitation coil (the primary signal, **B**). The second is from the eddy currents induced in the material which in turn produce their own magnetic field (the secondary signal, **ΔB**). The third contribution to **ΔB** is due to 'magnetostatic' induction of a polarization in the material by virtue of its magnetic susceptibility. This contribution is very small and is usually neglected in biomedical MIT as the relative permeability of biological materials is close to unity.

For a sinusoidally time-varying excitation at angular frequency ω, the skin depth of the electromagnetic field in the material is given by $\delta = \sqrt{2/(\omega\mu_0\mu_r\sigma)}$, where σ and μ_r are the conductivity and relative permeability of the sample and μ_0 is the permeability of free space. If δ is large compared with the thickness of the sample, which will normally be so for biological tissues,

$$\frac{\Delta\mathbf{B}}{\mathbf{B}} = P\omega\mu_0\left(\omega\varepsilon_0\varepsilon_r - j\sigma\right) + Q\left(\mu_r - 1\right), \tag{16.1}$$

where ε_r is the relative permittivity of the sample, ε_0 is the permittivity of free space and P and Q are geometrical constants[932]. Thus, the conduction currents induced in the sample give rise to a component of $\Delta\mathbf{B}$ which is proportional to frequency and conductivity and is imaginary and negative, meaning that it lags the primary signal by 90 degrees. Displacement currents cause a real (in-phase) component proportional to the square of the frequency. A non-unity relative permeability also gives rise to a real component, but with a value independent of frequency. The primary and secondary signals can be represented by the phasor diagram shown in figure 16.1.

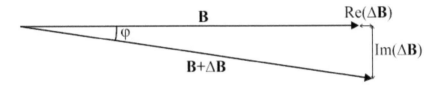

Figure 16.1   Phasor diagram representing the primary (**B**) and secondary ($\Delta\mathbf{B}$) magnetic fields detected. The total detected field ($\mathbf{B} + \Delta\mathbf{B}$) lags the primary field by an angle ϕ.

Because for biological tissues, $\Delta\mathbf{B}$ is much smaller in magnitude than $\mathbf{B}$ and is normally dominated by the conductivity term, the phase angle can be written

$$\phi \approx \frac{\Delta\mathbf{B}}{\mathbf{B}} \propto \omega\sigma \tag{16.2}$$

The phase angles concerned will be small, of the order of millidegrees (m°) to a few degrees. A higher frequency of excitation will increase the size of the signal.

For a metal sample, where the conductivity is high and the permittivity negligible, δ will be much smaller than the thickness of the sample and the behaviour of $\Delta\mathbf{B}/\mathbf{B}$ departs from the proportionality given in equation 16.1. Its value will be much larger than for the same volume of biological tissue and it will contain not just a negative imaginary part but also a negative real part as the sample tends to act as a 'screen'[384,1026].

16.3   MIT ARRAY DESIGN

16.3.1   EXCITORS AND SENSORS

A typical practical MIT system consists of an array of exciter and sensor coils mounted inside an outer screen (see figure 16.2). Each coil may be dedicated as an exciter or a sensor, or may be switched electronically to either mode.

The primary magnetic field generated by an exciter coil will be proportional to the current it carries and the number of turns on the coil. Increasing the number of turns or the coil dimensions, however, increases the inductance and dynamic impedance of the coil, which will then reduce the current if the coil is driven with a constant voltage. The optimal coil dimensions and turns will depend on the working frequency. Operating the coils at resonance allows more turns to be used and a higher magnetic field to be produced for a given drive voltage, but usually restricts the system to a single frequency of operation. However, Scharfetter[930] described an impedance matching circuit that enabled a coil to be driven efficiently at two distinct frequencies, 1 and 10 MHz. When multiple excitor coils are employed, they are normally isolated when inactive to ensure that they do not perturb the primary field of the active coil[573,1127]. With this in mind, Sharfetter et al[933] configured the excitation amplifiers as current sources and was then able to drive the coils simultaneously to reduce the data-acquisition time (see §16.4).

The magnetic-field sensors employed in low-conductivity MIT systems to date have been air-cored coils. The coils have been operated in voltage- rather than current-sensing mode to ensure they do not load the drive coil and perturb the primary magnetic field. They are normally operated at frequencies below their resonant frequency[887]. When a sensor coil is operated near resonance, the phase of the measurement is very sensitive to environmental changes such as temperature and mechanical disturbance. For this reason, operating coils at resonance has often been avoided. A differential amplifier is used to buffer and amplify the voltage induced in the sensor coil allowing rejection of electric-field coupling since a large component of this signal will be common-mode[574,1129]. The sensor buffer amplifier should be selected to have high input impedance and common-mode rejection ratio across the operating frequency range, and high unity-gain bandwidth to minimise the phase error across the amplifier and provide sufficient phase stability[1125].

Air-cored coils are by far the simplest and least expensive sensors for high-frequency, low-noise, detection of magnetic fields and a noise figure of $1\,pT/\sqrt{Hz}$, or better, is easily achievable. But coils have a limitation: the induced voltage is proportional to frequency, restricting broadband performance and sensitivity at lower frequencies. However, developments in magnetic-sensor technology such as magnetoresistive and spin-dependent-tunnelling devices and optical and atomic magnetometers are now providing sensors with competitive low-frequency and broadband characteristics.

Atomic and optically pumped magnetometers currently provide the most sensitive room-temperature magnetic-field sensors and have the potential for both miniaturization of the sensor and extension of the bandwidth into the MHz range. An atomic magnetometer has been proposed and experimentally demonstrated for application in MIT[1144] and more recently the same group developed a sensor demonstrating a noise floor of $<1\,pT/\sqrt{Hz}$ over a frequency range of 30 kHz to 1.1 MHz[226]. Miniaturized optically pumped sensors are now commercially available and provide a noise floor of $<20\,fT/\sqrt{Hz}$ but over a limited bandwidth of DC–1.4 kHz[928].

Commercially available sensors based on magnetoresistive effects, specifically anisotropic magnetoresistive and giant magnetoresistive sensors, provide a broadband response typically from DC to 5 MHz in small and convenient integrated-circuit packages. They have offered limited noise performance, $\approx 100\,pT/\sqrt{Hz}$, and have not been competitive with coil sensors for low-conductivity applications such as biomedical MIT, where the secondary magnetic fields to be detected are typically a few nT or less[1011]. Recently however, tunnelling magnetoresistive (TMR) sensors have appeared on the market, demonstrating low noise of $<5\,pT/\sqrt{Hz}$ for frequencies above 1 kHz and providing single- and 3-axis magnetometers in surface-mount packages. Given their small size, high sensitivity and broadband performance, TMR sensors may offer significant advantages over coils for high-conductivity, low-frequency MIT applications and for this reason they are now being evaluated for such applications[1112].

16.3.2 ARRAY CONFIGURATION

Most MIT systems described to date have employed cylindrical coil arrays with between 8 and 16 excitor/sensor coil pairs[574,623,933,1076,1132]. Cylindrical arrays are most appropriate for objects with

Figure 16.2   A practical MIT system, operating at 10 MHz (after[1126]). The 16 coils are mounted inside a cylindrical electromagnetic screen of aluminium. The circuit boards of the transceivers are enclosed in metal boxes fixed to the outside of the screen.

cylindrical geometry, for instance in the measurement of multi-phase flows within pipelines[1127], and allow simplicity of construction but not necessarily an optimal geometry for all applications.

Hemispherical arrays have been proposed for brain imaging[1177,1219] since this geometry provides superior sensitivity in comparison to a cylindrical array for this application[253]. Planar arrays have been developed for vital-signs monitoring, since they provide an 'open' geometry with measurements taken conveniently from one side of the subject, providing practical advantages. Hemispherical and planar-array MIT developments are discussed in more detail in §16.5.4.

The number of independent measurements produced by excitor/sensor array may be increased by mechanically translating or rotating either the sample or the array[1049,1133]. It is quite possible to take this concept to an extreme and to perform MIT with a single coil. Trakic *et al*[1049] employed a single coil as both excitor and sensor and rotated it around the sample to emulate a cylindrical array of 200 transmit-receive coils using time-division multiplexing. A hand-held MIT system has been demonstrated by Feldkamp and Quirk[275] in which a single coil was scanned around the sample and impedance changes determined by measurement of the resistive losses in the coil. Optical tracking of the device allowed the coil's position and orientation to be precisely determined (see §16.5.4).

16.3.3 SCREENING

Because of the small sizes of the signals to be measured in MIT, screening is important for two purposes: to reduce the sensitivity of the system to objects outside the imaging space and to reduce capacitive coupling between the coils. The scalar potential difference required to drive current through the excitation coil creates an electric field in the surrounding space that can induce a signal across the impedance of the sensing coil. This can either be by direct coupling or by indirect routes via the sample or external bodies[372,820]. Without careful design of the hardware, this unwanted capacitive signal can easily be much larger than the signal of interest from inductive coupling.

The outer screen can be of two types which function in different ways. A cylinder of high-permeability material (e.g. ferrite) acts as a magnetic-confinement screen by providing a low-reluctance return path for the field lines. Thus, no magnetic flux escapes from the cylinder to interact with external objects. This type of screen has been used in low-frequency MIT systems[822,1196]. A magnetic-confinement screen increases the measurement sensitivity to objects inside the coil array by up to a factor of 2[819].

A second type of outer screen, used in both low- and high-frequency systems is a highly conducting metal cylinder which functions as a so-called 'electromagnetic screen'[574,1126,1196]. Eddy currents are induced in the screen creating a magnetic field in opposition to the field from the coil. Provided the thickness of the screen is large compared with the skin depth of the fields in the metal, no magnetic flux exists outside the cylinder. In contrast to the magnetic-confinement screen, the eddy currents in an electromagnetic screen reduce the imaging sensitivity to objects inside the array by an amount depending on the stand-off distance of the coils from the screen[821]. This type of screen has a particular advantage, in that it 'attracts' electric field lines in a similar manner to a ground plane on a printed circuit board and significantly reduces capacitive coupling between the coils.

In addition to the outer screen, screening of the individual coils is often added. Griffiths *et al*[385] formed the coils as 'shielded turns', winding them from coaxial cable and terminating the core on the screen at the feed point. Korzhenevsky and Sapetsky[577] also formed coils from coaxial cable but used a different method of termination. Another way of screening a circular coil is to enclose it in a metal cylinder having radial cuts in the ends and longitudinal cuts in the sides (a technique used for inductive applicators in shortwave diathermy). In this way, the cuts are all perpendicular to the vector potential field from the coil and prevent eddy currents from flowing in the screen which would otherwise oppose the magnetic field from the coil. Manufacturing coils and screens from printed circuit board is an attractive method allowing high reproducibility and a low profile of construction. Again the breaks in the screen must be placed so as to prevent the flow of eddy currents (see figure 16.3. Many of the screening techniques used in MIT are not new and can be found in old texts on radio engineering (see[372,820] for further discussion and references).

Residual capacitive coupling and the effectiveness of screening can be determined by measuring the response of the real and imaginary signal components to changes in the conductivity of saline test samples[372]. Assuming the dimensions of the test samples are much less than the skin depth, the imaginary component should be linear with a negative gradient versus conductivity as expected theoretically (16.1). Placing a deionised water sample in the system provides a simple test, with zero change in the imaginary signal component expected if screening is effective[384].

16.3.4 CANCELLATION OF THE PRIMARY SIGNAL

Because the secondary signal has to be detected against the much larger primary signal, various methods have been tried for 'backing off' the primary signal, i.e. for subtracting the phasor **B** in figure 16.1, such that with no sample present all recorded signals should be zero. This allows the

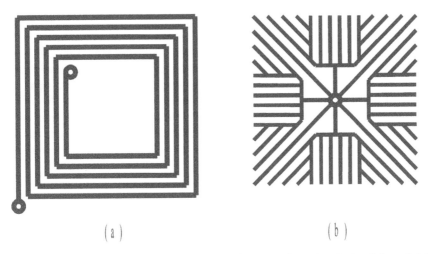

(a)   (b)

Figure 16.3   Designs of (a) spiral coil and (b) comb screen for printed circuit board fabrication (after[820]).

gain of the front-end amplifier to be increased and the overall signal-to-noise ratio to be improved. Another important consideration is that determining the MIT signal relies on the measurement of phase. In the presence of a large real component arising from the primary field, phase noise appearing in it produces noise in the imaginary component too. This most likely arises from short-term fluctuations in phase between the primary and reference signals which, in effect, cause part of the primary signal to be converted into an imaginary signal. Backing-off the primary field therefore greatly reduces the impact of phase noise on the measurement of the imaginary component.

In practice, perfect cancellation is of course never possible but usefully large cancellation factors can nevertheless be achieved. Three methods for primary-signal backoff have been employed: a third coil used to cancel the primary field at the sensor, before entering the electronics; orientating the sensor coil to reduce its flux linkage with, and sensitivity to, the primary field; and electronic generation of the backoff signal.

In single-channel measurements at 10MHz, Griffiths et al[385] used a third coil mounted separately, producing an antiphase signal that was then added to the signal from the sensing coil. Again in a single channel, operating in the band 40–370 kHz, Scharfetter et al[934] used a planar gradiometer as the sensor which provided a high rejection of the primary signal (cancellation factor 10^2–10^3). The residual signal was reduced further by a phase-compensation circuit (electronic backoff). The idea of using a third coil for backing off the primary signal was not new and had been used more than forty years earlier by Tarjan and McFee[1028]. They constructed what they termed a 'differential transformer' comprising two sensing coils positioned symmetrically on either side of the excitation coil on a cylindrical former. The sensing coils were connected such that the signals cancelled, in effect forming an 'axial gradiometer'. A related technique was used by Crowley and Rabson[216] for measuring the resistivity of semiconductor wafers. Also using axial gradiometers, Iker and Gencer[1065] achieved a cancellation factor of $\approx 10^3$ at 60 kHz, and combined it with electronic backoff, and Riedel et al[880] achieved a similar factor but at a higher frequency, 1 MHz.

Peyton et al[819] described the overlapping of excitation and sensing coils so that the net primary flux through, and hence primary signal from, the two sensing coils immediately adjacent to the excitation coil was close to zero. Watson et al[1124] proposed a method in which sensing and excitation coils were oriented at right angles to each other such that there was no net primary flux linkage between them. With a single channel, they achieved a primary-field cancellation factor of about 300 over the frequency range 1–10 MHz. They showed further that the noise level fell by a factor of over 40 when the primary field was in effect backed off. Scharfetter *et al*[936] computed sensitivity maps for a single excitation coil, first with a planar gradiometer as a sensor and then for Watson's 'right-angled-coil' method. The maps were found to be very similar in both form and magnitude. The authors then performed practical measurements at 500 kHz and showed the gradiometer to be much more efficient at rejecting interference from external sources, resulting in a signal-to-noise ratio higher by 20 dB than for the right-angled coil. In a subsequent paper[935] they described a new sensor based on a planar gradiometer oriented for zero net primary magnetic flux in both coils of the gradiometer. This was proposed in order to combine the advantages of gradiometers and the coil-orientation method, namely insensitivity to external sources of interference and a lower sensitivity to mechanical instability.

The main requirement for the backoff coil/gradiometer method is for good mechanical and temperature stability (see[932]). The disadvantage of the method is that whilst it is straightforward to implement for single-channel measurements, obtaining effective and stable backoff between multiple excitors and sensors is much more challenging. For cylindrical arrays, planar gradiometers and excitor coils may be oriented with their coil axes lying on a common plane to achieve multi-coil backoff[898,933]. Several authors[535,879,880,1065] mechanically scanned a single axial gradiometer in a plane above conducting objects (§16.5.4). From symmetry, an array of many such devices, with their excitor coil lying in a common plane (the axes of the gradiometers normal to the plane) would register no primary signal for any excitor/sensor combination. Similarly, Watson *et al*[1124] pointed out that a planar array of the 'right angled coils' would achieve zero primary flux linkage between any sensor and any exciter. In a practical implementation of this method, Igney *et al*[497] achieved modest primary-field cancellation factors in the range 19–50 across all excitor-sensor combinations.

Electronic backoff, on the other hand, can be programmed for all excitor/sensor combinations and the attractiveness of this arrangement is that a fully electronically scanned system with no moving parts is possible. However, small changes in the amplitude, phase or waveform of the compensation signal can upset the cancellation and great electronic stability is needed. An entirely electronic backoff, programmable in amplitude and phase, was employed by Yu *et al*[1195] in a 200 kHz industrial MIT system. The advantage of using a backoff coil, gradiometer or coil orientation however, is that the primary field is cancelled at the point of detection and is not dependent on the stability of the electronics. Any fluctuations in the excitation coil current or waveform will affect the sensing and backoff coils (or halves of the gradiometer) alike and will not affect the cancellation. For low-conductivity MIT, electronic back-off is likely to be most practical as a method to reduce further the residual primary-field signals after applying coil-based backoff methods.

An alternative approach to primary-field backoff for achieving precise measurement of the imaginary component of the MIT signal, is to reduce spurious phase fluctuations by using highly phase-stable measurement and distribution for both detected and reference signals. This has the advantage of being generally applicable to any array configuration, allowing the designer to select an array optimised in sensitivity and practicality for a specific application without the limitations imposed by signal backoff. Watson *et al*[1125] described the design of an ultra-phase-stable detector amplifier and found it to provide measurement stability equivalent to that of a gradiometer.

The use of coil backoff, electronic backoff and highly stable measurement and distribution need not be mutually exclusive, of course, and it is likely that the highest measurement precision will be achieved by utilising a combination or even all of these methods.

16.4   SIGNAL DEMODULATION

Various methods of phase-sensitive detection (PSD) have been employed in MIT. PSD may be performed by passing the signal through a pair of analog multipliers, multiplying with 0° and 90° reference signals and then low-pass filtering to provide the amplitudes of real and imaginary components respectively. The use of 'chopped' demodulation signals to increase phase measurement precision was described by Gough[374]. If the detected signal is digitised, then multiplication and filtering may be performed digitally providing significant advantages in both stability and flexibility. The digitized signal may alternatively be demodulated by performing a Fourier transform and selecting the real and imaginary magnitudes at the frequency of interest. Signal measurement by Fourier transform is particularly efficient for the demodulation of multi-tone (multifrequency) signals.

An alternative method of demodulation[576] is to measure the phase angle directly as it will be proportional to the sample conductivity (16.2). The method has been implemented by passing the signal and a reference waveform through zero-crossing detectors and feeding the resulting signals to an exclusive-OR gate; the output pulse width will then be proportional to the phase difference[574,1129].

Commercial lock-in amplifiers have provided an off-the-shelf solution incorporating a vector voltmeter (phase-sensitive detector), analog-to-digital conversion and digital filtering[535,880,882,1065,1126]. The appearance of multichannel digitisers with sufficiently high sampling rates, however, provide an alternative method allowing fast, efficient, parallel, multi-frequency demodulation. The MIT system described by Scharfetter et al[933] employed a fast multi-channel digitizer and FFT demodulation which achieved a frame rate of 1 Hz. Multi-frequency excitation, to obtain spectral bioimpedance information, was combined with a frequency encoding scheme for the excitation channels to enable simultaneous multi-frequency data collection from 8 exciters and 8 sensors.

Digitisation of the high-frequency MIT signal generates large data sets, and a very high data bandwidth between the digitizer and signal processing hardware is required. These requirements may be reduced by employing heterodyne downconversion of the signal to a much lower intermediate frequency before signal measurement[574,1129]. Modern embedded-systems devices, however, such as field-programmable gate arrays, allow efficient parallel multi-frequency demodulation to be implemented[810]. Direct digitization of the high-frequency signal, and processing by a dedicated multi-channel embedded-system processor, currently appears to be the most powerful and effective technique for MIT signal measurement.

Specifying the performance of MIT demodulators requires the noise and drift (the short and long-term variations) of the real and imaginary signal components to be measured and quoted for a specified measurement time constant. For MIT systems which do not employ back-off, measurement precision has been quoted in terms of phase noise and phase drift[1128], the signal phase being assumed directly proportional to the imaginary component (16.2). In this case, the magnitude of the measured signal should also be quoted to allow conversion between the phase and real and imaginary signal components.

16.5 WORKING IMAGING SYSTEMS AND PROPOSED APPLICATIONS

16.5.1 MIT FOR THE PROCESS INDUSTRY

MIT does not require physical contact with the sample. Measurements are therefore simpler to automate and the sensors can be isolated from physically extreme environments. MIT has the further significant advantage that it can operate through electrically isolating materials; air, plastic, rubber, glass, ceramics and concrete, for instance. These characteristics suggest that the technique has potential for a wide range of applications in industry for non-destructive evaluation, for process measurement and control and for security-threat detection.

The majority of practical MIT systems for the process industry have been designed for detecting metallic or ferromagnetic objects which, having either a high electrical conductivity or a high permeability, can produce large signals with an excitation frequency of 500 kHz or below. When the material to be imaged is entirely metallic and hence of very high conductivity, MIT has an advantage over EIT. In EIT the transimpedances would be very small indeed and difficult to measure.

Yu et al[1196] reported a system operating at 500 kHz employing a parallel excitation magnetic field generated by two pairs of large coils. Twenty-one sensing coils were arranged in a circle around the imaging volume. The assembly was situated within a magnetic-confinement screen and an electromagnetic screen. Imaging of metallic objects (copper bar and aluminium foil) was demonstrated. Subsequently, this research group described a system operating at 200kHz with a parallel excitation field and 24 detector coils[1195]. Metallic and ferromagnetic objects were identified from the phase information in the signals. Large signals were detected, $|\Delta \mathbf{B}/\mathbf{B}|$ being as much as 0.25.

Williams and Beck[1154] described an array of 12 excitation coils interleaved with 12 sensing coils arranged in a circle. The system operated at 5 kHz and employed phase-sensitive detection. In a further development of this type of 'multi-pole' design, Peyton et al[822] reported a 100 kHz system with 16 coils, each of which could serve either as an exciter or as a sensor. The coil assembly was housed within a magnetic-confinement screen and again it was shown possible to distinguish metallic from ferromagnetic objects from the sign of the signal.

Ramli and Peyton[857] proposed a 16-coil MIT linear array for detecting the positions and integrity of steel reinforcing bars embedded in concrete. Images were reconstructed by a simultaneous increment reconstruction technique (SIRT). Subsequently, Bissesseur and Peyton[117,118] developed an improved algorithm for this application involving a non-linear solution, parameterized for discrete conducting bars. Planar-array MIT systems have been proposed for the detection of corrosion and cracking in metallic plates[1190] and more recently for the detection of defects in conductive fibre-reinforced composites[664,873].

MIT is particularly suitable for use in high-temperature environments such as in the processing of molten metal. Of great commercial interest is the need for on-line monitoring of the flow regime in the pouring nozzle during the continuous casting of steel, control of which is critical to the quality of the final product. Pham et al[824] proposed MIT as a method for detecting the extent of solidification of molten metal flowing in a pipeline, exploiting the lower conductivity of the solid phase than the molten. Two-dimensional imaging of the conductivity distribution was demonstrated by an analytical method, but no practical measurements were reported. Experimental results were obtained by Binns et al[115] with a six-coil MIT system operating at 0.1–1 kHz. Initially, different molten flow regimes were simulated using Woods metal (a eutectic alloy with a melting point of 70°C) and glass beads simulating argon gas inclusions. For image reconstruction, a SIRT method was employed with a non-negativity constraint. This system subsequently underwent trials with molten steel[467,665] successfully acquiring flow images with a frame rate of 0.74 frames/s. More recently MIT has been used in combination with inductive flow tomography to monitor the flow structure in the pouring nozzle, specifically the two-phase steel/argon distribution, potentially allowing real-time control of the mould flow[1162].

16.5.2 SECURITY APPLICATIONS

Magnetic-induction-based metal detection has long been used for security applications for detecting threatening items such as guns or knives, and walk-through metal detection (WTMD) portals are now a ubiquitous feature of airports and government buildings. Modern WTMD portals typically provide information on the approximate location of the detected object to aid security staff. MIT has been proposed as a method of enhancing this capability by additionally providing classification of the object as a threat, or not. Potentially this could eliminate the requirement for passengers to remove all metal objects from their persons and could reduce the number of false positives from surgical implants such as hip replacements. Tian et al[1035] described a commercial WTMD portal utilising pulse-induction excitation, refitted with an array of 80 giant magnetoresistive sensors. The pulse responses produced by target objects allowed classification into ferrous/non-ferrous or a mixture of both. Principal component analysis applied to the transient response of the array as the object passed through allowed further discrimination between different types of objects and determination of the object's orientation. An MIT system comprising a 20×21 coil array operating at 10 kHz was proposed as a cost-effective small screening system for bags and parcels[1121]. An image of a tool within a plastic case placed on top of the array demonstrated impressive spatial resolution which would aid classification, but this is likely to be significantly degraded with increased lift-off of the object above the array.

Given that the major driver for the introduction of reliable object classification in security applications is the need to reduce inconvenient and expensive delays at airports, stations and crowded events, image reconstruction may not be the most practical and efficient method. An alternative approach which shows promise is the use of the magnetic polarizability tensor (MPT). The MPT is dependent on the excitation frequency and an object's size, shape, conductivity and permeability, allowing it therefore be used to define the characteristics of metallic and magnetic objects. The MPT has been applied to threat detection in a WTMD portal by treating the objects as infinitesimally small point dipole sources with an associated MPT value. A number of independent excitor/sensor measurements are collected as objects pass through the array, and the MPT is determined as an inverse problem. Further details of the inverse algorithms employed for MPT classification are covered in §16.6. The main potential advantages of this technique are classification accuracy and speed of calculation. This has been demonstrated in a WTMD system in which, for test data consisting of 67 objects and 835 scans, threats were discriminated from non-threat objects with a success rate of over 95%[671]. Measurements in this case were made at a single frequency. Because the MPT is frequency dependent, it is likely that multi-frequency measurements could further improve classification accuracy. Performing spectroscopic MPT measurements might be difficult in a walk-through metal-detection systems, given the speed of operation required, but the technique is also applicable to anti-personnel mine detection.

MPT spectroscopy has already been applied to the detection of 'minimum metal mines' in which metallic components are limited to a few, small, essential parts. These mines can be detected by more traditional methods but they are prone to a high false-positive rate due to contamination by metal 'clutter' in the vicinity, such as bullets, found in areas of previous conflict. However, the MPTs obtained from these types of mines have been found to be very different from those due to the clutter, so provide an improved specificity of mine detection[9].

16.5.3   PETROCHEMICAL INDUSTRY

The use of MIT for monitoring oil/gas/sea water flow in a pipeline has been proposed because sea water is conductive[1154]. Its conductivity is approximately 5 S/m which is many orders of magnitude lower than that of metals so much smaller eddy-current signals will be produced for a given excitation frequency. Consequently, frequencies in the megahertz range have been necessary in order to obtain measureable signals. The first practical results appeared in three papers from a Norwegian research group. They described a high-frequency, inductive 'dipstick' for sensing levels of sea water, oil and air in a gravitational separator[424] and in subsequent papers, the extension of the work to tomography was suggested[425]. They used finite-element modelling to investigate the effect of water droplet size and volume fraction on the eddy-current loss in a resonant coil[426]. Images of stratified flow of oil/gas/sea water mixtures, simulated using a cylindrical plastic tube filled with varying levels of saline solution were obtained by[1127] using a 16-channel MIT system.

As in the application to molten steel (§16.5.1), the measurement of multiphase flow can be enhanced by combining MIT with a second tomographic technique sensitive to flow velocity. An 8-channel MIT system, operating at 10 MHz and capable of measurement and image reconstruction at a rate of 2 frames per second was combined with electromagnetic velocity tomography in a prototype multiphase oil/gas/seawater flow metering system[663]. The system, tested on a flow loop, allowed the measurement of continuous water flowrate with a relative error of 1% for single water phase flow and 12% for a 65% water/oil flow, a value which was described by the authors as consistent with existing alternative flowmeters.

16.5.4 BIOMEDICAL MIT

The conductivities of biological tissues are of the same order of magnitude as that of sea water, or smaller still, so biomedical MIT again has tended to use high frequencies of excitation in order to obtain measurable signals. Even at a frequency of 10 MHz, the secondary signal due to 500 ml of muscle tissue is typically only about 1% of the magnitude of the primary signal, i.e. $\mathrm{Im}\,\Delta\mathbf{B}/\mathbf{B} \approx 0.01$ [385].

The first report of MIT for biomedical use was by Al-Zeibak and Saunders[34]. An excitation and a sensing coil operating at 2 MHz were scanned past a tank of tissue-equivalent saline solution, with immersed metallic objects, in a translate-rotate manner. Images were reconstructed by filtered backprojection and showed the outline of the tank and the internal features. Despite the fact that these images have been reproduced in many reviews of MIT since, questions have been raised about the origin of the signals. Using amplitude detection only, a change in total signal of about 70% was measured as the saline solution's conductivity was increased from zero to 1 S/m. Taking $\mathrm{Im}\,\Delta\mathbf{B}/\mathbf{B} \approx 0.01$, the proportional change in amplitude will be $|\mathbf{B}+\Delta\mathbf{B}|/|\mathbf{B}| = \sqrt{1^2+0.01^2} = 1.00005$, i.e. a change of only 0.005%. The reason for the large signals measured was most likely that the electric-field screening was inadequate, leaving significant capacitive coupling between the coils, and the system was, in effect, largely performing ECT, not MIT. The paper, however, has had the merit of stimulating a lot of interest in MIT.

Using a similar, two-coil, translate-rotate principle, Griffiths et al[385] measured volumes of tissue-equivalent saline solution at 10 MHz. The imaginary part of the signal, corresponding to the conductivity of the solution, agreed well with theoretical predictions and with subsequent more detailed modelling[730]. An image was reconstructed by filtered backprojection. The real part of the signal was much larger than predicted theoretically, and this was attributed to residual capacitive coupling which had been separated out by the phase-sensitive detection.

Korzhenevskii and Cherapenin[576] proposed a circular MIT array of 16 coils with direct phase measurement (see §16.4) and showed images reconstructed by weighted backprojection from simulated data. Subsequently, this research group reported the practical implementation of the method[574]. Like the 100 kHz multi-pole system described by Peyton et al[822] (§16.5.1), it employed multiple, electronically-switched excitation/sensing coil units arranged in a circle and housed within an electromagnetic screen. The excitation frequency of the system was 20 MHz but this was mixed down to 20 kHz (a process in which phase information is preserved) for signal distribution and demodulation. The coils were not individually screened but differential detection was used to minimise interference from capacitive coupling. Furthermore, any capacitive coupling affecting the real part of the signal would have had little effect on the measured phase. An image of a tank of tissue-equivalent saline solution clearly showed two embedded regions of higher and lower conductivity. The image was referenced to homogeneous saline, not to empty space.

In later publications[573,575] this group proposed that reduced spatial distortion could be achieved when images were reconstructed by an artificial neural network for some simple distributions of conductivity, and obtained the first in-vivo MIT images (figure 16.4). These appear to show internal anatomical structure, but a careful published validation of this MIT system on phantoms with similar distributions of conductivity to those of the anatomical sites studied would be beneficial to confirm the interpretation of these images.

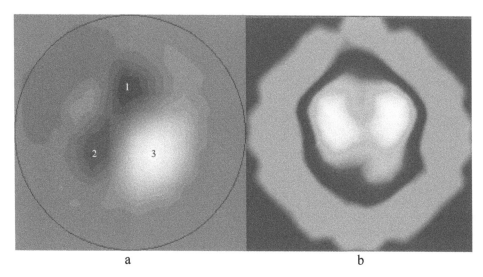

a   b

Figure 16.4   Human in-vivo images obtained with the Moscow 16-coil 20 MHz MIT system (after[577]. (a) Difference image of the thorax (inhalation-exhalation) reconstructed by weighted backprojection. The authors interpret features 1 and 2 as the left and right lungs and 3 as chest movement artefact. (b) Absolute image of the head (referenced to empty space) reconstructed by artificial neural network in which the two white (high conductivity) features are interpreted as the lateral ventricles of the brain.

Watson *et al*[1126] reported a 16-channel, electronically-switched MIT system similar to the design of[574] but operating at 10 MHz and employing phase-sensitive detection for signal demodulation. An MIT image of a human thigh in vivo was obtained and, from saline calibration images, mean thigh conductivity and permittivity calculated; the spatial resolution was insufficient to image internal structure. Details of the system are given in a later report[1127]. Figure 16.5 shows images, obtained with this system, of a saline bath simulating a brain with an immersed block of agar, simulating a hemorrhage. The conductivity contrast between the agar and the saline was a factor of 3.3, being similar to the contrast between blood and brain measured at 10 MHz[320]. The images were reconstructed with a single-step, linear algorithm as described by[730]. The simulated hemorrhage can be identified in both the absolute (row b) and difference (row c) images.

Tarjan and McFee[1028] measured an average value for brain conductivity inductively, using an axial gradiometer (§16.3), and subsequently a number of workers have identified the imaging of brain conductivity as a possible clinical application for MIT[575,932]. The advantage of MIT is that the magnetic field easily penetrates the skull whereas in EIT the skull acts as a resistive barrier. Two clinical applications for MIT which have been proposed are the detection and monitoring of cerebral edema, and the detection of cerebral hemorrhage.

Netz *et al*[769] suggested that brain edema might be detected more promptly from conductivity changes than is possible from imaging by CT or MRI. With a view to brain imaging, Merwa *et al*[709] described a finite element model for MIT, based on edge elements, combined with a realistic 3-dimensional tissue map of the head. Using the model, the group simulated a region of edema set at twice the conductivity of white matter. Taking a realistic signal-to-noise ratio, based on their practical, planar-gradiometer system, they calculated that if the region were 40 mm in diameter and located at the centre of the brain, it would be detectable with an operating frequency of 100 kHz[708].

The detection and classification of stroke is an area in which MIT might be able to contribute significantly to patient management. For ischemic stroke, where an artery has been occluded, prompt

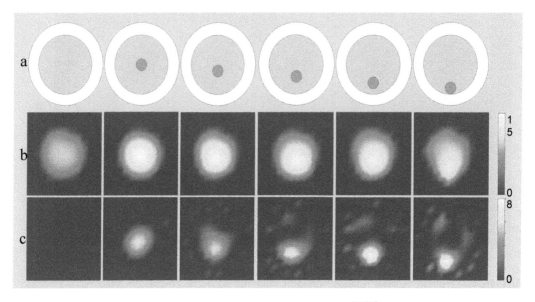

Figure 16.5 MIT images obtained with the 10 MHz system of[1126] for 4 cm diameter cylinder of agar, conductivity 1 S/m, in a 20 cm diameter bath of saline, conductivity 0.3 S/m. (a) Diagram indicating position of agar; the thickness of the air gap between the saline bath and the coils (white ring) was 3.5 cm; (b) absolute images reconstructed relative to empty space, 40 singular values; (c) difference images reconstructed from the difference in measurements with and without the agar present, 50 singular values. Only positive image values are displayed. (color figure available in eBook)

administration of thrombolytic drugs is effective, but this can only be carried out if cerebral hemorrhage has been definitively excluded. Currently used imaging techniques, MRI and CT, are expensive and may be in some cases inaccessible, and MIT has been proposed as an inexpensive, rapid imaging method for both initial diagnosis and perhaps more persuasively, as a means of providing continuous monitoring.

A hemispherical array MIT employing 15 axial gradiometer sensors was reported by Xu *et al*[1177]. Measurements were carried out with a phantom comprising a hemispherical bowl filled with saline, used to represent the brain, and agar blocks of higher conductivity representing the hemorrhage. The inverse problem was not solved, but 2D pseudo-images generated using an interpolation algorithm were presented. The use of gradiometer sensors allowed operation at lower frequencies, 40–120kHz, but limited the number of independent measurements to 15.

In a simulation study, Zolgharni *et al*[1220] investigated the feasibility of detecting cerebral hemorrhage with a 16-channel annular MIT system operating at 10 MHz. They used a realistic finite-element head model comprising 12 tissue types, and modelled the change in the detected signals produced by three different examples of stroke: a large (volume 49 cm^3) peripheral stroke, a small (8.2 cm^3) peripheral stroke and a small (7.7 cm^3) deeply located stroke. They found that the modelled strokes produced maximum phase changes of 71, 13 and 3.4 m° respectively, and that good visualisation of the strokes could be achieved when the measurement noise level was 1 m° (figure 16.6b). These simulations were, however, of time-differential imaging. As the authors pointed out, in practice this method is very unlikely to be useful for making an initial diagnosis, since no before-stroke reference will ever be available. With this limitation in mind, they then simulated frequency-difference imaging (1 and 10 MHz) allowing a snapshot static image to be obtained without the need for a before-stroke reference[1219]. The large peripheral stroke could be easily visualised when the measurement noise level was 1m° (figure 16.6c), but visualisation of the smaller strokes was poor. This was due to the low spatial resolution of the images and to the presence of confounding features caused by the fact that all the other tissues in the head increase in conductivity with frequency. It was also shown that an error in head position of just 5 mm or a size-scaling error of just 5%, relative to the position and size used for computing the sensitivity matrix, could cause similarly serious confounding features on the image (figure 16.6d). Millimetre accuracy in determining the shape and position of the head is therefore necessary, and developments of the hardware by which this might be achieved have already been reported [371,1025].

Xiao et al[1176] have proposed an improved method for detecting cerebral hemorrhage, using 7 frequencies instead of just 2. In simulations, they employed a 'frequency decomposition' that produced much clearer images of the hemorrhage with greatly reduced confounding features from the other tissues in the head. The method requires prior knowledge of the conductivity of each tissue as a function of frequency.

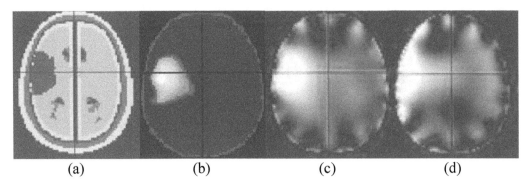

(a)   (b)   (c)   (d)

Figure 16.6 Simulated MIT data, using an anatomically realistic head model incorporating a large peripheral haemorrhage (after [1219,1220]. Left to right: (a) true tissue conductivity distribution, (b) LP stroke, time-differential image, (c) frequency-differential image 1–10 MHz, (d) same as image (b) but with a misregistration of the head relative to the coil array of 5 mm (left to right). Images (b) – (d) were reconstructed with 1 m° added phase noise.

A similar problem in the imaging of hemorrhage with MIT was highlighted by Chen et al[189]. Hemorrhages of varying volume ($10\,\text{cm}^3$ and $20\,\text{cm}^3$) and locations (shallow, medial and central) were simulated and reconstructed using a realistic head model for the 16-channel annular MIT system, operating at 10 MHz, described in Vaukhonen et al[1076]. In this case, however, they also modelled the effects on the detected signal produced by the deformation of the brain caused by the stress introduced by the hemorrhage. Using their combined bioelectrical/biomechanical model, they found that deformation of the brain tissues reduced the amplitude of the signal changes due to the hemorrhage and introduced artifacts which significantly decreased the ability to distinguish hemorrhages on the images.

Planar-array MIT systems have been proposed for the surface mapping of tissue conductivity and for non-contact vital-signs monitoring. Gencer and Tek[337] proposed an MIT system consisting of 49 excitation and 49 separate sensing coils arranged in two 7×7 planar arrays above the surface of a slab of conductor. From finite-element simulations they reconstructed images in three dimensions. The first step towards a practical implementation of the technique was reported by Iker and Gencer[1065] in which a single axial gradiometer operating at 60 kHz was scanned over a volume of tissue-equivalent saline solution (see also §16.3.4). Maps of signal strength were given. In a further paper, similar measurements were performed at 11.6 kHz and two-dimensional images of conductivity for volumes of saline were reconstructed[535]. Because only one position of the sensing coils was available for each position of the excitation coil (all being wound on the same former), the full data set described in[535] was not collected. Meaningful images of conductivity were nevertheless obtained, most likely because the subset of measurements collected were those with high sensitivity values.

Riedel *et al*[880] suggested the inductive measurement of wound conductivity using planar arrays. Performing impedance measurements of a wound from electrodes is very difficult as the surrounding skin is often uneven and in poor condition. A non-contacting inductive method might overcome these difficulties, but the fall-off in signal amplitude as the coil size decreases needs to be considered (see §16.9).

The 'open' geometry of planar arrays, with measurements taken conveniently from one side of the subject, may offer advantages in biomedical applications such as vital-signs monitoring where the coils could for instance be incorporated into a bed or a chair. Liebold *et al*[624] employed a 12-channel planar array placed within a bed to record cardiac and respiratory signals but did not perfrom imaging (figure 16.7). Steffen *et al*[1003] described a 14-channel multifrequency planar array, again designed to be placed within a bed, which could offer advantages for monitoring neonates, whose very sensitive skin could be damaged by the long-term application of electrodes.

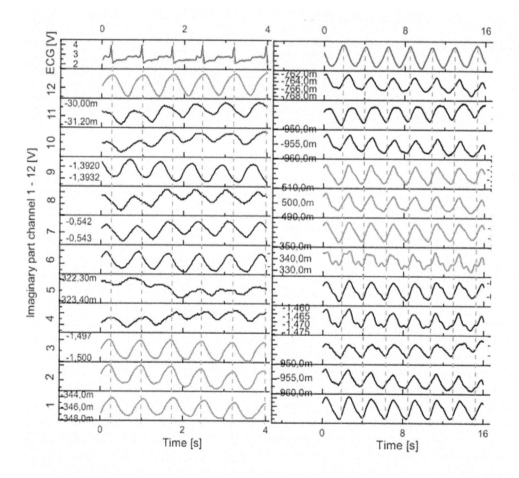

Figure 16.7   Signals obtained from a 12-channel planar MIT array[624]. The top traces in the left and right columns are ECG and plethysmograph respectively. The inductive signals in the left column were obtained during breath-holding and in the right column, with the subject breathing normally. (color figure available in eBook)

The MIT system described by Feldkamp[274] takes the planar-array concept further and reduces it to a hand-held, single-coil system which may be easily moved around the surface of an object. The system comprised a 4 cm diameter 10-turn coil consisting of 5 concentric loops with an excitation frequency of 12.5 MHz. Single-coil measurements of sample conductivity are achieved by measuring inductive losses through the change in the real part of the coil admittance. In-vivo images of the thoracic spine were reconstructed by collecting 132 measurements using a template to guide the operator. The resulting images (figure 16.8) showed features which were interpreted as rib articulations, spinal processes and the spinal canal, suggesting the system was capable of discriminating anatomical features at a depth of several centimetres. The concept was further developed through the addition of optical tracking: an optical body was attached to the hand held sensor and its location and orientation measured using an infrared 3D motion tracking device. This allowed the coil centre to be tracked with sub-millimetre accuracy while 1800 independent measurements were collected at a rate of 5 Hz[275]. This MIT method will likely be restricted to measurement of the conductivity of the surface of objects, or inclusions near the surface, but the concept of a hand-held, accurately tracked MIT device has merit in its flexibility.

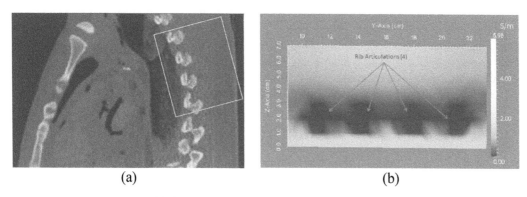

(a)   (b)

Figure 16.8   Results from[274]. (a) Parasagittal CT slice of upper torso with box in right hand side showing region scanned by MIT. (b) MIT image obtained using a hand-held single-coil MIT system

16.6   IMAGE RECONSTRUCTION

Early attempts at image reconstruction in MIT have involved the use of linear algorithms. Weighted backprojection has been used successfully for isolated objects imaged relative to free space where sensitive regions approximately correspond with 'flux tubes' between excitation and sensing coils[576,1196]. With a conductive background however, the sensitivity maps for low-contrast perturbations depart widely from the assumed flux tubes, the sensitivity increasing outwards towards the edge of the conductive region[937].

A single-step multiplication of the data by the pseudo inverse of the sensitivity matrix, computed by truncated singular value decomposition has been used widely in EIT. In a modelling study using this method, Morris and Griffiths[729] found the MIT images to be of poorer quality than the corresponding EIT images in a direct comparison with the same conductivity distributions. For process applications in MIT, the simultaneous increment reconstruction technique (SIRT) has been used extensively with good results for high-contrast objects[115,129,822,857]. This is a linear, iterative method in which the sensitivity matrix remains fixed. Lionheart[627] has pointed out that a given number of iterations of the SIRT (essentially the same as the Landweber method) can be implemented in a single matrix multiplication if the sensitivity matrix is inverted by singular value decomposition with the appropriate filter function. However, if pixel values are to be constrained (e.g. non-negative conductivity), the SIRT method is often still chosen, with the constraint applied at each iteration[115].

A prerequisite of all the above methods is efficient computation of the sensitivity matrix. This can be computed by assuming an initial conductivity distribution (e.g. uniformity) and solving the forward problem for all excitor/sensor combinations. Each voxel is then perturbed by a small amount (e.g. 1%) and the whole computation repeated for all such voxels in turn. As has been pointed out in the context of EIT, such a method is computationally very time-consuming and several authors have now described more efficient methods specifically for MIT. Gencer and Tek[337] derived a method for computing the sensitivity involving the impressed vector potential and a derivative of the scalar potential. Two papers have described rapid computation of the sensitivity matrix by what is in effect the Gezelowitz sensitivity formula extended to take account of changes in conductivity, permittivity and permeability and the fact that the electric field contains a magnetically induced component as well as that arising from the gradient of the scalar potential[485,631]. The methods require only two solutions of the forward problem for each coil pair, first exciting one coil and then the other. Ktistis et al[586] proposed an efficient method for approximating the sensitivity matrix by pre-computing one of the fields with the tissue absent from within the coil array, leaving just one field to be computed at each iteration of the conductivity distribution.

The artificial neural network method used by Korjenevsky[573] to produce in-vivo images (§16.5.2) is sometimes criticised for not being based on any underlying physical principles and depending for its accuracy on the training data. However, the method does not assume linearity, can be implemented with speed and may well prove valuable for practical MIT applications.

There is a general consensus that in order for MIT to advance significantly, the non-linear inverse problem will need to be solved in three dimensions. In contrast to the linear iterative methods, the Newton-Raphson or Gauss-Newton method will be used and the Jacobian (sensitivity matrix) recomputed at each iteration from the most recent estimate of the conductivity distribution[628]. Soleimani and Lionheart[992] employed the iterative Gauss-Newton method, with Tikhonov regularization applied at each step, to reconstruct images of inclusions of conductivity 0.8 S/m placed within a 0.2 S/m background. Relative to the initial image, 3 iterations produced an image with dramatically improved location of the inclusions and absolute conductivity values, although subsequent iterations provided little further improvement. This clearly demonstrated the advantage of nonlinear reconstruction, but the authors made the valid point that unless sufficient precision in both measurements and model is available, a non-linear algorithm is unlikely to perform better than a single-step one. Tamburrino et al[1021] described an interesting non-iterative, non-linear algorithm using the concept of a 'resistance matrix' for ERT and showed how it could be modified for MIT, but no illustrations of imaging were presented.

Many biological tissues display significant anisotropy in their electrical properties, skeletal muscle and brain white matter for instance. Image reconstruction may also therefore require tissue anisotropy to be considered, as prior information during forward modelling. Gürsoy and Scharfetter[398] developed an anisotropic forward solver and tensor image reconstruction algorithm and used them to investigate the feasibility of reconstructing anisotropic conductivity distributions. The use of isotropic reference sets – the initial estimate of conductivities used in reconstruction – resulted in significant image artefacts or outright failure to reconstruct anisotropic perturbations. The use of diffusion-tensor MRI as prior information for clinical applications was proposed as a strategy to improve MIT image reconstruction.

Because of the ill-posedness of the inverse problem, many workers have pointed out the certain advantages in introducing as much a priori information as possible in order to constrain the inverse solution, and this can be done in a number of ways. Non-negativity and maximum-conductivity constraints and regularization are all common examples of the use of a priori information, the first two because they disallow physically impossible values of conductivity and the third because it restricts the differences in conductivity between neighbouring voxels in the image to a physically realistic level. It is possible to correlate image voxels with their near neighbours in time as well as in space during image reconstruction. Wei and Soleimani[1132] describe such a 4D image reconstruction method and conclude that correlating multiple 3D data frames can improve both the spatial resolution and stability of reconstructed images as objects move through volumes of varying sensitivity within the MIT system. Another important piece of a priori information is the shape of the outer boundary (the skin surface). This needs to be known with high accuracy because it is where the largest conductivity contrast occurs – between tissue and air (§16.5.4).

A priori information can also be introduced by confining the solution to a certain class of problems or by introducing shape information determined by some other method. Bissesseur and Peyton[117] described non-linear, iterative, image reconstruction customised for their particular application in imaging discrete metal bars (§16.5.1). Casanova et al[181] demonstrated non-linear image reconstruction restricted to cylindrical regions within a larger cylindrical body and solved for their positions, radii and internal permeabilities.

A further approach to improving MIT image reconstruction is to combine MIT measurement with those obtained by other methods. Gürsoy et al[396] modelled a system comprising 16 coils and 16 electrodes operating in EIT, MIT and induced-current EIT (ICEIT) modes. They found that the images reconstructed using the combined data of the three modalities were superior, with a 20% improvement in spatial resolution compared to those reconstructed from MIT or EIT data sets exclusively.

For security applications requiring the rapid detection and classification of objects, reconstructing images may not be the most efficient use of the measured data and, as discussed in §16.5.2, an alternative approach involving the magnetic polarizability tensor is being explored. The "metal detector equation" (16.3) relates the voltage induced in a detector coil (V_{ind}) to the magnetic field produced by a current (I_R) flowing in an exciter coil ($\mathbf{H_E}$) and the field produced by the same current flowing in the detector coil ($\mathbf{H_D}$). The two fields contain the positional information. The MPT ($\mathbf{M}$) is a symmetric, complex, rank-2 tensor that depends on the excitation frequency and the object's size, shape and conductivity.

$$V_{ind} = \frac{j\omega\mu_0}{I_R}\mathbf{H_E} \cdot (\mathbf{MH_D}) \tag{16.3}$$

Equation 16.3 is used to compute a Jacobian and the unknown $\mathbf{M}$. The locations of one or more objects can then be solved as an inverse problem using a suitable inversion algorithm[687]. $\mathbf{M}$ can subsequently be decomposed into its eigenvalues and eigenfunctions. The eigenvalues are rotationally invariant representations of the MPT and are therefore useful in classifying objects[9].

16.7   SPATIAL RESOLUTION, CONDUCTIVITY RESOLUTION AND NOISE

The spatial resolution of an MIT system will depend on the number of independent excitor/sensor combinations. For an array of N transceivers (coil modules functioning as either exciter or sensor) fixed in position, $N(N-1)$ independent measurements will be possible. As these consist of reciprocal pairs, the number of independent measurements will be $N(N-1)/2$. For the 16-transceiver system of Korjenevsky et al[574], the number of independent measurements was 120. As the transceivers were arranged in a ring, producing a two-dimensional image, the theoretical maximum spatial resolution possible was approximately $1/\sqrt{120} \approx 9\%$ of the array diameter. This figure will of course be degraded by noise. Cylindrical objects in a saline bath, with positive and negative conductivity contrast, each with diameter 29% of the array diameter, were clearly resolved but the resolution limit of the system was not given.

For the simulated planar array of Gencer and Tek[337], low-contrast, single-voxel conductivity perturbations (10% of the array width) were clearly reconstructed in the surface layer. The resolution deteriorated to 30–40% of the array width at a depth in the slab of half the array width. A high signal-to-noise ratio of 80 dB was assumed. With industrial MIT systems, Peyton et al[819] and Borges et al[129] report a spatial resolution of 7-15% of the array diameter for metal bars in empty space.

It has been pointed out that since MIT is a contactless method, the array of coils could be shifted by a small amount, for instance by half of one coil spacing, doubling the number of independent measurements with a consequent increase in spatial resolution[337,896,1049]. Indeed, the optically tracked, hand-held coil of[275] enabled a very large number of measurements to be collected rapidly (§16.5.4). Although increasing the number of measurements increases the theoretical maximum spatial resolution, in practice image resolution is restricted by measurement noise and the ill-posed nature of MIT. Some regions within the object being imaged, especially centrally located ones, will display low sensitivity to some or all of the excitor/detector coil combinations, with large conductivity changes producing very little change in the signal at the detectors. Regularization therefore is applied during reconstruction to obtain stable images, resulting in measurement information being lost and spatial resolution reduced. The lower the signal-to-noise ratio, the more severe will be the regularization applied and the greater the image smoothing and loss of spatial resolution. Conductivity image resolution will thus depend on the interplay between measurement noise and the sensitivity across the volume of the specific MIT system.

The volume sensitivity of the MIT system is described by the sensitivity matrix, and it is therefore possible to derive generalised quality measurements of the imaging performance of a specific MIT array design by analysing the sensitivity matrix using singular value decomposition. Merwa and Scharfetter[710] proposed the use of the point spread function (PSF), derived from a sensitivity matrix, as a general quality measurement of image resolution and distortion for MIT. The PSF has been used extensively in microscopy and elsewhere to determine the response of an imaging system to point objects and quantify the degree of blurring or smoothing. This method was subsequently used to compare the performance of different detector coil types[711] and coil orientations[397] on MIT image resolution.

For biomedical MIT, equation 16.2 implies that the uncertainty in conductivity will depend on the phase noise in the system but that a higher noise level can be tolerated at higher frequencies. From numerical simulations Morris et al[730] proposed a phase measurement precision of 3 m° in order to resolve the internal conductivity features in some simple models of biological tissues at 10 MHz, while Zolgharni et al[1220] found that a precision of 1 m° was sufficient to image a large peripherally located cerebral hemorrhage, (§16.5.4). These figures reflect the very high measurement precision required of MIT. A phase difference 3 m° amounts to a time difference of only 1 picosecond. Light travels less than 1 mm in this time!

The phase noise and drift figures for six biomedical MIT systems are collated in table 16.1. The early MIT systems – the Moscow[574] and Cardiff Mk1[1126] – provided very limited phase noise and phase drift performance. A very significant improvement in the quoted precision can be seen in more recent systems; the Cardiff Mk2[1123] the Xi'an[623], the Phillips[703] and the Bath[1132]. The increase in the precision of these systems has been possible due to the continual improvements in commercially available components and instrumentation, allowing more powerful excitor amplifiers, more stable detector amplifiers and faster and more precise measurement electronics.

Table 16.1

Quoted phase noise and drift figures of six biomedical MIT systems developed between 2000–2012. To allow comparison of performance, an adjusted phase-noise figure showing the phase noise for a measurement integration time of 1 ms has been computed for each system, with adjusted phase noise equal to the quoted phase noise divided by the square root of the measurement integration time in ms. The Drift was the typical drift in phase observed over the Drift Time.

Year	MIT System	Operating Frequency (MHz)	Adjusted Phase Noise (m°)	Drift (m°)	Drift Time (min)
2000	Moscow	20	560	–	–
2002	Cardiff MK1	10	164	130	48
2009	Cardiff Mk2	10	7	8	60
2009	Xi'an	5.8	8	6	60
2009	Phillips	10	50	102	30
2012	Bath	10	10	25	300

It is still difficult to judge whether the noise figures achieved by the various hardware designs so far developed will be adequate for biomedical MIT imaging. It is likely that further progress in excitation amplifiers, sensors and measurement system design will provide a phase precision better than 1 m°. For practical clinical application however this level of precision may have to be achieved over extended measurement periods of hours to days, to allow time-differential measurements for continual monitoring applications. Precision will be needed over an extended frequency range to allow frequency-difference measurements and spectroscopy for tissue characterisation.

16.8 PROPAGATION DELAYS

Most theoretical modelling of MIT to date has used a quasi-static approximation to Maxwell's equations thereby neglecting the effects of wave propagation. Propagation delays are probably unimportant when metals and ferromagnetic materials are being imaged because the secondary signals from the eddy currents are large. In biomedical MIT, with its much smaller signals, propagation delays might be significant and will appear at the detecting coil as a phase lag which could be confused directly with the phase lag, ϕ, caused by the eddy current signal (figure 16.1). Using the formula for dipole radiation, the magnitude of the propagation delay has been estimated[373,375,382]. It was concluded that phase changes due to wave propagation were small compared with the total eddy current signals, but that in requiring a higher accuracy of MIT for imaging details of internal structure, these effects might need to be taken into account. This is an area of MIT that appears not to have been extensively investigated.

16.9 SCALING THE SIZE OF AN MIT ARRAY

A number of researchers have proposed reducing the size of an MIT array for imaging in small-animal studies or for performing spectroscopy on samples. Before considering MIT in this regard, it should be noted that in EIT, several groups have constructed miniature arrays of electrodes, some even measuring less than 1mm across, and have successfully used them to image or impedance-map human hairs, very thin wires, glass microspheres, vegetable matter or the growth of cell cultures[386,626,639,850,1193]. In addition, a single-channel array measuring just 3mm across, mounted on the tip of a probe, has been used clinically to investigate the early diagnosis of cancer in the human cervix, esophagus and bladder[368]. One reason that these developments have been successful is that reducing the size of an EIT array tends to increase the signal magnitudes, so an adequate, if not a better, signal-to-noise ratio can be obtained.

An EIT data set normally consists of a set of four-terminal measurements in which a current, I, is injected into a conducting volume via two electrodes in contact with its surface. A potential difference (voltage), V, is measured between another two electrodes sited elsewhere on the surface (figure 16.9a). The ratio, V/I, is the transimpedance and this is an important quantity in achieving a good signal-to-noise ratio as it determines the voltage output for a given current injected. It is easily shown[386] that if the complete 3-dimensional arrangement (electrodes, boundary of volume and its internal distribution of conductivity) is reduced in size by a linear scaling factor, s (where $s < 1$), the transimpedance will increase by a factor, s^{-1} (figure 16.9b).

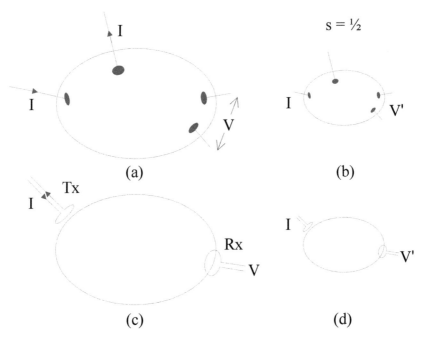

Figure 16.9   Schematic diagram of a 4-terminal measurement for (a) Galvanic coupling with a current, I, injected into a conducting volume via two electrodes causing a potential difference, V, to appear at the other two electrodes; (b) the arrangement in (a) reduced in size by a linear scaling factor, s, causing the potential difference to change to V'. (c) and (d): analogous measurement with magnetic coupling where the current is injected into an excitation coil, Tx, and an e.m.f. is induced in a sensing coil, Rx.

In MIT, the data collected is essentially again a set of transimpedances as current is injected into an excitation coil and the electromotive force (voltage) induced in a sensing coil is measured (figure 16.9c). We assume, as earlier, that the wavelength is long compared to the dimensions of the volume so a quasi-static approximation to Maxwell's equations may be used. We also assume that the skin depth is large compared with the dimensions of the volume.

Consider the scaling in three dimensions so that the conducting volume and the coils are all reduced by the linear scaling factor, s (figure 16.9d). At a superficial glance it might be thought that the transimpedance would scale in exactly the same way as for EIT, but this is not so because the governing equations for EIT and MIT are different special cases of Maxwell's equations.

The scaling of MIT can be illustrated by considering the derivation of equation 16.1 for a thin conducting disk positioned midway between an excitation coil and a sensing coil (figure 16.10) for which an analytical solution is available. There are two transimpedances that need to be calculated: the primary transimpedance due to direct coupling between the coils, and the secondary transimpedance due to the eddy currents induced in the disk.

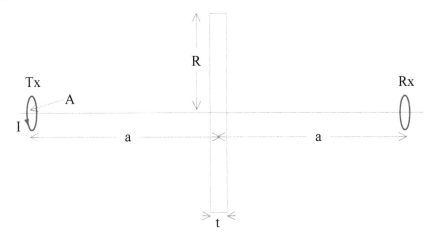

Figure 16.10   Diagram for calculating the e.m.f. induced in a small sensor coil (Rx) by the magnetic field produced by a current, I, flowing in a small excitation coil (Tx). A conducting disc of diameter, R, and thickness, t, is located coaxially and midway between the coils. Each coil consists of a single turn of area, A.

The primary magnetic field at the sensing coil is, $B_0 = \mu_0 m/(16\pi a^3)$, where m is the dipole moment of the excitation coil[385]. The magnetic flux through the sensing coil is $\Phi = B_0 A$ and the induced e.m.f. is $V = -j\omega\Phi$. Since the dipole moment of a planar loop is the product of its area and the current it carries, then $m = IA$ and the primary transimpedance is

$$Z_0 = \frac{V}{I} = -\frac{j\omega\mu_0 A^2}{16\pi a^3}. \tag{16.4}$$

Denoting the dimensions after scaling with the prime symbol, $A' = s^2 A$ and $a' = sa$, so $Z_0' = sZ_0$. The primary transimpedance is thus reduced by the linear scaling factor.

Similarly, from the expression for the secondary magnetic field, it follows that the secondary transimpedance due to the eddy currents in the disk is

$$Z_e = -\frac{\omega\mu_0 A^2 t R^4}{16\pi\delta^2 a^2(a^2 + R^2)^2} \tag{16.5}$$

where $\delta = \sqrt{2/(\omega\mu_0\mu_r\sigma)}$ is the skin depth[385]. The dimensions R and t must also be changed under the scaling such that $t' = st$ and $R' = sR$, so $Z_e' = s^3 Z_e$. The secondary transimpedance due to the eddy currents is reduced by the cube of the linear scaling factor.

The secondary transimpedance is often expressed as a proportion of the primary (equation 16.1). Because Z_e is real and Z_0 is imaginary, this proportion equals a phase shift in the total detected signal, $\phi = |Z_e/Z_0|$. After scaling, $\phi = s^2\phi$, so the phase shift due to the eddy currents in the volume is reduced by the square of the linear scaling factor. In the corresponding term in equation 16.1 the coefficient, P, is similarly reduced after the scaling.

Scharfetter et al[932] extended the derivation of Griffiths et al[385] to include a contribution to the secondary transimpedance due to magnetic polarization of the disk. This contribution is given by

$$Z_m = -\frac{j\omega A^2 \mu_0(\mu_r - 1)tR^2(8a^2 - R^2)}{32\pi(a^2 + R^2)^4} \tag{16.6}$$

After scaling, $Z'_m = sZ_m$, so unlike the eddy-current contribution, that due to magnetic polarization decreases just by the linear scaling factor, s. Because Z_m is imaginary, like Z_0, no phase shift is produced by the magnetic polarization but for convenience we retain the symbol, ϕ, when forming the ratio, $\phi = Z_m/Z_0$. ϕ remains unchanged after the scaling, as does the coefficient, Q, in the corresponding term in equation 16.1.

Table 16.2
Summary of the changes in transimpedance, Z, for a linear scaling factor, s. For MIT, the change in the ratio of secondary to primary transimpedance, ϕ, is also given.

		Z'/Z	ϕ'/ϕ
EIT		s^{-1}	–
MIT	Primary	s	–
	Secondary – eddy currents	s^3	s^2
	Secondary – magnetic polarization	s	1

The effect of scaling on the transimpedances is summarized in table 16.2. Only in EIT does a reduction in size ($s < 1$) result in an increase in transimpedance. In MIT, the transimpedance is always reduced, particularly severely in the eddy-current signal where it falls by a factor, s^3. As a proportion of the primary transimpedance, it falls by a factor, s^2, but as this amounts to a phase shift, phase-detection methods can aid its measurement. The transimpedance due to magnetic polarization falls less severely, by a factor, s, and as a proportion of the primary transimpedance it does not change at all ($\phi'/\phi = 1$). However, it is difficult to measure as it is in phase with the primary transimpedance so phase-detection methods are inapplicable.

These difficulties could be the reason why so few reports of MIT miniaturisation have appeared. Matoorian et al[691] proposed a 200 kHz MIT array with coils only a few millimetres across, for imaging caries in teeth. It is not known if practical results were ever obtained. Recently, Scharfetter and Gursoy[932] modelled magnetic-induction measurements of conductivity changes in the skin, for different levels of hydration, using a coil 60 μm in diameter. They concluded that a SNR of 49 dB might be feasible at an excitation frequency of 1 MHz but no practical results have yet been published.

In order to compensate for a fall in transimpedance upon miniaturisation, the excitation current could be increased but this may cause excessive heating in the reduced volume. Indeed, in some of the EIT studies quoted, the drive current has been reduced with this in mind. The above derivations have also assumed that the sensor in MIT is a coil. Because the sensitivity of a coil is proportional to its area, it will fall by a factor, s^2, upon miniaturisation. This reduction in sensor sensitivity may be addressed to an extent if, in future, different types of magnetic-field sensors with superior sensitivity per unit volume in comparison to coils become available. Wickenbrock et al[1144] have performed magnetic-induction measurements of conducting objects using an atomic magnetometer as a sensor and suggested that these highly sensitive devices have the potential to be miniaturised for use in MIT.

The use of higher excitation currents or more sensitive sensors does not however address the critical issue that the ratio of secondary to primary transimpedance Z_e/Z_0 falls by s^2 on reduction of scale, making signal measurement more practically difficult. One way of compensating for this would be to use a higher frequency since $Z_e/Z_0 \propto \omega$ (equations 16.4, 16.5, 16.6). Frequencies higher than those traditionally used for MIT (≤ 20 MHz) might be practicable because the self-resonance frequency of the coils would rise to an even higher value at the reduced size.

So far, only miniaturisation ($s < 1$) has been considered but the expressions apply equally to en­largement ($s > 1$) provided that the dimensions of the imaging volume remain small compared with the skin depth and the wavelength.

One obvious pitfall of MIT scaling for biomedical applications is in the use of animal models. A number of published papers have investigated MIT and single-channel magnetic induction spectroscopy for neurological applications using rodent and rabbit models. Since the volume of a rodent or rabbit brain is of the order of one hundredth of that of a human brain, the expected MIT signal sizes will be very small indeed and the use of such models may be invalid.

16.10 MULTIFREQUENCY MEASUREMENTS: MAGNETIC INDUCTION SPECTROSCOPY

For biomedical applications there is a strong incentive for developing multi-frequency MIT to match the large body of work in EIT spectroscopy for tissue characterisation and frequency-difference imaging, but without the drawback of having to attach electrodes.

Scharfetter et al[934] performed multifrequency inductive measurements in the band 40–370kHz (§16.5.4) and were able to obtain the conductivity spectrum of a sample of potato (figure 16.11). Later measurements were performed for vegetable material within a conducting saline background at frequencies up to 700kHz[931]. Further details of the hardware are given elsewhere[867,896]. In a subsequent paper[932], this group named the technique 'magnetic induction spectroscopy (MIS)', following the term electrical impedance spectroscopy (EIS) for multi-frequency electrode-based impedance measurement.

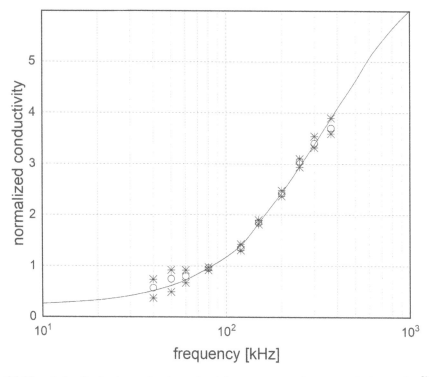

Figure 16.11   Inductively determined conductivity spectrum of potato (circles) (after[934]). The asterisks mark the error limits ±SD. For comparison, the solid line is the spectrum measured with needle electrodes.

More recently Barai et al[80] described the design of a single-channel MIS system with which impedance spectra of samples of potato, cucumber, tomato, banana, porcine liver and yeast suspensions were collected over the frequency range 200kHz–20MHz. The shapes of the MIS spectra were in good agreement with EIS spectra collected from the same samples, with a measurement precision of 0.05 S/m at 200 kHz, increasing to 0.01 S/m at 20 MHz. The authors also used a 'biomass density estimator' derived from the ratio of the MIS-derived conductivities at 400 kHz and 20 MHz to estimate the yeast concentration in a yeast/saline suspension and to track the ripening of bananas (figure 16.12). They concluded that, given the advantages of obtaining electrical impedance data in a non-contacting manner, MIS has considerable potential for bio-industrial process and product quality. Indeed, MIS systems are now being developed for the food industry, for assessing the quality of meat and fruit[169,798].

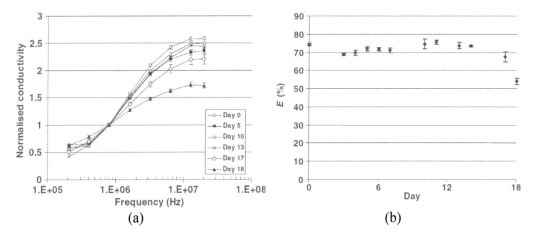

(a)   (b)

Figure 16.12 (a) Inductively determined conductivity spectrum of 5 bananas measured over a 19 day period ([80]). (b) Biomass density estimator, E, versus time. The points show the average for the 5 bananas with error bars showing ± 1 SD.

Electrical impedance spectroscopy has been investigated for detection of cervical intraepithelial neoplasia, potentially providing a 'low-cost' and easily applied alternative to Pap smear tests[10]. The use of electrodes, however, requires good contact with the cervix which might not always be practicable. MIS has been proposed as an alternative where direct electrical contact with the cervix is unnecessary. Wang et al[1114] constructed a single-channel MIS probe which consisted of an axial gradiometer surrounded by a ferrite cylinder and with either an air or a ferrite core. Measurements on saline solutions and in vivo on the human forearm and hand produced impedance spectra with an accuracy of 0.05–0.09 S/m and 0.04–0.28 S/m over the frequency range 50 kHz – 300 kHz, for the air-cored and ferrite-cored probes respectively. The ferrite core restricted the sensitivity to the surface of the sample, a requirement for the cervical screening application, but the thermal stability of the ferrite limited accuracy.

A multi-frequency MIT imaging system has been described by Scharfetter et al[933] following the publication of images reconstructed from measured data collected from agar/saline and potato/saline samples[165,898] (§16.4). Watson[1123] too has described the development of a multi-frequency MIT device.

In EIT the frequencies of interest have been below about 2 MHz, in the range where the β-dispersions of tissues mostly occur and changes due to different physiology and pathology are most pronounced. Useful contrast between different tissue types and states may be available at significantly higher frequencies for some applications, however. Because the signal in MIT/MIS is frequency-dependent, with a greater signal-noise-ratio available at higher frequencies, the optimal frequency range for biomedical applications, such as the detection and monitoring of cerebral edema and hemorrhage, is still to be determined. Operation at much higher frequencies, above 10 MHz and even in excess of 100 MHz, may be advantageous.

Gonzales and Rubinsky[365] carried out a theoretical study of single-channel inductive measurements for the detection of bulk and localised cerebral edema. They concluded that the phase shift of the signal had the potential to detect the formation of edema and hematoma and that the highest sensitivity would be achieved over the frequency range 10–100 MHz. This group subsequently carried out measurements on 54 human subjects, 46 of whom were healthy volunteers, 4 were classified as having cerebral edema, and 4 had hematoma[366]. They found that, using an operating frequency range 26–39 MHz, they could distinguish between healthy subjects and those with brain injury, while frequencies of 153–166 MHz could distinguish between those with edema and those with hematoma. Their method is now undergoing clinical trials in the detection and triage of acute stroke. Early results appear promising[542].

The naming of the technique 'volumetric impedance phase shift (VIPS)' rather than 'magnetic induction spectroscopy' may actually be an accurate reflection of the mode of coupling that was present. The magnitude of the phase shifts observed in phantom measurements[364] appear to be very high in comparison with other published MIT measurements, 20° phase shift being observed for a 350 ml sample of physiological saline at an excitation frequency of 10 MHz. This is over an order of magnitude higher than that observed by other groups[385,574], once scaled for conductivity and excitation frequency. Large phase shifts were also observed using small volumes injected into rats, with 1.4° phase shift observed at 8.5 MHz for just 26.5 ml of physiological saline injected into the abdominal cavity[363]. Such large phase shifts would not be expected from inductive coupling alone, especially given the scaling properties of MIT discussed in §16.9 of this chapter. Using a VIPS system of similar design, a very high phase shift of 13° (frequency not stated) was also observed by Li et al[620] in a rabbit model, for a typical brain volume of less than 15 ml. Both the VIPS systems discussed above appear to use unshielded excitation and detection coils. As discussed in §16.3.2, capacitive coupling can greatly exceed inductive coupling unless rigorous steps are taken to ensure effective screening and ground return paths. It is likely therefore that the VIPS systems employ mixed electric and magnetic coupling. This is possibly dominated by capacitive coupling between the coils, and analysis of such systems should therefore address the complex nature of the coil-to-coil coupling. Nevertheless, it appears from the results published so far that the technique, whatever the precise mode of coupling, might have considerable clinical value.

16.11   IMAGING PERMITTIVITY AND PERMEABILITY

It was demonstrated some years ago that samples of high permeability, such as ferrite, could readily be visualised by MIT (§16.5.1).

For biological tissues, very little work has been carried out in measuring permittivity and permeability as the signals are so small relative to the already-small conductivity signal. However, phase-sensitive detection provides a means of separating the conductivity signal, appearing in the imaginary part, from the permittivity and permeability signals in the real part, provided that hardware errors such as electric-field coupling can be reduced to a sufficiently low level. Researchers are now beginning to attempt such measurements. Because the permittivity signal is proportional to the square of frequency (equation 16.1), larger signals might be expected at high excitation frequencies, but this gain will be offset by the fall in relative permittivity with increasing frequency exhibited by all biological tissues. Measuring at 10 MHz, Watson et al[1130] obtained values for the relative permittivity of a water sample and an average for a human thigh in vivo.

Scharfetter et al[932] evaluated the stability and sensitivity requirements of an inductive sensor for measuring the magnetic susceptibility ($\chi_m = \mu_r - 1$) of liver with a view to detecting hepatic iron overload. The susceptibility of liver tissue is very small and ranges from a normal value of -9×10^{-6} to $+5 \times 10^{-6}$ in overload. The authors calculated that a very narrow receiver bandwidth (<1 Hz) would be needed to achieve the necessary signal-to-noise ratio for such measurements.

They furthermore calculated that for water ($\chi_m \approx 10 \times 10^{-6}$, $\varepsilon_r = 80$), the permittivity contribution to the signal at 50 kHz would be four orders of magnitude lower than that of magnetic susceptibility but that for liver tissue, the much higher relative permittivity ($\approx 10^4$) would reduce this difference. Single-channel practical measurements at 28 kHz (not imaging) combined with modelling were reported and showed that whilst the susceptibility of water could be measured inductively, for human measurements in vivo, the contribution of the permittivity outweighed that of the magnetic susceptibility and made the measurement more difficult[180]. The authors suggested the use of multiple frequencies to improve the accuracy of the measurement by exploiting the different frequency-dependences of the permittivity and permeability terms in equation 16.1.

Imaging permeability might also be applicable, and with somewhat larger signals than from the body's natural magnetism, by using magnetic material as a tracer. Forsman[282] introduced particles of iron oxide (Fe_2O_3) in a meal and was able to observe gastrointestinal motility with a magnetometer detection system (again not imaging). This offers the tantalising possibility of contrast enhancement in biomedical MIT.

16.12   CONCLUSIONS

Since the first MIT systems appeared in the early 2000s, considerable interest in MIT has developed worldwide with many new research groups entering the field. Presentations on MIT now feature regularly at international conferences and there is increasing collaboration and synergy between the fields of process tomography and biomedical imaging. The first commercial MIT systems have begun to appear.

Hardware development has been aided greatly by the advent of new, high-precision, high-frequency electronic devices such as direct digital waveform synthesisers and the latest generation of digital lock-in amplifiers and digitisers. These advances in technology have resulted in an improvement in measurement precision by almost 2 orders of magnitude in the wave of MIT systems produced since 2008, compared with earlier MIT technology (§16.7). The continual improvement in the performance of commercially available RF analog components and digital devices means that further significant increases in MIT speed and precision can be expected.

The number of in-vivo MIT images published to date remains disappointingly small, suggesting that measurement speed and precision are not the only factors limiting the successful practical application of MIT. The ill-posed nature of MIT image reconstruction leads to the same problems of low spatial resolution and image artefacts which have dogged EIT. There have been many developments in image reconstruction algorithms. Three-dimensional numerical modelling using realistic anatomical information is now routine, leading to better solution of the forward problem and computation of the sensitivity matrix. However, the incorporation of a priori information such as boundary shape and the appropriate use of frequency-difference measurements will need to be effectively employed to produce practically useful MIT images. Despite MIT's advantage in having coils fixed in space and in precisely known positions relative to each other, it would appear that the tissue boundary shape and its position relative to the coils needs to be known with millimetre accuracy if significant artefacts on the images are to be avoided (§16.5.4).

The development of a miniature magnetic-induction probe for performing 'virtual tissue biopsies', similar to the electrode-based probe of Gonzalez-Correa et al[368], but instead in a non-electrically-contacting way, is an attractive idea that has been raised over the years. However, the much smaller signals expected as the coil size is reduced (§16.9) mean that this may not be a realistic proposition. Indeed, any reduction in coil size, for instance for small-animal studies, is likely to present extreme SNR difficulties.

Because of the complexity and variability of the human body, and the importance of obtaining information of sufficient accuracy to be useful to clinicians, such as in detecting and monitoring cerebral edema or hemorrhage, the practical medical application of MIT still seems some distance away. One 'low-hanging fruit' which may soon be 'picked' however is bio-industrial MIS. Here, the materials to be measured are often much more homogeneous than in medical applications and are contained within simpler structures, such as cellular suspensions within process or storage tanks. Electrical impedance spectroscopy is known to provide valuable information on biomass concentrations and cellular structures. A method which allows this information to be collected with the sensors physically isolated from the samples, and through electrically insulating containers and packaging such as glass, ceramics and plastics, offers much potential for a wide range of industrial applications. These include food quality control, the monitoring of anaerobic digestion and fermentation processes, bioplastics production, and applications involving extreme environments such as metal and glass production.

Despite the many difficulties discussed above, MIT has made considerable progress in the past 10 years, most notably in detector development. Advances in modelling and instrumentation stimulated by the development of industrial process systems will likely influence and inform biomedical MIT/MIS development and vice versa and there are many research groups working on image reconstruction algorithms. It will be very interesting to watch the development of these areas over the next few years.

17 Electrical Impedance Imaging Using MRI

Oh In Kwon
Department of Mathematics, Konkuk University, Korea

Eung Je Woo
Department of Biomedical Engineering, Kyung Hee University, Korea

CONTENTS

DOI: 10.1201/9780429399886-17

17.1   INTRODUCTION

The electrical conductivity and permittivity are passive material properties that do not generate any signals by themselves[128,958,1165]. When an electrically conducting domain such as the human body is probed by either injecting or inducing electrical current, the conductivity and permittivity distributions inside the domain determine the produced magnetic field as well as electric field distributions following Maxwell's equations. Imaging of the conductivity and permittivity distributions can be, therefore, pursued by measuring not only the electric field or voltage but the magnetic field produced by a probing current. Noninvasive measurements of the magnetic field inside the human body can be done using the magnetic resonance imaging (MRI) technique.

In MRI, three different magnetic fields are applied to the imaging domain: dc main field, ac gradient field, and RF field. An exogenous electrical current at the same frequency of any one of these three magnetic fields produces a magnetic field at that frequency and perturbs the corresponding magnetic field of the MRI scanner. The perturbation in turn produces some changes or artifacts in the acquired MR signals. Extracting the effects of the magnetic field perturbation subject to the exogenous electrical current, images of the conductivity and permittivity distributions inside the domain can be reconstructed based on Maxwell's equations[525,598,948,959].

When a dc current of less than a few mA is externally injected into the imaging domain such as the human body, through a pair of electrodes, it generates a dc magnetic field distribution with a maximum of less than tens of nT inside the domain. The dc magnetic field outside the domain rapidly decays as the field point is moved far away from the domain. The generated weak dc magnetic field inside the domain perturbs the main dc magnetic field of the MRI scanner that is 3 T in many clinical MRI scanners. The weak dc magnetic field inside the domain subject to the externally injected current can be measured from the acquired MR phase images. In magnetic resonance electrical impedance tomography (MREIT), dc currents are injected into the domain between at least two pairs of surface electrodes along orthogonal directions as much as possible. The acquired data of the weak dc magnetic fields are used to reconstruct images of the internal dc conductivity distribution. Injecting currents in a form of pulses, low-frequency conductivity image reconstructions can be conducted at frequencies below a few hundreds Hz[116,955,958,1165].

In MRI, RF coils are used to produce RF magnetic fields inside the domain at the Larmor frequency of about 128 MHz at 3 T. The time-varying RF field induces a distribution of electrical current inside the domain, which is determined by the conductivity and permittivity distributions. The induced RF electrical current generates a secondary RF magnetic field to perturb the primary RF field applied by the RF coils. The same RF coils of the MRI scanner can be used to measure the magnetic field perturbation using the B1 mapping technique. In magnetic resonance electrical properties tomography (MREPT), images of the conductivity and permittivity distributions at the Larmor frequency are reconstructed from the acquired B1 map data[395,540,1093,1094].

During MRI scans, gradient fields in the audio frequency range are applied to the imaging domain for position encoding. The internal currents induced by the time-varying gradient fields are also affected by the conductivity and permittivity distributions. The secondary magnetic fields subject to the induced currents perturb the applied gradient fields. In addition, if external currents at the same frequencies of the applied gradient fields are injected, the magnetic fields produced by the injected currents should also perturb the applied gradient fields. In theory, therefore, the gradient field can be utilized for conductivity and permittivity imaging. However, conductivity and/or permittivity imaging using the gradient field has not been successfully demonstrated in practice and this topic is excluded in this chapter.

The electrical conductivity of an electrolyte is determined by concentrations and mobilities of charge carriers such as ions. Since water molecules and ions coexist in the same molecular environment in the electrolyte, there is a relation between ion mobility and diffusivity of water molecules. In a biological tissue including cells, extracellular fluid and extracellular matrix materials, both water diffusivity and ion mobility may become anisotropic. Noting that the MRI scanner can measure water diffusion coefficients inside the imaging domain along many selected directions, conductivity tensor image reconstructions can be attempted using the relation between water diffusivity and ion mobility with or without relying on Maxwell's equations. In these approaches, the conductivity tensor is assumed to be proportional to the water diffusion tensor. Once the water diffusion tensor image is obtained using the existing diffusion-weighted imaging (DWI) methods, the scale factor between two tensors could be determined for each pixel.

In diffusion tensor MREIT (DT-MREIT), the position-dependent scale factors between the water diffusion and conductivity tensors are determined from the measured internal magnetic flux density distributions subject to externally injected low-frequency currents[599,600]. The DT-MREIT technique requires two separate scans of MREIT and diffusion tensor imaging (DTI) and its image reconstruction algorithms are based on Maxwell's equations. DT-MREIT is capable of producing anisotropic low-frequency conductivity tensor images while MREIT is not practically feasible for recovering conductivity tensor images using a conventional MRI scanner.

Conductivity tensor imaging (CTI) is an electrodeless technique to produce low-frequency conductivity tensor images without requiring external current injections at low frequencies[539,917]. In CTI, the high-frequency conductivity is modelled as a weighted sum of the extracellular and intracellular conductivity values and the low-frequency conductivity is regarded as a property of the extracellular space only. The CTI technique adopts the MREPT method to obtain high-frequency isotropic conductivity images. The multi-b-value DWI method is then utilized to distinguish the influences from the extracellular and intracellular spaces. Instead of relying on Maxwell's equations except for the MREPT part, the CTI technique is based on the following two relations: the relation between the low-frequency and high-frequency conductivity values and the relation between the water diffusion and conductivity tensors. The CTI method can be easily implemented in a clinical MRI scanner and produce low-frequency conductivity tensor images without any additional hardware component.

In §17.2, we will review Maxwell's equations in an electrically conducting domain to summarize the relations between the material properties of conductivity and permittivity and the magnetic field. This will lead to introducing the governing equations of MREIT and MREPT. Measurement techniques, image reconstruction algorithms and potential clinical applications of MREIT and MREPT will be described in §17.3 and §17.4, respectively. In §17.5, the relation between the conductivity and the water diffusivity and the relation between the high-frequency and low-frequency conductivity values will be investigated to derive a basic formula describing the relation between the water diffusion and conductivity tensors. The required assumptions in the formula will be discussed with the limitations associated with the assumptions. Measurement techniques, image reconstruction algorithms and potential clinical applications of DT-MREIT and CTI will be described in §17.6 and §17.7, respectively.

The impedance imaging techniques using MRI can produce conductivity and/or permittivity images with a high spatial resolution of about 1 mm pixel size or less. The temporal resolution is, however, much slower than that of EIT since typical imaging times using an MRTI scanner could be a few minutes to tens of minutes. In the MRI-based impedance imaging methods described in this chapter, there are numerous research and development opportunities from pulse sequence designs and image reconstruction algorithms to experimental validations and clinical applications. With more successful clinical studies addressing specific unmet needs in diagnostic imaging, treatment planning and forward modelling of bioelectromagnetic phenomena, some of these methods could be adopted for routine clinical uses in the future.

17.2   CONDUCTIVITY, PERMITTIVITY, AND MAXWELL'S EQUATIONS

We denote the admittivity τ in an electrically conducting domain Ω as $\tau = \sigma + i\omega\varepsilon$ where σ and ε are the conductivity and permittivity, respectively, at the angular frequency ω. The magnetic permeability of a biological tissue is close to that of the free space μ_0 and assumed to be the constant $\mu_0 = 4\pi \times 10^{-7}$ N/A^2. We assume that there exist electric and magnetic fields inside the domain Ω by externally injecting dc or ac current or applying ac magnetic field. Distributions of σ and ε in Ω determine the distributions of voltage u, current density $\mathbf{J}$, electric field $\mathbf{E}$ and magnetic field $\mathbf{H}$ inside Ω. The relations among them are expressed by Maxwell's equations.

This section follows the derivations described in Seo et al[955]. Time-harmonic Maxwell's equations in Ω with no net free charge can be expressed as

$$\nabla \cdot \mathbf{E} = 0, \ \ \nabla \cdot (\mu_0 \mathbf{H}) = 0, \ \ \nabla \times \mathbf{E} = -i\omega\mu_0 \mathbf{H} \ \text{ and } \ \nabla \times \mathbf{H} = \tau \mathbf{E}. \tag{17.1}$$

Taking the curl operation to $\nabla \times \mathbf{H} = \tau \mathbf{E}$, we get

$$\nabla \times \nabla \times \mathbf{H} = \nabla \times (\tau \mathbf{E}) = \nabla\tau \times \mathbf{E} + \tau \nabla \times \mathbf{E} = \nabla\tau \times \mathbf{E} - i\omega\mu_0\tau\mathbf{H}. \tag{17.2}$$

From the vector identity $\nabla \times \nabla \times \mathbf{H} = -\nabla^2\mathbf{H} + \nabla\nabla \cdot \mathbf{H}$ and $\nabla \cdot \mathbf{H} = 0$,

$$-\nabla^2\mathbf{H} = \nabla\tau \times \mathbf{E} - i\omega\mu_0\tau\mathbf{H}. \tag{17.3}$$

The relation in (17.3) can be rewritten as

$$-\nabla^2\mathbf{H} = \frac{\nabla\tau}{\tau} \times (\nabla \times \mathbf{H}) - i\omega\mu_0\tau\mathbf{H} \tag{17.4}$$

which is the governing equation of MREIT and MREPT.

Since $\nabla \cdot \mathbf{H} = 0$, a magnetic vector potential $\mathbf{A}$ can be introduced as

$$\mu_0 \mathbf{H} = \nabla \times \mathbf{A} \tag{17.5}$$

and $\nabla \cdot \mathbf{A} = 0$ from the Coulomb gauge. Using Faraday's law, $\nabla \times (\mathbf{E} + i\omega \mathbf{A}) = 0$, we can define a scalar potential u satisfying the following relation:

$$-\nabla u = \mathbf{E} + i\omega \mathbf{A}. \tag{17.6}$$

Multiplying both sides of (17.6) by τ, the total current density $\mathbf{J}_t$ is expressed as

$$\mathbf{J}_t = -\tau \nabla u = \tau \mathbf{E} + i\omega \tau \mathbf{A} = \mathbf{J} + \mathbf{J}_e \tag{17.7}$$

where $\mathbf{J}_e$ is the eddy current. Since $\nabla \cdot \mathbf{J} = \nabla \cdot (\tau \mathbf{E}) = 0$ in Ω and $\nabla \cdot \mathbf{A} = 0$,

$$\nabla \cdot (\tau \nabla u) = -i\omega \nabla \cdot (\tau \mathbf{A}) = -i\omega \nabla \tau \cdot \mathbf{A}. \tag{17.8}$$

When the Faraday induction is negligible with $\mathbf{A} \approx \mathbf{0}$ at frequencies below 1 MHz, for example,

$$\nabla \cdot (\tau \nabla u) = 0 \tag{17.9}$$

which is the governing equation of EIT.

17.3 MAGNETIC RESONANCE ELECTRICAL IMPEDANCE TOMOGRAPHY (MREIT)

MREIT is an isotropic low-frequency conductivity imaging method using an MRI scanner with an added constant current source. As the first MRI-based conductivity imaging method, it triggered the ongoing research and development of high-resolution static imaging of conductivity distributions inside an electrically conducting object such as the human body. There are important clinical applications where external current injections are used for the purpose of disease treatments (transcranial direct current stimulation (tDCS), transcranial alternating current stimulation (tACS), deep brain stimulation (DBS), electroporation, etc.). MREIT technique is a valuable tool to directly visualize the internal low-frequency variables (conductivity, current density, electric field, etc).

In its early stage of technology development, a breakthrough was needed to remove the impractical requirement of rotating the imaging object inside the MRI scanner to measure all three components of the induced magnetic flux density $\mathbf{B} = (B_x, B_y, B_z)$ [959]. After the invention of the harmonic B_z algorithm described in this section, equivalent-isotropic low-frequency conductivity image reconstructions became possible using measured B_z data only – without rotating the imaging object. Recent progress in MREIT has focused on improving the image quality using smaller injection currents [260,330,331,357,496,915]. More clinical studies are needed to show its clinical usefulness as a tool for diagnostic imaging and treatment planning [600,914,915,918,1164].

17.3.1 GOVERNING EQUATIONS IN MREIT

We review the governing equations of MREIT in Box 17.13. Note that there is a complicated nonlinear relation between $\mathbf{H}$ and σ in (17.10) since $\mathbf{J}$ and u are nonlinear functions of σ as expressed in (17.11). Several MREIT image reconstruction algorithms will be introduced to recover σ from measured data of B_z using the governing equations in (17.10) and (17.11).

Box 17.13: Governing equations in MREIT

A dc or low-frequency current is injected into an imaging object Ω with its boundary $\partial\Omega$ between a pair of surface electrodes. Assume that $\frac{\sigma}{\omega\varepsilon} \ll 1$ and the Faraday induction is negligible.

- The magnetic field **H** and the conduction current density $\mathbf{J} = \sigma\mathbf{E}$ satisfy the following equation:

$$\nabla^2\mathbf{H} = -\frac{\nabla\sigma}{\sigma} \times (\nabla \times \mathbf{H}) = -\frac{\nabla\sigma}{\sigma} \times (\sigma\mathbf{E}) = -\frac{\nabla\sigma}{\sigma} \times \mathbf{J} \qquad (17.10)$$

where σ is the low-frequency conductivity and **E** is the electric field.
- The conduction current density **J** in Ω subject to the externally injected current is $\mathbf{J} = \sigma\mathbf{E} = -\sigma\nabla u$ and satisfies

$$\begin{cases} \nabla \cdot (\sigma\nabla u) & = & 0 & \text{in } \Omega \\ -\sigma\nabla u \cdot \mathbf{n} & = & g & \text{on } \partial\Omega \end{cases} \qquad (17.11)$$

where u is the voltage, **n** is the outward unit normal vector on $\partial\Omega$ and g is the current density on $\partial\Omega$ subject to the externally injected current.

17.3.2   MEASUREMENT TECHNIQUES IN MREIT

In this section, practical techniques to measure the magnetic flux density $\mathbf{B} = \mu_0\mathbf{H}$ will be described under the constraint of measuring only one component B_z of $\mathbf{B} = (B_x, B_y, B_z)$ using an MRI scanner with its main dc magnetic field in the z direction. The current-injection MRI technique was originally developed for magnetic resonance current density imaging (MRCDI)[525,948]. When a dc current is injected into the imaging domain Ω between a pair of surface electrodes, it produces distributions of current density $\mathbf{J} = (J_x, J_y, J_z)$ and magnetic flux density $\mathbf{B} = (B_x, B_y, B_z)$ in Ω. Since the z component B_z is aligned with the main dc magnetic field of the MRI scanner, it shifts the precession frequency of the magnetization. The amount of the frequency shift at every pixel can be extracted from the acquired MR phase image. In MRCDI, the current injection and MR signal acquisition are repeated by rotating the imaging subject twice to measure B_x and B_y in addition to B_z. Once all three components of **B** are measured, the current density can be computed as $\mathbf{J} = \frac{1}{\mu_0}\nabla \times \mathbf{B}$. In this section, however, only B_z will be measured to avoid the impractical subject rotations inside the MRI scanner.

Figure 17.1(a) shows the basic pulse sequence for the current-injection MRI technique in MRCDI and MREIT. Currents $I^{\pm}$ with the same amplitude and opposite polarities are sequentially injected into the domain to remove systematic phase artifacts of the MRI scanner. The width of the current pulses T_c is from the end of the first RF pulse to the beginning of the readout gradient. The induced magnetic flux densities $\pm B_z$ subject to $I^{\pm}$, respectively, cause additional dephasing of the spins and consequently extra phases are accumulated during T_c. Due to the spatially independent phase shifts using the pulse sequence, the corresponding k-space signals $S^{\pm}(k_x, k_y)$ can be represented as

$$S^{\pm}(k_x, k_y) = \int_{\mathbb{R}^2} \rho(x,y)e^{i\delta(x,y)}e^{\pm i\gamma T_c B_z(x,y)}e^{i2\pi(k_x x + k_y y)}dxdy \qquad (17.12)$$

where $\rho(x,y)$ is the proton density, $\delta(x,y)$ is the systematic phase artifact due to main field inhomogeneity, for example, and $\gamma = 26.75 \times 10^7 \text{rad/T} \cdot \text{s}$ is the gyromagnetic ratio of the proton[911,948]. Using the notation

$$\rho^{\pm}(x,y) = \rho(x,y)e^{i\delta(x,y)}e^{\pm i\gamma T_c B_z(x,y)}, \qquad (17.13)$$

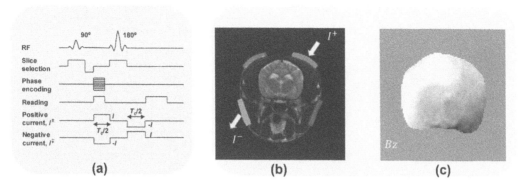

Figure 17.1   (a) Pulse sequence for MREIT. (b) Currents are injected into the imaging domain between a pair of surface electrodes. (c) The induced magnetic flux density B_z is extracted from the acquired MR signals. (color figure available in eBook)

the magnetic flux density B_z can be recovered as

$$B_z(x,y) = \frac{1}{2\gamma T_c} \tan^{-1} \left(\frac{Im(\rho^+/\rho^-)(x,y)}{Re(\rho^+/\rho^-)(x,y)} \right) \tag{17.14}$$

where $Im(\rho^+/\rho^-)$ and $Re(\rho^+/\rho^-)$ are the imaginary and real parts of ρ^+/ρ^-, respectively. From the analysis by Scott $et\ al$[948] and Sadleir $et\ al$[911], the noise standard deviation sd_{B_z} of the measured B_z data is inversely proportional to T_c and the SNR Υ of the MR magnitude image $|\rho^\pm|$ as

$$sd_{B_z} = \frac{1}{2\gamma T_c \Upsilon}. \tag{17.15}$$

Note that it is impossible to increase both T_c and $|\rho^\pm|$ or Υ due to the T_2 or T_2^* decay of MR signals.

The injection current nonlinear encoding (ICNE) method shown in figure 17.2(a) was proposed to extend the duration of the current pulses to the end of the reading gradient[806]. By combining the ICNE method with a multiecho technique, the ICNE-multiecho sequence produces multiple echoes with different phase signals influenced by different durations of the current pulses. The ICNE-multiecho method supplies multiple B_z data with different noise characteristics in each voxel influenced by different durations of the current pulses and effects of the T_2 or T_2^* decay. Figure 17.2(b) shows the ICNE-multiecho pulse sequence based on a gradient echo technique. Using the ICNE-multiecho method with N_E echoes, the acquired data of B_z^i from the ith echo for $i = 1, \cdots, N_E$ have different noise levels for each echo. The measured magnetic flux density data B_z^i can be summed to compute the B_z data with the weighting factors w_i as follows:

$$B_z = \sum_{i=1}^{N_E} w_i B_z^i. \tag{17.16}$$

The optimal weighting factor can be determined at every pixel by minimizing the noise variation of the pixel as

$$w_i(\mathbf{r}) = \frac{T_{ci}^2 e^{-2T_{ci}/T_2^*(\mathbf{r})}}{\sum_{j=1}^{N_E} T_{cj}^2 e^{-2T_{cj}/T_2^*(\mathbf{r})}}, \quad i = 1, \cdots, N_E \tag{17.17}$$

Figure 17.3 shows an example of the MREIT data acquisition from a cylindrical phantom with 13 cm diameter and 16 cm height. To acquire six echoes, the multi-spin-echo ICNE pulse sequence

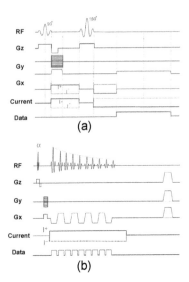

(a)

(b)

MREIT MR pulse sequence	
Sequence	Spoiled multi gradient echo
TR/TE	35/1.88 ms
Number of excitation	10
Echo Space (ES)	2.2 ms
Number of echoes	5-15
Flip angle	6.65
Slice thickness	2-5 mm
Matrix size	128 x 128
Current amplitude	2-10 mA

(c)

Figure 17.2   (a) ICNE pulse sequence. (b) ICNE-multiecho pulse sequence based on the gradient echo technique. (c) Typical pulse sequence parameters.

was adopted. The phantom was filled with a saline of 0.12 S/m conductivity and included two cylindrical agar-gel objects with 1.79 and 1.14 S/m conductivity values. Figure 17.3(a) shows the magnitude images for each echo time T_{Ej} for $j = 1, \cdots, 6$. The same current was repeatedly injected during each echo time. Figure 17.3(b) shows the images of $\left| \nabla B_{zj} \right|$ for $j = 1, \cdots, 6$. Due to the variations in the magnitude images and also current injection pulses, the qualities of the B_z images are different for each echo. Figure 17.3(c) and (d) show the determined weighting factor to optimize the conductivity distribution and the recovered conductivity map using the measured B_{zj} data for $j = 1, \cdots, 6$[556].

17.3.3   IMAGE RECONSTRUCTION ALGORITHMS IN MREIT

In this section, we describe MREIT image reconstruction algorithms based on the governing equations in (17.10) and (17.11). We chose three algorithms: harmonic B_z algorithm[959], projected current density algorithm[806], and direct harmonic B_z algorithm[766]. These three algorithms reconstruct conductivity images pixel by pixel utilizing only B_z data without adopting a global error minimization approach such as the least square error method. Note that there exist other types of MREIT image reconstruction algorithms such as the J-substitution algorithm, sensitivity matrix algorithms and others[765,1108,1165].

17.3.3.1   Harmonic B_z Algorithm

The harmonic B_z algorithm was the first method to produce equivalent-isotropic low-frequency conductivity images using a set of measured B_z data without rotating the subject inside the MRI scanner. Assuming a distribution of isotropic low-frequency conductivity σ in Ω, the following relation between σ and B_z can be derived from the governing equation in (17.10):

$$\mu_0 \left(\frac{\partial \sigma}{\partial x} \frac{\partial u}{\partial y} - \frac{\partial \sigma}{\partial y} \frac{\partial u}{\partial x} \right) = \nabla^2 B_z. \qquad (17.18)$$

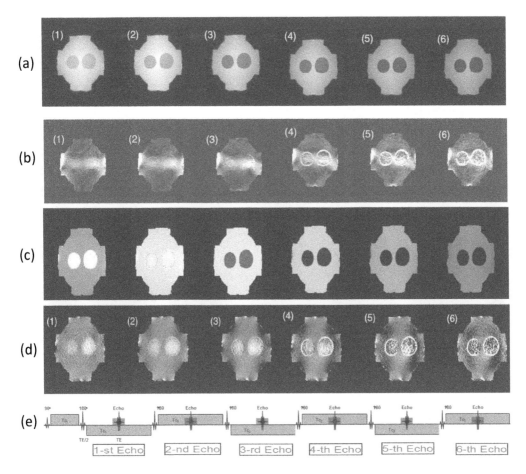

Figure 17.3 Multiecho MREIT phantom experiment. (a) Magnitude images for each echo time T_{E^j} for $j = 1, \cdots, 6$. (b) Images of $\left| \nabla B_{z^j} \right|$ for each echo time T_{E^j} for $j = 1, \cdots, 6$. (c) Estimated weighting factors. (d) Recovered conductivity images for each echo time T_{E^j} for $j = 1, \cdots, 6$. (e) Diagram for the mutiecho spin MR pulse sequence. (color figure available in eBook)

With at least two independent injection currents along orthogonal directions, for example, horizontal and vertical directions, the harmonic B_z algorithm recovers the conductivity σ by iteratively solving the following update formula at the kth iteration:

$$\mathbf{U}^k \mathbf{s}^{k+1} = \mathbf{b} \tag{17.19}$$

where

$$\mathbf{U}^k = \begin{pmatrix} \frac{\partial u_1^k}{\partial y} & -\frac{\partial u_1^k}{\partial x} \\ \frac{\partial u_2^k}{\partial y} & -\frac{\partial u_2^k}{\partial x} \\ \vdots & \vdots \\ \frac{\partial u_N^k}{\partial y} & -\frac{\partial u_N^k}{\partial x} \end{pmatrix}, \quad \mathbf{s}^{k+1} = \begin{pmatrix} \frac{\partial \sigma^{k+1}}{\partial x} \\ \frac{\partial \sigma^{k+1}}{\partial y} \end{pmatrix},$$

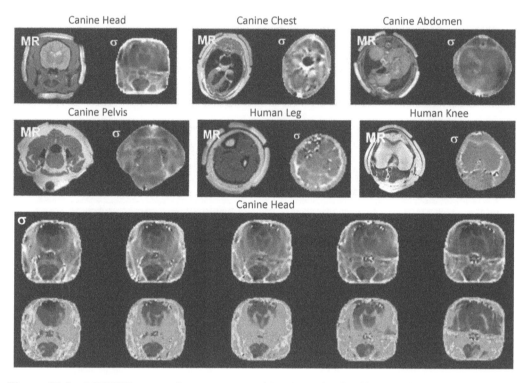

Figure 17.4 MREIT images of postmortem and *in vivo* animal subjects and *in vivo* human subjects. (color figure available in eBook)

and

$$\mathbf{b} = \frac{1}{\mu_0} \begin{pmatrix} \nabla^2 B_{z,1} \\ \nabla^2 B_{z,2} \\ \vdots \\ \nabla^2 B_{z,N} \end{pmatrix}$$

where $N \geq 2$ is the number of injection currents. To solve (17.19), a regularization parameter λ is introduced as

$$\mathbf{s}^{k+1} = \left(\mathbf{U}^{k^T}\mathbf{U}^k + \frac{\lambda}{|\mathbf{U}^{k^T}\mathbf{U}^k|}\mathbf{I}_2 \right)^{-1} \mathbf{U}^{k^T}\mathbf{b} \tag{17.20}$$

where $\mathbf{I}_2$ is the 2×2 identity matrix. To obtain σ^{k+1} from $\tilde{\nabla}\sigma^{k+1} = \mathbf{s}^{k+1}$, the following Poisson equation is solved for σ^{k+1}:

$$\begin{cases} \tilde{\nabla}^2\sigma^{k+1} = \tilde{\nabla} \cdot \mathbf{s}^{k+1} & \text{in} \quad \Omega_t \\ \sigma^{k+1} = \sigma_0 & \text{on} \quad \partial\Omega_t \end{cases} \tag{17.21}$$

where Ω_t denotes an imaging slice and σ_0 is the known conductivity values on its boundary $\partial\Omega_t$. Figure 17.4 shows equivalent-isotropic low-frequency conductivity images reconstructed using the harmonic B_z algorithm from postmortem and *in vivo* animal subjects and *in vivo* human subjects.

17.3.3.2 Projected Current Density Algorithm

Once the equivalent-isotropic low-frequency conductivity images are obtained using the harmonic B_z algorithm, the current density $\mathbf{J} = \sigma\mathbf{E} = -\sigma\nabla u$ can be numerically computed by solving the governing equation in (17.11). In this approach, however, the computed current density is influenced by the isotropic conductivity assumption of the harmonic B_z algorithm. In the projected current density method, current density image reconstructions are conducted directly from measured B_z data without any assumption on the conductivity distribution in the imaging domain.

The projected current density $\mathbf{J}^P$ is an optimal current density that can be recovered from the measured B_z data using the following relation[806]:

$$\mathbf{J}^P = \mathbf{J}^0 + \left(\frac{\partial \psi}{\partial y}, -\frac{\partial \psi}{\partial x}, 0 \right) \tag{17.22}$$

where $\mathbf{J}^0$ is a computed current density from a three-dimensional model of the imaging domain Ω with a presumed homogeneous conductivity distribution subject to the same current injection as in the imaging experiment to measure the B_z data. One may compute $\mathbf{J}^0 = -\nabla \alpha$ by solving the following equation for α:

$$\begin{cases} \nabla^2 \alpha &= 0 & \text{in } \Omega \\ \nabla \alpha \cdot \mathbf{n} &= g & \text{on } \partial \Omega. \end{cases} \tag{17.23}$$

The scalar variable ψ in (17.22) is computed from the measured B_z data by solving the following two-dimensional Poisson equation:

$$\begin{cases} \tilde{\nabla}^2 \psi &= \frac{1}{\mu_0} \nabla^2 B_z & \text{in } \Omega_t \\ \psi &= 0 & \text{on } \partial \Omega_t \end{cases} \tag{17.24}$$

where $\tilde{\nabla}^2$ is the two-dimensional Laplacian and Ω_t is a chosen imaging slice. Since the projected current density method has no restriction in the conductivity distribution, the recovered projected current density $\mathbf{J}^P$ could be utilized in a subsequent isotropic or anisotropic conductivity image reconstruction.

Figure 17.5 shows an example of projected current density image reconstructions from a biological tissues phantom[557]. The ICNE multi-gradient-echo pulse sequence was used to acquire the B_z data. The imaging parameters were as follows: $T_R = 35$ ms, echo spacing $E_s = 3.7$ ms, number of echoes $N_E = 15$, $NEX = 50$, flip angle = 6.65°, $FOV = 260 \times 260$ mm^2 and image matrix = 128×128. Note that the projected current density can be recovered separately for each current injection.

17.3.3.3　Direct Harmonic B_z Algorithm

The projected current density $\mathbf{J}^P$ consists of the background current density $\mathbf{J}^0$ and the curl of a solution of the two-dimensional harmonic equation. By investigating the relation between the projected current $\mathbf{J}^P$ and the conductivity distribution, the direct harmonic B_z algorithm reconstructs the conductivity image from the measured B_z data without iteration.

From (17.18), the current density $\mathbf{J} = -\sigma \nabla u$ satisfies the following relation:

$$\tilde{\nabla} \times (J_x, J_y) = -\tilde{\nabla} \sigma \times \tilde{\nabla} u = \frac{\tilde{\nabla} \sigma}{\sigma} \times (J_x, J_y) = \tilde{\nabla} \log \sigma \times (J_x, J_y) \tag{17.25}$$

where $\tilde{\nabla} \times (J_x, J_y) = \frac{\partial J_y}{\partial x} - \frac{\partial J_x}{\partial y}$. Instead of the current density $\mathbf{J}$ in (17.25), the projected current density $\mathbf{J}^P$ satisfies

$$\tilde{\nabla} \log \sigma \times (J_x^P, J_y^P) \approx -\frac{1}{\mu_0} \nabla^2 B_z. \tag{17.26}$$

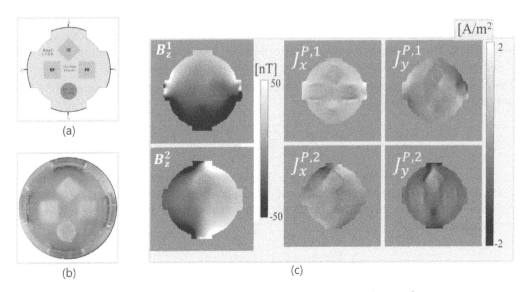

Figure 17.5 (a) and (b) Biological tissue phantom. (c) Measured B_z^1 and B_z^2 data subject to for horizontal and vertical current injections, respectively, and corresponding projected current densities $\mathbf{J}^{P,1}$ and $\mathbf{J}^{P,2}$. (color figure available in eBook)

For independent current injections I_i for $i = 1, \cdots, N$, the relation in (17.26) produces the following linear system of equations[766]:

$$
\begin{pmatrix}
J_{y,1} & -J_{x,1} \\
J_{y,2} & -J_{x,2} \\
\vdots & \vdots \\
J_{y,N} & -J_{x,N}
\end{pmatrix}
\begin{pmatrix}
\frac{\partial \log \sigma}{\partial x} \\
\frac{\partial \log \sigma}{\partial y}
\end{pmatrix}
=
\begin{pmatrix}
\frac{\partial J_{y,1}}{\partial x} - \frac{\partial J_{x,1}}{\partial y} \\
\frac{\partial J_{y,2}}{\partial x} - \frac{\partial J_{x,2}}{\partial y} \\
\vdots \\
\frac{\partial J_{y,N}}{\partial x} - \frac{\partial J_{x,N}}{\partial y}
\end{pmatrix}.
\tag{17.27}
$$

In the direct harmonic B_z algorithm, the computed projected current densities $J_{x,i}^P$ and $J_{y,i}^P$ using the measured $B_{z,i}$ data are used in places of $J_{x,i}$ and $J_{y,i}$, respectively, to reconstruct a conductivity image $\log \sigma$ or σ. This method provides a single-step solution of the iterative method in (17.20) and (17.21).

17.3.4 CLINICAL APPLICATIONS OF MREIT

Though MREIT can be used for diagnostic imaging, the requirement of external current injections along at least two orthogonal directions hinders its clinical acceptance mainly due to the inconvenience of the added hardware to a clinical MRI scanner and also the electrode attachment process during MRI scans. Though the amplitude of injection currents could be reduced to below 1 mA, the external injection current may still stimulate nerves and muscles depending on a chosen imaging area. Since a new electrodeless method of conductivity tensor imaging (CTI) has been lately developed and validated, we restrict the clinical applications of MREIT to such cases where currents are already externally injected for the purpose of disease treatments.

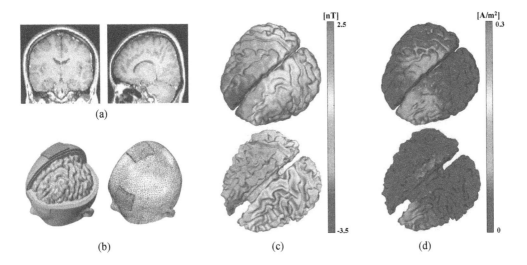

Figure 17.6   Current density imaging during tDCS. (a) Anatomical MR coronal and sagittal images. (b) Three-dimensional head model using anatomical MR images. (d) Projected current density images in the gray and white matter regions, which were recovered using the simulated magnetic flux density B_z in (c). (color figure available in eBook)

17.3.4.1   MREIT for Transcranial Direct Current Stimulation (tDCS)

Transcranial direct current stimulation (tDCS) is a noninvasive neurostimulation method delivering dc currents of 0.5 to 2 mA into the brain through surface electrodes. Clinical applications of tDCS may include enhancements of motor or memory functions and treatments of diseases such as Alzheimer, Parkinson disease, stroke, epilepsy, pain syndrome, obsessive-compulsive disorder (OCD), general anxiety disorder (GAD), post-traumatic stress disorder (PTSD), and schizophrenia[166,591,781,782,1201]. Since the internal current flow caused by an externally injected current is different for each individual, MREIT could be a promising tool to estimate the current density distribution in the brain during tDCS using the following steps:

- Acquire anatomical MR images of a head including its entire brain.
- Construct an anatomically correct model of the head including attached electrodes.
- During tDCS, acquire B_z data subject to tDCS currents. Systematic phase artifacts could be removed using separately acquired data without any injection current.
- Reconstruct projected current density images in the brain using the B_z data.

Figure 17.6 shows an example of current density imaging during tDCS. A realistic three-dimensional head model including attached surface electrodes was constructed in figure 17.6(b) using three-dimensional anatomical MR images. Figure 17.6(a) shows the anatomical MR coronal and sagittal images. The projected current density image of $\mathbf{J}^P$ in (17.22) was reconstructed using the measured B_z data induced by the injected tDCS current. Figure 17.6(d) shows the projected current density images in the gray matter and white matter regions. In this example, the simulated data of B_z in figure 17.6(c) were used.

17.3.4.2 MREIT for Deep Brain Stimulation (DBS)

Deep brain stimulation (DBS) is an effective and increasingly popular treatment method for a variety of movement disorders including dystonia[581,582], tremor[102] and Parkinson disease[101,625]. Currently, DBS is conducted in an open-loop mode without feedback, and the stimulation parameters such as the rate, intensity and duration of stimulating currents and electrode positions are predetermined and fixed without knowing actual current density distributions in the brain.

MREIT may provide a way to visualize the internal current density distribution during a DBS treatment session inside an MRI scanner by measuring B_z data induced by DBS currents. Figure 17.7 shows the results of an animal experiment using a DBS electrode inserted into the brain of a dog. Figure 17.7(a) shows the magnitude images of the canine head on seven imaging slices. The thickness of each slice was 3 mm with no gap between them. The DBS electrode was positioned in the second and third imaging slices. The DBS current was continuously injected in the bipolar and monopolar mode. Figure 17.7(b) shows the projected current density images of J_y^P and J_x^P in the bipolar mode. Note that the current density is localized within the third and fourth imaging slices where the DBS electrode was positioned. Figure 17.7(c) shows the projected current density images of J_y^P and J_x^P in the monopolar mode.

17.3.5 FUTURE WORK IN MREIT

MREIT was the first MRI-based low-frequency conductivity imaging method experimentally validated on various conductivity phantoms, animals and human subjects using conventional clinical MRI scanners. It provides equivalent-isotropic low-frequency conductivity images and projected current density images from measured B_z data without rotating the imaging object inside the MRI scanner. It, however, requires external injection currents to induce measurable data of induced magnetic flux density B_z using the adopted MRI scanner.

The injection current amplitude has been decreased from tens of mA to a value smaller than 1 mA. This was achieved by developing and optimizing current sources, pulse sequences, RF coils and other hardware components as well as novel algorithms for preprocessing, image reconstruction and postprocessing. However, MREIT has not been adopted yet in clinical settings and lacks of clinical studies. The major reasons include the followings:

- A current source must be added to an existing clinical MRI scanner and it should be controlled in a synchronized way with the spectrometer.
- Multiple electrodes with lead wires must be attached around a chosen imaging area inside the MRI scanner.
- The external injection current may stimulate the nerve and muscle depending on a chosen imaging area.
- Anisotropy cannot be properly handled in reconstructed equivalent-isotropic low-frequency conductivity images.

Supported by Maxwell's equations, the basic theory of MREIT is concrete and software tools have been developed to facilitate its clinical application studies[916]. As a diagnostic imaging tool providing new contrast information, the above-mentioned technical hurdles could be removed by continuing efforts in research, clinical studies and commercialization.

For clinical applications using low-frequency current stimulations, MREIT could be a valuable tool to quantitatively visualize internal current density distributions produced by external stimulation currents. The projected current density method could be further developed and optimized for these clinical applications.

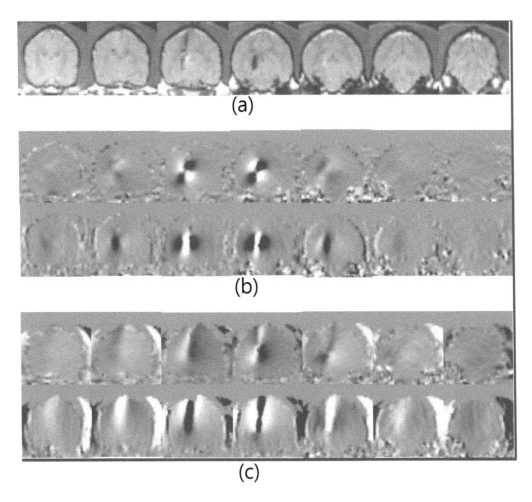

Figure 17.7 Current density imaging during DBS. (a) Magnitude images on seven imaging slices. The slice thickness was 3 mm with no gap between them. (b) Projected current density images of J_y^P and J_x^P in the bipolar mode. (c) Projected current density images of J_y^P and J_x^P in the monopolar mode.

17.4　MAGNETIC RESONANCE ELECTRICAL PROPERTIES TOMOGRAPHY (MREPT)

MREPT produces images of conductivity and permittivity at the Larmor frequency of 128 MHz at 3 T field strength [540,1094]. Using the B1 mapping technique of the MRI scanner, it does not require external current injections. Without using any additional hardware components and electrodes, MREPT can be easily implemented in an existing clinical MRI scanner as a software module.

There have been two technical challenges in MREIT. One is the conventional technical problem of acquiring high-quality B1 maps. As the performance of the clinical MRI scanner has been improved, the quality of B1 maps has been improved. The other is the requirement of the assumption about piecewise constant conductivity and permittivity distributions needed for image reconstructions. Recent progress in image reconstruction algorithms significantly relaxed this requirement to recover spatially-varying conductivity and permittivity images and also edges of almost piecewise constant regions.

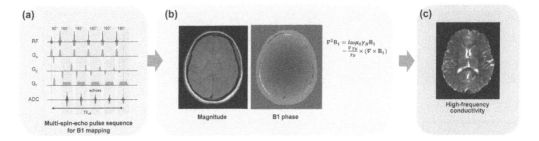

Figure 17.8 (a) Multi-spin-echo MREPT pulse sequence diagram. (b) Measured magnitude and phase images. (c) Reconstructed high-frequency conductivity image. (color figure available in eBook)

At the Larmor frequency of 128 MHz at 3 T, the conductivity of a biological tissue does not provide much structural information since cellular membranes become almost transparent to such high-frequency currents. For the same reason, the tissue anisotropy disappears at high frequencies above 1 MHz. High-frequency conductivity and permittivity images of MREPT, therefore, provide the information about ion concentrations and amount of total fluids in the extracellular and intracellular spaces. Numerous clinical studies of MREPT have been conducted and more are expected in the future to show its clinical usefulness as a new tool for diagnostic imaging. Beyond clinical applications, MREPT could be a useful technical tool to estimate a distribution of specific absorption rate (SAR) inside the human body especially when a high-field MRI scanner is used.

17.4.1 GOVERNING EQUATION IN MREPT

We review the Governing equation in MREPT in Box 17.14.

Box 17.14: Governing Equations in MREPT

An MRI scanner excites the imaging domain Ω using its transmitting RF coils around the Larmor frequency of, for example, 128 MHz at 3 T field strength. The scanner receives RF signals using its receive RF coils and the signals include the effects of an internal current density distribution induced by the RF excitation.

- The magnetic field $\mathbf{H}$ and the admittivity $\tau = \sigma + i\omega\varepsilon$ in Ω satisfy the following equation:

$$\nabla^2\mathbf{H} = -\frac{\nabla\tau}{\tau}\times(\nabla\times\mathbf{H}) + i\omega\mu_0\tau\mathbf{H}. \tag{17.28}$$

The positively rotating field is expressed as $H^+ = \frac{1}{2}(H_x + iH_y)$ in terms of the x and y components of the B1 field of the MRI scanner. For a measurable positively rotating field H^+, the governing equation in (17.28) is expressed as

$$-\nabla^2 H^+ = \frac{1}{2}\left((1,i,0)\times\nabla\times\mathbf{H}\right)\cdot\frac{\nabla\tau}{\tau} - i\omega\mu_0\tau H^+. \tag{17.29}$$

Here, a difficulty arises in dealing with the first term of the right side of (17.29). In the first attempt to recover τ from H^+ measured with a single transceiver RF coil, Katscher *et al*[540] assumed the piecewise constant conductivity and permittivity distribution, that is, the conductivity and permittivity are locally homogeneous. Under this local homogeneity assumption of $\nabla\tau = 0$, the admittivity

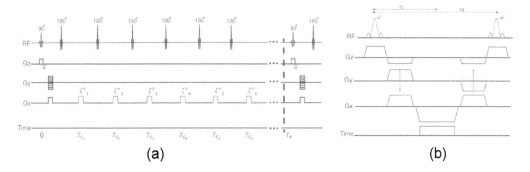

Figure 17.9   Pulse sequences to measure B_1^+ phase maps: (a) Multi-spin-echo pulse sequence and (b) bSSFP pulse sequence.

τ is recovered from (17.29) as

$$\tau(\mathbf{r}) = \sigma(\mathbf{r}) + i\omega\varepsilon(\mathbf{r}) = \frac{\nabla^2 H^+(\mathbf{r})}{iw\mu_0 H^+(\mathbf{r})}. \tag{17.30}$$

Seo *et al*[956] provided a rigorous mathematical error analysis of the reconstruction formula in (17.30). Recently developed algorithms significantly reduced the error caused by the local homogeneity assumption using complementary information from transmit and receive sensitivity distributions of multiple RF coils[960].

17.4.2   MEASUREMENT TECHNIQUES IN MREPT

To reconstruct a conductivity image in MREPT, a conventional spin-echo pulse sequence is commonly used to measure the phase of H^+. As shown in figure 17.9(a), a multi-spin-echo pulse sequence is useful to enhance the weak phase signal and reduce its noise and artifacts. Using a multi-spin-echo pulse sequence, the signals $\tilde{S}^k$ for $k = 1, \cdots, N_E$ are acquired where N_E is the number of echoes. Focusing on the conductivity image reconstruction, we may reduce the amount of noise in the processed transceiver phase $\phi_1^+ + \phi_1^-$. Since the phase noise is inversely proportional to $|\tilde{S}^k|$, the measured phase can be averaged using the following weight for the kth echo:

$$w_k = \frac{|\tilde{S}^k|^2}{\sum_{k=1}^{N_E} |\tilde{S}^k|^2}. \tag{17.31}$$

Since the spin-echo pulse sequence suffers from eddy current effects and requires a long acquisition time, a gradient echo, UTE/ZTE or balanced steady-state free precession (bSSFP) pulse sequences have been also successfully used in experimental studies[342,554,1004]. The gradient echo pulse sequences are advantageous in reducing the scan time but suffer from off-resonance effects. The bSSFP pulse sequence is promising to measure the phase signals since its scan time is short with a high SNR and eddy current effects are automatically compensated. However, the bSSFP pulse sequence should overcome the problems caused by the B_0 inhomogeneity and banding artifacts.

Using the bSSFP pulse sequence, we denote the steady-state total phase offset during each repetition time T_R as $\phi_0(\Delta B, \vec{G}, T_R)$ that is a resonance offset induced by the static and gradient field inhomogeneities, ΔB and $\vec{G}$, respectively. The steady-state transversal components of the magnetization are as follows[800]:

$$M_x = M_0(1-E_1)\frac{E_2\sin\alpha\sin\phi_0(\Delta B,\vec{G},T_R)}{d}$$
$$M_y = M_0(1-E_1)\frac{\sin\alpha(1-E_2\cos\phi_0(\Delta B,\vec{G},T_R))}{d} \tag{17.32}$$

where $E_1 = e^{-T_R/T_1}$ and $E_2 = e^{-T_R/T_2^*}$, T_1 and T_2^* denote the longitudinal and transverse relaxation times, M_0 is the proton density and

$$\begin{aligned}d \;:=\; & (1-E_1\cos\alpha)(1-E_2\cos\phi_0(\Delta B,\vec{G},T_R))\\ & -E_2(E_1-\cos\alpha)(E_2-\cos\phi_0(\Delta B,\vec{G},T_R)).\end{aligned}$$

Using an MRI scanner with multiple receive coils, the complex MR images $\mathbf{S}_l^k$ for $k=1,\cdot,N_C$ from N_C receive coils can be averaged using the complex channel coefficients c_l^k for $k=1,\cdots,N_C$ as

$$\tilde{\mathbf{S}}^k = \Sigma_{l=1}^{N_C} c_l^k \mathbf{S}_l^k. \tag{17.33}$$

Note that $\tilde{\mathscr{B}}_1^- = \Sigma_{l=1}^{N_C} c_l^k \mathscr{B}_{1,l}^- = |\tilde{\mathscr{B}}_1^-|e^{\tilde{\phi}_1^-}$ in (17.33) is the combined received B1 field from all receive channels. To satisfy the condition of $\nabla|\mathscr{B}_1| \approx \mathbf{0}$ as much as possible, we set

$$|\tilde{\mathbf{S}}^k| = V_1 M_0^k \sin\left(V_2\mathbf{a}\left|\mathscr{B}_1^+\right|\right)\left|\tilde{\mathscr{B}}_1^-\right| = 1 \tag{17.34}$$

where V_1 and V_2 are the system dependent constants and M_0^k is a quantity by the proton density, and $\mathbf{a}$ is the nominal flip angle of the RF excitation.

To determine the channel coefficients satisfying (17.34), the following optimization problem can be solved for c_l^k [611]:

$$\hat{c}_l^k = \text{argmin}_{c_l^k} \sum_{\mathbf{r}\in\Omega}\left(\|\sum_{l=1}^{N_C} c_l^k \mathbf{S}_l^k(\mathbf{r})-1\|^2 + \lambda\sum_{l=1}^{N_C}|c_l^k|^2\right) \tag{17.35}$$

where Ω is a region of interest and λ is a regularization parameter. Using $\hat{c}_l^k$ from (17.35), $\tilde{\phi}_1^-$ can be computed as

$$\tilde{\phi}_1^- = \arg\tilde{\mathscr{B}}_1^- = \arg\sum_{l=1}^{N_C}\hat{c}_l^k\mathscr{B}_{1,l}^- \tag{17.36}$$

and $\phi_1^+ + \tilde{\phi}_1^-$ replaces $\phi_1^+ + \phi_{1,l}^-$ to reconstruct the high-frequency conductivity σ.

17.4.3 IMAGE RECONSTRUCTION ALGORITHMS IN MREPT

In a homogeneous region where $\nabla\tau = 0$, the governing equation in (17.28) is simplified to the Helmholtz equation $\nabla^2\mathbf{H} = i\omega\mu_0\tau\mathbf{H}$. The circularly polarized component of the magnetic field, H^+, can be used to compute the admittivity in the homogeneous region based on the following simplified algebraic equation[539,1094]:

$$\tau(\mathbf{r}) = \sigma(\mathbf{r})+i\omega\varepsilon(\mathbf{r}) = \frac{\nabla^2 H^+(\mathbf{r})}{i\omega\mu_0 H^+(\mathbf{r})}. \tag{17.37}$$

At the boundary of the locally homogeneous region, the reconstruction formula in (17.37) fails and produces boundary artifacts[956].

To remove the assumption of the local homogeneity, a new image reconstruction formula based on the convection-reaction equation was derived as follows[395]:

$$\beta^{\pm} \cdot \nabla \ln(\gamma_H) - \nabla^2 \mathscr{B}_1^{\pm} + i\omega\mu_0\gamma_H\mathscr{B}_1^{\pm} = 0 \qquad (17.38)$$

where $\beta^{\pm} = (\frac{\partial \mathscr{B}_1^{\pm}}{\partial x} \mp i\frac{\partial \mathscr{B}_1^{\pm}}{\partial y} + \frac{1}{2}\frac{\partial B_z}{\partial z}, \pm i\frac{\partial \mathscr{B}_1^{\pm}}{\partial x} + i\frac{\partial \mathscr{B}_1^{\pm}}{\partial y} \pm i\frac{1}{2}\frac{\partial B_z}{\partial z}, -\frac{1}{2}\frac{\partial B_z}{\partial x} \mp i\frac{1}{2}\frac{\partial B_z}{\partial y} + \frac{\partial \mathscr{B}_1^{\pm}}{\partial z})$. The phase-based formula becomes

$$\left(\nabla\phi^{tr} \cdot \nabla\left(\frac{1}{\sigma_H}\right)\right) + \frac{\nabla^2\phi^{tr}}{\sigma_H} - 2\omega\mu_0 = 0 \qquad (17.39)$$

where $\phi^{tr} = \phi^+ + \phi^-$. To stabilize the formula in (17.39), an artificial regularization term is added as follows:

$$-c\nabla^2\left(\frac{1}{\sigma_H}\right) + \left(\nabla\phi^{tr} \cdot \nabla\left(\frac{1}{\sigma_H}\right)\right) + \frac{\nabla^2\phi^{tr}}{\sigma_H} = 2\omega\mu_0 \qquad (17.40)$$

where c is a small diffusion coefficient.

17.4.4 CLINICAL APPLICATIONS OF MREPT

Though no clinical MRI scanner is currently equipped with an embedded MREPT software yet, numerous clinical studies using MREPT have been conducted and more are expected in the future. MREPT has a key advantage to MREIT that it does not require external current injections using additional hardware components including a synchronized constant current source and electrodes. Though MREPT is limited to provide only high-frequency conductivity and permittivity images without much structural information, it can be readily used in clinical studies using a conventional clinical MRI scanner. MREPT could become a new tool for diagnostic imaging as its clinical usefulness is further demonstrated from animal studies with various disease models and also from carefully designed clinical studies.

17.4.4.1 Imaging of Conductivity Changes Caused by Radiation Therapy (RT)

Ionizing radiation applied to a biological tissue alters its conductivity by changing concentration of charge carriers and/or causing damages to cells. MREPT could be a useful tool to monitor short-term and also long-term effects of a radiation therapy (RT). In a feasibility study, Park et al[807] acquired T_2-weighted images, diffusion-weighted images and MREPT conductivity images from mouse brains before and after radiation exposure. Figure 17.10 (a) and (c) show MR magnitude images of a mouse brain before and after radiation exposure, respectively. There was no significant morphological change in the anatomical MR and T_2 images of the brain after irradiation. Figure 17.10(b) and (d) show conductivity images of the mouse brain before and after irradiation, respectively. The overall conductivity values of the brain tissues increased immediately after irradiation. The increase in the gray matter conductivity values was relatively higher than those of the white matter region and thalamic region. The MREPT conductivity image showed a significantly higher sensitivity than other MR contrasts to irradiation. Future clinical studies are needed to evaluate the utility of MREPT as a new imaging method to monitor the efficacy of RT during and after the therapy.

17.4.4.2 Conductivity Imaging of Ischemic Stroke

High-frequency conductivity imaging was tried in a mouse model of ischemic stroke. Phase-based MREPT imaging was conducted to assess changes in tissue structure and heterogeneity due to ischemic stroke. Figure 17.11 shows MR magnitude images, before and 3, 6 and 12 hours after induction. The reconstructed conductivity images are also shown in the lower row of figure 17.11. The

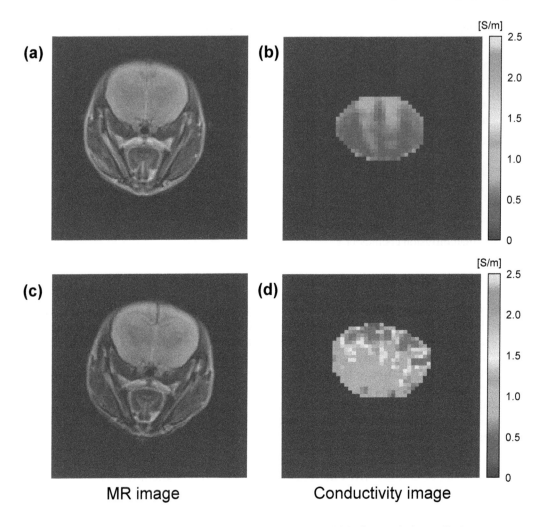

Figure 17.10   MREPT imaging experiment of a mouse model before and after radiation exposure. (a) and (c) are MR magnitude images of a mouse brain before and after irradiation, respectively. (b) and (d) are high-frequency conductivity images before and after irradiation, respectively. (color figure available in eBook)

conductivity values in the ischemic stroke region were found to be higher than those of the normal tissue.

17.4.4.3   Conductivity Imaging of Lower Extremity

Figure 17.12 shows *in vivo* MREPT images of a human lower extremity on four imaging slices. The MR magnitude images in the upper row show the anatomical structure of the human knee and calf. The colour-coded conductivity images in the lower row show different contrast among the bones, muscles, subcutaneous adipose tissues and conductive fluids. The cancellous bone in the knee and the cortical bone in the calf show a quite homogeneous conductivity distribution inside them. The subcutaneous adipose tissues around both the knee and calf show slightly higher conductivity values than the conductivity value of the muscle. This could have stemmed from the conductive fluids and blood inside the adipose tissues in their *in vivo* wet states. The conductivity of the muscle appeared to be smallest among all tissues of the knee and calf. The prominent conductivity contrast

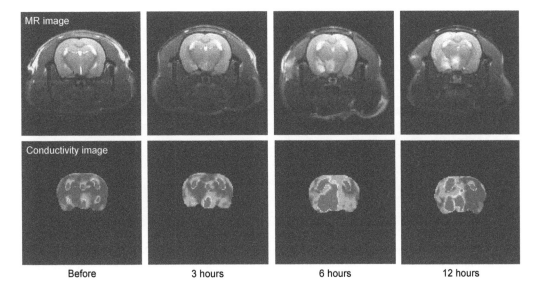

Figure 17.11   MREPT imaging experiment of a mouse model with ischemic stroke. MR magnitude images and recovered high-frequency conductivity images are shown in the upper and lower row, respectively. (color figure available in eBook)

of the synovial fluid in the knee and crural fascia and intermuscular septum in the calf suggests that MREPT conductivity imaging could be used to clearly detect changes the such conductive body fluids.

17.4.5   FUTURE WORK IN MREPT

Most phase-based MREPT methods require the assumption of a piecewise constant conductivity distribution, which produces boundary artifacts in reconstructed high-frequency conductivity images[956]. Though the method proposed by Gurler and Ider[395] is not based on the local homogeneity assumption, it requires other assumptions of $\sigma \ll \omega \varepsilon$, $\nabla |B_1^+| = 0$ and $\nabla |B_1^-| = 0$. The artificial diffusion constant term c in (17.40) is sensitive to the noise level of the B1 phase signals. Future improvements in MREPT image reconstruction algorithms are needed to enhance the quality of high-frequency conductivity images in MREPT. The major goal is either how to deal with the local homogeneity assumption or how to handle the governing equation in (17.28) without relying on the local homogeneity assumption. In terms of the MREPT measurement method, a new multi-channel multi-echo method is expected to improve the quality of measured B1 map data. To validate the clinical usefulness of MREPT, properly designed multi-center clinical studies using clinical 3 T MRI scanners should be conducted for several carefully chosen clinical application areas, especially tumor imaging.

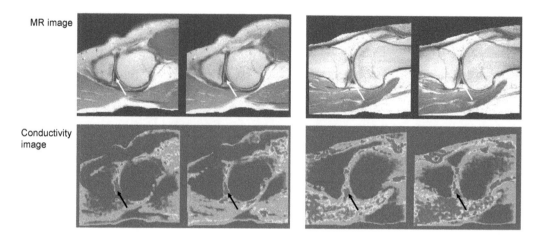

Figure 17.12 MREPT imaging experiment of a human lower extremity. MR magnitude images and colour-coded high-frequency conductivity images are shown in the upper and lower row, respectively. (color figure available in eBook)

17.5 FREQUENCY DEPENDENCE AND DIRECTION DEPENDENCE OF CONDUCTIVITY

MREIT and MREPT provide equivalent-isotropic low-frequency conductivity images and isotropic high-frequency conductivity images, respectively. Most physiological functions occur at low frequencies below 10 kHz, and some biological tissues such as the white matter and muscle exhibit anisotropic properties at low frequencies. It is, therefore, highly desirable to produce anisotropic low-frequency conductivity images of the human body. Before introducing anisotropic low-frequency conductivity image reconstruction methods, we explain the frequency dependence and direction dependence of a tissue conductivity in this section (see also §3.2).

17.5.1 FREQUENCY DEPENDENCE OF CONDUCTIVITY

Previous studies measuring conductivity spectra of various biological tissues show that their conductivity values increase with frequency[320,322,323]. Though the frequency dependence in the conductivity of a biological tissue is related with the dispersion characteristics[683,689], we restrict our description to the frequency dependence of the simplified tissue model shown in figure 17.13. In this model, the cellular membrane is assumed to be a thin insulating membrane with negligible leakage. The thin cellular membrane is transparent to high-frequency currents whereas it blocks dc or low-frequency currents. In figure 17.13(a), therefore, the macroscopic high-frequency conductivity σ_H of the tissue model can be expressed as follows

$$\sigma_H = \alpha\sigma_e + (1-\alpha)\sigma_i \qquad (17.41)$$

where α is the volume fraction of the extracellular space, σ_e is the conductivity of the extracellular space and σ_i is the conductivity of the intracellular space. At dc or low frequencies where displacement currents are negligible, the cell behaves as an insulator as shown in figure 17.13(b). The low-frequency conductivity σ_L is expressed as follows:

$$\sigma_L = \alpha\sigma_e \qquad (17.42)$$

since $\sigma_i = 0$ in effect. Note that $\sigma_H > \sigma_L$ and the typical conductivity spectrum of a biological tissue shows gradually increasing conductivity values from σ_L to σ_H as frequency is increased.

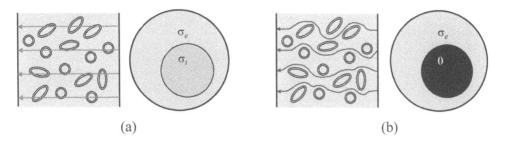

Figure 17.13   Simplified models of a biological tissue: (a) high-frequency model and (b) low-frequency model. (color figure available in eBook)

17.5.2   DIRECTION DEPENDENCE OF CONDUCTIVITY

Some biological tissues such as the white matter and muscle exhibit anisotropy in their conductivity values at low frequencies, and the anisotropy disappears at high frequencies, above approximately 1 MHz. Figure 17.14(a) shows a simplified model of a microscopic environment in a biological tissue including conduction of ions, polarization of molecules such as water and formation of charge double layers across the membranes. Note that the ions inside the cells are restricted whereas the ions outside the cells are hindered in their motions produced by an externally applied electric field.

Figure 17.14(b) shows a macroscopic view of the model in Figure 17.14(a) assuming two different cellular structures. At low frequencies, the ions in the extracellular space cannot penetrate the cell membranes and must move around the cells. When the cellular structure holds a directional property, the low-frequency conductivity exhibits the same directional property or anisotropy. At high frequencies, the anisotropy disappears since motions of the ions are parallel or antiparallel to the applied electric field and the cells are transparent to the resulting high-frequency ac currents.

The anisotropic conductivity of a biological tissue with a complicated cellular structure may have many directional components. In this chapter, we will express the anisotropic conductivity of a biological tissue as a conductivity tensor of 3×3 positive definite matrix. Though the expression of an anisotropic conductivity as a tensor with three directional components could be an over-simplification for a tissue with a complicated cellular structure, it provides a practically useful and mathematically manageable way of handling the tissue anisotropy.

For a particle in a homogeneous electrolyte without cells, we consider the following Einstein relation:

$$m = \frac{1}{k_B T} d \tag{17.43}$$

where m is mobility, d is diffusivity, k_B is the Boltzmann constant and T is the absolute temperature. For a particle surrounded by a cellular structure, we express the relation using tensors as

$$\mathbf{M} = \frac{1}{k_B T} \mathbf{D} \tag{17.44}$$

where $\mathbf{M}$ is the mobility tensor and $\mathbf{D}$ is the diffusion tensor. Since water molecules and ions exist in the same microscopic environment, we assume that they have the same directional property as

$$\mathbf{M}_{ion} \propto \mathbf{M}_{water}, \ \mathbf{D}_{ion} \propto \mathbf{D}_{water} \ \text{and} \ \mathbf{M}_{ion} \propto \mathbf{D}_{water}. \tag{17.45}$$

The above assumption will play a key role in the following section where we derive a relation between the conductivity tensor and water diffusion tensor.

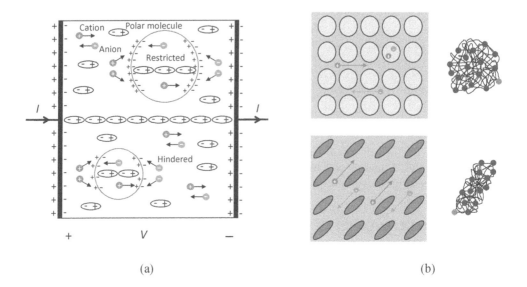

(a)   (b)

Figure 17.14   Simplified models of a microscopic environment in a biological tissue. (a) Conduction of ions, polarization of molecules and formation of charge double layers across membranes. (b) Directional properties of ion conduction as well as diffusion affected by two different cellular structures. (color figure available in eBook)

17.5.3   CONDUCTIVITY TENSOR AND WATER DIFFUSION TENSOR

Conductivity image reconstructions in MREIT and MREPT are based on Maxwell's equations. To overcome the limitations of MREIT and MREPT and produce anisotropic conductivity tensor images, we now pay attention to other physical principles of water diffusion to acquire some additional information. This approach could be practically useful since numerous DWI methods are readily available in clinical MRI scanners. In this section, the relation between ion mobility and water diffusion in (17.45) will be used to establish a connection between the water diffusivity tensor and conductivity tensor.

Tables 17.1 and 17.2 summarize the charge carriers and the symbols of the physical quantities, respectively, considered in this section. The conductivity of a homogeneous electrolyte can be expressed as a sum of the products of concentrations and mobilities of ions[689]. The macroscopic high-frequency conductivity σ_H in (17.41) of a voxel can be expressed as

$$\begin{aligned}
\sigma_H &= \alpha \sigma_e + (1-\alpha)\sigma_i \\
&= \alpha \sum_{j=1}^{M} q A_v z^{(j)} \gamma_e^{(j)} c_e^{(j)} m_e^{(j)} + \\
&\quad (1-\alpha)\sum_{j=1}^{N} q A_v z^{(j)} \gamma_i^{(j)} c_i^{(j)} m_i^{(j)}
\end{aligned} \tag{17.46}$$

where α is the extracellular volume fraction. The mobility of the jth charge carrier can be expressed in terms of the first charge carrier ($j = 1$):

$$m_e^{(j)} = k^{(j)} m_e^{(1)} \quad \text{and} \quad m_i^{(j)} = k^{(j)} m_i^{(1)} \tag{17.47}$$

where $k^{(j)}$ is a constant for each j with $k^{(1)} = 1$. The mobility-weighted effective concentrations $\bar{c}_e$ and $\bar{c}_i$ can be defined as

$$\begin{aligned}
\bar{c}_e &= \sum_{j=1}^{M} q A_v z^{(j)} \gamma_e^{(j)} c_e^{(j)} k^{(j)} \\
\text{and} \quad \bar{c}_i &= \sum_{j=1}^{N} q A_v z^{(j)} \gamma_i^{(j)} c_i^{(j)} k^{(j)}.
\end{aligned} \tag{17.48}$$

The high-frequency conductivity is expressed as follows:

$$\sigma_H = \alpha \bar{c}_e m_e^{(1)} + (1-\alpha)\bar{c}_i m_i^{(1)}. \tag{17.49}$$

From the Einstein relation, we have

$$m_e^{(1)} = \frac{1}{k_B T} d_e^{(1)} \quad \text{and} \quad m_i^{(1)} = \frac{1}{k_B T} d_i^{(1)} \tag{17.50}$$

where k_B is the Boltzmann constant and T is the absolute temperature. The diffusion coefficients $d_e^{(1)}$ and $d_i^{(1)}$ are determined by Stoke's radius of the particle and medium viscosity. Assuming that the reference charge carrier and water molecule exist in the same microscopic environment, we set

$$m_e^{(1)} = K d_e^w \quad \text{and} \quad m_i^{(1)} = K d_i^w \tag{17.51}$$

for a constant K.

We now assume that

$$\bar{c}_i = \beta \bar{c}_e \tag{17.52}$$

where β is a position-dependent constant. Since there is no currently available method to estimate β, we may use a fixed constant for β. For the human brain tissues, $\beta = 0.41$ was estimated by adopting reference values of intracellular and extracellular ion concentrations of four predominant ions including Na^+, Cl^-, K^+ and Ca^{2+} [539,917]. From (17.49) to (17.52), $\bar{c}_e$ is expressed as

$$\bar{c}_e = \frac{1}{K} \left(\frac{\sigma_H}{\alpha d_e^w + (1-\alpha) d_i^w \beta} \right) \tag{17.53}$$

which is independent of the choice of the reference charge carrier. Since low-frequency conduction currents flow only in the extracellular space, the low-frequency conductivity σ_L of a presumed isotropic macroscopic voxel is expressed as

$$\sigma_L = \alpha \bar{c}_e m_e^{(1)} = K \alpha \bar{c}_e d_e^w = \frac{\alpha \sigma_H}{\alpha d_e^w + (1-\alpha) d_i^w \beta} d_e^w = \eta d_e^w \tag{17.54}$$

where η is a position-dependent scale factor.

For an anisotropic macroscopic voxel, $m_e^{(1)}$ in (17.54) could be replaced by $\mathbf{M}_e^{(1)}$ to get the low-frequency conductivity tensor $\mathbf{C}$ as

$$\mathbf{C} = \alpha \bar{c}_e \mathbf{M}_e^{(1)} = K \alpha \bar{c}_e \mathbf{D}_e^w = \frac{\alpha \sigma_H}{\alpha d_e^w + (1-\alpha) d_i^w \beta} \mathbf{D}_e^w = \eta \mathbf{D}_e^w \tag{17.55}$$

where we assume $\mathbf{M}_e^{(1)} = K \mathbf{D}_e^w$. The mean diffusivity values can be estimated as

$$d_e^w = \frac{D_{e,11}^w + D_{e,22}^w + D_{e,33}^w}{3} \quad \text{and} \quad d_i^w = \frac{D_{i,11}^w + D_{i,22}^w + D_{i,33}^w}{3} \tag{17.56}$$

where $d_{e,jj}^w$ and $d_{i,jj}^w$ for $j = 1,2,3$ denote the diagonal components of $\mathbf{D}_e^w$ and $\mathbf{D}_i^w$, respectivey. The formulae in (17.54) and (17.55) provide a new framework for electrodeless low-frequency conductivity imaging using MRI.

Table 17.1

Charge carriers in a biological tissue. P is the last charge carrier to be considered with its net charge of $\pm qz^{(j)}$ C.

Index (j)	1	2	3	4	$\cdots$	M or N
Charge carrier	Na^+	K^+	Cl^-	Ca^{2+}	$\cdots$	$P^{\pm z^{(j)}}$

Table 17.2

Symbols of physical quantities included in the derivation of the CTI formula in (17.55).

Quantity	Symbol	Units
Absolute value of the charge of the electron	q	1.6×10^{-19} C
Avogadro's number	A_v	6.02×10^{23} 1/mol
Extra- and intra-cellular conductivities	σ_e and σ_i	S/m
Extra- and intra-cellular activity coefficients	$\gamma_e^{(j)}$ and $\gamma_i^{(j)}$	$0 < \gamma_e^{(j)}, \gamma_i^{(j)} < 1$
Extrac- and intra-cellular concentrations	$c_e^{(j)}$ and $c_i^{(j)}$	mol/m^3
Extra- and intra-cellular mobility	$m_e^{(j)}$ and $m_i^{(j)}$	m^2/Vs
Extra- and intra-cellular water diffusion coefficients	d_e^w and d_i^w	m^2/s
Extra- and intra-cellular mobility tensors	$\mathbf{M}_e^{(j)}$ and $\mathbf{M}_i^{(j)}$	m^2/Vs
Extra- and intra-cellular water diffusion tensors	$\mathbf{D}_e^w$ and $\mathbf{D}_i^w$	m^2/s
Volume fraction of extracellular space	α	$0 < \alpha < 1$
High-frequency and low-frequency conductivities	σ_H and σ_L	S/m
Conductivity tensor	$\mathbf{C}$	S/m

17.6 DIFFUSION TENSOR MAGNETIC RESONANCE ELECTRICAL IMPEDANCE TOMOGRAPHY (DT-MREIT)

DT-MREIT was suggested to determine the scale factor η in (17.55) using the measurement method of MREIT. In DT-MREIT, the conductivity tensor $\mathbf{C}$ is recovered from a measured diffusion tensor $\mathbf{D}$ using the position-dependent scale factor η. Since DT-MREIT finds the scale factor η using measured B_z data subject to external injection currents, it requires two separate scans of MREIT and DWI.

17.6.1 GOVERNING EQUATIONS IN DT-MREIT

We review the governing equations of MREIT in Box 17.15. There is a complicated nonlinear relation between $\mathbf{H}$ and $\mathbf{C}$ in (17.59) since $\mathbf{J}$ and u are nonlinear functions of $\mathbf{C}$ as expressed in (17.57). Image reconstruction algorithms in DT-MREIT will be described using these governing equations.

Box 17.15: Governing equations in DT-MREIT

A dc or low-frequency current is injected into the imaging domain Ω with its boundary $\partial\Omega$ between a pair of surface electrodes. Assume that $\frac{\sigma}{\omega\varepsilon} \ll 1$ and the Faraday induction is negligible. DWI and MREIT scans are sequentially conducted to acquire diffusion-weighted image data B_z data subject to the external injection current.

- The conduction current density $\mathbf{J}$ in Ω subject to the externally injected current is $\mathbf{J} = -\mathbf{C}\nabla u$ and satisfies

$$\begin{cases} \nabla \cdot (\mathbf{C}\nabla u) &= 0 \quad \text{in} \quad \Omega \\ -\mathbf{C}\nabla u \cdot \mathbf{n} &= g \quad \text{on} \quad \partial\Omega \end{cases} \tag{17.57}$$

where $\mathbf{C}$ is the low-frequency anisotropic conductivity tensor, u is the voltage, $\mathbf{n}$ is the outward unit normal vector on $\partial\Omega$, and g is the current density on $\partial\Omega$ subject to the externally injected current.

- The conductivity tensor $\mathbf{C}$ is expressed as

$$\mathbf{C} = \eta\mathbf{D} \tag{17.58}$$

where η is a position-dependent scale factor and $\mathbf{D}$ is the water diffusion tensor.

- The magnetic field $\mathbf{H}$ and the conduction current density $\mathbf{J}$ satisfy the following equation:

$$\nabla^2 \mathbf{H} = -\nabla \log \eta \times \mathbf{J} = \nabla \log \eta \times (\mathbf{C}\nabla u) \tag{17.59}$$

17.6.2 MEASUREMENT TECHNIQUES IN DT-MREIT

In DT-MREIT, two separate scans of DWI and MREIT are conducted. Since the MREIT data acquisition was described (§17.3), we describe only the DWI data acquisition in this section. Figure 17.15 shows a single-shot spin-echo echo-planar (SS SE-EPI) MR pulse sequence diagram, where strong diffusion-sensitizing gradients (DSGs) are applied on either side of the 180°-pulse. The magnitude intensity is not affected by phases of stationary spins produced by DSGs since any phase accumulation from the first gradient is reversed by the second one. The magnitude intensity, however, decreases because spins diffuse to different locations between the first and second gradients.

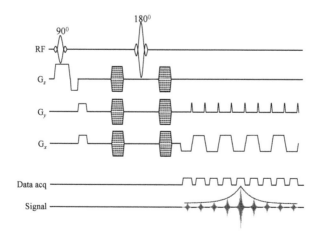

Figure 17.15 Single-shot spin-echo echo-planar (SS SE-EPI) MR pulse sequence diagram for DWI.

The signal intensity ρ in DWI is given by

$$\rho^j = \rho_0 \exp(-b\mathbf{g}_j^T \mathbf{D}\mathbf{g}_j), \quad j = 1, \cdots, N_d$$

where ρ_0 is the signal obtained without any diffusion-sensitizing gradient, $\mathbf{g}_j$ is the jth normalized diffusion-sensitizing gradient vector and b denotes a b-value expressed as

$$b = \gamma^2 \delta^2 G^2 \left(\Delta - \frac{\delta}{3} \right).$$

Here, δ is the duration of the applied gradient, Δ is the duration between the paired gradient pulses and G is the amplitude of the gradient pulse. The signal loss in ρ^j compared with ρ_0 is interpreted as the apparent diffusion coefficient (ADC) along the jth direction due to Brownian motion. The diffusion tensor $\mathbf{D}$ in (17.58) can be obtained from acquired DWI data using more than three DSGs with different directions.

17.6.3 IMAGE RECONSTRUCTION ALGORITHMS IN DT-MREIT

A projected current density $\mathbf{J}^P$ can be directly recovered from the measured B_z data induced by an externally-injected low-frequency current. Then, $\mathbf{J}^P$ reflects the anisotropic conductivity property of a biological tissue.

Using the estimated diffusion tensor $\mathbf{D}$, the projected current $\mathbf{J}_i^P$ satisfies the following relation:

$$\nabla \times (\mathbf{D}^{-1}\mathbf{J}_i^P) = -\nabla \times (\eta \nabla u_i) = -\nabla \eta \times \nabla u_i \tag{17.60}$$

and it implies

$$\nabla \times (\mathbf{D}^{-1}\mathbf{J}_i^P) = -\frac{\nabla \eta}{\eta} \times (\eta \nabla u_i) = \nabla \log \eta \times (\mathbf{D}^{-1}\mathbf{J}_i^P). \tag{17.61}$$

The electric field leads to

$$\mathbf{E}_i = (E_{i,x}, E_{i,y}, E_{i,z}) = \mathbf{D}^{-1}\mathbf{J}_i^P \tag{17.62}$$

and the identity (17.61) can be written as

$$\tilde{\nabla} \times (E_{i,x}, E_{i,y}) = -\tilde{\nabla}\eta \times \tilde{\nabla} u_i = \tilde{\nabla} \log \eta \times (E_{i,x}, E_{i,y}) \tag{17.63}$$

where $E_{i,x}$ and $E_{i,y}$ are the x- and y-components of $\mathbf{E}_i$, respectively. The relation (17.63) can be discretized into the following matrix equation:

$$\mathbf{A}\mathbf{x} = \mathbf{b} \tag{17.64}$$

where

$$\mathbf{A} = \begin{pmatrix} E_{1,x} & E_{1,y} \\ \\ E_{2,x} & E_{2,y} \end{pmatrix}, \qquad \mathbf{b} = \begin{pmatrix} -\tilde{\nabla} \times (E_{1,x}, E_{1,y}) \\ \\ -\tilde{\nabla} \times (E_{2,x}, E_{2,y}) \end{pmatrix} \tag{17.65}$$

and the unknown variable $\mathbf{x} = \left(\frac{\partial \log \eta_{ext}}{\partial x}, \frac{\partial \log \eta_{ext}}{\partial y} \right)^T$. By solving (17.64), $\log \eta$ is recovered by

$$\log \eta(\mathbf{r}) = -\int_{\Omega_t} \tilde{\nabla}\Phi_2(\mathbf{r} - \mathbf{r}') \cdot \tilde{\nabla} \log \eta(\mathbf{r}')d\mathbf{r}' + \int_{\partial\Omega_t} \frac{\partial \Phi_2(\mathbf{r} - \mathbf{r}')}{\partial \mathbf{n}} \log \eta(\mathbf{r}')dl_{\mathbf{r}'} \tag{17.66}$$

where $\Phi_2(\mathbf{r} - \mathbf{r}') = \frac{1}{2\pi} \log |\mathbf{r} - \mathbf{r}'|$ is the two-dimensional fundamental solution of the Laplace equation.

From the estimated scale parameter η and the diffusion tensor $\mathbf{D}$, the low-frequency conductivity tensor map can be approximately recovered as $\mathbf{C} = \eta\mathbf{D}$ in (17.58).

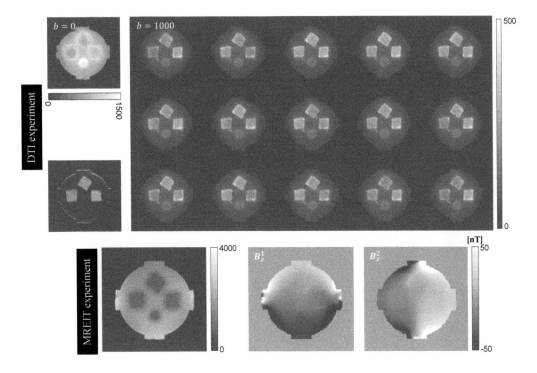

Figure 17.16　Biological phantom experiment of DT-MREIT. For DWI, the number of diffusion gradients was 15 for a b-value of $b = 1000$ s/mm^2. The colour-coded fractional anisotropy (FA) map shows different anisotropic properties of the three pieces of chicken breast. For MREIT, MR magnitude image and B_z^1 and B_z^2 images were acquired. (color figure available in eBook)

17.6.4　DT-MREIT IMAGING EXPERIMENTS

In a phantom experiment of DT-MREIT, DWI data were acquired using the SS SE-EPI pulse sequence. DSGs were applied in 15 directions with b-value of 1000 s/mm^2. The imaging parameters were as follows: repetition time T_R = 3000 ms, echo time T_E = 73 ms, slice thickness = 5 mm, number of excitations (NEX) = 2, field of view (FOV) = 180×180 mm^2 and acquisition matrix size = 64×64. The upper row of figure 17.16 shows the multi-b-value diffusion-weighted images with the applied gradients in 15 directions and the colour-coded FA map. The left chicken breast object shows diffusion along the x-direction (red), the middle-top object shows diffusion along intermediate directions (red and green) and the right one shows diffusion along the y-direction (green). The lower row of figure 17.16 shows the T_2 weighted MR magnitude image and measured magnetic flux density images corresponding to the transversal and vertical injection currents.

Figure 17.17 shows the results of DT-MREIT image reconstructions. The images in figure 17.17(a) are the diffusion tensor images with b = 1000 s/mm^2. In the three chicken breast regions with muscle fibres oriented in the x-, y- and xy-directions, the diffusion tensor map shows different anisotropic diffusion effects in each region. The cylindrical gel region was isotropic. The recovered low-frequency conductivity tensor images using the DT-MREIT image reconstruction algorithm are displayed in figure 17.17(b).

The reconstructed conductivity values of the tissues show their anisotropic characteristics depending on muscle fiber orientations. The conductivity value σ_{11} was higher in the left tissue and σ_{12} was higher in the top-middle tissue. Figure 17.17(c) shows the selected ROIs (ROI-1,$\cdots$, ROI-5).

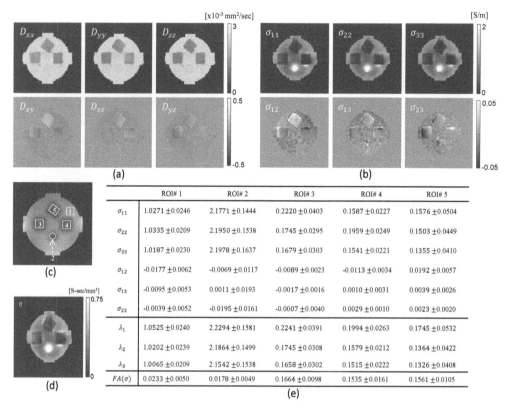

Figure 17.17 Biological phantom experiment of DT-MREIT. (a) and (b) are reconstructed images of water diffusion tensor and conductivity tensor, respectively. (c) and (d) show chosen ROIs and the image of η, respectively. Mean and standard deviation values of the pixels in the ROIs are shown in (e).

Figure 17.17(d) shows the image of the scale factor η. Figure 17.17(e) shows the conductivity values, eigenvalues and FA values of the reconstructed conductivity tensor $\mathbf{C}$ in the ROIs. The maximum eigenvalues from the background saline and agar gel regions were 1.05 and 2.22 S/m, respectively, which were close to the true values of 1 and 2 S/m. For the anisotropic muscle regions, the ratio between the maximum and minimum eigenvalues was 1.38.

DT-MREIT is capable of producing a cross-sectional conductivity tensor image of an anisotropic object such as the human body. It is based on Maxwell's equations and the assumption of $\mathbf{C} = \eta\mathbf{D}$. This method could be useful in clinical applications where electrical currents are injected for therapeutic purposes, e.g., tDCS, DBS and electroporation, where images of anisotropic conductivity, current density and electric field help design and validate such a treatment method.

17.7 CONDUCTIVITY TENSOR IMAGING (CTI)

As explained in §17.5, low-frequency conductivity tensor images can provide information about tissue structure via direction-dependent mobility of a charge carrier in addition to information about its concentration. Although DT-MREIT can produce an image of conductivity tensor distribution, it requires low-frequency current injections through surface electrodes. In both DT-MREIT and MREIT, therefore, we should use an external current source synchronized with a spectrometer and attach multiple electrodes on the subject. For low-frequency conductivity tensor imaging using a conventional clinical MRI scanner without any added hardware component,[917], developed a method of electrodeless low-frequency conductivity tensor imaging based on the following two assumptions.

- Information about the concentration of a charge carrier is embedded in the high-frequency conductivity and the concentration is independent of frequency.
- Mobility of a charge carrier is proportional to that of a water molecule when they exist in a same microscopic environment.

The idea is to obtain information about the concentration using MREPT and information about the direction-dependent mobility using DWI.

17.7.1 GOVERNING EQUATIONS OF CTI

We review the governing equations of MREIT in Box 17.16. Note that the governing equations of CTI include that of MREPT and the relations between conductivity and water diffusivity described in §17.5. Once the high-frequency conductivity σ_H is obtained, the key ingredients of CTI are the extracellular volume fraction α, extracellular water diffusivity d_e^w and intracellular water diffusivity d_i^w that should be extracted from the acquired multi-b-value DWI data. Figure 17.18 shows the CTI process from data acquisitions to image reconstructions.

Box 17.16: Governing equations in CTI

Two separate scans of MREPT and multi-b-value DWI are conducted on a subject Ω placed inside an MRI scanner.

- The high-frequency conductivity σ_H is recovered from the following governing equation of MREPT with $\tau_H = \sigma_H + i\omega\varepsilon_H$:

$$\nabla^2 \mathbf{H} = -\frac{\nabla \tau_H}{\tau_H} \times (\nabla \times \mathbf{H}) + i\omega\mu_0\tau_H\mathbf{H}. \qquad (17.67)$$

- The low-frequency isotropic conductivity is recovered from the following relation:

$$\sigma_L = \frac{\alpha\sigma_H}{\alpha d_e^w + (1-\alpha)d_i^w\beta}d_e^w = \eta d_e^w \qquad (17.68)$$

where α is the extracellular volume fraction, d_e^w is the extracellular water diffusivity, d_i^w is the intracellular water diffusivity, β is the ratio of the ion concentrations in the extracellular and intracellular spaces and η is a position-dependent scale factor.

- The low-frequency anisotropic conductivity tensor is recovered from the following relation:

$$\mathbf{C} = \frac{\alpha\sigma_H}{\alpha d_e^w + (1-\alpha)d_i^w\beta}\mathbf{D}_e^w = \eta\mathbf{D}_e^w \qquad (17.69)$$

where $\mathbf{D}_e^w$ is the extracellular water diffusion tensor.

17.7.2 MEASUREMENT TECHNIQUES

Since CTI is an electrodeless method using a conventional clinical MRI scanner without any added hardware component, its measurement technique deals with usual MRI data collection methods including RF coils and pulse sequences. There could be, however, many different implementations in practice depending on an adopted MRI scanner. In this section, we describe a few examples of

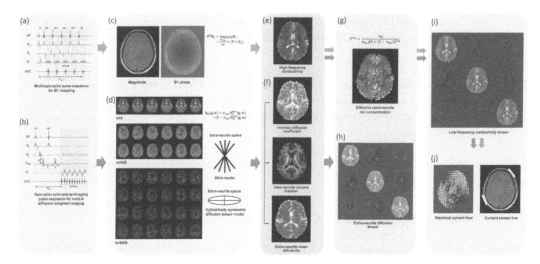

Figure 17.18   CTI image reconstruction:(a) pulse sequence for B1 maps, (b) pulse sequence for multi-b DWI, (c) B1 maps, (d) multi-b DWI data set, (e) σ_H image, (f) images of α, d_e^w and d_i^w, (g) η image, (h) $\mathbf{D}_e^w$ image, (i) $\mathbf{C}$ image, and (j) images of tensor and current stream lines. (color figure available in eBook)

CTI pulse sequences. Since CTI requires two separate MRI scans to sequentially acquire B1 maps and multi-b-value DWI data, a pair of pulse sequences will be included for each example. Note that the primary concern is the quality of the acquired B1 map and multi-b-value DWI data. The choices of b-values and gradient directions in DWI scans are important to successfully extract the extracellular volume fraction α, extracellular water diffusivity d_e^w and intracellular water diffusivity d_i^w.

Figure 17.19 shows the first example of CTI pulse sequences implemented in a 9.4 T research MRI scanner (Agilent Technologies, USA) with a single-channel mouse body coil. The MREPT scan to acquire B1 maps can be conducted using the multi-echo spin-echo sequence in figure 17.19(a) with an isotropic voxel resolution of 0.5 mm. The imaging parameters could be as follows: TR/TE = 2200/22 ms, number of signal acquisitions = 5, field-of-view (FOV) = 65×65 mm², slice thickness = 0.5 mm, flip angle = 90°, and image matrix size = 128×128. Six echoes were acquired with the total scan time of 23 min. The multi-b-value DWI scan needs to be separately performed using the SS SE-EPI sequence in figure 17.19(b). The imaging parameters could be as follows: TR/TE = 2000/70 ms, number of signal acquisitions = 2, FOV = 65×65 mm², slice thickness = 0.5 mm, flip angle = 90°, and image matrix size = 128×128. The number of directions of the diffusion-weighting gradients is 30 with b-values of 50, 150, 300, 500, 700, 1000, 1400, 1800, 2200, 2600, 3000, 3600, 4000, 4500 and 5000 s/mm². With Δ = 53.8 ms and δ = 6 ms, the resulting diffusion time (T_D) is $(\Delta - \delta/3)$ = 51.8 ms.

Figure 17.20 shows the second example of CTI pulse sequences implemented in a 3 T clinical MRI scanner (Magnetom Skyra, Siemens Healthcare, Germany) equipped with a 16-channel head coil. For high-frequency isotropic conductivity image reconstructions using MREPT, the multi-echo spin-echo (MSE) pulse sequence with multiple refocusing pulses is adopted to acquire B1 phase maps. The imaging parameters could be follows: TR/TE = 1500/15 ms, flip angle = 90°, bandwidth = 250 Hz/Px, number of echoes = 6, number of slices = 30, slice thickness = 4 mm, acquisition matrix = 128×128 and FOV = 260×260 mm². The voxel size was 2×2×4 mm³. The imaging time to cover the full brain with the interleaved multi-slice mode was 19 minutes. For accelerated multi-b-value DWI data acquisitions, multi-band echo-planar imaging (EPI) could be conducted with a quadrature transmit coil and a 16-channel receive head coil. A multi-band factor of 3 (MB = 3) was used with

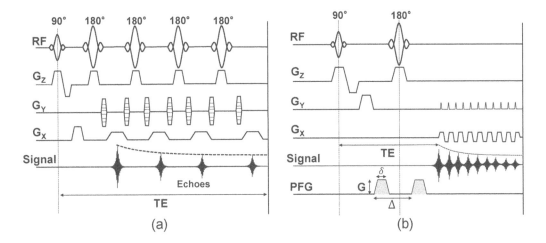

Figure 17.19   CTI pulse sequences on a 9.4 T MRI scanner with a single-channel RF coil. (a) Multi-echo spin-echo MR pulse sequence for B1 mapping. (b) Single-shot spin-echo echo planar imaging sequence for multi-b DWI. (color figure available in eBook)

the three-band RF excitation and axial spin-echo EPI (SE-EPI) readout with phase encoding (PE) in the anterior-posterior (AP) direction. The diffusion gradient was applied in 12 directions with 12 b-values of 0, 50, 150, 300, 500, 700, 1000, 1400, 1800, 2200, 2600 and 3000 s/mm^2. The acquisition parameters were as follows: TR/TE = 2000/80 ms, flip angle = 90°, number of slices = 30, slice thickness = 4 mm, acquisition matrix = 128×128 and FOV = 260×260 mm^2. This multi-b-value DWI scan took 9 minutes.

Figure 17.21 shows an example of the acquired DWI data set using 12 b-values of 0, 50, 100, 300, 600, 800, 1000, 3000, 4000, 5000, 6000, 7000 and 8000 s/mm^2. Note that the signal intensity varies differently depending on a chosen ROI. In the next section, signal intensity at every pixel with respect to the b-values will be analyzed to extract the intermediate variables of α, d_e^w and d_i^w. At small b-values, the acquired DWI data could have been influenced by flow effects. Partial volume effects and flow effects to α, d_e^w and d_i^w need to be investigated. The choices of b-values and gradient directions are important in CTI. One may devise new pulse sequences and adopt better multi-channel RF coils to improve the SNR and reduce artifacts and scan time.

17.7.3   IMAGE RECONSTRUCTION ALGORITHMS IN CTI

Since the image reconstruction algorithm for σ_H is already explained in §17.4, we will describe the method to extract d_e^w, d_i^w and α. After explaining the methods to estimate β and $\mathbf{D}_e^w$, we will show two examples of CTI image reconstructions from a conductivity phantom and human subject.

17.7.3.1   Extraction of d_e^w, d_i^w and α

DWI measures the effects of Brownian motions of water molecules in a voxel to provide quantitative information about the tissue microstructure and its functions[608]. Denoting the jth direction of the gradient as a unit vector $\mathbf{g}_j$, the acquired DWI signal $S_{b,j}$ of a voxel is expressed as

$$S_{b,j} = S_0 e^{-bD_{b,j}} = S_0 e^{-b\mathbf{g}_j^T \mathbf{D}_b \mathbf{g}_j} \tag{17.70}$$

where S_0 is the pixel value without diffusion gradient, $D_{b,j}$ is the diffusion coefficient for the chosen b-value along the $\mathbf{g}_j$ direction and $\mathbf{D}_b$ is the symmetric 3×3 diffusion tensor at the b-value.

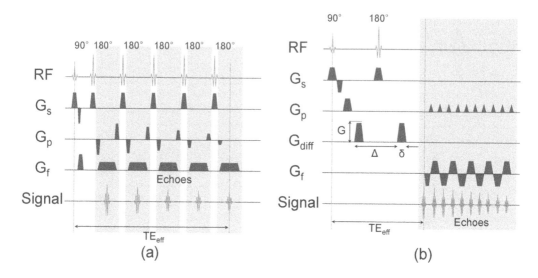

Figure 17.20 CTI pulse sequences on a 3 T clinical MRI scanner with a 16-channel head coil. (a) Multi-echo spin-echo pulse sequence for B1 mapping. (b) Multi-band spin-echo echo-planar imaging sequence for multi-b DWI.

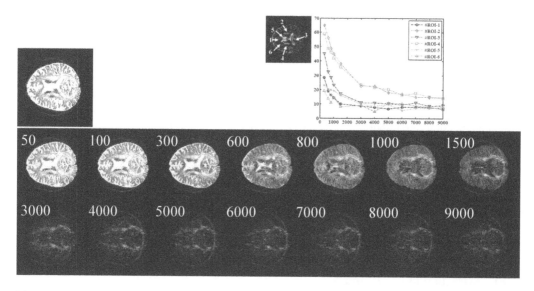

Figure 17.21 Example of a multi-b DWI data set using b-values of 0, 50, 100, 300, 600, 800, 1000, 3000, 4000, 5000, 6000, 7000 and 8000 s/mm^2. Signal intensity variations in six different ROIs are plotted on the upper right panel. (color figure available in eBook)

The diffusion weighted images from all gradient directions for each b-value can be averaged to get the image S_b as

$$S_b = \frac{1}{N_D} \sum_{j=1}^{N_D} S_0 e^{-b \mathbf{g}_j^T \mathbf{D}_b \mathbf{g}_j} \tag{17.71}$$

where N_D is the number of gradient directions. Using the Taylor series expansion, the image S_b exhibits the exponential decay property at each voxel:

$$\begin{aligned} S_b &\approx S_0 \left(1 - \frac{b}{N_D} \sum_{j=1}^{N_D} \mathbf{g}_j^T \mathbf{D}_b \mathbf{g}_j \right) \\ &\approx S_0 e^{-\frac{b}{N_D} \sum_{j=1}^{N_D} \mathbf{g}_j^T \mathbf{D}_b \mathbf{g}_j} = S_0 e^{-b \bar{D}_b} \end{aligned} \tag{17.72}$$

where $\bar{D}_b$ is the average diffusion coefficient of the voxel over all the gradient directions at the chosen b-value.

Since the macroscopic voxel includes both extracellular and intracellular spaces, the plot of S_b at each voxel as a function of b-value can be fitted to a decay curve of multiple exponentials. To estimate d_e^w, d_i^w and α, a three-compartment model can be used. The intracellular space is one compartment consisting of both intracellular water and solid components. It should be noted that the effective mean diffusivity of water molecules in the intracellular compartment is smaller than that of free water diffusion. The extracellular space is subdivided into two compartments: the extracellular water (ECW) compartment of free water molecules and the extracellular matrix (ECM) compartment of a mixture of water molecules and solid components. The data in (17.72) is then fitted to the following equation for each voxel:

$$S_b = S_0 \cdot \left(v_{ecm} e^{-b d_{ecm}^w} + v_{ecw} e^{-b d_{ecw}^w} + v_i e^{-b d_i^w} + v_o \right) \tag{17.73}$$

where v_{ecm}, v_{ecw} and v_i are the volumes of the extracellular matrix, extracellular water and intracellular compartments, respectively, with v_o denoting an offset value. Note that d_{ecm}^w and d_{ecw}^w are the water diffusion coefficients of the extracellular matrix and extracellular water compartments, respectively. For the extracellular isotropic free water compartment, we set $d_{ecw}^w = 3 \times 10^{-3}$ mm²/sec[1203]. The extracellular volume fraction α is estimated at each voxel as

$$\alpha = \frac{v_{ecm} + v_{ecw}}{v_{ecm} + v_{ecw} + v_i}. \tag{17.74}$$

The extracellular diffusion coefficient d_e^w is estimated as

$$d_e^w = \frac{v_{ecm}}{v_{ecm} + v_{ecw}} d_{ecm}^w + \frac{v_{ecw}}{v_{ecm} + v_{ecw}} d_{ecw}^w. \tag{17.75}$$

The method described above could be improved in future studies. As we understand more about water diffusion in a biological tissue, one may devise better methods to extract these intermediate variables. Recently, a model-based approach commonly adopted in nerve tractography ([1203]) was investigated to visualize the low-frequency dominant anisotropic conductivity tensor map in Jahng et al[507].

17.7.3.2   Estimation of $\mathbf{D}_e^w$

From (17.70), we get

$$\frac{1}{b} \ln \left(\frac{S_{b,j}}{S_0} \right) = -\mathbf{g}_j^T \mathbf{D}_b \mathbf{g}_j. \tag{17.76}$$

Since there are six unknowns in the 3×3 symmetric diffusion tensor $\mathbf{D}_b$, we may use at least six gradient directions to determine $\mathbf{D}_b$ for each b-value. This conventional water diffusion tensor

$\mathbf{D}_b$, however, includes the effects of both extracellular and intracellular spaces. To estimate $\mathbf{D}_e^w$ in (17.69), we need to extract the effects of the extracellular space only. Though we may use a model similar to (17.73) for this goal, the problem to extract $\mathbf{D}_e^w$ from a three-compartment model is ill-posed.

We may adopt the following model including 13 unknowns[201,669]:

$$\frac{S_{b,j}}{S_0} = (1 - \xi)e^{-b\mathbf{g}_j^T \mathbf{D}_F^w \mathbf{g}_j} + \xi e^{-b\mathbf{g}_j^T \mathbf{D}_S^w \mathbf{g}_j} \tag{17.77}$$

where $\mathbf{D}_F^w$ and $\mathbf{D}_S^w$ are the fast and slow diffusion tensors, respectively and ξ is a weight factor. Before separating $\mathbf{D}_F^w$ and $\mathbf{D}_S^w$ in (17.77), we first estimate ξ using the averaged signal in (17.72). Noting that the signal from the fast component is negligible at a high b-value, the weight ξ is calculated from (17.72) at two high b-values of b_1 and b_2 as

$$\xi = \frac{1}{S_0}\left(S_{b_1} + b_1 \frac{S_{b_1} - S_{b_2}}{b_2 - b_1}\right) \tag{17.78}$$

where we may set $b_1 = 4,500$ and $b_2 = 5,000$ sec/mm^2, for example. For a high b-value, (17.77) is simplified to

$$S_{b,j} \approx \xi S_0 e^{-b\mathbf{g}_j^T \mathbf{D}_S^w \mathbf{g}_j}. \tag{17.79}$$

The slow diffusion tensor $\mathbf{D}_S^w$ with six unknowns is extracted by fitting the measured data at $b = 4,500$ sec/mm^2, for example, with at least six gradient directions to (17.79) using the weighted least square method. The fast diffusion tensor $\mathbf{D}_F^w$ with six unknowns is then obtained from the measured data at $b = 700$ sec/mm^2 (Tuch *et al* 2001) with at least six gradient directions from

$$S_{b,j} - S_0 \xi e^{-b\mathbf{g}_j^T \mathbf{D}_S^w \mathbf{g}_j} \approx (1 - \xi)S_0 e^{-b\mathbf{g}_j^T \mathbf{D}_F^w \mathbf{g}_j}. \tag{17.80}$$

Assuming that water diffusion in the extracellular space is less restricted and hindered compared to that in the intracellular space, we set the extracellular diffusion tensor as $\mathbf{D}_e^w \approx \mathbf{D}_F^w$[953]. To improve the method to estimate $\mathbf{D}_e^w$ in future studies, one may incorporate a model-based approach in nerve tractography as suggested earlier.

17.7.3.3 Estimation of β

From (17.52), β can be expressed as

$$\beta = \frac{\bar{c}_i}{\bar{c}_e} = \frac{\sum_{j=1}^N z^{(j)}\gamma_i^{(j)}c_i^{(j)}k^{(j)}}{\sum_{j=1}^M z^{(j)}\gamma_e^{(j)}c_e^{(j)}k^{(j)}} \tag{17.81}$$

where $k^{(j)}$ is the mobility ratio between the jth charge carrier and the reference charge carrier ($j = 1$). According to the Stokes-Einstein relation, mobility m of a charge carrier depends on the medium viscosity ς and hydration diameter d_h, that is, $m = \frac{1}{3\pi\varsigma d_h}$. Therefore, $k^{(j)}$ can be expressed as

$$k^{(j)} = \frac{m_e^{(j)}}{m_e^{(1)}} = \frac{m_i^{(j)}}{m_i^{(1)}} = \frac{d_h^{(1)}}{d_h^{(j)}}. \tag{17.82}$$

Substituting (17.82) into (17.81) and assuming that $\gamma_i^{(j)} \approx \gamma_e^{(j)}$, β is expressed as

$$\beta = \frac{\sum_{j=1}^N z^{(j)}\frac{c_i^{(j)}}{d_h^{(j)}}}{\sum_{j=1}^M z^{(j)}\frac{c_e^{(j)}}{d_h^{(j)}}}. \tag{17.83}$$

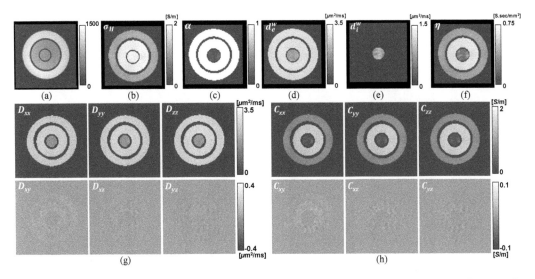

Figure 17.22 Images of a conductivity phantom including a giant vesicle suspension: (a) T_2 weighted image and images of (b) σ_H, (c) α, (d) d_e^w, (e) d_i^w and (f) η. (g) and (h) are images of reconstructed water diffusion tensor and conductivity tensor, respectively.

We may consider four major ions of Na^+, Cl^-, K^+, and Ca^{2+}. We set the values of c_e for Na^+, Cl^-, K^+, and Ca^{2+} as 154, 129, 3.10, and 1.30 mmol/liter, respectively[433]. The values of c_i are 19.67, 3.30, 89.93, and 1×10^{-3} mmol/liter for Na^+, Cl^-, K^+, and Ca^{2+}, respectively (Hansen *et al* 1985). The hydration diameter values are $d_h^{Na^+} = 716$, $d_h^{Cl^-} = 664$, $d_h^{K^+} = 661$, and $d_h^{Ca^{2+}} = 824$ pm[1095]. The value of β is then calculated as 0.41. Since there is no method to estimate β at every voxel to the best of our knowledge, this value may be used for all voxels.

17.7.4 CTI IMAGING EXPERIMENTS

17.7.4.1 Conductivity Phantom with Giant Vesicle Suspension

Experimental validation of the CTI method requires a conductive phantom containing cell-like materials with thin insulating membranes. We may use electrolytes with known electrical conductivity values outside and inside the cell-like materials for error estimation of a reconstructed conductivity tensor image. Employing giant vesicles as cell-like materials, CTI experiments of a conductivity phantom including a giant vesicle suspension were conducted using a 9.4 T MRI scanner with a spatial resolution of 0.5 mm[539].

Figure 17.22(a) and (b) show a conductivity phantom including three compartments of two different electrolytes and giant vesicle suspension. Using the pulse sequences in figure 17.19, CTI images of the phantom in figure 17.22(c)–(f) were reconstructed. The recovered values of the high-frequency conductivity σ_H and low-frequency conductivity C_{xx}, C_{yy} and C_{zz} were in good agreement with the values measured using an impedance analyzer.

17.7.4.2 *In Vivo* Human Brain

Conductivity tensor imaging of *in vivo* human brain was conducted using a 3 T clinical MRI scanner with the pulse sequences shown in figure 17.20. To stabilize the CTI for human brain, various multi-compartment models can be adopted to stably determine the microstructural parameters such as d_e^w, d_i^w and α. The recently proposed multi-compartment microscopic diffusion imaging based on the

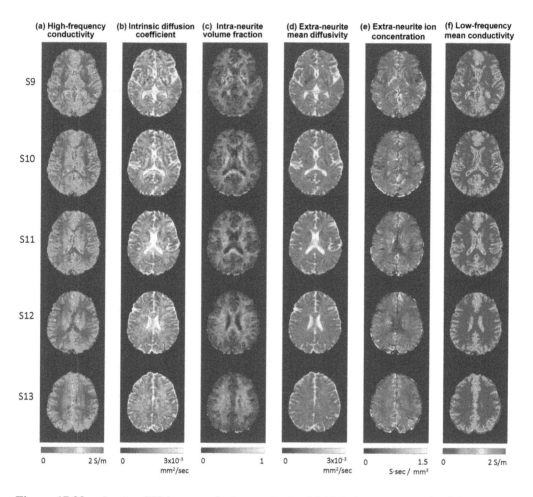

Figure 17.23 *In vivo* CTI images of a human brain: (a) high-frequency conductivity (σ_H), (b) intrinsic diffusivity, (c) intracellular volume fraction (α), (d) extracellular mean diffusivity (d_e^w), (e) effective extracellular ion concentration ($\bar{c}_e$) and (f) low-frequency mean conductivity (σ_L) in five different imaging slices of S9 to S13. (color figure available in eBook)

spherical mean technique (MC-SMT), which is a model to estimate the parameters, d_e^w, d_i^w and α, does not assume the neurite orientation distribution and estimates the diffusion coefficient and volume fraction in the intra- and extra-neurite spaces[528]. In the MC-SMT, the brain area is separated into intra and extra-neurite compartments surrounded by single or double lipid layer membranes and it is able to overcome several limitations of the three pool model in (17.73).

Figure 17.23 shows reconstructed CTI images of a human head using MC-SMT technique. Recovered values of σ_H and **C** were in good agreement with the values measured from extracted tissue samples[539]. Comparing to the high-frequency conductivity image, the low-frequency conductivity image with smaller values reflects the ion concentrations and direction-dependent mobilities in the extracellular space.

Figure 17.23(a) shows the reconstructed high-frequency conductivity images, σ_H, from the acquired transceiver phases of the B1 maps. Figure 17.23(b-e) show the intrinsic diffusion coefficients (λ), the intracellular volume fraction (v_{int}), the extracellular mean diffusivity ($\bar{\lambda}^{ext}$), and the reconstructed extracellular ion concentration distribution at each imaging slice, respectively.

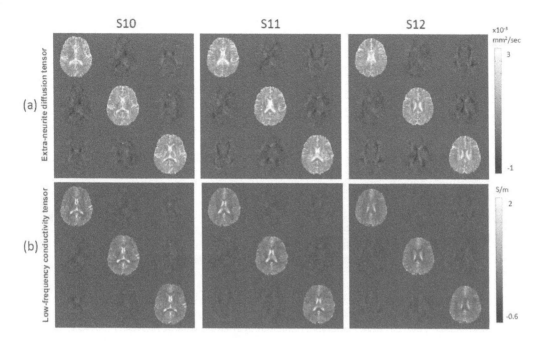

Figure 17.24 Images of (a) water diffusion tensor using $b=800$ s/mm^2 and (b) low-frequency conductivity tensor in three different imaging slices of S10 to S12.

Figure 17.23(f) shows the low-frequency mean conductivity from the 9-th to 13-th imaging slices. Comparing to the high-frequency conductivity map, the low-frequency conductivity map reflects the ion concentrations and diffusion coefficient in ECS and shows the different electrical properties, especially in the white matter region. Figure 17.24(a) and (b) are plots of the diffusion tensor and conductivity tensor images, respectively.

17.7.5 CLINICAL APPLICATIONS OF CTI

Since CTI can be readily implemented in an existing clinical MRI scanner without adding any hardware components, there is no technical barrier for its clinical applications. There must be, however, various validation studies to show its usefulness in chosen clinical applications. Numerous experimental studies with animal models and human subjects are needed in future studies. In this section, we introduce the following three potential clinical applications of CTI:

- Forward modelling of bioelectromagnetism
- Treatment planning of electrical stimulation therapy
- Diagnostic imaging

Figure 17.25(a) shows six different anatomical models of a human head constructed from conventional MRI images using image segmentation methods. Figure 17.25(b)–(d) plots CTI images of the entire brain, which can be combined with the anatomical models in (a). This approach enables us to construct a patient-specific head model with both geometry and conductivity tensor information. Figure 17.25(e) shows a three-dimensional finite element mesh in two different views. This finite element mesh can be used for numerous forward modelling studies in bioelectromagnetism.

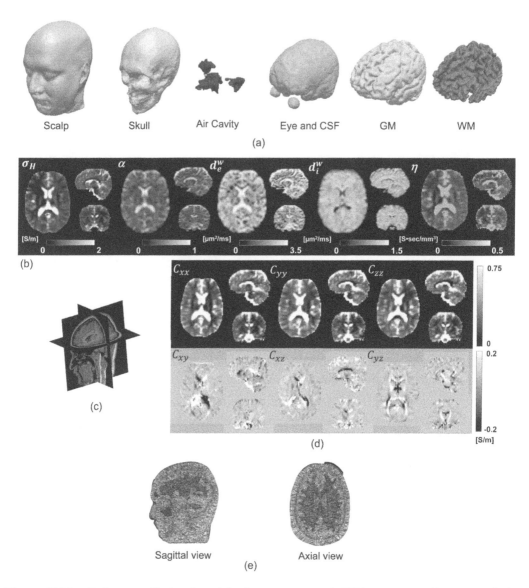

Figure 17.25   Patient-specific head model using reconstructed CTI images as well as anatomical images of a human head. (a) Segmented anatomical models of the head. (b) Images of σ_H, α, d_e^w, d_i^w and η. (c) Slice positions. (d) Conductivity tensor images. (e) Patient-specific finite element mesh for numerous forward modelling studies. (color figure available in eBook)

The patient-specific model in figure 17.25(e) can be use for treatment planning of electrical stimulation therapy such as transcranial direct current stimulation (tDCS), transcranial alternating current stimulation (tACS), deep brain stimulation (DBS), electroporation and RF ablation. When a DC current of I mA is injected into the head between two electrodes $\mathcal{E}_A$ and $\mathcal{E}_C$, the voltage u inside

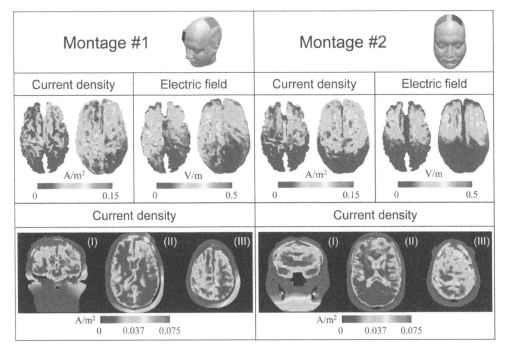

Figure 17.26 Effects of two different electrode montages on electric field and current density distributions in the white matter and gray matter regions. The reconstructed conductivity tensor images in figure 17.25(d) were used to compute the field distributions. The images in (I) are the current density distributions in the coronal slices of the heads at the position of the anode. The images in (II) and (III) are the current density distributions in the axial slices of the heads at 75 and 25 mm, respectively, from the skull. (color figure available in eBook)

the head denoted as Ω with its boundary $\partial\Omega$ satisfies the following partial differential equation:

$$\begin{cases} \nabla \cdot \mathbf{C}\nabla u = 0 \quad \text{in } \Omega \\ \int_{\mathscr{E}_A} \mathbf{C}\frac{\partial u}{\partial \mathbf{n}} \, ds = I \\ u = 0 \quad \text{on } \mathscr{E}_C \\ \mathbf{C}\frac{\partial u}{\partial \mathbf{n}} = 0 \quad \text{on } \partial\Omega \backslash (\mathscr{E}_A \cup \mathscr{E}_C) \end{cases} \tag{17.84}$$

where $\mathbf{n}$ is the outward unit normal vector on the boundary $\partial\Omega$. We can numerically calculate the voltage u, electric field $-\nabla u$ and current density $-\mathbf{C}\nabla u$ using the finite element method, for example. Figure 17.26 shows current density and electric field distributions inside the brain subject to two different electrode montages commonly used in tDCS.

Figure 17.27 shows CTI images of a canine head. Currently, different animal models of tumor, cirrhosis, abscess, inflammation, ischemia and others are being investigated for potential clinical applications of CTI as a diagnostic imaging tool.

17.7.6 FUTURE WORK IN CTI

Unlike other low-frequency conductivity imaging methods of MREIT and DT-MREIT, CTI does not require external injecting currents. This allows CTI to be readily applicable to *in vivo* human imaging studies without causing any adverse effects. Without adding any special hardware components, CTI can be implemented in a conventional clinical MRI scanner for diagnosis and treatment planning.

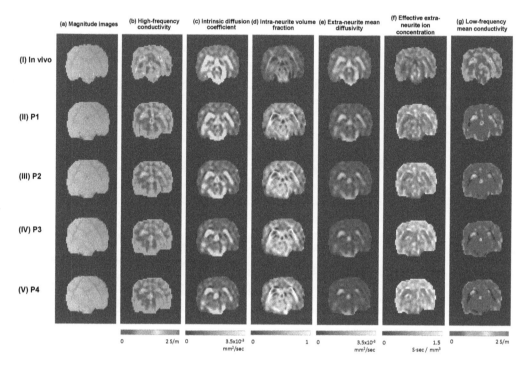

Figure 17.27 CTI images of a canine head using a 3 T MRI scanner: (a) MR magnitude image, (b) high-frequency conductivity (σ_H), (c) intrinsic diffusivity, (d) intracellular volume fraction (α), (e) extracellular mean diffusivity (d_e^w), (f) effective extracellular ion concentration (α) and (g) low-frequency mean conductivity (σ_L). (color figure available in eBook)

Currently, there is no method to experimentally determine β at every voxel. In future studies, it will be worthwhile to pursue a method to experimentally estimate β for each voxel to separately quantify $\bar{c}_e$ and $\bar{c}_i$. Depending on how many of $d_e^w, d_i^w, \alpha, \beta$ and $\mathbf{D}_e^w$ are experimentally determined with enough accuracy or assumed to be constant, we may interpret the reconstructed image of $\mathbf{C}$ as a conductivity tensor or conductivity weighted image, that is, CTI or CWI.

Since the image contrast in CTI or CWI is based on ensemble averages of microscopic motions of charge carriers in a structured tissue, macroscopic CTI image parameters may lead to new methods to extract quantitative information about the tissue microstructure and its functions. It would be meaningful to quantitatively examine the images of σ_H, $\frac{d_e^w}{\chi d_e^w + (1-\chi)d_i^w \beta}$, $\frac{\mathbf{D}_e^w}{\chi d_e^w + (1-\chi)d_i^w \beta}$ and $\bar{c}_e$ together with the image of $\mathbf{C}$. Future clinical studies may be pursued in the following areas.

- Changes in concentrations of charge carriers are reflected in $\bar{c}_e$ from which pathological changes of tumor, cirrhosis, inflammation and bleeding can be examined.
- Structural changes are reflected in $\frac{d_e^w}{\chi d_e^w + (1-\chi)d_i^w \beta}$ or $\frac{\mathbf{D}_e^w}{\chi d_e^w + (1-\chi)d_i^w \beta}$, which can be related with demyelination and ion channel openings, for examples, as well as pathological changes.
- Changes in cell size and density are reflected in α and σ_L, which can provide information about cell swelling, for example.
- Conductivity tensor images can be used in patient-specific models for electromagnetic source imaging and electrical stimulation techniques such as tDCS, tACS, DBS, electroporation and RF ablation.

Table 17.3

Comparison of four electrical impedance imaging methods using MRI.

	MREIT	MREPT	DT-MREIT	CTI
Pixel value	Equivalent isotropic conductivity	Conductivity and permittivity	Conductivity tensor	Conductivity tensor
Frequency	$< 1\,\text{kHz}$	$128\,\text{MHz}$ at $3\,\text{T}$	$< 1\,\text{kHz}$	DC
Electrode	≥ 4	None	≥ 2	None
Injection current	$\leq 2.5\,\text{mA}$	None	$\leq 2.5\,\text{mA}$	None
Contrast information	Ion concentration Ion mobility Hindered conduction	Ion concentration Ion mobility Dipole polarization	Ion concentration Ion mobility Hindered conduction Anisotropy	Ion concentration Ion mobility Hindered conduction Anisotropy

17.8 SUMMARY

Electrical impedance imaging using MRI has been studied for more than two decades. Currently, there are four methods summarized in table 17.3. Based on Maxwell's equations, the physical and mathematical frameworks of MREIT, DT-MREIT and MREPT have been validated through theoretical analyses, numerical simulations and imaging experiments of conductivity phantoms, animals and human subjects. MREIT and DT-MREIT require external current injections inside an MRI scanner, which could be inconvenient in practice. Depending on a chosen part of the imaging area, the injection current may stimulate nerves and/or muscles, which restricts clinical applications of the methods. MREPT is an electrodeless method without requiring external current injections. MREPT is expected to find clinical applications where high-frequency isotropic conductivity images provide enough information. CTI is also an electrodeless method without using any external injection currents to produce low-frequency anisotropic conductivity tensor images. Since CTI recovers the high-frequency isotropic conductivity as an intermediate variable, it may function as a dual-frequency method. CTI is, however, based on the assumptions about the relation between the conductivity tensor and water diffusion tensor, and needs to be further validated through more rigorous experimental studies. In addition to clinical applications, applications in electrochemistry, biology, food science and material science can also be pursued.

18 Geophysical ERT

Alistair Boyle
Systems and Computer Engineering, Carleton University, Ottawa, Canada

Paul Wilkinson
British Geological Survey, UK

CONTENTS

18.1 INTRODUCTION

In geophysics, electrical measurement techniques to estimate near-surface impedances were initially developed in the context of mineral prospecting by Conrad Schlumberger in 1911[39,941] and have been widely used in subsurface investigations ever since. Geoelectric imaging is used world-wide in industry, consultancy and academia, and is the subject of considerable ongoing research and development. Terminology in geophysics has evolved and what was initially referred to as "vertical electrical sounding" (a one-dimensional layered Earth problem) has been refined to Electrical Resistivity Tomography (ERT). Electrical Resistivity Imaging (ERI) is also used in some literature as an alternative to the use of ERT. Resistivity (units: Ωm), the reciprocal of conductivity (units: S/m), is generally the preferred unit when discussing geological phenomena.

It is important to recognize that, mathematically, biomedical EIT and geophysical ERT solve the same equations; the Calderón Problem. Nonetheless, there are some important distinctions in geophysics for specific applications. In particular, ERT is usually performed at much lower frequencies

DOI: 10.1201/9780429399886-18

(1 Hz to 10 kHz) and over greater distances (50–100 m electrode arrays are common), often on an open domain (the Earth's surface) or in bore-holes, and encounters resistivities which can vary over orders of magnitude across nearby regions. In biomedical EIT, time difference EIT is often preferred, but for geophysical settings, a static reconstruction of the actual resistivity distribution is often required. There are also monitoring applications in which, similar to EIT, changes in resistivity are of interest.

18.2   COMMON APPLICATIONS

As in biomedical EIT, the basic DC geoelectrical resistivity measurement configuration comprises four electrodes connected to a resistivity meter, with two passing current (labelled A,B; C1,C2; or C+,C-) and two measuring a potential difference (M,N; P1,P2; or P+,P-). Typically many electrodes (up to hundreds) are connected to the meter simultaneously, and user-defined measurement schedules control which configurations of electrodes are measured in which order. It is often the case that multiple potential difference measurements can be made simultaneously on different pairs of potential electrodes for each current injection ("multi-channel" acquisition). Connections to the electrodes are typically made using multi-core cables with regularly spaced take-outs, and the electrodes are usually stainless steel rods or plates (figure 18.1). Distances between electrodes range from tens of centimetres to hundreds of metres depending on the desired survey area and depth of investigation.

Figure 18.1   Example of an Electrical Resistivity Tomography field survey (BGS UKRI 2019). (color figure available in eBook)

Geoelectrical resistivity meters mostly employ either a switched DC signal, where the current switches from positive to zero to negative to zero to positive over a few seconds, or low-frequency AC signal between 1 Hz and tens of Hz. Signal processing is either performed by the instrument or the full waveform may be recorded for later analysis. Switched DC systems tend to require more power (of the order of 100 W), whilst low-frequency AC systems, using lock-ins or digital signal processing, are usually more efficient (of the order of 10 W). Field systems are usually powered from automotive batteries, with typical applied voltages of tens to hundreds of volts driving currents of tens to hundreds of milliamps between the current electrodes. The ratio of the applied voltage and the injected current is usually dominated by the contact impedance at the electrodes, which depends on the surface area of the electrodes, the resistivity of the surrounding earth material, and the degree of galvanic contact between the two. Contact resistances can be lowered by using larger electrodes with greater surface area. Normally in ERT applications the electrode dimensions are much smaller than the spacing between electrodes. Most geoelectrical inversion algorithms treat the electrodes as point-like. In instances where contact resistances are too high, it is more common to reduce them by treating the ground in the immediate vicinity of the electrode, often using water, saline solution, or conductive slurries or gels, rather than using larger electrodes. Reducing the contact resistance increases the signal-to-noise or, conversely, may justify reducing the required power. Some systems can run from generators and apply greater voltages for use in highly resistive environments. It is also possible to use physically separated current and potential bipoles with high power systems to enable larger scale (multi-kilometre) surveys.

Unlike biomedical applications, in geoelectrical surveying it is rarely possible to place electrodes over an entire boundary surrounding the region of interest. Most investigations are undertaken using either lines or grids of electrodes on the ground surface of the area to be surveyed, or, less commonly, lines of electrodes deployed in boreholes. The positions of these electrodes must be accurately recorded (either by GPS or other surveying techniques, e.g. tape). Geoelectrical surveys usually comprise sets of measurements made in one or more standard configurations (figure 18.2) with a range of bipole lengths and spacings. It is sometimes possible to deploy electrodes sufficiently far from the survey area to be practically at infinity, which enables the use of pole-pole and pole-dipole measurements.

For quick interpretation of raw data, the measured transfer resistance R can be multiplied by a geometric factor k (figure 18.2, right-hand column) based on the assumption of a homogeneous flat half-space, giving a quantity known as the apparent resistivity ρ_a. This provides a simple check of the range of resistivities present in the subsurface and the approximate locations of structures of interest. The apparent resistivity also normalizes measurements so that they are more equally weighted in the reconstruction (see §18.12).

For an electrode at the surface of a half-space, current I causes a radial potential $\phi = \rho I / 2\pi r$ in a medium with homogeneous resistivity ρ at radial distance r from the electrode. Note that there is a singularity at the electrode when $r = 0$. A dipole created by current flowing between source electrodes A and B (figure 18.2) causes a potential ϕ measured at electrodes M and N as ϕ_M and ϕ_N. A surface ERT measurement V is the difference between these potentials at the two measurement electrodes $\Delta\phi$

$$\phi_M = \frac{\rho I}{2\pi}\left(\frac{1}{AM} - \frac{1}{MB}\right) \qquad \phi_N = \frac{\rho I}{2\pi}\left(\frac{1}{AN} - \frac{1}{NB}\right)$$

$$V = \Delta\phi = \frac{\rho I}{2\pi}\left(\frac{1}{AM} - \frac{1}{MB} - \frac{1}{AN} + \frac{1}{NB}\right) \tag{18.1}$$

where each distance AM, MB, AN, NB is between the corresponding stimulus electrode A or B and a measurement electrode M or N. The model may be applied for any arbitrary pair-wise electrode placement. Buried electrodes may be handled by placing "mirror image" electrodes at an equal distance above the flat surface to correct for boundary effects [88,644].

Equation (18.1) may be rearranged to find the apparent resistivity $\rho \to \rho_a$ as an estimate of the homogeneous resistivity given a single measurement

$$\rho_a = \frac{V}{I}k = Rk \qquad \text{for } k = \frac{2\pi}{\left(\frac{1}{AM} - \frac{1}{MB} - \frac{1}{AN} + \frac{1}{NB}\right)} \tag{18.2}$$

where $R = V/I$ is the measured transfer resistance as the ratio of a measured difference in potentials V resulting from an applied current I. A geometric factor k accounts for the electrode configuration.

Geometric Factor k

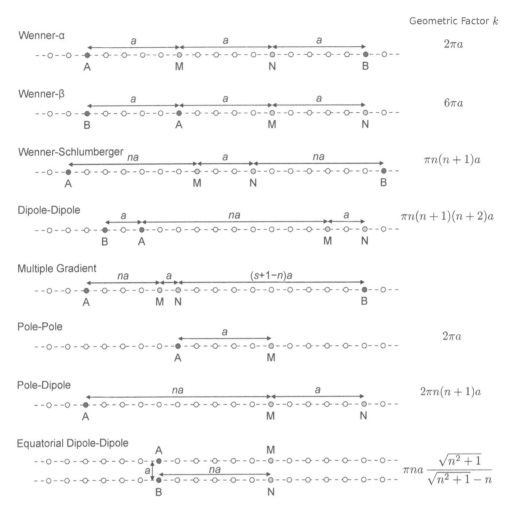

Figure 18.2 Commonly used ERT survey configurations showing current electrodes (A,B) in blue ● and potential electrodes (M,N) in green ○. The factors n and s are typically integer multipliers ≥ 1 of the electrode spacing a and specify the lengths and spacings of the bipoles; geometric factors k from equation (18.2) are shown adjacent to selected configurations where a simple expression exists (BGS UKRI 2019). (color figure available in eBook)

The resistivity is "apparent" because inhomogeneity in the medium, electrodes with finite size, or a surface that is not flat introduce errors. The geometric factor k may also be calculated as a normalization factor from an arbitrary geometry by calculating the measurements for a homogeneous $1\,\Omega$m model $k = 1/\mathscr{F}(1)$ using a FEM forward model $\mathscr{F}$ with $1\,A$ (unit) stimulus.

More quantitative interpretation requires solving the inverse problem. How this is parametrized depends on the geological setting, the placing of the electrodes, and the types of measurements made. In the simplest case, where the subsurface structure is horizontally layered, Wenner measurements (figure 18.2: Wenner$-\alpha$ and Wenner-β) can be made on a line of electrodes with a range of bipole lengths a and centred on the same point. This is known as vertical electrical sounding, and can be inverted using a 1D-model where the resistivity varies only as a function of depth. More commonly, a survey line will be placed perpendicular to the strike[1] of the subsurface structure. Then

---

[1]Strike (compass heading) and dip (steepest angle) define a unit vector orientation of the layered rock, for example sedimentary layers rotated by faulting or other deformation processes relative to a flat horizontal layer.

the assumption can be made that the resistivity varies with depth and the distance along the line, but not in the perpendicular direction. The inverse model is then a 2D parametrization (although it represents a 3D structure). Unlike biomedical EIT, the current flow is treated as 3D from point electrodes, so the resulting model is referred to as a 2.5D solution (see §18.10). Lastly, measurements can be taken on a 2D surface grid (see §18.3). Often there will not be sufficient numbers of electrodes available to emplace the whole grid at once, so multiple linear surveys are carried out along parallel lines. For better results with 3D inversion, it is good practice to include at least some lines in the perpendicular direction (referred to as tie-lines) and along diagonal directions. If more than one parallel line of electrodes can be laid out at once, then true 3D measurements (such as equatorial dipole-dipole) can also be used.

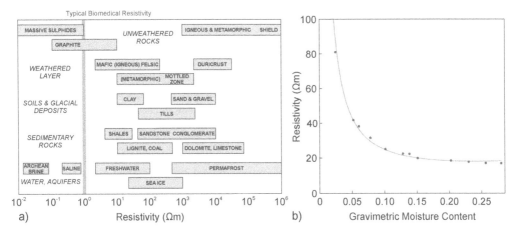

Figure 18.3   Material and moisture content affect resistivity; (a) approximate resistivity ranges for surface waters, rocks and soils; (b) example of petrophysical relationship between resistivity and moisture content of a porous material showing data and fitted model (BGS UKRI 2019). (color figure available in eBook)

Geoelectrical surveys are widely used because the resistivity of the earth is sensitive to a wide range of properties of interest to geoscientists. Predominantly the resistivity varies strongly with lithology (rock type) over a range of many orders of magnitude (figure 18.3a). It also depends on the weathering of the rock, its fracturing, its degree of saturation with fluids, the ionic content of saturating pore fluids, its temperature, and the presence of contaminants and pollution. Consequently the results of geoelectrical surveys are often transformed into other subsurface properties of interest (such as moisture content) via petrophysical relationships, which relate the results to physical and chemical properties of rocks and their contained fluids. These can be determined experimentally by recording the resistivity as a function of the target parameters in representative samples, and fitting petrophysical models (figure 18.3b), such as Archie's Law for sands, gravels and sedimentary rocks, or the Waxman-Smits or Dual Water models where significant clay mineralization is present[55,354,1131].

Their dependence on a wide variety of physical parameters (e.g. lithology, porosity, water saturation, and permeability) means that geoelectrical surveys are rarely undertaken without significant ground-truth being acquired (or already available), since interpretation is otherwise difficult and open to question. Types of commonly used ground truth, prior information and calibration include geological maps and ground models, topographic surveys, trial pits, borehole logs, water level loggers (piezometers), point sensors (measuring local electrical conductivity, moisture content and temperature), tilt sensors, accelerometers, other geophysical survey methods (e.g. ground penetrating radar, seismics, microgravity), and laboratory testing and calibration of material samples to develop petrophysical relationships between resistivity and parameters of interest (e.g. moisture content).

There are many and varied applications of geoelectrical imaging, including geological mapping and ground model development; hydrology and hydrogeology (e.g. marine, rivers, aquifers); natural hazard detection and mitigation (e.g. landslides, sinkholes); geotechnical and engineering hazards (e.g. slope stability); detection and mapping of voids (e.g. caves, tunnels, mineshafts); mineral prospecting and resource assessment; contaminated land / brownfield site investigation (e.g. landfills, leachate plumes, groundwater pollution); and archaeology. Several review papers from the last decade or so are available which cover the methods and applications in detail[114,646,672,817,876,980].

18.3   RESEARCH APPLICATIONS

Ongoing research and development into geoelectrical methods has seen advances in instrumentation and inversion algorithms, which have led to a rapid expansion in applications beyond standard, widely used 2D resistivity surveys. Some examples of recent developments are the increased use of large numbers of electrodes in surface grids for full 3D imaging; time-lapse/4D data acquisition and inversion for monitoring processes; optimal selection of measurements and electrode locations to maximise image resolution; subterranean electrodes to improve image resolution at depth; joint inversion of complementary geophysical data; accommodating changes in boundaries and electrode locations over time; and capacitively coupled electrodes for highly resistive environments. 3D geoelectrical imaging relies on a combination of a 3D inverse model, typically finite-element or finite-volume and parametrized into hexahedral or tetrahedral model cells, and data measured on a grid of electrodes at the surface. As noted in the previous section, it is common to collect such data using a small linear array, comprising a few tens of electrodes, making multiple parallel linear surveys and combining the data for inversion. In this case, the parallel surveys should not be separated by more than two or three times the electrode spacing along the line[341], and some perpendicular tie-lines should be included, particularly at the ends of the parallel surveys, to ensure overlapping sensitivity. Without these overlapping regions, each survey is effectively independent. Such independent acquisitions schemes can cause directional artefacts in the resulting images in the regions between survey lines[341]. These can be avoided if several parallel lines of electrodes, or ideally the entire grid, can be emplaced simultaneously permitting the collection of more general planar configurations as well as collinear measurements aligned in different directions (figure 18.4).

Since laying out arrays with larger numbers of electrodes for 3D surveys can be time-consuming, it is often desirable to install the electrodes in a semi-permanent manner, especially if repeat surveys of the same site are needed to monitor changes. This usually involves burying the array a few centimetres below the ground surface to protect the electrodes and wiring from mechanical damage. Sometimes the galvanic contact with the ground can be poor if the surface is dry or rocky, in which case the immediate vicinity of the electrodes can be treated using conductive slurries, gels or grouts to temporarily reduce the contact resistances. The array can then be connected periodically to a survey resistivity meter, or permanently to one of a growing number of geoelectrical monitoring systems (figure 18.5a). These are used to collect repeated data sets, which are inverted to produce models of the changes in resistivity over time (figure 18.5b). Several types of time-lapse inverse method are available[534]. These include inverting directly for the changes in the data, or imposing constraints between the current model and a baseline. Both of these approaches require a well-characterised baseline data set and background model. Another approach is to invert all the data sets as a 4D model with temporal constraints between subsequent time steps similar to the spatial constraints between adjacent model cells. This places no emphasis on any particular time step, and so does not require extra effort to characterise the baseline data and background model.

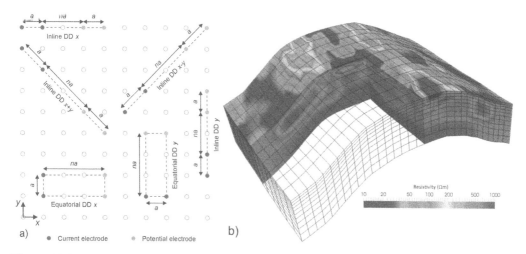

Figure 18.4 a) Examples of full 3D survey measurement types comprising inline and equatorial dipole-dipole configurations with different orientations; b) Example of a 3D resistivity model comprising hexahedral cells generated from a geoelectrical survey on a grid of 12×32 electrodes[183] (BGS UKRI 2019). (color figure available in eBook)

Since the most common causes of changes in the ground resistivity are variations in the degree of saturation and quality of the groundwater, there has been rapid growth in the use of time-lapse / 4D ERT to monitor hydrological processes. Example fields of application include landslide hydrology, earthwork stability, dam integrity, CO_2 sequestration, landfills, contaminated ground, nuclear waste decommissioning, leak detection, permafrost, aquifer exploitation, agriculture and soil/plant science, geothermal systems, and tracer tests. Details of these and other applications can be found in recent review papers[114,646,876,977,1141,1146].

Geophysical monitoring encounters a range of seasonally varying surface temperatures which directly affect ionic mobility, and therefore resistivity, by approximately 2%/°C near 25°C. Biomedical EIT does not generally encounter large temperature variation courtesy of the body's tight thermal regulation. On freezing, water solidifies into crystalline ice which drastically increases resistivity. Below ground, temperatures vary as a time delayed sinusoid that decays with depth and is balanced by steady heat radiated from below. Homogeneous models of this heat conduction are often sufficient to correct for seasonal variation in long-term monitoring data[183] particularly when validated against thermal depth arrays (e.g. thermistors).

As with biomedical EIT, a fundamental limitation of the technique is that image resolution decreases rapidly as the distance from the electrodes is increased. Since most geophysical applications use surface electrodes, this typically means that resolution decreases with depth. There has been considerable interest in the last 10–15 years in using optimal experimental design techniques to maximise resolution in more poorly resolved regions of the model, ideally without increasing survey time. Several approaches have been explored to maximise the resolution of the reconstructed image including reconstructing comprehensive data sets from a linearly independent complete subset; maximising the sum of the Jacobian sensitivity matrix elements; maximising the sum of the model resolution matrix elements; minimizing the average of the point spread function; and maximising the determinant of the normal matrix (see[1150] and references therein). The methods involving the model resolution matrix have received the majority of the research effort, and have been applied to 2D and 3D survey design, including investigations with subsurface electrodes (see[650] and references therein). The resulting surveys require care to use in practice due to the types of multichannel instrumentation available and the effects of electrode polarization[1150], but they have been shown

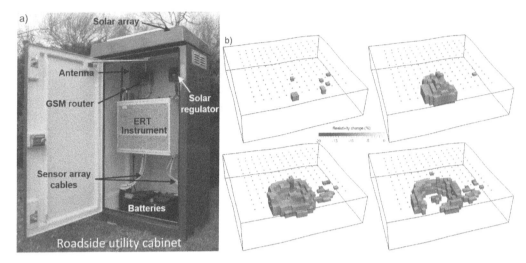

Figure 18.5   a) Permanently installed geoelectrical monitoring system, comprising ERT instrument, communications, batteries, solar panel and connections to electrode arrays; b) Relative change images from 4D inversion of ERT monitoring data taken from a grid array (green dots) above a simulated utility pipe leak, showing regions of model with resistivity changes $< -7.5\%$ (BGS UKRI 2019). (color figure available in eBook)

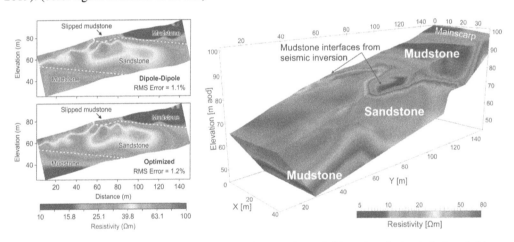

Figure 18.6   Comparison of standard dipole-dipole (above left) and optimized (below left) survey images of a landslide. Both images capture the geological structure but in the optimized image the sandstone/mudstone interfaces are better resolved; (right) 3D ERT image of the same landslide where the structure of the interface between the upper mudstone layer and the sandstone was obtained from a seismic refraction inversion and incorporated as structural information in the ERT inversion (BGS UKRI 2019). (color figure available in eBook)

to produce images with measurably better fidelity, especially in poorly resolved regions such as the base, edges and corners of the models (figure 18.6 left images). Similar techniques have also been used to produce adaptive measurement schemes for time-lapse geoelectrical monitoring[1152], and to optimize the placement of electrodes in arbitrary arrangements for non-standard 2D and 3D surveys[1063,1104].

A different way to improve the resolution at depth is to implant electrodes beneath the ground in the vicinity of the imaging region, and make measurements using combinations of all surface, all

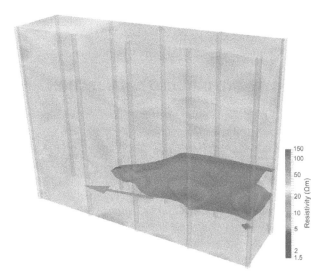

Figure 18.7 Isosurface (blue) showing regions of resistivity model changed by $< -20\%$ due to the injection of a conductive tracer in a confined aquifer. Each borehole (grey cylinder) contained 16 equally spaced electrodes at 0.5 m depth intervals (BGS UKRI 2019). (color figure available in eBook)

subsurface, and subsurface-to-surface electrodes[1147]. This is most commonly done by installing electrodes in boreholes[596], but other approaches include direct-push electrodes[648,827] (typically pushed or percussed less than 30 m into unconsolidated sediments using small-diameter rods) or placing electrodes in void spaces below ground (e.g. on tunnel walls:[616,1059]). Figure 18.7 shows an example of using resistivity measurements between boreholes to investigate the flow and dispersal of a tracer test in a confined aquifer[1151].

The fidelity of inverse images can also be improved by jointly inverting data sets from different geophysical techniques, since their resolution and sensitivity to subsurface features are often complementary. Various approaches have been taken, depending on the types of data available. If the techniques depend on the same property, but cover complementary regions of the model with different sensitivities, they can be inverted together directly[1118]. In cooperative joint inversions, data are inverted separately for different parameters but derived information, such as structure, are exchanged[137,1061] see figure 18.6 right image. Other approaches couple the inversions together via the spatial structure of the parameter distributions ([205] and references therein) or via petrophysical relationships that relate the multiple geophysical parameters to some common property such as the porosity ([454] and references therein;[726] and references therein).

A particular type of challenge that can cause artefacts in ERT image reconstruction is that of changes to the boundary on which the electrodes are installed. If these changes can be measured and recorded, they can be incorporated into the inversion and their adverse effects minimised e.g. mapping fissures at length scales smaller than the electrode spacing[329] or measuring the displacements of electrodes during ground movements such as landslides[328,1061] or shrink-swell[1062]. In particular, the application of ERT to landslide monitoring has led to inverse methods that incorporate electrode positions as model parameters to be reconstructed[144,649,1149] in much the same way as in lung function EIT imaging.

For certain applications, e.g. on dry, frozen or paved surfaces, galvanic contact impedances can be so high that they prevent injection of sufficient current for a reliable signal. In such cases, capacitively coupled systems can be used that employ non-grounded electric dipoles. They operate in a quasi-electrostatic regime typically at audio frequencies (10-20 kHz), in which the DC resistivity inversions can be applied[593]. Capacitive resistivity systems have been used for rapid acquisition of data on resistive surfaces using towed arrays[594,770], for assessing the strength of heritage stonework[998], and for monitoring of freeze-thaw cycles in permafrost[762].

18.4 COMPLEX RESISTIVITY AND INDUCED POLARIZATION

In biomedical applications of EIT, the quadrature component of the voltage is usually very small, and below the resolution of the instruments. In geophysics, the phase shift between the measured voltage and applied current for certain materials can be appreciable (up to several hundred milliradians). The phase shifts result from the reversible accumulation of charges, and the effects can be observed and measured in either the time- or frequency-domain. In the time-domain the technique is known as Induced Polarization (IP) and is measured by integrating the decaying residual voltage after current switch-off and normalizing by the initial DC voltage. This gives a dimensionless measure of the apparent chargeability in mV/V (or per-mille). In the frequency domain, the technique is referred to as either Complex Resistivity, if working at a single frequency, or Spectral Induced Polarization (SIP), if measuring over a range of frequencies. IP techniques were first applied in the exploration industries, but more recently have been used in environmental applications ([112] and references therein).

IP measurements should ideally be made using non-polarizing electrodes with shielded cables. Nonetheless, in many practical circumstances good data can be collected using the same multi-core cables and stainless steel electrodes as used in resistivity surveys. Having the current and potential electrodes on separate cables will improve the data quality if capacitive coupling in the multi-core cables is significant, e.g. when contact impedances are high[219]. Consequently, IP surveys are often carried out using the same equipment as for resistivity surveys, but they take longer due to the signal-to-noise ratio usually being significantly smaller. IP effects can arise from several mechanisms[544], but the predominant ones are electrode polarization and membrane polarization. Electrode polarization is caused by charge build up when electrolytic current flowing though the pore water is impeded by conductive mineral grains, through which the current has to flow electronically. Membrane polarization is largely caused by clay minerals which possess a negative surface charge that attracts positive ions and impedes electrolytic flow through narrow pores. Chargeability effects in the ground can be caused by materials like metals, metallic ores, clays, landfill waste, and hydrocarbons.

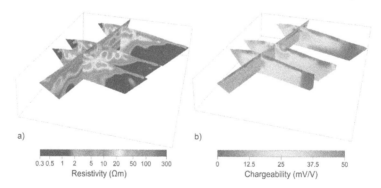

a)

0.3 0.5 1   2   5   10 20   50 100   300
Resistivity (Ωm)

b)

0   12.5   25   37.5   50
Chargeability (mV/V)

Figure 18.8   a) Resistivity image of a coastal embankment with a core containing household waste. b) The waste materials in the core have a high chargeability signature (BGS UKRI 2019). (color figure available in eBook)

Different inverse methods have been used and researched for induced polarization data, depending on the type. For time-domain measurements in terms of apparent chargeability, the data are inverted with the apparent resistivities to give a resistivity and a chargeability model of the ground (figure 18.8). If the full voltage waveforms are recorded, parameters for empirical spectral models can be extracted, such as Cole-Cole-based models which do not attempt to describe the underlying physical processes[112,1085]. In the frequency domain, the data are inverted for a complex resistivity ground model, either directly or by decoupling the real and imaginary components[519]. If data are available at several frequencies (SIP), then spectral model parameters can be fitted to the observed dispersion data.

18.5   LOGARITHMIC PARAMETRIZATION

Two numerical problems can occur if geophysical resistivity reconstructions are attempted blindly. First, non-physical negative resistivities can be reconstructed which can then skew nearby regions of the reconstruction. Second, the resistivities tend to span orders of magnitude which makes it difficult to apply effective regularization to the reconstruction. Both issues are addressed by converting to a log parametrization. A transformation to log units allows an unconstrained reconstruction, for example using iterative Gauss-Newton methods, to solve a constrained problem. The transform must be injective, which is to say that there is a one-to-one mapping of values in the new and old parameter spaces.

It is more common to use resistivities for geophysical applications, but in the following we use the conductivity parameters familiar to biomedical EIT. To restrict solutions to positive valued conductivities ($\sigma > 0$), we can apply a log parametrization $\mathbf{p}$ where

$$\mathbf{p} = g \ln \sigma \qquad \longleftrightarrow \qquad \sigma = \exp\left(\frac{\mathbf{p}}{g}\right) \qquad \text{for } g > 0, \quad 0 < \sigma < \infty \qquad (18.3)$$

with natural log $\ln : g = 1$ and base-10 $\log_{10} : g = 1/\ln(10)$. In a Gauss-Newton iteration the transformation is first used to convert the conductivity to the log parameter space, update the parameters using the Jacobian, then take the inverse transform to check data misfit (line search and stopping criteria) using the forward model.

The Jacobian for the new log parameter space is computed from the chain rule

$$\mathbf{J}_{\mathbf{p},i,j} = \frac{\partial \mathbf{b}_i}{\partial \mathbf{p}_j} = \frac{\partial \mathbf{b}_i}{\partial \sigma_j} \frac{\partial \sigma_j}{\partial g \ln \sigma_j} \qquad\qquad \mathbf{J}_\mathbf{p} = \mathbf{J}_\sigma \frac{\text{diag}(\sigma)}{g} \qquad (18.4)$$

for the i-th measurement $\mathbf{b}_i = V_i$ and j-th conductivity element σ_j, so that the original Jacobian columns are scaled by the conductivity σ at which the Jacobian was calculated and g is, as above, selects the type of log scaling. In this parametrization for resistivity ρ, one can either directly substitute $\sigma \to \rho$ resulting in an inverted weighting and regularization scheme, or translate the equations $\sigma = 1/\rho$ to get the equivalent numerical result in terms of resistivity.

18.6   ABSOLUTE RECONSTRUCTION

The term "absolute reconstruction" is somewhat misleading, as it does not refer to the resistivity being greater than zero, which can be achieved by using the log parametrization (see §18.5). Instead, the term refers to a reconstruction using a single set of measurements rather than a time or frequency difference reconstruction. (Again these equations are written in terms of conductivity σ here, but are typically written in terms of resistivity ρ in geophysical contexts.) The most common approach is to use a Gauss-Newton iterative update, for which the left-hand side of the update $\delta\sigma$ looks identical to the Gauss-Newton single-step solution used in time-difference EIT. On the other hand, the right-hand side differs by requiring terms that are most often dropped from the single-step solution, as they are set to zero. To arrive at a solution, one iterates through the following two equations

$$\delta\sigma = \left(\mathbf{J}^\mathsf{T}\mathbf{W}\mathbf{J} + \lambda^2\mathbf{Q}\right)^{-1}\left(\mathbf{J}^\mathsf{T}\mathbf{W}\left(\mathbf{b} - \mathscr{F}(\sigma_n)\right) + \lambda^2\mathbf{Q}(\sigma_n - \sigma_*)\right)$$

$$\sigma_{n+1} = \sigma_n + \alpha_n\,\delta\sigma \qquad\qquad (18.5)$$

for Jacobian $\mathbf{J}$, inverse measurement covariance $\mathbf{W}$, regularization $\mathbf{Q}$ scaled by hyperparameter λ, prior conductivity estimate σ_*, measured data $\mathbf{b} = [V_1 V_2 \ldots V_n]^\mathsf{T}$, current conductivity estimate σ_n, and forward model of the measured data at that conductivity estimate $\mathcal{F}(\sigma_n)$. The prior conductivity estimate is often set to the initial conductivity estimate ($\sigma_* = \sigma_0$). The update direction $\delta\sigma$ gives a search direction. A line search may be used to find an "optimal" step size $0 < \alpha_n < 1$ or various heuristics are available, for example ($\alpha_n = 1$). Line searches are expensive to compute, since they require multiple calculations of the forward model, but can significantly improve convergence rates and solution accuracy.

The Jacobian should be recalculated at each σ_n when conductivities change significantly: current density through the domain changes as conductivities are modified, resulting in changes to the regional sensitivity $\mathbf{J}$. This is also the key reason why absolute reconstructions are most often a necessary first step in any time-difference solution for ERT datasets: a Jacobian calculated on a homogeneous conductivity can be wildly incorrect when conductivities vary over the ranges commonly observed in the near surface.

We can observe that the forward model $\mathcal{F}(\sigma_n)$ now enters the picture, where it would otherwise cancel out in time-difference EIT. The fact that the forward model plays a key role in the absolute reconstruction indicates that model errors such as electrode placement, electrode movement, boundary or contact impedance are now critical factors in a successful reconstruction.

Absolute solutions are much more computationally expensive to reconstruct than typical EIT single-step time-difference reconstructions. The majority of the computational time is spent recalculating the Jacobian used in the update $\delta\sigma$ and forward solutions for the line search α at each iteration.

Iterations are halted at "stopping criteria" which can be an iteration limit, when the total misfit is sufficiently reduced, or when progress slows. The hyperparameter is sometimes modified as iterations proceed, which introduces "trust region" type iterative solutions.

An initial conductivity estimate σ_0 is required as a starting point. As with most iterative algorithms, the starting point can determine what sorts of solutions can be found as the algorithm may become trapped in local minima as it traverses away from the initial estimate. The most common approach is to take the mean of the apparent resistivities, which works by using the scaling from a homogeneous $1\,\Omega$m resistivity used in the forward model $\mathcal{F}(1) = 1/k$ geometric factor k, so that

$$\rho_0 = \frac{1}{n} \sum_{i=1}^{n} \frac{V_i}{I_i} k_i \tag{18.6}$$

where the single scalar resistivity ρ_0 value gives a best-fit homogeneous estimate of conductivity $\sigma_0 = 1/\rho_0$ from the measurements $\mathbf{b}_i = V_i$. The initial estimate can be thrown off by outlier apparent resistivities, which can be treated by discarding or de-weighting based on an error model (see §18.11).

18.7 TIMELAPSE INVERSION

Timelapse inversion in ERT differs from time-difference EIT, in that the changes can be non-linear enough to require iterative updates: large changes in conductivity that no longer match extrapolated linearized measurement changes. The absolute conductivity Gauss-Newton solution (18.5) is modified so that the data and conductivity represent differences $\mathbf{b}_\Delta = \mathbf{b}_t - \mathbf{b}_{t=0}$ and $\sigma_\Delta = \sigma_t - \sigma_{t=0}$ from a reference time $t = 0$, often the solution of an absolute reconstruction. This expansion leads to a timelapse Gauss-Newton iterative formulation over the data misfit $\mathbf{d}$ where

$$\mathbf{d} = (\mathbf{b}_t - \mathbf{b}_{t=0}) - (\mathscr{F}(\sigma_t) - \mathscr{F}(\sigma_{t=0})) = \mathbf{b}_\Delta - (\mathscr{F}(\sigma_{t=0} + \sigma_{\Delta,n}) - \mathscr{F}(\sigma_{t=0}))$$
$$\delta\sigma_\Delta = (\mathbf{J}^\mathsf{T}\mathbf{W}\mathbf{J} + \lambda^2 \mathbf{Q})^{-1} \left(\mathbf{J}^\mathsf{T}\mathbf{W}\mathbf{d} + \lambda^2 \mathbf{Q}(\sigma_{\Delta,n} - \sigma_{\Delta,*})\right)$$
$$\sigma_{\Delta,n+1} = \sigma_{\Delta,n} + \alpha_n\,\delta\sigma_\Delta \tag{18.7}$$

and, as with time-difference EIT, it is often reasonable to initially assume no change in conductivity $(\sigma_{\Delta,*} = \sigma_{\Delta,0} = 0)$ [602]. The same absolute reconstruction code can often be used to calculate a timelapse solution with a few trivial modifications because the changes are entirely related to the data misfit term $\mathbf{d}$. An alternate approach is to regularize the temporal changes simultaneously through a Kronecker expansion of the Gauss-Newton update (18.7) over multiple frames [448,534,647].

If one assumes no line search ($\alpha_n = 1$), one iteration, and a "no change" initial estimate, then the above formulation simplifies to the familiar time-difference EIT equation for a single-step Gauss-Newton solution

$$\sigma_\Delta = (\mathbf{J}^\mathsf{T}\mathbf{W}\mathbf{J} + \lambda^2 \mathbf{Q})^{-1}\mathbf{J}^\mathsf{T}\mathbf{W}(\mathbf{b}_t - \mathbf{b}_{t=0}) \tag{18.8}$$

though these simplifications are often inappropriate in the presence of large underlying variations in conductivity. In biomedical EIT, where the body is largely constrained to a relatively narrow range of conductivities due to intra- and extra-cellular fluids and the body's regulation of salinity, these assumptions are much more reasonable.

18.8 THE USE OF ELECTRODE MODELS

In geoelectrical applications, the electrodes tend to be small in comparison to the separations between them. This means that they can usually be modelled as point sources. Rucker *et al* [903] showed that for rod electrodes, a Point Electrode Model (PEM) is adequate for electrode lengths of up to 20% of the unit electrode separation. By contrast, in most biomedical applications a Complete Electrode Model (CEM) has to be used to account for the finite size of the electrode. Larger electrodes are sometimes used, e.g. to reduce contact resistance (often in highly resistive environments, or in small-scale laboratory work where point-like electrodes would be too small to provide good galvanic contact). In these cases, the finite size of the electrodes must be taken into account. If the electrode geometry is simple, such as a ring [878] or ellipsoid [498], the effects can be calculated analytically. For more general cases, CEMs have been developed for the 3-D [903] finite element formulation of the forward and inverse problems, although the 2.5-D problem has not been tackled convincingly. For induced polarization/complex resistivity, the CEM is often found to be an important ingredient in reconstructions because the electrode contact can strongly influence measurements. Simpler approaches, which are sufficient if the contact impedance is small, are to model the electrode as a point source embedded in a highly conductive region [1083] or as an extended perfect conductor [1181]. A small number of studies have used CEM techniques to analyse field and laboratory data [99,144,1103,1105].

18.9 MODELLING OPEN DOMAINS

A major difference between the biomedical and geoscientific applications of EIT is that the former make measurements on a bounded surface (the patient), whereas the later are frequently applied to open domains (in field applications, although laboratory tank experiments and sample investigations are bounded). Open domains typically consist of the ground surface, at which Neumann boundary conditions are applied, and a far subsurface boundary effectively at infinity. An early approach to treat this boundary was to impose a Robin condition

$$\frac{\partial \phi}{\partial \mathbf{n}} + \alpha \phi = 0 \tag{18.9}$$

on a nearer, artificial subsurface boundary with outward normal $\mathbf{n}$ such that the potential ϕ has a $1/r$ behaviour at a distance r from the electrode[904]. The parameter $\alpha = (\mathbf{n} \cdot \mathbf{r})/r^2$ with $r = ||\mathbf{r}||$ for a surface current electrode[238] and $\alpha = (r'^3 \mathbf{n} \cdot \mathbf{r} + r^3 \mathbf{n} \cdot \mathbf{r}')/(r^2 r'^2 (r + r'))$ for a buried current electrode[111], where $\mathbf{r}$ is the vector from the electrode to the boundary element and $\mathbf{r}'$ from the electrode's mirror image above the surface to the boundary with $r' = ||\mathbf{r}'||$. The parameter α is, in part, related to $\cos\theta = (\mathbf{n} \cdot \mathbf{r})/r$ for the angle θ between $\mathbf{r}$ and $\mathbf{n}$. These calculations make use of the analytic half-space model of ERT (18.1).

A single boundary condition for all electrode source positions is often sufficient when boundaries are far from sources and sources are close to each other (avoiding updated boundary conditions for each electrode stimulation pair) by taking the average electrode position[904]. For many geophysical settings with apparent topological relief, the overall surface variation from a flat surface is sufficiently small to make these approximations useful. For example, many hillsides tend to lie at a roughly uniform angle of repose over hundreds of meters (a typical ERT survey length), despite having a significant slope. Rotating the model domain to match this average slope will remove the majority of the topological error in the boundary condition.

Another approach, which has become more common as computer power has increased, is to move the artificial subsurface boundary to a suitably large distance from the electrodes by using progressively larger model cells, such that simply applying either Dirichlet[814] or Neumann[112] boundary conditions is a reasonable approximation to the boundary actually being at infinity. More recently, infinite elements have been used to impose the far boundary condition, improving accuracy and reducing computational load in finite element formulations of the geoelectrical resistivity problem[122,1197].

18.10   2.5D CALCULATIONS

Two and a half-dimensional (2.5D) solutions are forward problems that are solved in three dimensions under the assumption that one dimension has uniform conductivity from positive to negative infinity with collinear electrodes. In real impedance imaging systems the electrodes are not of infinite length in a particular direction, even when the conductivity distribution is uniform in that direction. The finite size of the electrodes indicates that one cannot correctly approximate three-dimensional solutions with a two-dimensional model unless the electrodes extend to the boundaries, generating a uniform current distribution in the third dimension.

The approach taken in the 2.5D method is to use a 2D FEM, apply a Fourier transform in one dimension, and integrate over the spatial frequencies to obtain a correction, while simultaneously solving the FEM for the other two dimensions[141,238].

As before, the 2.5D method is presented in terms of conductivity, though resistivity $\sigma = 1/\rho$ is more common in geophysics. By subscripts, the dimensionality of the variables and partial derivatives are denoted, so that a three-dimensional potential ϕ_{xyz} is caused by current at the boundary applied to a three-dimensional conductivity distribution σ_{xyz}. When the conductivity is constant in one dimension z, the partial derivative is zero $\partial \sigma_{xyz}/\partial z = 0$ for constant σ_z and σ_{xyz} is denoted σ_{xy}. For such a conductivity, the potential will vary in the z-dimension.

The z-dependence of the potential ϕ_{xyz} is Fourier-transformed into the spatial frequency domain $\tilde{\phi}_{xy\tilde{k}}$ using the cosine transform and its inverse

$$\tilde{\phi}_{xy\tilde{k}} = \int_0^\infty \phi_{xyz} \cos(\tilde{k}z) dz \qquad \leftrightarrow \qquad \phi_{xyz} = \frac{2}{\pi} \int_0^\infty \tilde{\phi}_{xy\tilde{k}} \cos(\tilde{k}z) d\tilde{k} \tag{18.10}$$

for a potential that is reflected across the xy-plane so that the potential is an even function $\phi(z) = \phi(-z)$. The conductivity-potential relationship for uniform conductivity in the z-dimension is Fourier-transformed

$$-\nabla \cdot (\sigma_{xy} \nabla \phi_{xyz}) = \frac{\partial \rho}{\partial t} \delta_{xyz} \qquad \rightarrow \qquad -\nabla \cdot (\sigma_{xy} \nabla \tilde{\phi}_{xy\tilde{k}}) + \tilde{k}^2 \sigma_{xy} \tilde{\phi}_{xy\tilde{k}} = \tilde{Q} \delta_{xy} \qquad (18.11)$$

for a scalar wave number $\tilde{k}$ and steady state current density in the spatial frequency domain with $\tilde{Q}\delta_{xy} = \frac{I}{2\partial t} \delta_{xy} \simeq \frac{I}{2A}$ using an approximated of a shunt electrode with constant current[238], conducting current I over an electrode area A. The result is a shunt model in the z-dimension and FEM electrode model (PEM, CEM, etc.) in the xy-dimensions.

We note that (18.11) takes the same general form in regular and Fourier space, albeit the spatial frequency domain in place of the z-dimension, with an additional dissipation term dependent on the square of the spatial wave number $\tilde{k}^2$ which suggests an efficient implementation. When assembling matrices for many $\tilde{k}$, only the value of $\tilde{k}$ changes, so the additional 2.5D computations are a small incremental cost relative to the 2D solution.

The potentials found for the forward solution at many $\tilde{k}$ are inverse Fourier transformed and an adaptive quadrature numerical integration is then typically used to accumulate the inverted solutions at appropriate $\tilde{k}$. The first solution $\tilde{k} = 0$ is the two-dimensional solution. For a sufficient summation of $\tilde{k}$, the solution will converge to the three-dimensional solution.

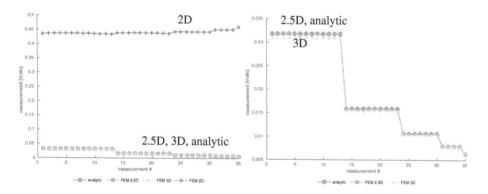

Figure 18.9 2D versus 3D measurements; a 16 CEM electrode half-space model with Wenner stimulus pattern, models for 2D and 2.5D used the 2D model (half-space, collinear electrode array, equally spaced electrodes), 3D measurements used an equivalent 3D model, and the analytic model used geometry from the 2D model, (a) 2D measurements are significantly different than 3D measurements and the half-space (PEM) analytic model, while (b) 3D, 2.5D and analytic model are in close agreement. (color figure available in eBook)

The difference between the 2D and the 3D or 2.5D solutions can be significant as illustrated in figure 18.9, where in 2D widely separated bipoles exhibit simulated measurements in error by as much as 71.5 times their true value. The 2D and 2.5D simulations use the same 2D model (half-space, collinear electrode array, equally spaced electrodes), while the 3D simulations use a 3D model. Both models are 16 electrode CEM half-space models, sharing the same linear electrode arrangement (5 m spacing, 0.1 m diameter), with homogeneous conductivity ($\sigma = 1$). In figure 18.9, the analytic solution (18.1) uses the 2D model geometry and estimates PEM electrodes at the centre of the CEM electrode positions.

The agreement between the analytic, 2.5D and 3D solutions illustrates that the FEM models extend far enough to approximate a half-space without introducing significant truncation errors.

The difference between the analytic and 2.5D, and the 3D solution are due to differences in the modelled shape of the electrodes. We do not explore the source of the analytic versus 3D error further here, though in principle it is straightforward to eliminate the possibilities such as PEM versus CEM, mesh density or electrode shape.

18.11   DATA QUALITY MEASURES

Errors in ERT data arise from a number of sources, both systematic and random. It is good practice to try to minimise these errors, and to obtain estimates of data quality to remove outliers and weight the data in the inversion. Sources of systematic error include uncertain electrode locations (through misidentification, poor measurement, or ground movement); electrode polarization effects; damage to cables or electrodes; and cross-talk between cables[601]. Location uncertainties have become less problematic in surface surveys as electrode positions can be readily measured to high accuracy by Global Navigation Satellite System receivers, although positions of borehole electrodes can be harder to determine accurately[1148]. Misidentification of electrodes can be checked by measuring sequences of short-offset Wenner or dipole-dipole configurations, which tend to yield negative voltages where electrode cables have been swapped. Ground movements can be accommodated by updated measurement or estimation of the electrode positions[1061,1149]. Electrode polarization errors occur when electrodes are used to measure potential soon after passing current but can be mitigated by careful measurement sequencing[1150]. Cable and electrode damage are an issue in long-term monitoring installations. In such cases electrodes and cables tend to be installed in protective piping or buried at shallow depth (figure 18.10a), but damage can still occur due to e.g. ground movement, animals chewing through cables (figure 18.10b&c) or human activity[596].

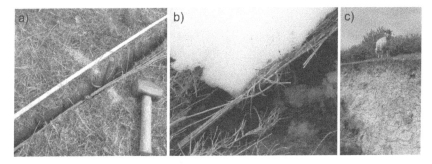

Figure 18.10   a) Burying electrodes and cables at a landslide monitoring site. b) Broken cables. c) Livestock and ground motion, both potential culprits for damage to ERT monitoring installations (BGS UKRI 2019). (color figure available in eBook)

Sources of random error include telluric currents induced by fluctuations in the Earth's magnetic field; self-potential effects from natural ground sources such as ore bodies acting as batteries and streaming potentials due to groundwater motion; and anthropogenic sources of electromagnetic noise such as power lines. They can be reduced in switched DC systems by subtracting slowly varying backgrounds and averaging fast variations, or by filtering in low-frequency AC systems. Signal-to-noise can be maximised by keeping galvanic contact resistances as low as possible, although electrode surface changes due to corrosion and scale formation can cause contact resistances to increase over time in monitoring installations[597], and weather tends to cause higher contact resistances (due to drier, harder ground conditions) with worse data quality in summer months than winter for warmer climates.

Data quality is typically assessed using stacking errors (from repeated measurements made over several cycles during acquisition), repeat errors (from repetitions of the same survey), and reciprocal errors (by performing the same survey in reciprocal configurations, i.e. with current and potential dipoles exchanged)[1057]. Stacking errors are usually smaller than repeat or reciprocal errors since the ground conditions have less time to change during stacking cycles than during the repeat period of entire surveys which can take hours due to low excitation frequencies and large numbers of electrodes. Reciprocal errors tend to be preferred over repeat errors since exchanging the dipoles also changes certain factors, such as electrode locations, magnitudes of injected current and received voltage, and degree of electrode polarization, which helps to identify some types of systematic error. Per-electrode distributions of errors can be a useful measure to identify problematic electrodes[227] as can distributions of contact resistances. Poor quality data can be discarded prior to inversion by filtering out problematic electrodes and setting limits on error estimates; magnitudes and polarities of apparent resistivity; magnitudes of geometric factors, return voltages, and contact resistances; and sensitivities to errors in position.

Once suspect data have been removed, it is often helpful to try to improve error estimates in individual measurements by constructing an error model[763]. Individual error estimates can be poor when based on differences between low numbers of repeat or reciprocal surveys (often only two). Better estimates can be produced by assuming that the error is a function of transfer resistance, binning the data on a logarithmic scale, and fitting a low-order polynomial or other simple function to the averaged data (figure 18.11). Measurement accuracy is assumed to vary smoothly by transfer resistance, and binning error estimates can provide enough data to build statistical significance. More robust error estimates, based on this fitting, enable the selection of more appropriate measurement weights in the conductivity reconstruction. Such approaches tend to lead to more robust inverse models with fewer artefacts, and have recently been extended to cover time-lapse monitoring data[617] and to account for correlations due to bad electrodes[1057].

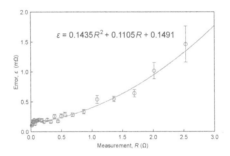

Figure 18.11   Averaged transfer resistance measurements and reciprocal error estimates, binned on a logarithmic scale, and fitted quadratic error model (BGS UKRI 2019). (color figure available in eBook)

18.12   DATA WEIGHTING

Apparent resistivities $\rho = \rho_a$ are transformed ERT *measurements* and do not represent reconstructed image units. The apparent resistivity transformation is a method of normalizing the measurement data **b** according to a geometric factor k and is essentially a method of re-weighting the data so that the magnitude of the measurements relative to a homogeneous model are all treated equally in the inversion. Without this or a similar correction, the small valued measurements will essentially be ignored because they contribute little to the overall measurement misfit in the reconstructed image. Similar to the log parametrization (see §18.5) the apparent resistivity (18.2) may be formulated as a weighting matrix in the reconstruction, which modifies the data misfit and Jacobian terms

$$\rho_i = \frac{V_i}{I_i}k \qquad\qquad\qquad \rho = \mathbf{G}_\rho\,\mathbf{b}$$

$$\mathbf{J}_{\rho,i,j} = \frac{\partial \rho_i}{\partial \sigma_j} = \frac{\partial \rho_i}{\partial \mathbf{b}_i}\frac{\partial \mathbf{b}_i}{\partial \sigma_j} \qquad\qquad \mathbf{J}_\rho = \mathbf{G}_\rho\,\mathbf{J}_\sigma \qquad (18.12)$$

for the i-th measurement $\mathbf{b}_i = V_i$ and j-th conductivity element σ_j, where the measured transfer resistance V_i/I_i scaled by the geometric factor k_i. which can be represented as the diagonal measurement normalization matrix $\mathbf{G}_{\rho,(i,i)} = k_i/I_i$. The conductivity Jacobian $\mathbf{J}_\sigma$ is modified by the same normalization matrix $\mathbf{G}_\rho$ after applying the chain rule.

The log of apparent resistivity ($\ln \rho$) is used in ERT reconstructions when the measurements differ from the homogeneous model by orders of magnitude. For example, this range of measurements can occur during a freeze-thaw cycle where liquid ground water (very low resistivity) freezes (very high resistivity). Apparent resistivity can become negative for small magnitude measurements on slightly misplaced electrodes, making log scaling unusable for some datasets[1148].

The apparent resistivity conversion results in a right matrix multiplication of $\mathbf{G}_\rho$ with the conductivity Jacobian $\mathbf{J}_\sigma$ and measurements $\mathbf{b}$. For a time difference Gauss-Newton update, the inverse measurement covariance matrix $\mathbf{W}$ can fulfill the same role as the apparent resistivity measurement transformation

$$\Delta\sigma = (\mathbf{J}_\rho^\mathsf{T}\mathbf{J}_\rho + \lambda^2\mathbf{Q})^{-1}\mathbf{J}_\rho^\mathsf{T}(\Delta\mathbf{b}_\rho) = (\mathbf{J}_\sigma^\mathsf{T}\mathbf{G}_\rho^2\mathbf{J}_\sigma + \lambda^2\mathbf{Q})^{-1}\mathbf{J}_\sigma^\mathsf{T}\mathbf{G}_\rho^2(\Delta\mathbf{b}) \qquad (18.13)$$

$$= (\mathbf{J}_\sigma^\mathsf{T}\mathbf{W}\mathbf{J}_\sigma + \lambda^2\mathbf{Q})^{-1}\mathbf{J}_\sigma^\mathsf{T}\mathbf{W}\Delta\mathbf{b} \qquad \text{with } \mathbf{W} = \mathbf{G}_\rho^2 \qquad (18.14)$$

for change in conductivity $\Delta\sigma$, apparent resistivity Jacobian $\mathbf{J}_\rho$, regularization $\mathbf{Q}$ with hyperparameter λ, using apparent resistivity time difference measurements $\Delta\mathbf{b}_\rho$. The inverse measurement covariance $\mathbf{W}$ typically serves to weight the measurements in the reconstruction, often by an estimated noise variance. If an apparent resistivity reconstruction uses a uniform noise estimate $\mathbf{W} = \mathbf{I}$, then using apparent resistivity is exactly equivalent to $\mathbf{W} = \mathbf{G}_\rho^2$, by which we can see that apparent resistivity is an objective method of applying a data weighting to measurements, possibly modified by available measurement error estimates.

18.13   AVAILABLE HARDWARE

A sample of available ERT hardware is listed in Table 18.1. The table identifies some key characteristics of commonly available commercial ERT systems currently in use world-wide, including system name and the "source," a company or organization selling and supporting the system. The number of electrodes for some systems are listed as "independent" (indep) and "maximum" (max) numbers of electrodes for systems supporting switching infrastructure between electrode cables (often through external switching boxes): a system supporting 2 cables of at most 256 electrodes would be listed as "256 indep (512 max)." The number of channels "# chan" indicates the number of simultaneous measurements available in the system. Systems supporting switched DC ($\pm$DC) and AC (sinusoidal) stimulus, as well as systems with Induced Polarization (IP) and Self-Potential (SP) measurement capabilities are identified with bullets •. The typical system power requirements and operating frequencies are also listed, where some systems can operate in high or low power modes. There is a remarkable difference in the power requirements of switched DC versus AC systems. For fields where information was not publicly available, question marks "?" have been indicated.

Table 18.1
ERT hardware

System	Source	Electrodes	# chan	±DC	AC	IP	SP	Power (W)	Freq (Hz)
ALERT	BGS[1]	256 indep (512 max)	10	•		•	•	200	0.5–2
PRIME	BGS	1024 indep	7		•	•		10	0.1–200
GEOMON 4D	GSA[2]	unlimited	> 1?	•		•	•	?†	?
GeoTom/ A_ERT	Geolog[3]	100	4		•	•		10	0.5–25
WGDM-9/ WERT-120	Langeo	120	?	•		•	•	7.2k	0.017–1
SuperSting	AGI	224	8	•		•	•	200	0.07–5
ZETA	Zonge Intl.	30 indep (7680 max)	?	?	?	?	?	?	?
Terrameter	ABEM	81 indep (16384 max)	12	•		•	•	250	< 300
IRIS	Syscal	120	10	•		•	•	250/1.2k	0.125–4
4PL	Lippmann	100	1		•	•		10	0.26–30
DAS-1	MPT	64 indep (16384 max)	8	•	•	•	•	250	≤ 225*
IRIS	FullWaver	2/node	1/node	•		•	•	10k	< 50
Flashres	ZZRI	64 indep	61	•				250	0.1–1
Geotection	HGI	∞	180	•			•	960	0.05–5
OhmMapper	Geometrics	?	?	?	?	?	?	?	?
V-fullWaver	II	?	?	?	?	?	?	?	?

BGS: British Geological Survey; GSA: Geological Survey of Austria; MPT: Multi Phase Technologies; ZZRI: ZZ Resistivity Imaging; II: Iris Instruments; † 235 W solar, 25 W fuel cell; * 0.016–13.5 DC, ≤ 225 AC; 1 [595], 2 [1013], 3 [468], ?: unpublished ∞: unlimited electrodes (multiplexed in groups of 180);

18.14 AVAILABLE SOFTWARE

A sample of available ERT software is listed in Table 18.2. The variety of software for ERT reconstructions reflect a mature (if specialized) market with a relatively rich array of commercial offerings. The most common commercial software is likely Res2DInv and Res3DInv (Geotomo), though it is difficult to externally judge uptake by academic and commercial entities. All codes support the point electrode model (PEM); these have not been indicated in the table.

18.15 DISCUSSION

In general, one can observe that the similarities between geophysical ERT and biomedical EIT outweigh the differences. Both EIT and ERT face similar resistivity/conductivity reconstruction issues directly because they are tackling the same mathematical problem. ERT has a long history but many of the major advances in reconstruction and instrumentation have been driven by wider technological advances: the numerical techniques available due to faster, better computational resources, and the availability of improved electronic components. Its long history means that ERT has benefited from a degree of world-wide commercial adoption: a widely recognized electrical tool in geophysical site investigations. EIT has similarly benefited from these enabling technologies, and is perhaps at the cusp of widespread commercial adoption.

Table 18.2
ERT software

Software	Author	License	2.5D	3D	IP	CEM
BERT/pyGIMLI	Günther & Rücker	GPLv3+	•	•	•	•
E4D	Johnson/PNNL	BSD		•	•	
R2/R3t	Binley	csf	•	•		
cR2	Binley	csf	•		•	
Res2D/3DInv,	Loke/GeoTomo	$	•	•	•	
ZondRes2dp/3d,	Zonge Intl.	$	•	•	•	
EarthImager2D/3D	AGI USA	$	•	•	•	
ERTLab	MPT	$		•	•	
Aarhus Workbench	HG-AU	$	•		•	
VOXI	Geosoft	$		•	•	
DCIP2D/3D	UBC-GIF	a/$	•	•	•	
DC_2D/3DPro	Kim/KIGAM	a/$	•	•		
ResInvM3D	Pidlisecky	SEG		•		
IP4DI	Karaoulis	BSD	•	•	•	
ELRIS2d	Acka		•			
EIDORS	Adler	GPLv2/3	•	•	•	•
SimPEG	Cockett	MIT	•	•	•	
V-fullWaver	Iris Instruments	?	?	?	?	?

$: commercial; : source available, copyright retained; a: academic; csf: closed source/freeware; GPLv3+: GPL v3 and Apache v2; SEG: SEG Open Source; PNNL: Pacific Northwest National Laboratory, United States Department of Energy; MPT: Multi Phase Technologies; HG-AU: Hydrogeophysics Group, Aarhus University; UBC-GIF: University of British Columbia Geophysical Inversion Facility; KIGAM: Korea Institute of Geoscience and Mineral Resources

There are many examples of parallel research in the two fields of EIT and ERT, but the language used to describe biomedical and geophysical EIT/ERT problems and their solutions differ enough that interdisciplinary research, or even discovering related research, is challenging. In addition, access controls on the literature limit discoverability because many institutions have either biomedical or geophysical library access to journals but not both. Recent movements toward open access publishing are breaking some of these barriers.

The technical language of EIT and ERT differ in their specifics, including mathematical conventions, and the translation to a familiar framework can be challenging. Despite the technical language barriers, there is a richness in exploring geophysical techniques and looking for opportunities to apply these to biomedical problems; opportunities to open new avenues of research or imagine new areas of application.

ACKNOWLEDGEMENTS

The contributions of Wilkinson are published with permission of the Executive Director of the British Geological Survey.

19 Industrial Process Tomography

Thomas Rodgers, William Lionheart and Trevor York
University of Manchester, UK

CONTENTS

19.1 INTRODUCTION

In this chapter we explore some of the differences between biomedical applications of electrical impedance and magnetic induction tomography and applications in industrial process monitoring. The very first attempt at electrical resistivity imaging (or at least electrical prospecting) was probably the Sclumberger brothers resistivity experiments in their mother's copper bath tub in Chateau

DOI: 10.1201/9780429399886-19

de Crevecoeur, Normandy in 1912[39]. However it was not until biomedical EIT became popular, largely from the work of Barber and Brown in the 1980s, that the use of low frequency electric currents were seriously studied, from the 1990s, as a way to non-invasively monitor vessels containing conductive fluids in industrial processes such as mixing and filtration. Much of the initial work was done by a group lead by Maurice Beck at UMIST in Manchester. The term Electrical Resistance Tomography (ERT) was coined for this technique (in slight contrast to Electrical *Resistivity* Tomography, as used in Geophysics, chapter 18). This technique followed closely the method used for test tanks (phantoms) by biomedical EIT researchers. A typical configuration being a single ring of sixteen circular or rectangular metal electrodes around a cylindrical vessel. Pairs of electrodes were excited with an alternating current while the magnitude of the voltage was measured on other electrodes. By contrast to biomedical applications using adhesive electrodes on the skin industrial process monitoring had the advantage of a rigid tank with regularly placed electrodes.

While ERT was aimed at imaging conductivity contrasts in conductive fluids a related method, electrical capacitance tomography (ECT) was devised for dielectric objects such as particles in a pneumatic conveyor. In this case larger electrodes were used to maximize the capacitance, typically with shielding around the outside of the vessel. An electrode was excited with MHz pulses while the displacement current was measured to the other electrodes that were grounded. Other than being a voltage driven system with measured currents, mathematically the problem of recovering a distribution of permittivity is very similar to recovering conductivity. The contrasts under investigation were typically quite small and hence the linear approximation was useful although the number of electrodes used was typically fewer than 16.

The terminology used to describe tomographic imaging methods varies. For example X-ray, neutron, gamma-ray or microwave tomography all refers to the waves or articles that pass through the object. Electrical resistance/impedance/capacitance tomography refers to the measurement that is made externally, where as Electrical resistivity/permittivity/conductivity tomography refers to the physical parameter being imaged. Imaging of conductivity, permittivity and indeed magnetic permeability of an object can also be attempted using an array of coils to excite the object with an alternating magnetic field and measure the response of the field to the object. In a context in which the length scale of the object being imaged being much smaller than a wavelength we can regard the coils as being inductively coupled, rather than a radio transmitter and receiver. This is commonly called "Magnetic Induction Tomography" using the convention that it is a measurement of magnetic inductance (chapter 16). The potential advantage is that the coils need not be in contact with the object measured and indeed with a larger gap than ECT. In common with ECT, MIT has the problem of also measuring changes caused by objects outside the coil array. The technique is especially good at locating and characterizing high conductivity objects.

The promise of electrical methods were that, while having a very low spatial resolution, they had a high temporal resolution and could be deployed *in situ* in an industrial plant or in a laboratory test rig that duplicated industrial conditions. Methods such as gamma ray tomography were also available but added radiological hazards to an already hazardous environment. The hardware for electrical imaging methods was also relatively low cost. Often the goal of industrial process imaging is more straightforward than in medical imaging. The detection of an undersirable state as a warning that maintenance is needed, the measurement of a flow or the amount of mixing or separating to provide feedback to alter process inputs or controls for example. Typically making the process more efficient, saving money or increasing quality of the output directly leads to lower costs or increased profits.

Progress in industrial process tomography is recorded in the proceedings from a number of international meetings. From 1992 to 1995 European activity was coordinated through the European Concerted Action on Process Tomography.. This was followed by two international conferences, "Frontiers in Industrial Process Tomography", that were organized by the Engineering Foundation in 1995 and 1997. The "World Congress on Industrial Process Tomography" was first held in Buxton, UK, in 1999, and there have been subsequent meetings in Hannover, Germany (2001); Banff, Canada (2003); Aizu, Japan (2005); Cape Town, South Africa (2012); Jeju, South Korea (2014); Dresden, Germany (2015); Iguassu Falls, Brazil (2016) and Bath, UK (2018). The previous edition of this book[484] included a comprehensive chapter by Trevor York reviewing electromagnetic methods in industrial process tomography circa 2004. See also an earlier review "Status of Electrical Tomography in Industrial Applications"[1191]. An introduction to process tomography can be found also in Williams and Beck[1154]. The current chapter has a more modest aim to highlight the contrast and commonality between biomedical and industrial applications and to provide some contemporary notes on the ways in which these techniques are currently used.

19.2 DATA ACQUISITION

Electrical tomography systems comprise sensors, measurement electronics, switching electronics, signal conditioning, analog-to-digital conversion, communications and a computer hosting control and data processing, including inversion, analysis and display algorithms. In many applications it is beneficial to obtain measurements of more than one parameter. For instance, in an oil pipeline the flow may be oil continuous, in which case capacitance tomography would be appropriate; or water continuous, in which case resistance tomography would be appropriate. To take full advantage of such an approach it is necessary that the signals from the various sensors are synchronized, and for this reason efforts have been applied to a multi-modal instrument[491]. Indeed a common trend throughout imaging in medicine, geophysics and industry is synthesis of data from multiple physical measurements.

Typically, electrodes are located in rings around regions that are to be interrogated. For capacitance and inductance systems the electrodes are frequently non-invasive, lying outside the vessel wall, as well as non-intrusive – touching but not penetrating the materials in the vessel. For resistance measurements the electrodes are usually invasive but not intrusive. However, for vessels with conducting walls it is necessary in all cases for the electrodes to be located inside the vessel. It is becoming apparent that there are a number of applications, notably for batch processes, which allow electrodes to be placed above and below the vessel as well as around the circumference. This allows a richer variety of electrode configurations to be considered which should, in turn, lead to higher quality results.

Signals from the sensors are routed to the measuring electronics by a multiplexer, which is usually implemented using solid state switches. Parasitics that are associated with these switches are particularly important, affecting switching speed and noise, and selection of appropriate devices is an important decision. When the initial signals have been amplified and buffered programmable gain and offset are usually employed, to accommodate a wide range of signals, with demultiplexing and filtering in analog hardware. The characteristics of the electrical modalities are summarized in table 19.1.

Electrical tomography has motivated applications for process design and validation, on-line monitoring and control. This can, for instance, lead to improved product quality and process efficiency, with accompanying improved profits through reduced time and waste. There are also important consequences for environmental issues and the reduction of exposure hazards for plant operators. Typical fields of application in the early years of development included two-phase flow, fluidized beds, mixing and environmental monitoring[4,5,6,7].

19.2.1 ELECTRICAL RESISTANCE TOMOGRAPHY (ERT)

ERT measurements correspond to EIT considered elsewhere in this book. Early systems were strongly influenced by developments in the medical field, notably the Sheffield Applied Potential

Tomography system[1156]. Electrodes are relatively small and placed in contact with the conducting materials. Commonly industrial applications use a sinusoidal current source to a pair of electrodes, at a frequency of some tens of kilohertz, and to measure the resulting electric potentials between other pairs of electrodes. This adjacent strategy provides high sensitivity near the vessel walls, but is poor in the centre of the region. Alternatively, other strategies can be adopted, for instance to inject current between opposite electrodes. An ERT system that resulted from work done at UMIST has been developed into a commercial instrument by Industrial Tomography Systems Ltd. (http://www.itoms.com).

Table 19.1
Comparison of electrical tomography techniques

Method	Arrangement	Measurand	Material properties	Typical material
ECT	Capacitance plates	Capacitance C	$\varepsilon_r, 10^0 - 10^2$ $\sigma < 10^{-1}$ S/m	Oil De-ionized water Non-metallic powders Polymers Burning gases
ERT	Electrode array	Resistance (impedance) $R(Z)$	$\sigma, 10^{-1} - 10^7$ S/m $\varepsilon_r < 10^2$	Water/saline Biological tissue Geological materials Semiconductors
EMT	Coil array	Self/mutual inductance (L/M)	$\sigma, 10^2 - 10^7$ S/m $\mu_r < 10^4$	Metals Some minerals Magnetic materials Ionized water

19.2.2   ELECTRICAL CAPACITANCE TOMOGRAPHY (ECT)

Distributions of electrical permittivity are determined from measurements of current around the boundary of a vessel. For capacitance measurements, electrodes must have a large surface area in order to provide sufficient signal. Electrodes are often located outside the vessel, such that the technique is non-invasive as well as non-intrusive. In contrast to ERT, an a.c. voltage signal is usually applied to a drive electrode and the resulting current on the remaining electrodes is measured. Typical excitation frequencies, to provide sufficient sensitivity, are about 1 MHz. The main difference between the various systems that have been described is the use of either sinusoidal or square pulse excitation, often referred to as 'charge-discharge'. The sinusoidal approach provides a readily analysed measurement in which the phase as well as the magnitude can be exploited, but this is at the cost of increased complexity to generate and demodulate the signals.

The biggest challenge is to detect changes of the order of femtofarads (10^{-15} F) in the presence of standing capacitances that are typically 10 times larger, and stray capacitances that are of the order of 100 pF. Electrodes and cables must be well screened and careful attention must be given to the layout of printed circuit boards. The use of 'driven' guards, in an attempt to confine the excitation signal to a single plane, has enjoyed some success. Process Tomography Ltd (PTL) manufactures capacitance tomography systems based on the so-called charge–discharge system developed at UMIST.

19.2.3 MAGNETIC INDUCTION TOMOGRAPHY (MIT)

MIT (or Electromagnetic tomography, EMT) is treated in detail in chapter 16. EMT seeks to determine the distribution of electrical permeability or conductivity from boundary measurements of mutual inductance. The region of interest is interrogated with a time varying magnetic field. Non-conducting, magnetic materials, such as ferrite, increase the measured signal. High conductivity, non-magnetic materials, for instance non-ferrous metals, decrease the signal, and low conductivity materials, such as saline, produce a small change in the quadrature component. For low conductivity materials there is an increase in the measured signal for increases in excitation frequency, and consequently values of 1–20 MHz are popular. Due to the skin effect excitation frequencies are limited to a maximum of about 100 kHz for high conductivity materials.

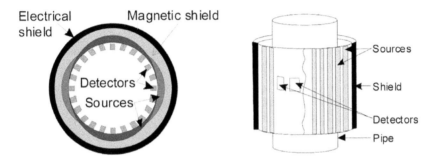

Figure 19.1   Diagram of a Magnetic Introduction Tomography (MIT or EMT) system (a) Parallel-field system, (b) current-strip source system [1192].

A major difference between MIT and the other electrical methods is in the operation of the sensor array.

- *Use of coils.* Coils can give enormous flexibility in the design of arrays. For example, coils can be superimposed allowing excitation and detection elements in virtually the same positions, and measurements combined to cancel the background signal. For some systems a parallel field is established using two orthogonal excitation coils, in which varying magnitudes are used to generate a rotating field.
- *Screening.* Magnetic screening is generally accepted as being difficult compared with electrical screening. If the external environment is defined, the screening is not required, as external conductive or magnetic objects will have a constant effect, which can usually be subtracted during calibration. Otherwise magnetic shielding is required, typically a high permeability material to provide a low reluctance return path for the interrogating field.

19.2.4 ELECTRICAL IMPEDANCE TOMOGRAPHY

As suggested above, most electrical tomography systems that have been described for industrial applications are only single modality, and measure resistance, capacitance or inductance to yield information on resistivity, permittivity or permeability distributions. This somewhat undesirable situation has arisen due to the contrasting requirements of circuitry to optimize the measurement of each component. However, commercial instruments to determine complex impedance from four-point measurements of amplitude and phase are readily available for other application areas, and this approach should be considered for electrical tomography.

19.2.5 INTRINSICALLY SAFE SYSTEMS

Many industrial processes operate in hazardous environments. For instance, the use of solvents presents a potentially explosive atmosphere. In order to exploit the benefits of electrical tomography in such cases, it is essential to provide certified safe equipment. York et al[1192] describe the design of the world's first, certified, Intrinsically Safe (I.S.) electrical tomography system. This has been designed for a research project that is seeking to monitor progress during pressure filtration of agrochemical products, as described in §19.3.3 below, but could be readily applied in other application areas which may or may not involve tomographic processing.

Across the process sector many organic solvents and products are common, which are flammable in air or other gas mixtures. To allow electrical apparatus to be applied within such an environment, a branch of engineering has been developed to classify the risk and reduce the probability of an ignition source being present. This methodology is a legal requirement in Europe and industrial nations elsewhere. There are a number of ways of ensuring that a flammable atmosphere is isolated from any significant energy source, but by its nature, electrical tomography requires energy to be injected into a potentially flammable atmosphere.

I.S. certification relies on constructing apparatus in such a manner that the maximum electrical energy that can be provided to the flammable atmosphere during normal operation or in the case of worst case failure will be less than the minimum ignition energy of the flammable gas mixture into which it is placed. It was decided to simplify the problem by locating only passive electrical components in the hazardous, "Zone 0", environment found within the pressure filter. In other words, only the electrodes and connecting wires are located in the hazardous area. All active, electronic, components and power supplies are located on the safe side of the barrier. By limiting the dimensions of the electrodes and the maximum capacitance and inductance of the interconnecting wires, it has been possible to define the equipment within the hazardous area as "simple apparatus" and allow up to 50 m of co-axial cable to be connected to each of the electrodes. For an intrinsically safe system it is necessary to define a boundary between the hazardous and non-hazardous areas. In the system described by York et al, all of the interface electronics and the control computer are mounted remotely in the plant switch room, which is a safe area some 50 m from the filter.

19.3 PREVIOUS INDUSTRIAL APPLICATIONS OF ELECTRICAL TOMOGRAPHY

Previous reviews have summarized early applications of electrical tomography[98]. Table 19.2 directs the reader to some interesting recent reports. The following sections summarize developments in a number of contrasting application areas. Much of the material has been extracted, with approval, from earlier papers by the researchers involved. Rather than presenting exhaustive lists, it is intended that reference to recent publications will direct the reader to related earlier work. Criteria for selection of the applications presented here include progress towards industrial benefit, contrasting modalities, sensing challenge and on-going effort.

19.3.1 APPLICATIONS OF ELECTRICAL RESISTANCE TOMOGRAPHY TECHNOLOGY TO PHARMACEUTICAL PROCESSES

Ricard et al[877] evaluate the applicability of ERT to pharmaceutical process development. A 3.5 litre, 150 mm diameter, glass reactor, located at GSK, has been fitted with 64 electrodes arranged in four planes, as shown in figure 19.2.

The platinum electrodes were deposited in liquid layers and have high chemical resistance with a thermal expansion coefficient that matches the walls of the reactor. A P2000 ERT system from Industrial Tomography Systems Ltd. was used to acquire measurements and reconstruct images.

Table 19.2

Recent reported applications of electrical tomography

Process	Modality	Status
Bead milling [947]	ECT	Industrial tests
Hydrocyclone monitoring [124]	ERT	Industrial tests
Monitoring pressure filtration [380,1198]	ERT	Industrial tests
Pneumatic conveying [59]	ECT	Industrial tests
Density flowmeter [232]	ECT	Industrial tests
Nylon polymerization [249]	ERT	Industrial tests
Onset of crystallization in steel production [964]	EMT	Industrial tests
Nuclear waste site characterization [223]	ERT	Field tests
Waste storage ponds [113]	ERT	Field tests
Subsurface resistivity [332]	ERT	Field tests
Leaks in buried pipes [520]	ERT	Field tests
Flame monitoring [1122]	ECT	Laboratory tests
Fluidized beds [489,1140]	ECT	Laboratory tests
Multi-phase flow [643,801]	ERT	Laboratory tests
Bubble column dynamics [103,1138]	ECT, ERT	Laboratory tests
Pneumatic conveying [512,797]	ECT	Laboratory tests
Mixing in a stirred vessel [489,682]	ERT	Laboratory tests
Foam density distribution [198,801]	ERT, ECT	Laboratory tests
Powder flow in a dipleg [1117]	ECT	Laboratory tests
Belt conveyor [1153]	ECT	Laboratory tests
Blast furnace – hearth wall thickness [1116]	ERT	Laboratory tests
Dust explosions [831]	ECT	Laboratory tests
Solid rocket propellant [1046]	ECT	Laboratory tests
Metal solidification [824]	EMT	Theoretical
Paste extrusion [1138]	ERT	Laboratory tests
Flow of molten steel [665]	EMT	Industrials
Pneumatic conveying [512]	ECT	Laboratory tests
Imaging wet gas [1184]	ECT	Laboratory tests
Slurry transport [802]	ERT	Laboratory tests

Using the adjacent current strategy for the 64 electrode ERT sensor described above, there are effectively 1264 non-intrusive electrical conductivity probes so that a much higher data density is obtained when recording the distribution of a tracer compared with the traditional method of inserting conductivity probes.

The tracer distribution images obtained from the mixing time experiments were compared with computational fluid dynamics (CFD) results, as shown in figure 19.3. The tracer is seen to cover a large proportion of the surface before being ingested into the bulk. After it reaches the impeller a well mixed zone emerges. The final layer to be mixed lies between the well mixed impeller zone and the surface. The results suggest that there is some advantage to adding material close to the baffle and working with a liquid height equal to the impeller diameter.

Conclusions from this work suggest that ERT shows promise for on-line control of process mixing performance, as well as efficiency evaluation and optimization of reactor geometries. Results show successful modelling and analysis of pharmaceutical mixing processes. ERT is capable of offering superior mixing time information for vessel characterization purposes compared with existing techniques, and can also provide valuable data for CFD validations. The authors plan for the work to evolve to an increased level of process complexity with the study of multiphase, solid/liquid systems.

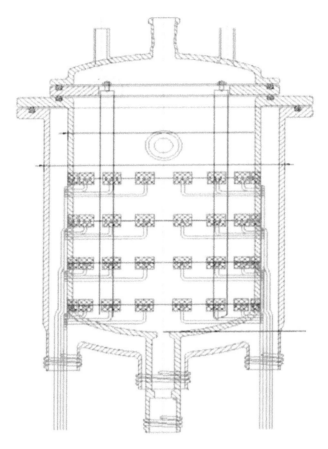

Figure 19.2   Overall design of the ERT reactor.

19.3.2 IMAGING THE FLOW PROFILE OF MOLTEN STEEL THROUGH A SUBMERGED POURING NOZZLE

Continuous casting is a process by which molten steel is formed into semi-finished billets, blooms and slabs. Liquid steel from the basic oxygen steelmaking (BOS) or electric arc furnace (EAF) process, and subsequent secondary steelmaking, is transferred from a ladle, via a refractory shroud, into the tundish. The tundish acts as a reservoir, both for liquid steel delivery and removal of oxide inclusions. Primary solidification takes place in the water-cooled copper mould, and casting powder is used on the surface to protect against re-oxidation and serve as a lubricant in the passage of the strand through the mould. Exiting the mould, the strand consists of a solid outer shell surrounding a liquid core. This is continuously withdrawn through a series of supporting rolls and banks of water sprays, where further uniform cooling and solidification take place. The resulting cooled and solidified strand is finally divided by cutting torches into pieces as required for removal and further processing.

In continuous casting, control of molten steel delivery through the pouring nozzle is critical to ensure the stability of the meniscus and to create the optimum flow patterns within the mould. These factors influence the surface quality and also the cleanliness of the cast steel product. Knowledge of the flow regime enable improved control of conditions in this area of the caster. The authors describe the application of electromagnetic tomography (EMT) to the imaging of the flow profile of molten steel through a submerged pouring nozzle.

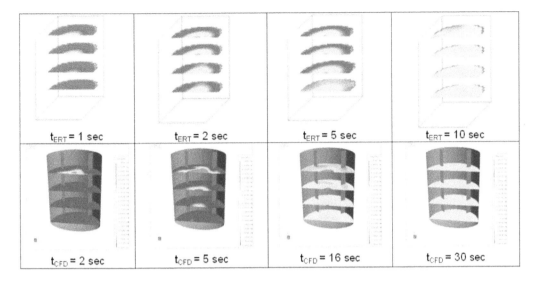

t_{ERT} = 1 sec	t_{ERT} = 2 sec	t_{ERT} = 5 sec	t_{ERT} = 10 sec
t_{CFD} = 2 sec	t_{CFD} = 5 sec	t_{CFD} = 16 sec	t_{CFD} = 30 sec

Figure 19.3   Comparison between ERT and CFD tracer plots at selected timesteps.

Tomography is important in this application because it demonstrates the ability to measure real flows, but the steel producers are not really interested in images. Full scale industrial implementation would require a simpler system, with fewer coils and a GO/NO-GO output. An important practical point is that the sensor cannot totally enclose the nozzle, as it must be possible to withdraw it quickly if something goes wrong.

19.3.3   THE APPLICATION OF ELECTRICAL RESISTANCE TOMOGRAPHY TO A LARGE VOLUME PRODUCTION PRESSURE FILTER

Pressure filtration is a generic process operation applied across the chemical industry for rapid, cost-effective separation and drying of a solid phase from a liquid slurry. Existing instrumental techniques are inadequate for providing both diagnostic information and measured variables on which to apply closed-loop control. This results in sub-optimal process settings, which are designed to accommodate the worst-case conditions. The effect of this is pervasive; at the very least there will be an extended pressure filtration cycle time, which implies an inefficient use of the asset. In addition, there may be yield loss when processing an unstable intermediate product, poor energy usage during elevated temperature drying, or additional environmental impact through excessive use of wash solvents.

To address these issues, Electrical Resistivity Tomography (ERT) is being developed to provide real-time information on end point of filtration and drying; imperfections in the filter cake; and solvent displacement of the mother liquor. As the filters operate in potentially explosive environments, it is necessary to employ intrinsically safe equipment as described in §19.2.5.

A suitable 36 m³ asset was identified on the Syngenta Huddersfield site (figure 19.4).

The scale of the unit is readily appreciated from consideration of the doors on the left-hand side of the photograph. This unit is of metallic construction, with a non-conductive filter cloth. As the vessel was not originally designed to accept ERT electrodes, an additional series of challenges soon became apparent:

- *Electrode geometry*: It was agreed with the plant management that the pressure rating of the vessel could not be jeopardized by attempting to machine into the wall of the unit. This led to the alternative option of mounting the 24 electrodes in a planar arrangement above the filter cloth.

Figure 19.4   36 m^3 subject pressure filter.

- *Electrode design*: To locate the electrodes above the filter cloth, it was necessary to design an assembly that could be easily removed during routine cloth replacement and which would be small enough to not affect the normal operation of the filter.
- *Materials of construction*: In common with the majority of processes operated within the chemical industry, the materials of construction of the subject process unit were carefully selected to prevent erosion and corrosion. The demonstration filter is predominantly hastelloy-C276, an alloy of nickel, with a mesh fabricated from polypropylene. These materials, together with PTFE, PVDF and viton, for the O-ring elastomer, were used exclusively in the electrode assembly.
- *Cable routing*: The pressure vessel had no provision for additional flanges through which the 24 electrode cables could exit. Surprisingly, for such a large vessel, the best solution involved routeing the 24 cables through two 1 cm diameter air balance ports.
- *Operational constraints*: As the demonstration unit was also a manufacturing asset, access to get into the filter to fit the electrodes was severely restricted to an existing time window during the planned annual maintenance period. The effect of this was to limit the electrode installation time to a four-day period each year. The usable resource was further constrained as safety procedures dictated that to ensure a breathable atmosphere within the vessel only two people could enter the unit at any one time.

Results show the effect of the slurry, acetic acid and water washes can be seen and the tomographic measurements clearly track the process. The tomography measurements lag behind the level measurements and it is reasonable to assume that this is due to the time for the liquid to pass through the cake.

Over three years results were repeatable and the electrodes are transparent to the process. The main challenge is to deliver 3D images and this is being impeded by the proliferation of metal current sinks in the vessel.

19.3.4   A NOVEL TOMOGRAPHIC FLOW ANALYSIS SYSTEM

Hunt *et al*[494] describe a flow analysis system, Tomoflow R100 ECT, which uses twin-plane tomographic data to derive detailed pictures of the velocity and concentration structure within complex two-phase flows. Initial results have been obtained using electrical capacitance tomography (ECT), but other modalities may also be used.

Twin-plane sensors are used in conjunction with guard electrodes to create two image "planes" that are separated axially along the flow. Each "plane" is in fact a cylinder of finite length made up of 812 pixels on a 32×32 square. For eight-electrode systems, the cross-sectional flow is divided into 13 zones each containing approximately 62 pixels. Results show a peak if the flow structures are coherent over the sensor length and contains information about the time domain statistics of the flow–primarily convection and dispersion. The simplest assumption is that the time delay at the peak of the correlogram corresponds to the transit time of flow structures between the two planes.

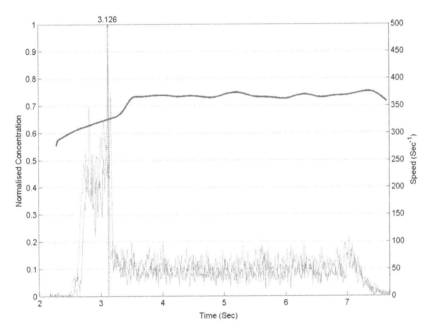

Figure 19.5   Concentration (left-hand scale) and velocity (right-hand scale) against time in centre zone.

Figure 19.5 shows the concentration and velocity against time for the central zone of a 13-zone map for a typical test with data acquisition of 200 frames per second. The dashed line shows the concentration in the first plane, light grey shows the concentration in the second plane and the black line shows velocity.

This work demonstrates the feasibility of making a flowmeter for blown and gravity-fed solids. A few technical challenges remain, for instance calibration and varying moisture content of materials, but these are likely to be solved in the near future. The main obstacle to implementing a full scale commercial integrated flowmeter is availability of capital on the 3–5 years scale to fund the large engineering programme to launch the product. This would involve engineering design, integration of electronics, manufacturing route, marketing, distribution and servicing. The technical risk is small, but the commercial risk is difficult to evaluate as there is not a current market because such flowmeters do not exist.

19.3.5 APPLICATION OF ELECTRICAL CAPACITANCE TOMOGRAPHY FOR MEASURE-MENT OF GAS/SOLIDS FLOW CHARACTERISTICS IN A PNEUMATIC CONVEYING SYSTEM

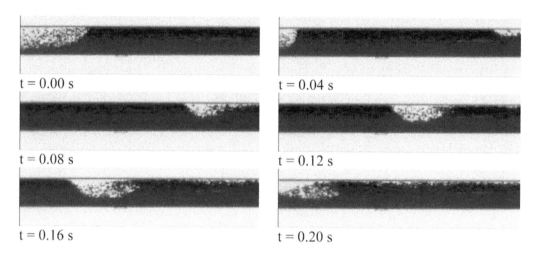

t = 0.00 s   t = 0.04 s

t = 0.08 s   t = 0.12 s

t = 0.16 s   t = 0.20 s

Figure 19.6   Video images of slug flow in a horizontal pipe.

Applications of pneumatic conveying (i.e. the use of air for transporting granular materials, such as flour, coal, lime, plastic pellets, granular chemicals etc.) along pipelines date back as early as the mid-19th century. In 'dilute' (or 'lean') phase conveying, the particles are usually transported in the form of a suspension with the solids concentrations typically below 10%. For "dense-phase" transport the pipe is filled with particles at one or more cross-sections. Previous studies show that the predominant mechanism for solids transport is due to flow instabilities referred to as 'slugs' and 'plugs'. Jaworski and Dyakowski[512] report the study of pneumatic conveying using twin-plane ECT, supported by high-speed video and pressure measurements.

Figure 19.6 illustrates shows a series of six photographs illustrating the passage of two consecutive slugs in the horizontal pipe. These images clearly illustrate some of the parameters of interest that are associated with such slugs, such as height and density of slug and slope of leading and trailing edges. Figure 19.7 shows a time series of cross-sectional tomographic images corresponding to the slug flow shown in figure 19.6. The first seven images show the transition between a half-filled and fully-filled pipe that corresponds to the passage of the slug front. Similarly, the last four images show the passage of the slug's tail through the measurement plane.

The use of a twin plane system allows the shape of the slugs to be reconstructed, as shown in figure 19.7. The pixels lying along a vertical line passing through the centre are selected from each frame. These are combined to give a longitudinal cross-section of the slug, as shown in figure 19.8. Difficulties associated with such images include limited spatial resolution in the cross-sectional images, averaging of the concentration of solids along the length of the electrodes and smearing of boundaries between phases.

The solids mass flow rates that are calculated from the tomographic data underestimate those obtained by weighing of the material by about 20–30%. Several issues for further research were identified by the authors:

- The electrodes are of finite length and therefore it is not obvious which electrode distance should be taken for calculating the velocity of flow structures.
- For improved accuracy, cross-correlation analysis should be performed on the pixel-by-pixel basis rather than for the whole cross section.

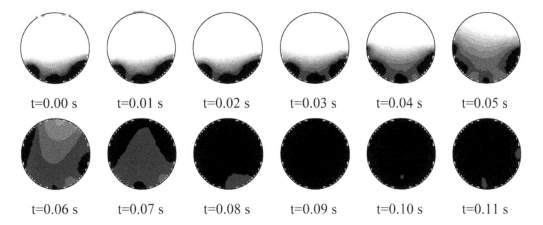

images between t=0.12 s and t=0.29 s are omitted as they show fully filled pipe (black)

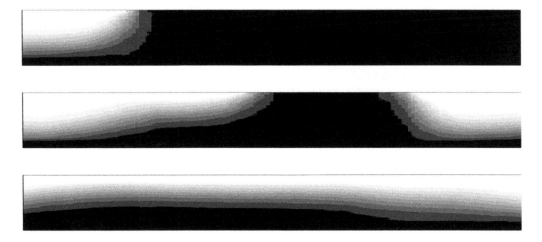

Figure 19.7   ECT images of slug flow in a horizontal pipe.

Figure 19.8   Axial reconstruction of horizontal slug flow.

- The technique may be inappropriate for flow regimes which are close to blocking the system. In this case, long plugs of almost stationary material fill the sensor and render the cross-correlation techniques ineffective.
- A more accurate estimate of the solids mass flow will use an improved model of the relationship between material density and dielectric permittivity.

19.4 RECENT INDUSTRIAL APPLICATIONS OF ELECTRICAL TOMOGRAPHY

Previous reviews have summarized some applications of electrical tomography[98,965,1135,1191]. The following sections summarize developments in a number of contrasting application areas in which electrical tomography has had success. Rather than presenting exhaustive lists it is intended that reference to some recent publications will direct the reader to related work. Criteria for selection of the applications presented here include progress towards industrial benefit, contrasting modalities, sensing challenge, and on-going effort.

19.4.1 APPLICATION OF ELECTRICAL RESISTANCE TOMOGRAPHY TO THE MEASUREMENT OF BATCH MIXING

One area where ERT has been used with success for many years is with the monitoring of mixing with batch vessels. Typically, these systems involve the use of the standard 16 electrode ring arrangement (as one plane) in multiple planes, but on a number of industrial scale systems, a linear probe has been used as this requires much less retrofitting of existing equipment. This section will provide a few examples of different applications within batch mixing vessels, where key metrics are required, e.g., mixing time, chemical concentration, phase location/concentration.

Mixing time is often used to assess quantitatively the blending performance of stirred tanks, historically conductivity probes have been used to detect as many different local values of the mixing time as there are probes in the reactor. The mixing time over the whole tank can be obtained by combining all these local measurements.

Using the adjacent current strategy with no repeat measurements a standard 16 electrode plane provides effectively 104 non-intrusive electrical conductivity probes so that a much higher data density is obtained when recording the distribution of a tracer compared to the traditional method of inserting conductivity probes.

The majority of industrially relevant semi-batch reactions undergo a change in level during the progression of the process, due to liquid being added to the vessel. This changes the volume of the fluid being measured. Rodgers et al[891] examined a batch level change with a tracer addition to measure the mixing time in these systems in a 0.91 m diameter vessel. During the experiment, the liquid height changed by about 16.7%. The results were compared with an initial static reference and a new adaptive reference technique with a moving finite element model taking into account the change in the fluid volume, figure 19.9. This technique is very effective in reconstructing the correct conductivity distribution with these two simultaneous changes.

Very recently Alberini et al[36] have looked at using ERT measurements to accurately calculate the concentration of several chemical components in a 0.2 m diameter vessel. This has been achieved by coupling the equations for reactions inside the vessel with a machine learning algorithm to undertake the reconstruction in terms of pH and concentration, rather than an inverse problem to find the conductivity. This allows the close monitoring of different regions of the fluid and would be highly useful for the control of reacting systems.

As well as these single-phase mixing additions it is possible to track multiphase dissolution processes. Rodgers et al[892] investigated the addition of a surfactant (viscosity of around 20-30 Pa·s) in vessels ranging from diameter 0.15 to 0.91 m. This process can be seen in figure 19.10, where the conductivity rises sharply at the point of the surfactant ball. At around 10 seconds it can even be seen that the single large lump has been broken into two smaller pieces. Small mm size strands are sheared off the larger lumps, leading to areas of higher conductivity around the larger pieces; however, these are too small to be directly resolved with the ERT.

By around 60 seconds the surfactant has been broken by the agitator into only the mm size high conductivity strands and after about 150 seconds the rate of increase in conductivity slows, which

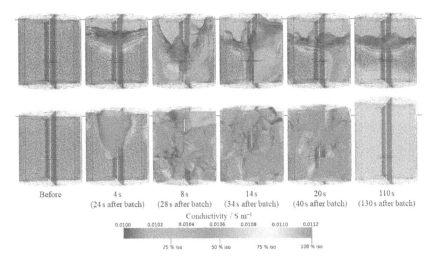

Before | 4 s | 8 s | 14 s | 20 s | 110 s
(24 s after batch) | (28 s after batch) | (34 s after batch) | (40 s after batch) | (130 s after batch)

Conductivity / S m⁻¹

0.0100   0.0102   0.0104   0.0106   0.0108   0.0110   0.0112

25 % Iso   50 % Iso   75 % Iso   100 % Iso

Figure 19.9 Variation of the conductivity distribution with time for a batch level change with an addition of a small volume high conductivity salt solution tracer (0.02% vessel volume, 11.5 S/m) part way into the batch addition (same conductivity of original vessel contents, 0.01 S/m). Top row is the solution with the reference taken as the initial liquid height. Bottom row is the solution with the adaptive reference taking into account the change in liquid volume. (color figure available in eBook)

corresponds to the later stages of the dissolution where the smaller pieces of surfactant are now slowly dissolving. As the dissolution time of the small strands is much slower than the mixing time the dissolution increases the bulk conductivity fairly evenly throughout the whole vessel during these later stages.

Jamshed et al[508,509] have monitored the distribution of gas bubbled through an agitated vessel as well as measuring the mixing time in these multi-phase systems. These were undertaken on vessels of diameter 0.61 and 0.91 m with the 16-electrode circular plane system with high gas phase fractions, up to 40% gas. Measured values of the variation of the gas hold-up with ERT measurements agreed well with results from a much smaller vessel with diameter 0.08 and much lower gas fraction[725].

Forte et al[284] undertook ERT measurements in a vessel of diameter 0.14 m with both a gas fraction and the addition of solids; however, in this case they use a 16-electrode linear probe. Gas fraction results measured, even with this different modality, agree with the other work in this area. An example of gas distributions measured with ERT is given in figure 19.11; in this case the gas concentration is calculated from the reconstructed conductivity by using the Maxwell equation.

19.4.2 APPLICATION OF ELECTRICAL RESISTANCE TOMOGRAPHY TO INLINE FLOW MEASUREMENT

As pipes and piping systems are used to convey fluids (liquids or gases) from one location to another, being able to have multi-dimensional information of the features and condition of the flow inside these usually opaque systems will have major benefits in process design, monitoring, and control. Therefore, it is not surprising that a there are a large number of ERT publications focused on this area. Various different configurations have been considered, including vertical, horizontal, or inclined pipe, most of which use the standard 16 electrode circular plane configuration with most work using pipes ranging from diameter 0.02 to 0.15 m: for example figure 19.12.

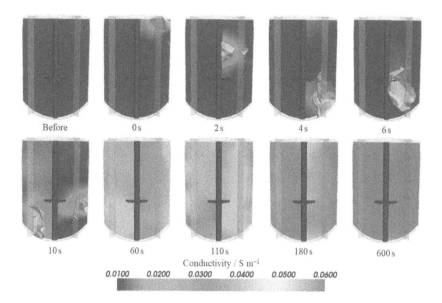

Figure 19.10   Variation of conductivity distribution with time for the addition of an approximately 0.17 m diameter ball of surfactant (very high conductivity) in a 0.91 m diameter vessel (water at 0.01 S/m). The white iso-surface approximately represents the surface of the surfactant lump (based on the normal to the conductivity gradient in vessel) until it has broken to small mm diameter strands which slowly dissolve. (color figure available in eBook)

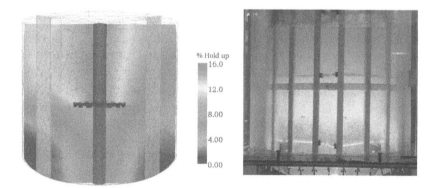

Figure 19.11   Measurement of the local gas concentration calculated from the conductivity using the Maxwell equation compared to the visual distribution. The volume average gas fraction (hold-up) measured by the ERT is 5.9% while the global value measured directly from the vessel is 6%. (color figure available in eBook)

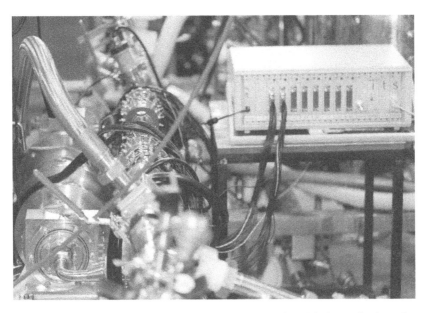

Figure 19.12   Typical set up for inline pipe measurement using 16 electrode ring planes. (color figure available in eBook)

One common multi-phase system examined in pipe flow is gas-liquid flow, where the regime of flow is measured at a range of gas and liquid fractions. This is commonly undertaken using two or more circular ERT planes such as Deng *et al*[236] where they use cross correlation measurements to visualise bubbles in a vertical pipe. ERT has also been used to look at solid-liquid flow in pipes, in a very similar way to gas-liquid flow. Giguere *et al*[344] has examined a number of reconstruction algorithms to determine the optimum one for imaging the interface between settled solids in a pipe and also the solids concentration. They found that the generalized iterative algorithm works well for this and ERT has the ability to produce quantitative concentration measurements. This was also confirmed more recently by Kotz *et al*[580].

Recently this use of ERT as has been built into the commercial, CombiMeter consisting of an electromagnetic flow meter by Krohne Altometer with an ITS (Industrial Tomography Systems) ERT system to measure the density. This has been tested on the CEMEX aggregate dredger Sand Falcon in a 0.7 m diameter pipeline and gave results within 1-3% of the typically used radioactive type[1134].

Forte *et al*[283] examined mixing in a 0.025 m diameter pipe using a Kenics KM static mixer and compared this to planar laser-induced fluorescence imaging showing that although there was less detail in the resolution that the extent of mixing could still be measured well.

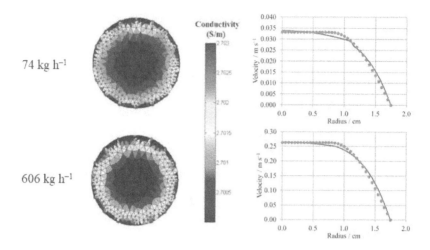

Figure 19.13 Variation of conductivity of fluid in a pipe with change in flowrate. This change is conductivity is caused by a change in shear-rate due to the radial position in the pipe. This allows the conductivity distribution to be reconstructed into a radial velocity profile, and thus fluid flowrate. On the graphs, the red points are those calculated from the ERT measurements and the line is the theoretical velocity profile from direct measurement of the fluid rheology. (color figure available in eBook)

It has also been shown possible to measure velocity profiles in pipelines using ERT,[871], using the rheological properties of the fluid (i.e., viscosity and conductivity change with shear rate) to directly measure the velocity profile within a pipe. Figure 19.13 shows an example of this, where the change in conductivity can be used to calculate the radial change in shear rate, and thus integrated to produce the velocity profile and hence the flow rate. Although this technique shows the use of ERT without a tracer and only using structural changes, it is commented that as a practical technique it is unlikely to be useful as the variation of fluid conductivity with position in the pipe is of the same order of magnitude as changes due to temperature, so calibration is vital.

Similar to this, Machin *et al*[666] inject a small heat tracer into the material within the pipe and track this change in conductivity. The key advance that is provided in this analysis is using a mixture of linear sensors of different sizes and circular sensors. The linear sensors have different widths to "look" different distances into the pipe, and then the circular sensor allow examination of the whole pipe diameter. This allows determination of the velocity profile, and thus the rheological properties of the fluid. This system has recently been developed into the commercial Stream Sensing Rheometer.

19.4.3 APPLICATION OF ELECTRICAL RESISTANCE TOMOGRAPHY TO CLEANING-IN-PLACE

One of the newer areas for ERT use, but one that has shown good success, is monitoring cleaning-in-place. One of the first applications of this was by Chen *et al*[187] who looked at the removal of milk fouling. In this case they only used two electrode plates and measured the electrical resistance as the fouling was removed. Henningsson *et al*[461] used a standard circular plane sensor to examine the washing out of yogurt. They were able to successfully monitor the displacement of the yogurt and compare the amount remaining with simulations. The sensor used in this case was produced with a hygienic design allowing their use onsite in food process lines of 0.06 m diameter.

Hou *et al*[488] and Wang *et al*[1119] both examined the cleaning of shampoo from 0.038 m diameter pipes using the standard 16 electrode circular plane. The use of multiple planes allowed

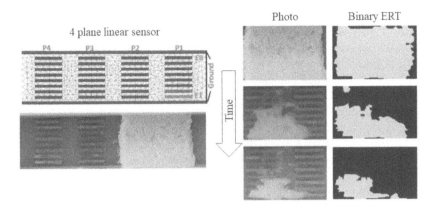

Figure 19.14   Multi-plane linear electrode system developed for the monitoring of surface deposits. The images produced from ERT with a binary metric (i.e., pixel either has deposit or no deposit) matches well with visual inspection of the same surface area. (color figure available in eBook)

monitoring of a section of pipe, or connection, e.g., T-junction. The cleaning time as measured by the ERT planes agreed well with that visually measured by inspection through specially designed transparent pipe work.

Ren *et al*[872] developed a new ERT sensor design specifically to examine surface deposits. They use a set of 8 linear electrodes across the pipe surface, in 4 planes, to look at the removal of milk fouling. Measurements are taken not only in each plane, but also between the planes to capture as much of the pipe wall as possible (even between the planes). The results produced match well with images captured through a viewing window, i.e., the remaining material, and also with measurement of the material removed from the surface, monitored as concentration in the cleaning water. The sensor was sensitive to even small amounts of fouled material remaining on the pipe wall, less than 0.05% of the initial fouling. Figure 19.14 shows an example of the 4-plane linear sensor, with two planes being used to monitor the amount of fouling remaining with ERT.

19.5   SUMMARY

Recent years have seen the beginning of a migration of the application of electrical tomography systems from the university laboratory into industrial environments. Simple sensors and compact electronic hardware are particularly well suited to on-site measurements for on-line process monitoring and control. Both resistance and capacitance modalities are now available commercially and true impedance tomography systems are beginning to emerge. Cost is low compared, for example, with X-ray tomography or magnetic resonance imaging, and would reduce considerably in mass production, especially with shrewd use of custom silicon. Many challenges have been addressed successfully by prototype solutions that accommodate metal walls, elevated temperature and pressure, reactive chemicals and restricted access. Image resolution is still disappointing to those familiar with nucleonic hard-field systems, but new programmes of work are delivering mathematically driven solutions that promise significant improvements. However, the limited number of measurements means that without dramatic technological developments the problem will remain severely underdetermined. Multi-modal systems are beginning to emerge which will provide synchronized measurements from a variety of sensors, not necessarily electrical, and these will benefit from research into appropriate methods of data fusion. Miniaturized tomography systems have not been considered here, but progress is being made with the development of sensors that may eventually prove invaluable for process intensification[1155].

As the technology is in its third decade of evolution, it is incumbent to offer reasons for the slow uptake by industry. Although reconstructed images are relatively coarse, compared for instance with those from X-ray or magnetic resonance, this is not perceived to be a limiting factor. In many cases the low resolution is more than adequate to provide invaluable information in a wide variety of processes. In fact it is not uncommon for the images to be superfluous to process operators as suggested by some of the case studies above. Single parameters (e.g. void fraction, mass flow rate, mixing time) that are better determined from knowledge of the physical distribution of materials provided by tomographic measurements are often the 'only' requirement. Similarly, although extreme applications would benefit from imaging rates of thousands of frames/s, for instance monitoring flame propagation in an internal combustion engine, there are many applications with much more modest requirements that can be easily satisfied with current technology. An important factor discouraging the uptake of the technology for production plant is the potential disruption to normal operation. The continuous application to production pressure filtration plant described above is, perhaps, the most advanced in this respect and has successfully overcome many challenges, but this has only been possible following a significant and mutually sympathetic programme of collaboration.

20 Devices, History and Conferences

CONTENTS

20.1 EIT CONFERENCES

In the more than 40 years since the earliest published biomedical impedance images, the conference series on biomedical applications of EIT has been the premier venue for EIT research. Table 20.1 lists these conferences and links to the proceedings (*Proc*) or associated journal special issues (*Issue*) (if available).

DOI: 10.1201/9780429399886-20

Table 20.1

Conferences on Biomedical Applications of EIT

Year	Conference	Proc	Issue
2021	National University of Ireland, Galway, Ireland (Online) (21th Int. Conf. Biomedical Applications of EIT)	Z:4635480	–
2019	University College London, UK (20th Int. Conf. Biomedical Applications of EIT)	Z:2691704	P:41(6)
2018	Edinburgh University, UK (19th Int. Conf. Biomedical Applications of EIT)	Z:1210246	P:40(focus)
2017	Dartmouth College, NH, USA (18th Int. Conf. Biomedical Applications of EIT)	Z:601555	P:39(focus)
2016	Stockholm, Sweden (16th Int. Conf. Electrical Bioimpedance and 17th Int. Conf. Biomedical Applications of EIT)	Z:55753	P:38(6)
2015	Neuchtel, Switzerland (16th Int. Conf. Biomedical Applications of EIT)	Z:17752	P:37(6)
2014	Gananoque, Ontario, Canada (15th Int. Conf. Biomedical Applications of EIT)	Z:17749	P:36(6)
2013	Heilbad Heiligenstadt, Germany (15th Int. Conf. Electrical Bioimpedance and 14th Int. Conf. Biomedical Applications of EIT)	[1]	P:35(6)
2012	Tianjin, China (13th Int. Conf. Biomedical Applications of EIT)	Z:18168	P:34(6)
2011	University of Bath, UK (12th Int. Conf. Biomedical Applications of EIT)		P:33(5)
2010	Gainsville, Florida (14th Int. Conf. Electrical Bioimpedance and 11th Int. Conf. Biomedical Applications of EIT)		P:32(7)
2009	Manchester, UK (10th Int. Conf. Biomedical Applications of EIT)		P:31(8)
2008	Dartmouth College, Hanover, NH, USA (9th Int. Conf. Biomedical Applications of EIT)		P:30(6)
2007	Graz, Austria (13th Int. Conf. Electrical Bioimpedance and 8th Int. Conf. Biomedical Applications of EIT)		P:29(6)
2006	Seoul, Korea (7th Int. Conf. Biomedical Applications of EIT)		P:28(7)
2005	London, UK (6th Int. Conf. Biomedical Applications of EIT)		P:27(5)
2004	Gdansk, Poland (12th Int. Conf. Electrical Bioimpedance and 5th Int. Conf. Biomedical Applications of EIT)		P:26(2)
2003	Manchester, UK (4th Int. Conf. Biomedical Applications of EIT)	Z:17924	P:25(1)
2002	Pingree Park, CO, USA (1st Mummy Range Workshop on Electrical Impedance Imaging)		P:24(2)
2001	London, UK (3th Int. Conf. Biomedical Applications of EIT)		P:23(1)
2000	London, UK (2nd Int. Conf. Biomedical Applications of EIT)		P:22(1)

Z:# indicates Zenodo DOI:10.5281/zenodo.#; P:#(#) indicates volume(issue) of *Physiological Measurement*

---

[1] IOP Conference series 434(1)

Table 20.1
Conferences on Biomedical Applications of EIT

1999	London, UK	P:21(1)
	(1st Int. Conf. Biomedical Applications of EIT)	
1998	Barcelona, Spain	[2]
	(X. Int. Conf. Electrical Bioimpedance)	
1995	Heidelberg, Germany	P:17(4A)
	(7th European Community Workshop on EIT)	
1994	Ankara, Turkey	P:16(3A)
	(6th European Community Workshop on EIT)	
1993	Barcelona, Spain	P:15(2A)
	(5th European Community Workshop on EIT)	
1992	Royal Society, London, UK	[3]
	(EC Conf. Biomedical Applications of EIT)	
1991	York, UK	P:13(A)
	(4th European Community Workshop on EIT)	
1990	Copenhagen, Denmark	–
	(3rd European Community Workshop on EIT)	
1987	Lyon, France	P:9(4A)
	(2nd European Community Workshop on EIT)	
1986	Sheffield, UK	P:8(4A)
	(1st European Community Workshop on EIT)	

Z:# indicates Zenodo DOI:10.5281/zenodo.#; P:#(#) indicates volume(issue) of *Physiological Measurement*

20.2 HISTORICAL PERSPECTIVE

The first published impedance images appear to have been those of Henderson and Webster in 1976[458] and 1978[459]. Using a rectangular array of 100 electrodes on one side of the chest earthed with a single large electrode on the other side, they were able to produce a transmission image of the tissues. Low conductivity areas in the image were claimed to correspond to the lungs. Shortly after, an impedance tomography system for imaging brain tumours was proposed by Benabid *et al* in 1978[100]. They reported a prototype impedance scanner which had two parallel arrays of electrodes immersed in a saline filled tank which was able to detect an impedance change inserted between the electrode arrays.

The first clinical impedance tomography system, then called applied potential tomography (APT), was developed by Brian Brown and David Barber and colleagues in the Department of Medical Physics in Sheffield. They produced a commercially available prototype, the Sheffield mark 1 system[161], which has been widely used for performing clinical studies, and is still in use in many centres today. This system made multiple impedance measurements of an object by a ring of 16 electrodes placed around the surface of the object.

The first published tomographic images were from this group in 1982 and 1984[84]. They showed images of the arm in which areas of increased resistance roughly roughly corresponded to the bones and fat. As EIT was developed, images of gastric emptying, the cardiac cycle and the lung ventilation cycle in the thorax were obtained and published. The Sheffield EIT system had the advantage that 10 images/s could be obtained, the system was portable, and the system was relatively inexpensive compared to ultrasound, CT and MRI scanners.

---

[2] Ann NY Acad Sci 873(1), 1999.

[3] DH Holder, "Clinical and Physiological Applications of Electrical Impedance Tomography", CRC Press, 1993.

Around the same time, a group in Oxford proposed that EIT could be used to image the neonatal brain[218]. They developed a clinical EIT system and obtained preliminary EIT images in two neonates. Their system used 16 electrodes placed in a ring around the head, but in contrast to the Sheffield system, the current was applied to the head by pairs of electrodes which opposed each other in the ring in a polar drive configuration. This maximized the amount of current which entered the brain and therefore maximized the sensitivity of the EIT system to impedance changes in the brain.

In the 1990s a group at the University of Göttingen developed a system originally for microgravity and supported by the German Space Agency (DLR). At least two hundred of these Goe MF II systems were built and delivered to Carefusion and Draeger medical and then to a large number of clinical researchers. Because of the availability, much of the clinical and preclinical EIT research before 2010 was done on the Goe MFII.

Since the first flush of interest in the mid to late 1980s, about thirty groups have developed their own EIT systems and reconstruction software, and publications on development and clinical applications have been produced by perhaps another twenty or so. Initial interest in a wide range of applications at first has now settled into the main areas of imaging lung ventilation, cardiac function, gastric emptying, brain function and pathology, and screening for breast cancer. Convincing pilot and proof of principle studies have been performed in these areas. In 1999, FDA approval was given to a method of impedance scanning to detect breast cancer. In the 2010s, at least three companies have built commercial EIT systems and obtained medical device approval for clinical use: Draeger Medical, Swisstom (later SenTec), and Timpal. At the time of writing, a small but growing number of clinical teams were using EIT routinely.

20.3 EIT HARDWARE

In the following section, we describe EIT systems which have been built. Information was obtained from publications or directly from the groups or companies involved. Not all requests for information were answered, in which case no entry is provided in this book.

20.3.1 COMMERCIAL SYSTEMS

Each vendor of a commercial EIT system was invited to contribute text and an image to this chapter. This information is presented below.

20.3.1.1 Draeger Medical

The PulmoVista 500 (figure 20.1), developed and manufactured by Drägerwerk AG & Co. KGaA, Lbeck, Germany, and launched in 2011, is the first commercially available EIT device, which was intended for continuous regional ventilation monitoring in the clinical environment.

By now, the results of more than 100 clinical studies have already been published in which PulmoVista 500 was used to assess the patient's regional distribution of ventilation, end-expiratory lung volume and its changes over time. Those studies, conducted in a large range of patient populations and during various therapeutic interventions, included a total of about 2.500 patients have used.

First studies, published between 2013 and 2015, investigated the effects of using high-flow nasal cannula and changing the body position on end-expiratory lung volume[808,885], the effects of Positive End-Expiratory Pressure (PEEP) and Pressure Support on ventilation distribution[694], the influence of tidal volume on ventilation inhomogeneity[94], the feasibility of EIT based titration of PEEP[536,651], the physiological effects of the Open Lung Approach in patients with Acute

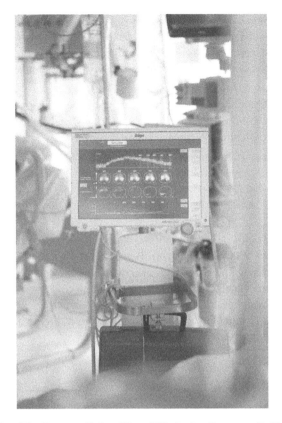

Figure 20.1   The Draeger PulmoVista 500. (color figure available in eBook)

Respiratory Distress Syndrome[199] and the effects of sighs on regional lung strain and ventilation heterogeneity in patients with Acute Respiratory Failure[695].

The principles of the electrical design of PulmoVista 500 are based on the Goe-MF II system of the EIT Group Göttingen and Dräger's EIT Evaluation Kit 2, a prototype version. However, PulmoVista 500 has been enhanced as a 16 channel hybrid of a serial and parallel measurement device, in order to allow – like the most common parallel EIT systems – conducting measurements at frame rates of up to 50 fps.

The operating frequency of the feed current and the resulting measurements is automatically set within a band of 80 to 130 kHz to the value, which has been determined to represent the lowest level of electromagnetic background noise. Depending on the selected frame rate, the pulse length of the feed current ranges between 2.5 and 12.5 ms. In order to gain the highest possible signal to noise ratio, the amplitude of the feed current is dynamically set between 80 to 100% of maximum patient auxiliary current according to the IEC standards, which depends on the set operating frequency of the feed current. At frequencies greater than 100 kHz, the maximum amplitude is limited to 9 mA rms.

Tomographic images are generated using a Finite Element Method (FEM) based linearized Newton-Raphson reconstruction algorithm, which converts the 208 voltages of a single frame into an ellipsoid EIT image.

PulmoVista 500 is used in combination with 16-electrode silicone rubber belts, which are available in 9 different sizes to cover a patient range from newborns up to morbidly obese patients. The minimum chest circumference starts at 36 cm, corresponding to a weight of about 3.5 kg, while the largest belt allows measurements of patients with a chest circumference up to 150 cm, corresponding to a patient's weight well above 200 kg. For the measurements, the reusable belt is connected to a patient cable, which can be separated for effective cleaning and disinfection. The patient interface, consisting of the electrode belt and the patient cable, is then connected to the device's EIT module via a trunk cable.

For calculating regional compliance and for interpreting trended EIT parameters together with the data from the ventilator, PulmoVista 500 provides a serial MEDIBUS interface allowing data import from about 20 types of ventilators and anesthesia machines.

As advanced respiratory monitoring increasingly includes esophageal and transpulmonary monitoring, PulmoVista 500 also provides a connection to a PressurePod via USB, which allows importing airway, esophageal and also gastric pressure waveforms. From airway pressure and esophageal pressure, the transpulmonary pressure – the pressure that actually distends the lung tissue – can then be calculated.

From the EIT images and impedance waveforms, various numerical parameters are calculated and displayed by PulmoVista 500 (see chapter 8): Regional tidal variations, quantifying regional ventilation distribution; Regional changes of end-expiratory lung impedance, representing regional lung volume changes; Regional changes of the respiratory system compliance; Estimation of alveolar overdistension and alveolar collapse for assessing PEEP trials; and Regional ventilation delay, for assessing cyclic opening and closing, Pendelluft and regional inspiratory dynamics transpulmonary pressures and various derived parameters

20.3.1.2 Sentec

Figure 20.2   LuMon^{TM} System: monitor and belts (left) LuMon^{TM} Monitor, (middle) Belt for adults/children (right) Belt for neonates/infants. (color figure available in eBook)

SenTec EIT refers to the technology based on the principles of chest electrical impedance tomography developed and commercialized by SenTec (Switzerland). The current CE-marked EIT system is the LuMon^{TM} System, while the predecessor was the BB^2 System (originally developed by Swisstom AG which was acquired by SenTec AG in 2018). SenTec EIT is also present in the Pioneer Set, a research platform mainly devoted to veterinary research.

The LuMon^{TM} System consists of the LuMon Monitor, a compact (<4kg, 12.1 in. display size) monitor (figure 20.2a), linked through a connector to a textile belt figure 20.2b and c, respectively for adults/children and neonates/infants), whose contact with the skin is enhanced by a specifically designed contact agent. The LuMon System is available in two configurations, one for adults/children and one for neonates/infants, each with specifically configured software but with the same underlying technology. The textile belts contain 32 electrodes embedded in a structured fabric and are designed to avoid restricting work of breathing, e.g. by following the ribs in order to avoid any impairment of the ribcage movement. As the ribcage anatomy is age-dependent, the neonatal belts follow the ribs on a transverse plane and the adult ones on an oblique plane. The belts are for single-patient use, in order to avoid cross-contamination, and feature an adhesive-free, textile belt/skin interface, which is particularly important for patients with sensitive and fragile skin such as preterm neonates. By means of very weak feed currents (0.7 – 3.7 mArms; 200 kHz ±10%), frames of tomographic images are generated with a temporal resolution of about 50 frames per second.

The reconstruction algorithm is based on GREIT[18] with the special feature that data processing makes use of thorax and lung models derived from computed tomography anatomical data. This allows to contextualize measurements with respect to individual patients as characteristics such as weight, height, gender, underbust girth are taken into account in the models. Focusing on values inside the lung contours for the reconstruction algorithm and the displayed images not only allows for a better interpretability and contextualization, but also for the introduction of novel indicators or concepts, e.g. Silent Spaces, that represent those regions, within the lung, that exhibit little or no ventilation. Interpretation of data is also enhanced by a sensor, embedded either in the belt or in the connector, that continuously detects the patient position, whether prone, supine, lateral or inclined, thereby helping to assess the influence of gravity on lung mechanics and ventilation distribution. Moreover the contact quality between skin and electrodes is continuously verified, which helps to infer when measurements are more or less reliable, and in case of impaired contact of up to 6 electrodes, an automatic compensation can still safeguard the quality of the signal. Several features are available on the display, including: a Global Dynamic Image showing impedance changes within the thorax in real-time; a Plethysmogram, i.e. a waveform representing changes of lung impedance over time; the Respiratory Rate based on the last breaths detected from the Plethysmogram; trends of End-Expiratory Lung Impedance, End-Inspiratory Lung Impedance and Aeration, representing respectively end-expiratory, end-inspiratory and mean lung volume; a Stretch Image, representing tidal distribution, and connected Silent Spaces, divided into Dependent and Non-Dependent; the Center of Ventilation, which is shown taking patient position into account with respect to gravity; different analysis modes, either breath-based (BB) or time-based (TB-I and TB-II), to assess the above mentioned features.

There are several studies on SenTec EIT, either on validation of the system or on clinical applications, with a considerable number of patients, including neonates. To name just a few, validity studies confirmed, among others, the advantage of using anatomical models[1033], the ability to detect objects[1159], the linear relationship between volume and impedance[739], the usefulness of Silent Spaces in assessing lung recruitment[999]. Clinical applications included, among others, detecting impaired ventilation after anesthesia[1145], assessing the effect of PEEP application during robot-assisted surgery[969], assessment of surfactant therapy[531] and detection of complications[852] in neonates.

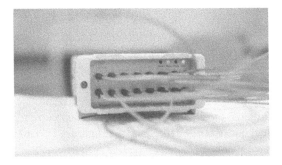

Figure 20.3   The Sciospec EIT system (color figure available in eBook)

20.3.1.3   Sciospec

Sciospec offers electrical impedance tomography solutions both as standard instruments and as customer specific systems. For standard instruments the channel count ranges from 8 to 256. By default these systems have a pseudo differential current source for excitation signals up to 10 mA, a tightly synchronized data acquisition block for truly simultaneous sampling of all channels and an ultra-low leakage reed relay injection matrix for selection of any combination of channels for positive and negative injection. Some modules also support fast switching semiconductor switches as alternative injection matrix allowing for higher frame rates with the trade-off of increased parasitics. By default, all these systems allow for multifrequency (single sine sweep) measurements and any to any injection sequence configurations, enabling almost any electrode configuration – from standard ring arrangements over grid structures as used in planar EIT all the way to custom configurations. Integration times can be tuned in between maximum signal quality or maximum frame rate – e.g. for 32 measuring electrodes using 16 injections per frame can yield up to 73 fps with the low parasitic reed relay switches and up to 280 fps with faster semiconductor switches. While all instruments offer single ended potential measurement, most of them also allow for differential measurements between two electrodes including options to choose between different pairs via software (figure 20.3).

Frontend connections are available through a selection of cable connectors with banana plugs, miniature DSUB or coaxial standards like MCX being standard options. Sciospec also has edge card type connection options that allow for highly flexible connectivity through custom made connector cards (e.g. sensor carriers or custom-made cable sets).

The instruments offer communication through Ethernet, high speed USB, isolated full speed USB with WIFI and Bluetooth as options. Through the integration of the optional medical grade isolation module low level GPIOs are another option for controlling the instrument through UART, SPI and parallel interface and fast hardware synchronization via sync ports. The Sciospec-COM-Interface API is available on any of these interfaces and allows for full instrument control from Java, C, LabView, Matlab etc. A standard software for PC-based control & data analysis including image reconstruction for standard setups is also included. Sample code for EIDORS integration is available.

To complete the EIT product range options like medical grade power supplies, electrical safety AC coupling frontends, EIT phantoms, EIT tanks, cables, sensor adapters, temperature control and IO modules are available. For more advanced non-standard setups Sciospec offers customizations and all EIT technology is also available as OEM modules for the integration into application specific systems.

20.3.2   RESEARCH SYSTEMS

20.3.2.1   Sheffield Mk 2 System

The first EIT systems were built at the University of Sheffield. The Mk 1 and Mk 2 Applied Potential Tomography (APT) serial systems utilized 16 electrodes and operated at a single frequency[161,983]. The Mk 2 system was important in that it included many of the features used in serial systems going forward. The Mk 2 systems applied currents to adjacent electrodes and collected data at 25 frames/s. It used a digitally-generated sinusoidal excitation signal of 20.83 kHz, produced using a 12-bit DAC and a 48-entry ROM look-up table clocked at 1 MHz. The applied current was produced using a floating-load voltage-to-current converter like that shown in figure 4.11. Direct measurement of the applied current, performed using an in-line resistor and an instrumentation amplifier, was used to account for the presence of variations in phase and amplitude of the applied current with variations in the load impedance at the electrodes. Two 1-to-16 multiplexers (Analog Devices DG506) were used to direct the currents to a single pair of electrodes at a given time. A current amplitude of 5 mA peak-to-peak was used.

Differential voltage measurements were made between adjacent pairs of electrodes. The electrode voltages were a.c.-coupled to a set of 16 instrumentation amplifiers (Burr-Brown INA110), providing parallel measurement of all the differential voltages. The instrumentation amplifier outputs were transformer-coupled to programmable-gain amplifiers (PGAs), with gains from 1 to 256 in powers of 2. PGA output voltages were processed by synchronous, phase-sensitive voltmeters. Only the real component of the measured voltages was used in image reconstruction due to the greater impact of stray capacitance on the accuracy of the reactive measurements.

A common-mode feedback circuit was used to reduce the common-mode voltage applied to the instrumentation amplifiers in the voltage measurement circuit. Since all differential voltages were measured simultaneously, the common-mode voltage could not be minimized for all voltage measurements but, rather, the circuit reduced the common-mode voltage seen by all instrumentation amplifiers. The circuit worked using a pair of electrodes located away from the electrodes used to collect image data. One electrode was used to sense the common-mode voltage and the second electrode was driven with a compensating voltage which acted to drive the common-mode voltage to zero. The gain of the feedback loop was kept sufficiently low (32 dB) in order to avoid oscillation problems.

20.3.2.2   Goettingen GoeMFII System

The Goe-MF II system was been developed by the EIT Group Goettingen and is based on the Goe-MF type which has been originally build for application of EIT in microgravity supported by the German Space Agency (DLR). Both systems use an adjacent drive current injection and 16 multiplexed electrodes corresponding to the classical Sheffield Mark I pattern. Operating frequency range is 5 kHz to 500 kHz. Electrodes were connected by low capacity coaxial cables (RG179, 63 pF/m) to a 16 channel multiplexer (4 times MAX307) additionally multiplexing the driven shields of current feeding and the shields of sensing cables. The Goe-MF used early digitization and digital I/Q demodulation of incoming signals, but required computer equipment of medical grade electrical safety for control and measurement. For this reason all safety barriers were integrated into the Goe-MF II on the analogue side of the signal chain. This is achieved by transformer coupling of the driving voltage for the floating current source and the analogue signal after passing the differential input amplifier stage. Coupling of the input signal uses coaxial magnetic shielding of an inner transformer core by a bigger outer one with permalloy caps. Total gain is adjusted by a variable gain amplifier (AD 603) before filtering by an elliptic filter (LTC 1560) allows. An additional isolated analogue input for auxiliary signals and a trigger input are implemented. The toroidal transformer supplying all electronics has segmental windings to secure low stray capacitance between patient side and line power.

The current source is driven by the voltage output of a 14-bit DAC 40 module while the filtered signal from the multiplexer is digitized at 10 MHz sample rate and 14 bit resolution by an A4D1 module. Both modules are placed on a SBC 62 DSP board (Innovative Integration) equipped by am TMS 320 digital signal processor with 1600 MIPS computing power and additional FPGA for control. All digital signal processing is performed in the SBC62 and the demodulated data are transferred via USB to external computer for further image reconstruction and control. The maximum frame rate in this configuration is 44 Hz. The software has implemented a spectral view of the received voltages to set the injection frequency to a band of less environmental electrical noise.

The Goe-MF II system is identical to EIT systems that have been formerly delivered by Carefusion. Also Draeger Medical distributed the hardware for scientific application named "Draeger Evaluation Kit I". The Goe-MF II was CE-certified at that time by Carefusion and approved as a laboratory device for space application by DLR. A further development funded by DLR was a multiple plane system using the hardware of three synchronized Goe-MF II devices collecting data simultaneously from three electrode planes. This allows the assessment of cranio-caudal changes in thoracic impedance distribution (e.g. during parabolic flights[298]).

20.3.2.3   École Polytechnique de Montréal

Involvement in EIT began in 1985 to examine the possibility of using data from cardiac-driven conductivity changes to improve accuracy of ECG inverse problem models. Three exploratory 16-channel EIT systems were built during the next decade[125,392]. The third system introduced the concept of Smart Active Electrodes (SAEs) for each ECG electrode, containing a current source, a boot-strapped follower, a microcontroller and analog gates, interconnected in daisy-chain. For recording on reclining or supine subjects, a module (Scan-Head) was developed that in an N-length daisy-chain of SAEs laid-out on a single printed-circuit board. Scan-Heads designs for N=16 to 32 were developed. From 1996 to 2012, five generations and configurations of EIT systems were developed:

- SigmaTome I and SigmaTome II: designed for monitoring lung function in a clinical setting. The second iteration was built with surface-mount components and faster communication[440].
- Scan-Head multiplexer for 3-D Imaging: designed to connect four 16-channel Scan-Heads to a Base-Station. It was used to acquire data frames using four rows of 16 electrodes.
- Combined Induced-Current and Applied-Current EIT System: designed to compare Induced-Current EIT to conventional EIT. The system comprised: a SigmaTome system, a PC, an Interface module, a Coil-Driver module, a Coil-Array and a Phantom[1024].
- Test tools and methods for optimizing EIT Systems: a method for designing EIT phantoms[326], and a later version with active elements to vary contact impedance. Based on these tools, the group developed methods to modelling and optimizing EIT hardware[440], and accounting for system hardware imperfections[441].
- SigmaTome DF and SigmaTome MF: a 16-channel, dual frequency system using wideband front-end electronics, and a 16-channel, 8-tone carrier, system using specialized processors developed primarily for telecommunications infrastructure with very high integration density and processing speed[889].
- System front-end design for concurrent acquisition of electroencephalograms and EIT data: A 24 electrode Scan-Head was designed to measure both EIT and EEG signals using the same set of 24 EEG electrodes[393].

20.3.2.4   Russian Academy of Sciences Breast Imaging System

A number of serial instruments were produced by this group for imaging the thorax and breast[194,195,196]. The breast imaging system[194,196] was particularly noteworthy because it used a large planar electrode array rather than a ring with 8 to 32 electrodes as was used by other systems. While frame rate was important in the design of systems for imaging ventilation or cardiac events, speed was less of a driving factor in this system since its goal was to measure static breast tissue impedance. The system used a single source and voltmeter, and supported 256 electrodes arranged in a round, planar matrix. This system required approximately 20 s to collect the data for a single image. A three op amp voltage-to-current converter driven by a DAC was used as the single source. A 1-to-256 multiplexer directed current to one electrode on the array and a second remote electrode was placed on the wrist of the patient completed the circuit. The system was able to produce excitation signals up to 110 kHz, with higher frequencies resulting in better coupling to the patient but greater losses due to stray capacitance. Due to these considerations, an excitation frequency of 50 kHz was generally used with a current amplitude of 0.5 mA. A voltage threshold detector was used at the output of the current source to enable the detection of electrodes with poor contact to the patient.

Difference voltages were measured between all non-current carrying electrodes on the array and a second remote electrode that was placed on the other wrist of the patient. A 256-to-1 multiplexer was used to attach one electrode at a time to an instrumentation amplifier input, with the second input permanently tied to the remote electrode. To produce an image, 255 voltage measurements were made for each applied current, resulting in a total of 65,280 voltage measurements when all 256 electrodes were in contact with the patient. The instrumentation amplifier had a programmable gain that was adjusted based on the physical distance of the electrode from the drive electrode, with gain increasing with distance. The electrodes were d.c. coupled to the instrumentation amplifier through the multiplexer and, as a result, the d.c. potential due to the electrode/patient interface appeared at the amplifier input. The system utilized a compensation system in which a DAC fed the bias adjustment on the instrumentation amplifier to compensate for the contact potential. This correction was performed for each electrode prior to the measurement of the a.c. voltage due to the applied current. The instrumentation amplifier output, after lowpass filtering, was sampled and quantized by a 14-bit ADC, and digital synchronous detection was used to measure only the real part of the electrode voltage.

20.3.2.5 CRADL System

Active electrode serial EIT systems are being developed as part of the Continuous Regional Analysis Device for neonate Lung (CRADL) project[1173,1174]. These systems, designed to provide continuous monitoring of infant lung function, represent the state-of-the art in serial EIT systems. The CRADL v2.0 system[1173] has 16 electrodes, each implemented using a fully-integrated ASIC designed in 0.35 μm high-voltage CMOS technology. The active electrodes are built into a wearable belt with the necessary inter-electrode connections. The belt connects to a central hub that performs waveform generation, voltage measurement, and control operations.

Each active electrode IC includes an instrumentation amplifier for voltage buffering, a fully-differential current driver, and five analog switches. The inputs to the IA on a given active electrode IC are the voltage on that electrode (-) as well as that on the adjacent electrode (+). A switch at the IA output allows the output voltage or a ground to be connected to a 16-input summer. The choice of switch settings for the set of 16 ICs enables the summer output to be the difference voltage between any pair of selected electrodes. A differential analog sinusoidal waveform is distributed to all the ICs and an analog switch in each IC allows it to be connected to the input of the differential current driver on the desired electrode. When the switch is engaged on a desired electrode, the positive output of the current driver is sent to that electrode and the negative output is connected to a bus that connects to all other active electrode ICs. Engaging an analog switch on the desired electrode will apply the negative current to that electode. In this way, the positive and negative current drives can be connected to any pair of electrodes. The current driver itself is constructed using differential difference transconductance amplifiers and OTAs and incorporates feedback from a current sense resistor. Common-mode feedback is used to minimize the common-mode voltage on the load between the positive and negative current carrying electrodes. The ICs also include a Micro-Electro-Mechanical Systems (MEMS) sensor that detects changes in the shape of the thorax and the orientation of the electrode. This information can be to perform boundary shape correction when reconstructing images from the data.

The central hub generates a single tone sinusoid using a look-up table followed by a 12-bit, 32 MSps DAC and distributes the analog tone to the active electrode ICs. Sinusoid frequencies of 125 kHz, 250 kHz, 500 kHz, and 1 MHz can be produced. The output from the 16 input summer described above comes to the central hub, is processed by a programmable-gain IA and digitized using a 12-bit, 32 MSps ADC. Digital demodulation recovers the real and reactive voltages. The digital processing in the central hub is performed using a Xilinx Artix 7 FPGA. The overall system operates with frame rate of up to 122 fps. Tests with a parallel RC load show an SNR of 54.3 dB at 125 kHz and 48.8 dB at 1 MHz.

20.3.2.6 Rensselaer Polytechnic Institute ACT 3 System

This group developed a series of adaptive current tomograph (ACT) systems, with the primary application being the imaging of the thorax[210,252,773]. The ACT 3[210,252] system was a 32-channel, multiple current source system that was capable of producing real-time images of conductivity and permittivity at a rate of roughly 20 images/s. The system was fully parallel, having 32 current sources and 32 voltmeters. A grounded thirty-third electrode was placed away from the measurement electrodes to provide a path for residual common-mode current due to the applied currents not summing exactly to zero.

A 10-bit digital sinusoidal reference waveform at 28.8 kHz was generated using a PROM look-up table and distributed to each channel over a backplane. An amplitude-scaled analog sinusoid waveform was produced from this digital sinusoid using a four-quadrant MDAC that was constructed using two bipolar two-quadrant MDACs (Analog Devices DAC10) and a 16-bit audio DAC (Analog Devices AD1856)[210]. This configuration, though expensive from a hardware viewpoint, provided 16 bits of amplitude control without introducing amplitude-dependent phase shifts in the resulting analogue sinusoidal waveform. Voltage-to-current conversion was performed using a Howland-type current source that was implemented using an instrumentation amplifier (Analog Devices AMP05). The current source circuit included a digital potentiometer (Dallas Semiconductor DS1867) that allowed adjustment of the output impedance of the source. An NIC negative capacitance circuit, including a digital potentiometer to enable automatic adjustment, was placed in parallel with the current source output to perform capacitance cancellation.

Single-ended real and reactive voltages on all the electrodes were measured using 32 phase-sensitive voltmeters. Each electrode voltage was sampled and quantized by a 12-bit ADC (Analog Devices AD678), and processed by a digital matched filter voltmeter that was implemented in an Analog Devices ADSP-2100 digital signal processor to obtain real and reactive voltage values. The voltage waveforms were sampled five times per cycle over multiple cycles, with the number of cycles dependent on the desired precision/image rate trade-off. With an imaging rate of approximately 20 images/s, 160 samples were collected per measurement, yielding an effective precision of 15 bits. Integrating over 640 samples yielded a precision of 16 bits and an imaging rate of approximately seven images/s.

The ACT 3 system included an automated calibration system for adjusting the digital potentiometers in the current sources and NICs to optimize the output impedance. The calibration system also determined calibration constants for the applied current amplitudes and the voltmeters. Frequent calibration of the current sources was needed to maintain a small value of common-mode current.

20.3.2.7 KHU Mark2.5 System

Investigators at Kyung Hee University have built a number of serial and parallel EIT instruments[791,792,1143]. The KHU Mark2.5 system[1143] is an improved version of the KHU Mark2 system[791], adding automatic self-calibration, interface signals to external devices such as patient monitors and ventilators, and improved system software. A key feature of these systems is that they include switches that enable them to operate in a hybrid serial-parallel mode in which the number of electrodes exceeds the number of sources. With 16 sources, they can operate in a fully parallel mode for 16 electrodes or a serial-parallel mode for 32 or more electrodes. Additionally, the system can apply multiple tones simultaneously, enabling simultaneous data collection at multiple frequencies for spectroscopy or frequency-difference imaging.

The electronics for a single source is implemented on a impedance measurement module (IMM). An Altera EP3C10F256C8N FPGA controls the IMM functions and drives an Analog Devices AD9783 dual 16-bit DAC to generate the excitation signals. Each DAC can produce up to 3 sinusoids of different frequencies (with frequency ratios of 1, 5 and 10) and, by summing the DAC outputs before the voltage-to-current converter (current source), up to 6 frequencies can be applied simultaneously. Additionally, two 10-bit DACs are used to provide d.c. compensation that removes any residual d.c. component at the current source output. A Howland current source is used with a Dallas DS 1267 digital potentiometer in one branch for trimming its output resistance. A Maxim MAX4545 T-bar switch enables one of four GIC circuits to be placed in parallel with the output of the Howland source to provide output capacitance mitigation. Each GIC is designed to operate within a particular subrange of the full 10 – 500 kHz frequency range of the instrument, with a digital potentiometer providing fine tuning for the GIC within that range.

The electrode voltage is processed by a differential amplifier, bandpass filter, and variable gain amplifier before being sampled using an Analog Devices AD9235 12-bit, 65 MSps ADC. The differential amplifier allows the voltage difference between a pair of electrodes to be measured. For single frequency operation, the ADC is set to acquire 1000 samples per cycle of the excitation sinusoid. The samples are fed to the FPGA which performs digital matched filtering to recover the real and reactive voltages. The total integration period for the matched filtering can be adjusted to obtain the desired SNR while trading off frame rate. The maximum frame rate for the system is 100 frames per second.

The digital potentiometers for the Howland source and GIC circuits are adjusted to maximize the output impedance using an automated calibration procedure. The droop method[210] is used to measure the output impedance during the procedure. While the adjustment of the Howland source and GIC primarily affects the output resistance and capacitance, respectively, the adjustments are not orthogonal, requiring a two-dimensional search to find the optimal settings. Resistive loads and switches are built into the system to enable automatic calibration of the current sources, voltmeters, and output impedances.

20.3.2.8   Dartmouth Broadband, High Frequency System

This group has developed several multiple-source systems for breast cancer detection that incorporate both current and voltage sources[406,444]. The system described in[405,406] has 64 electrodes and operates over a frequency range from 10 kHz to 12.5 MHz. It is an applied voltage system and uses a quasi-active electrode design that brings the electronics close to the electrodes to minimize the stray capacitance effects. Eight wedge-shaped measurement module circuit boards, each implementing the front end for four of the 64 total electrodes, are placed in a circular arrangement in which the breast to be imaged is at the center, avoiding the need for cables.

Each measurement module has two ADSP-21065L digital signal processors (DSPs), each of which provides digital control, signal generation and processing for one pair of channels and interfaces with the system controller. The sinusoidal waveform for each channel is generated from a 256 entry deep wavetable stored in an FPGA. One FPGA provides the waveforms for a pair of channels. Digital values are sent to an Analog Devices AD9754 14-bit current output DAC followed by an Analog Devices AD8061 amplifier that converts the differential DAC outputs to a single-ended voltage. This voltage is applied to the electrode through a current-sensing resistor and d.c. blocking capacitors. The voltages on both sides of the resistor are sampled and quantized using Analog Devices AD7677AST 1 MSps differential ADCs. The digital values are read by the DSP which performs digital matched filtering over 256 samples to produce the real and reactive values of the applied voltage and current. The system can acquire 182.5 frames per second when operating at a single frequency and applying 15 spatial patterns per image. For increased precision, multiple blocks of 256 samples can be used in the matched filter when performing voltage and current measurements, incurring a corresponding decrease in frame rate.

The control module communicates with the DSPs on the measurement modules using a serial peripheral interface (SPI) at 33 Mbps. This module maintains the system timing, including synchronizing the signal generators on the individual channels.

In addition to care in the mechanical design to minimize stray capacitance and timing skew, an extensive set of calibration procedures are used with the system. Calibration measurements determine the conversion factor for converting ADC values to voltages, the values of the current sense resistors, the values of the d.c. blocking capacitors, and the values of stray and parasitic impedances.

Bibliography

1. Dräger Manufacturer Brochure. `https://www.draeger.com/en_uk/Products/PulmoVista-500`. [online].
2. Swisstom BB^2 Product Information. `http://www.swisstom.com/wp-content/uploads/BB2_Brochure_2ST100-112_Rev.000_EIT_inside.pdf`. [online].
3. Timpel Enlight: Technology. `http://www.timpel.com.br/?#AboutUs`. [online].
4. *Proc. of 1st European Concerted Action on Process Tomography (ECAPT) Workshop*, 1992.
5. *Proc. of 2nd European Concerted Action on Process Tomography Workshop*, 1993.
6. *Proc. of 3rd European Concerted Action on Process Tomography Workshop*, March 1994.
7. *Proc. of 4th European Concerted Action on Process Tomography Workshop*, April 1995.
8. J.-F. P. Abascal, S. R. Arridge, R. H. Bayford, and D. S. Holder. Comparison of methods for optimal choice of the regularization parameter for linear electrical impedance tomography of brain function. *Physiological measurement*, 29(11):1319, 2008.
9. O. A. Abdel-Rehim, J. L. Davidson, L. A. Marsh, M. D. O'Toole, and A. J. Peyton. Magnetic polarizability tensor spectroscopy for low metal anti-personnel mine surrogates. *IEEE Sensors Journal*, 16(10):3775–3783, May 2016.
10. S. Abdul, B. Brown, P. Milnes, and J. Tidy. The use of electrical impedance spectroscopy in the detection of cervical intraepithelial neoplasia. *International Journal of Gynecologic Cancer*, 16(5):1823–1832, 2006.
11. R. A. Adams. *Sobolev Spaces*. Academic Press, 1975.
12. W. Adey, R. Kado, and J. Didio. Impedance measurements in brain tissue of animals using microvolt signals. *Experimental neurology*, 5(1):47–66, 1962.
13. A. Adler. Accounting for erroneous electrode data in electrical impedance tomography. *Physiological measurement*, 25(1):227, 2004.
14. A. Adler. EIDORS version 3.10. In *Proc 17th Int conf Biomedical Applications of EIT (EIT 2019)*, page 63, July 2019.
15. A. Adler, M. Albaghdadi, U. Siddiqui, K. B. Worms, U. Gordon, N. Racheli, and K.-H. Kuck. Detection of leak from left atrial appendage occlusion using dielectric imaging. *IEEE Transactions on Biomedical Engineering*, 2020.
16. A. Adler, M. B. Amato, J. H. Arnold, R. Bayford, M. Bodenstein, S. H. Böhm, B. H. Brown, I. Frerichs, O. Stenqvist, N. Weiler, et al. Whither lung eit: where are we, where do we want to go and what do we need to get there? *Physiological measurement*, 33(5):679, 2012.
17. A. Adler, R. Amyot, R. Guardo, J. H. T. Bates, and Y. Berthiaume. Monitoring changes in lung air and liquid volumes with electrical impedance tomography. *Journal of Applied Physiology*, 83(5):1762–1767, Nov. 1997.
18. A. Adler, J. H. Arnold, R. Bayford, A. Borsic, B. Brown, P. Dixon, T. J. C. Faes, I. Frerichs, H. Gagnon, Y. Gärber, B. Grychtol, G. Hahn, W. R. B. Lionheart, A. Malik, R. P. Patterson, J. Stocks, A. Tizzard, N. Weiler, and G. K. Wolf. GREIT: a unified approach to 2d linear EIT reconstruction of lung images. *Physiological Measurement*, 30(6):S35–S55, June 2009.
19. A. Adler and A. Boyle. Electrical impedance tomography: Tissue properties to image measures. *IEEE Transactions on Biomedical Engineering*, 64(11):2494–2504, 2017.
20. A. Adler, J. X. Brunner, D. Ferrario, and J. Solà i Caros. Method and apparatus for the non-invasive measurement of pulse transit times (PTT), 05 2013.
21. A. Adler, M. Faulkner, K. Aristovich, S. Hannan, J. Avery, and D. S. Holder. Cerebral perfusion imaging using EIT. In *Proc 17th Int conf Biomedical Applications of EIT (EIT 2017)*, page 44, June 2017.
22. A. Adler, R. Gaburro, and W. Lionheart. *Electrical Impedance Tomography*. Springer Science & Business Media, 2016.
23. A. Adler, P. O. Gaggero, and Y. Maimaitijiang. Adjacent stimulation and measurement patterns considered harmful. *Physiological measurement*, 32(7):731, 2011.

24. A. Adler, B. Grychtol, and R. Bayford. Why is EIT so hard, and what are we doing about it. *Physiological measurement*, 36(6):1067–1073, 2015.

25. A. Adler and R. Guardo. A neural network image reconstruction technique for electrical impedance tomography. *IEEE Transactions on Medical Imaging*, 13(4):594–600, 1994.

26. A. Adler and R. Guardo. Electrical impedance tomography: regularized imaging and contrast detection. *IEEE transactions on medical imaging*, 15(2):170–179, 1996.

27. A. Adler, R. Guardo, and Y. Berthiaume. Impedance imaging of lung ventilation: do we need to account for chest expansion? *IEEE transactions on Biomedical Engineering*, 43(4):414–420, 1996.

28. A. Adler and W. R. B. Lionheart. Uses and abuses of EIDORS: an extensible software base for EIT. *Physiological Measurement*, 27(5):S25–S42, apr 2006.

29. A. Adler and W. R. B. Lionheart. Minimizing EIT image artefacts from mesh variability in finite element models. *Physiological measurement*, 32(7):823, 2011.

30. A. Adler, M. Proença, F. Braun, J. X. Brunner, and J. Solà. Origins of cardiosynchronous signals in EIT. *Proceedings of the 18th International Conference on Biomedical Applications of Electrical Impedance Tomography*, page 73, 06 2017.

31. A. Adler, N. Shinozuka, Y. Berthiaume, R. Guardo, and J. H. T. Bates. Electrical impedance tomography can monitor dynamic hyperinflation in dogs. *Journal of Applied Physiology*, 84(2):726–732, Feb. 1998.

32. M. Akhtari, H. Bryant, A. Mamelak, E. Flynn, L. Heller, J. Shih, M. Mandelkern, A. Matlachov, D. Ranken, E. Best, M. DiMauro, R. Lee, and W. Sutherling. *Brain Topography*, 14(3):151–167, 2002.

33. M. Akhtari, H. Bryant, A. Mamelak, L. Heller, J. Shih, M. Mandelkern, A. Matlachov, D. Ranken, E. Best, and W. Sutherling. *Brain Topography*, 13(1):29–42, 2000.

34. S. Al-Zeibak and N. H. Saunders. A feasibility study of in vivo electromagnetic imaging. *Physics in Medicine and Biology*, 38(1):151–160, Jan. 1993.

35. N. A. Aladjalova. *Slow electrical processes in the brain*. Elsevier, 1964.

36. F. Alberini, D. Bezchi, I. Mannino, A. Paglianti, and G. Montante. Towards real time monitoring of reacting species and ph coupling electrical resistance tomography and machine learning methodologies. *Chemical Engineering Research and Design*, 2021.

37. G. Alessandrini. Stable determination of conductivity by boundary measurements. *Applicable Analysis*, 27(1-3):153–172, 1988.

38. G. Alessandrini and J. Isakov V. Powell. Local uniqueness of the inverse conductivity problem with one measurement. *Trans Amer Math Soc*, 347:3031–3041, 1995.

39. L. Allaud and M. Martin. *Schlumberger: The History of a Technique*. John Wiley & Sons, 1977.

40. C. M. V. Allen, G. E. Lindskog, and H. G. Richter. Collateral respiration. transfer of air collaterally between pulmonary lobules. *Journal of Clinical Investigation*, 10(3):559–590, Aug. 1931.

41. M. Alsaker, S. J. Hamilton, and A. Hauptmann. A direct D-bar method for partial boundary data electrical impedance tomography with a priori information. *Inverse Probl. Imaging*, 11(3):427–454, 2017.

42. M. Alsaker and J. Mueller. A D-bar algorithm with a priori information for 2-dimensional electrical impedance tomography. *SIAM J. Imaging Sci*, 9(4):1619–1654, 2016.

43. M. Alsaker and J. Mueller. EIT images of human inspiration and expiration using a D-bar method with spatial priors. *Journal of the Applied Computational Electromagnetics Society (ACES)*, 34(2):325–330, 2019.

44. M. Alsaker, J. Mueller, and R. Murthy. Dynamic optimized priors for D-bar reconstructions of human ventilation using electrical impedance tomography,. *Journal of Computational and Applied Mathematics*, 362:276–294, 2019.

45. M. Alsaker and J. L. Mueller. Use of an optimized spatial prior in D-bar reconstructions of EIT tank data. *Inverse Problems and Imaging*, 12(4):883–901, 2018.

46. T. D. Ambrisko, J. Schramel, K. Hopster, S. Kästner, and Y. Moens. Assessment of distribution of ventilation and regional lung compliance by electrical impedance tomography in anaesthetized horses undergoing alveolar recruitment manoeuvres. *Veterinary Anaesthesia and Analgesia*, 44(2):264–272, Mar. 2017.

47. T. D. Ambrisko, J. P. Schramel, A. Adler, O. Kutasi, Z. Makra, and Y. P. S. Moens. Assessment of distribution of ventilation by electrical impedance tomography in standing horses. *Physiological Measurement*, 37(2):175–186, Dec. 2016.

48. T. D. Ambrisko, J. P. Schramel, U. Auer, and Y. P. S. Moens. Impact of four different recumbencies on the distribution of ventilation in conscious or anaesthetized spontaneously breathing beagle dogs: An electrical impedance tomography study. *PLOS ONE*, 12(9):e0183340, Sept. 2017.

49. A. M. Ambrosio, T. P. Carvalho-Kamakura, K. K. Ida, B. Varela, F. S. Andrade, L. L. Facó, and D. T. Fantoni. Ventilation distribution assessed with electrical impedance tomography and the influence of tidal volume, recruitment and positive end-expiratory pressure in isoflurane-anesthetized dogs. *Veterinary Anaesthesia and Analgesia*, 44(2):254–263, Mar. 2017.

50. H. Ammari, O. Kwon, K. J. Seo, and E.J. Woo. T-scan electrical impedance imaging system for anomaly detection, ppreprint (submitted to siam j). *Math Anal*, 2003, 2003.

51. J. W. Anast, G. L. Andriole, T. A. Bismar, Y. Yan, and P. A. Humphrey. Relating biopsy and clinical variables to radical prostatectomy findings: can insignificant and advanced prostate cancer be predicted in a screening population? *Urology*, 64(3):544–550, 2004.

52. M. S. Andersen. Optimization methods for tomography. In P. C. Hansen, J. S. Jørgensen, and W. R. B. Lionheart, editors, *Computed Tomography: Algorithms, Insight and Just Enough Theory*, chapter 13. SIAM, Philadelphia, 2021.

53. N. C. Andreasen, R. Rajarethinam, T. Cizadlo, S. Arndt, V. W. Swayze, L. A. Flashman, D. S. O'Leary, J. C. Ehrhardt, and W. T. C. Yuh. Automatic atlas-based volume estimation of human brain regions from MR images. *Journal of Computer Assisted Tomography*, 20(1):98–106, Jan. 1996.

54. M. Arad, S. Zlochiver, T. Davidson, Y. Shoenfeld, A. Adunsky, and S. Abboud. The detection of pleural effusion using a parametric EIT technique. *Physiological Measurement*, 30(4):421–428, 03 2009.

55. G. E. Archie. The electrical resistivity log as an aid in determining some reservoir characteristics. *Transactions of the American Institue of Mining, Metallurgical, and Patroleum Engineers*, 146(1):54–62, 1942.

56. K. Aristovich, M. Donegá, C. Blochet, J. Avery, S. Hannan, D. J. Chew, and D. Holder. Imaging fast neural traffic at fascicular level with electrical impedance tomography: proof of principle in rat sciatic nerve. *Journal of neural engineering*, 15(5):056025, 2018.

57. K. Y. Aristovich, G. S. dos Santos, B. C. Packham, and D. S. Holder. A method for reconstructing tomographic images of evoked neural activity with electrical impedance tomography using intracranial planar arrays. *Physiological measurement*, 35(6):1095, 2014.

58. K. Y. Aristovich, B. C. Packham, H. Koo, G. S. dos Santos, A. McEvoy, and D. S. Holder. Imaging fast electrical activity in the brain with electrical impedance tomography. *NeuroImage*, 124:204–213, Jan. 2016.

59. A. Arko, R. C. Waterfall, M. S. Beck, T. Dyakowski, P. Sutcliffe, and M. Byars. Development of electrical capacitance tomography for solids mass flow measurement and control of pneumatic conveying systems. In *Proc. 1st World Congress on Industrial Process Tomography*, pages 140–146, Apr. 1999.

60. S. Arridge. Optical tomography in medical imaging. *Inverse Problems*, 15:R41–93, 1999.

61. S. Arridge, P. Maass, O. Öktem, and C.-B. Schönlieb. Solving inverse problems using data-driven models. *Acta Numerica*, 28:1–174, 2019.

62. S. R. Arridge and M. Schweiger. A gradient-based optimisation scheme for optical tomography. *Optics Express*, 2:213–226, 1998.

63. Y. Asfaw and A. Adler. Automatic detection of detached and erroneous electrodes in electrical impedance tomography. *Physiological measurement*, 26(2):S175, 2005.

64. M. Assenheimer, O. Laver-Moskovitz, D. Malonek, D. Manor, U. Nahaliel, R. Nitzan, and A. Saad. The T-Scan technology: electrical impedance as a diagnostic tool for breast cancer detection. *Physiological measurement*, 22(1):1, 2001.

65. K. Astala, J. Mueller, L. Päivärinta, A. Perämäki, and S. Siltanen. Direct electrical impedance tomography for nonsmooth conductivities. *Inverse Problems and Imaging*, 5(3):531–549, 2011.

66. K. Astala, J. Mueller, L. Päivärinta, and S. Siltanen. Numerical computation of complex geometrical optics solutions to the conductivity equation. *Applied and Computational Harmonic Analysis*, 29(1):391–403, 2010.

67. K. Astala and L. Päivärinta. A boundary integral equation for Calderón's inverse conductivity problem. In *Proc. 7th Internat. Conference on Harmonic Analysis, Collectanea Mathematica*, 2006.

68. K. Astala and L. Päivärinta. Calderón's inverse conductivity problem in the plane. *Annals of Mathematics*, pages 265–299, 2006.

69. K. Astala and L. Päivärinta. Calderón's inverse conductivity problem in the plane. *Annals of Mathematics*, 163(1):265–299, 2006.

70. K. Astala, L. Päivärinta, J. M. Reyes, and S. Siltanen. Nonlinear fourier analysis for discontinuous conductivities: Computational results. *Journal of Computational Physics*, 276:74–91, 2014.

71. R. Aster, B. Borchers, and C. Thurber. *Parameter Estimation and Inverse Problems*. Academic Press, 2004.

72. U. Auer, J. P. Schramel, Y. P. Moens, M. Mosing, and C. Braun. Monitoring changes in distribution of pulmonary ventilation by functional electrical impedance tomography in anaesthetized ponies. *Veterinary Anaesthesia and Analgesia*, 46(2):200–208, Mar. 2019.

73. J. Avery, K. Aristovich, B. Low, and D. Holder. Reproducible 3d printed head tanks for electrical impedance tomography with realistic shape and conductivity distribution. *Physiological measurement*, 38(6):1116, 2017.

74. J. Avery, T. Dowrick, M. Faulkner, N. Goren, and D. Holder. A versatile and reproducible multi-frequency electrical impedance tomography system. *Sensors*, 17(2):280, 2017.

75. J. Avery, T. Dowrick, A. Witkowska-Wrobel, M. Faulkner, K. Aristovich, and D. Holder. Simultaneous EIT and EEG using frequency division multiplexing. *Physiological measurement*, 40(3):034007, 2019.

76. S. B. Ayati, K. Bouazza-Marouf, and D. Kerr. In vitro localisation of intracranial haematoma using electrical impedance tomography semi-array. *Medical engineering & physics*, 37(1):34–41, 2015.

77. Comsol. The femlab reference manual. 2000.

78. A. P. Bagshaw, A. D. Liston, R. H. Bayford, A. Tizzard, A. P. Gibson, A. T. Tidswell, M. K. Sparkes, H. Dehghani, C. D. Binnie, and D. S. Holder. Electrical impedance tomography of human brain function using reconstruction algorithms based on the finite element method. *NeuroImage*, 20(2):752–764, 2003.

79. M. F. Bakker, S. V. de Lange, R. M. Pijnappel, R. M. Mann, P. H. Peeters, E. M. Monninkhof, M. J. Emaus, C. E. Loo, R. H. Bisschops, M. B. Lobbes, et al. Supplemental MRI screening for women with extremely dense breast tissue. *New England Journal of Medicine*, 381(22):2091–2102, 2019.

80. A. Barai, S. Watson, H. Griffiths, and R. Patz. Magnetic induction spectroscopy: non-contact measurement of the electrical conductivity spectra of biological samples. *Measurement Science and Technology*, 23(8):085501, June 2012.

81. C. B. Barber, D. P. Dobkin, and H. Huhdanpaa. The quickhull algorithm for convex hulls. *ACM Trans Math Software*, 22:469–483, 1996.

82. D. Barber and B. Brown. Recent developments in applied potential tomography-apt. *in Information Processing in Medical Imaging, ed S L Bacharach (Amsterdam: Nijhoff)*, pages 106–121, 1986.

83. D. C. Barber and B. H. Brown. Applied potential tomography. *Journal of Physics E: Scientific Instruments*, 17(9):723, 1984.

84. D. C. Barber, B. H. Brown, and I. L. Freeston. Imaging spatial distributions of resistivity using applied potential tomography–apt. In *Information Processing in Medical Imaging*, pages 446–462. Springer, 1984.

85. D. C. Barber and A. D. Seagar. Fast reconstruction of resistance images, Clin. *Phys. Physiol. Meas*, 8(4):47–54, 1987.

86. B. Barceló, E. Fabes, and J. K. Seo. The inverse conductivity problem with one measurement: uniqueness for convex polyhedra. *Proceedings of the American Mathematical Society*, 122(1):183–189, 1994.

87. T. Barceló, D. Faraco, and A. Ruiz. Stability of Calderón inverse conductivity problem in the plane. *Journal de Mathématiques Pures et Appliqués*, 88(6):522–556, 2007.

88. R. Barker. Signal contribution sections and their use in resistivity studies. *Geophys. J. Int.*, 59(1):123–129, 1979.

89. R. Barrett, M. Berry, and T. F. Chan. *Templates for the Solution of Linear Systems: Building Blocks for Iterative Methods*. Society for Industrial and Applied Mathematics, 1994.

90. M. Bauer, A. Opitz, J. Filser, H. Jansen, R. H. Meffert, C. T. Germer, N. Roewer, R. M. Muellenbach, and M. Kredel. Perioperative redistribution of regional ventilation and pulmonary function: a prospective observational study in two cohorts of patients at risk for postoperative pulmonary complications. *BMC Anesthesiology*, 19(1), July 2019.

91. S. Baumann, D. Wozny, S. Kelly, and F. Meno. The electrical conductivity of human cerebrospinal fluid at body temperature. *IEEE Transactions on Biomedical Engineering*, 44(3):220–223, Mar. 1997.

92. R. H. Bayford, A. Gibson, A. Tizzard, A. T. Tidswell, and D. S. Holder. Solving the forward problem for the human head using ideas (integrated design engineering analysis software) a finite element modelling tool. *Physiological Measurements*, 22:55–63, 2001.

93. T. Becher, M. Dargvainis, D. Hassel, N. Weiler, I. Alkatout, H. Ohnesorge, and I. Frerichs. Residual alveolar overdistension and collapse at electrical impedance tomography-guided positive end-expiratory pressure in patients with and without ards. volume 7(Suppl 3), page 000931.

94. T. Becher, M. Kott, D. Schädler, B. Vogt, T. Meinel, N. Weiler, and I. Frerichs. Influence of tidal volume on ventilation inhomogeneity assessed by electrical impedance tomography during controlled mechanical ventilation. *Physiological Measurement*, 36(6):1137–1146, May 2015.

95. T. Becher, T. Meinel, D. Bläser, G. Zick, N. Weiler, and I. Frerichs. Assessment of tidal recruitment and overdistension by regional analysis of respiratory system compliance at different tidal volumes. In *Proc 15th Int conf Biomedical Applications of EIT (EIT 2014)*, page 71, 04 2014.

96. T. Becher, B. Vogt, M. Kott, D. Schädler, N. Weiler, and I. Frerichs. Functional regions of interest in electrical impedance tomography: A secondary analysis of two clinical studies. *PLOS ONE*, 11:1–16, 03 2016.

97. T. Becher, A. Wendler, C. Eimer, N. Weiler, and I. Frerichs. Changes in electrical impedance tomography findings of ICU patients during rapid infusion of a fluid bolus: A prospective observational study. *American Journal of Respiratory and Critical Care Medicine*, 199(12):1572–1575, June 2019.

98. M. Beck, T. Dyakowski, and R. Williams. Process tomography - the state of the art. *Transactions of the Institute of Measurement and Control*, 20(4):163–177, Oct. 1998.

99. L. Beff, T. Günther, B. Vandoorne, V. Couvreur, and M. Javraux. Three-dimensional monitoring of soil water content in a maize field using electrical resistivity tomography. *Hydrology and Earth System Sciences*, 17(2):595–609, 2013.

100. A. Benabid, L. Balme, J. Persat, M. Belleville, J. Chirossel, M. Buyle-Bodin, J. de Rougemont, and C. Poupot. Electrical impedance brain scanner: principles and preliminary results of simulation. *T.-I.-T. journal of life sciences*, 8(1-2):59, 1978.

101. A. L. Benabid, S. Chabardes, J. Mitrofanis, and P. Pollak. Deep brain stimulation of the subthalamic nucleus for the treatment of parkinson's disease. *The Lancet Neurology*, 8(1):67–81, 2009.

102. A. L. Benabid, P. Pollak, D. Gao, D. Hoffmann, P. Limousin, E. Gay, I. Payen, and A. Benazzouz. Chronic electrical stimulation of the ventralis intermedius nucleus of the thalamus as a treatment of movement disorders. *Journal of neurosurgery*, 84(2):203–214, 1996.

103. M. A. Bennett, S. P. Luke, X. Jia, R. M. West, and R. A. Williams. Analysis and flow regime identification of bubble column dynamics. In *Proc. 1st World Congress on Industrial Process Tomography*, pages 54–61, Apr. 1999.

104. E. Beretta and E. Francini. Lipschitz stability for the electrical impedance tomography problem: the complex case. *arXiv*:1008.4046 [math.AP], Aug. 2010.

105. M. Bertero and P. Boccacci. *Introduction to Inverse Problems in Imaging*. IOP Publishing Ltd, London, 1998.

106. T. B. Bevers, M. Helvie, E. Bonaccio, K. E. Calhoun, M. B. Daly, W. B. Farrar, J. E. Garber, R. Gray, C. C. Greenberg, R. Greenup, et al. Breast cancer screening and diagnosis, version 3.2018, NCCN clinical practice guidelines in oncology. *Journal of the National Comprehensive Cancer Network*, 16(11):1362–1389, 2018.

107. J. Bickenbach, M. Czaplik, G. Polier, G. Marx, N. Marx, and M. Dreher. Electrical impedance tomography for predicting failure of spontaneous breathing trials in patients with prolonged weaning. *Critical Care*, 21(1), July 2017.

108. J. Bikowski. *Electrical Impedance Tomography Reconstructions in Two and Three Dimensions; from Calderón to Direct Methods*. PhD thesis, Colorado State University, Fort Collins, CO, 2008.

109. J. Bikowski, K. Knudsen, and J. L. Mueller. Direct numerical reconstruction of conductivities in three dimensions using scattering transforms. *Inverse Problems*, 27:19pp, 2011.

110. J. Bikowski and J. Mueller. 2D EIT reconstructions using Calderón's method. *Inverse Problems and Imaging*, 2(1):43–61, 2008.

111. Z. Bing and S. Greenhalgh. Finite element three dimensional direct current resistivity modelling: accuracy and efficiency considerations. *Geophys. J. Int.*, 145(3):679–688, 2001.

112. A. Binley. Tools and techniques: Electrical methods. In *Treatise on Geophysics*, volume 11, pages 233–259. Elsevier, Oxford, UK, 2 edition, 2015.

113. A. Binley, W. Daily, and A. Ramirez. Detecting leaks from waste storage ponds using electrical tomo-graphic methods. In *Proc. 1st World Congress on Industrial Process Tomography*, pages 6–13, Apr. 1999.

114. A. Binley, S. Hubbard, J. Huisman, D. Revil, Robinson, K. Singha, and L. Slater. The emergence of hydrogeophysics for improved understanding of subsurface processes over multiple scales. *Water Resources Research*, 51(6):3837–3866, 2015.

115. R. Binns, A. R. A. Lyons, A. J. Peyton, and W. D. N. Pritchard. Imaging molten steel flow profiles. *Measurement Science and Technology*, 12(8):1132–1138, July 2001.

116. Ö. Birgül, B. M. Eyüboğlu, and Y. Z. Ider. Current constrained voltage scaled reconstruction (CCVSR) algorithm for MR-EIT and its performance with different probing current patterns. *Physics in Medicine & Biology*, 48(5):653, 2003.

117. Y. Bissesseur and A. J. Peyton. Image reconstruction for electromagnetic inductance tomography em-ploying a parameterized finite-element-based forward model. In H. McCann and D. M. Scott, editors, *Process Imaging for Automatic Control*, volume 4188, pages 261–272. SPIE, Feb. 2001.

118. Y. Bissesseur and A. J. Peyton. A forward model for planar array emt imaging of cylindrical conductors embedded in a non-conducting medium. In *Proc. 2nd International Symposium Process Tomography*, pages 17–24, Sept. 2002.

119. B. Blad, B. Persson, and K. Lindström. Quantitative assessment of impedance tomography for temper-ature measurements in hyperthermia. *International journal of hyperthermia*, 8(1):33–43, 1992.

120. P. Blankman, D. Hasan, G. Erik, and D. Gommers. Detection of 'best' positive end-expiratory pressure derived from electrical impedance tomography parameters during a decremental positive end-expiratory pressure trial. *Critical Care*, 18(3):R95, 2014.

121. P. Blankman, D. Hasan, M. S. van Mourik, and D. Gommers. Ventilation distribution measured with EIT at varying levels of pressure support and neurally adjusted ventilatory assist in patients with ALI. *Intensive Care Medicine*, 39(6):1057–1062, Apr. 2013.

122. M. Blome, H. Maurer, and K. Schmidt. Advances in three-dimensional geoelectric forward solver techniques. *Geophys. J. Int.*, 176(3):740–752, 2009.

123. M. Bodenstein, M. David, and K. Markstaller. Principles of electrical impedance tomography and its clinical application. *Critical care medicine*, 37(2):713–724, 2009.

124. J. Bond, J. C. Cullivan, N. Climpson, I. Faulkes, X. Jia, J. A. Kostuch, D. Payton, M. Wang, S. J. Wang, R. M. West, and R. A. Williams. Industrial monitoring of hydro-cyclone operation using electrical resistance tomography. In *Proc. 1st World Congress on Industrial Process Tomography*, pages 102–107, Apr. 1999.

125. G. Bonneau, G. Tremblay, P. Savard, R. Guardo, A. R. LeBlanc, R. Cardinal, P. L. Page, and R. A. Nadeau. An integrated system for intraoperative cardiac activation mapping. *IEEE transactions on biomedical engineering*, (6):415–423, 1987.

126. K. Boone, A. Lewis, and D. Holder. Imaging of cortical spreading depression by EIT: implications for localization of epileptic foci. *Physiological measurement*, 15(2A):A189, 1994.

127. L. Borcea. A nonlinear multigrid for imaging electrical conductivity and permittivity at low frequency. *Inverse Problems*, 17:329–359, 2001.

128. F. Bordi, C. Cametti, and A. Di Biasio. Passive electrical properties of biological cell membranes determined from maxwell?wagner conductivity dispersion measurements. *Journal of electroanalytical chemistry and interfacial electrochemistry*, 276(2):135–144, 1989.

129. A. R. Borges, J. E. de Oliveira, J. Velez, C. Tavares, F. Linhares, and A. J. Peyton. Development of electromagnetic tomography (emt) for industrial applications. part 2: Image reconstruction and software framework. In *Proc. 1st World Congress on Industrial Process Tomography*, pages 219–225, Apr. 1999.

130. J. B. Borges, G. Hedenstierna, J. S. Bergman, M. B. P. Amato, J. Avenel, and S. Montmerle-Borgdorff. First-time imaging of effects of inspired oxygen concentration on regional lung volumes and breathing pattern during hypergravity. *European Journal of Applied Physiology*, 115(2):353–363, Oct. 2015.

131. J. B. Borges, F. Suarez-Sipmann, S. H. Bohm, G. Tusman, A. Melo, E. Maripuu, M. Sandström, M. Park, E. L. V. Costa, G. Hedenstierna, and M. Amato. Regional lung perfusion estimated by electrical impedance tomography in a piglet model of lung collapse. *Journal of applied physiology (Bethesda, Md. : 1985)*, 112(1):225–236, 2012.

132. A. Borsic. *Regularization Methods for Imaging from Electrical Measurements*. PhD thesis, Oxford Brookes University, 2002.

133. A. Borsic and A. Adler. A primal–dual interior-point framework for using the L1 or L2 norm on the data and regularization terms of inverse problems. *Inverse Problems*, 28(9):095011, 2012.

134. A. Borsic, B. M. Graham, A. Adler, and W. R. B. Lionheart. In vivo impedance imaging with total variation regularization. *IEEE Transactions on Medical Imaging*, 29(1):44–54, 2010.

135. A. Borsic, R. Halter, Y. Wan, A. Hartov, and K. Paulsen. Electrical impedance tomography reconstruction for three-dimensional imaging of the prostate. *Physiological measurement*, 31(8):S1, 2010.

136. A. Borsic, L. Wrb, and C. N. McLeod. Generation of anisotropic-smoothness regularization filters for EIT. *IEEE Transactions of Medical Imaging*, 21:596–603, 2002.

137. A. Bouchedda, M. Chouteau, A. Binley, and B. Giroux. 2-D joint structural inversion of cross-hole electrical resistance and ground penetrating radar data. *J. Applied Geophys.*, 78(0):52–67, 2012.

138. J. Bourne, editor. *Critical Reviews in Biomedical Engineering*, volume 24. 1996.

139. G. Boverman, D. Isaacson, T.-J. Kao, Saulnier, G. J., and J. C. Newell. Methods for direct image reconstruction for EIT in two and three dimensions. In *Proceedings of the 2008 Electrical Impedance Tomography Conference*, Dartmouth College, in Hanover, New Hampshire, USA, June 16 to 18 2008.

140. G. Boverman, T.-J. Kao, R. Kulkarni, B. S. Kim, D. Isaacson, G. J. Saulnier, and J. C. Newell. Robust linearized image reconstruction for multifrequency EIT of the breast. *IEEE Transactions on Medical Imaging*, 27(10):1439–1448, Oct. 2008.

141. A. Boyle. *Geophysical applications of electrical impedance tomography*. PhD thesis, Carleton University, Ottawa, Canada, 2016.

142. A. Boyle and A. Adler. Integrating circuit simulation with EIT FEM models. In *Proc 17th Int conf Biomedical Applications of EIT (EIT 2018)*, page 20, June 2018.

143. A. Boyle, K. Aristovich, and A. Adler. Beneficial techniques for spatio-temporal imaging in electrical impedance tomography. *Physiological measurement*, 41(6):064003, 2020.

144. A. Boyle, P. B. Wilkinson, J. E. Chambers, P. I. Meldrum, S. Uhlemann, and A. Adler. Jointly reconstructing ground motion and resistivity for ERT-based slope stability monitoring. *Geophys. J. Int.*, 212(2):1167–1182, Feb. 2018.

145. O. Brabant, A. Wallace, P. Buss, and M. Mosing. Construction of a finite element model in two large species for EIT application. In *Science Week of the Australian and New Zealand College of Veterinary Scientists*, 2018.

146. R. Bragos, J. Rosell, and P. Riu. A wide-band AC-coupled current source for electrical impedance tomography. *Physiological Measurement*, 15:A91–A99, 1994.

147. F. Braun, M. Proença, A. Adler, T. Riedel, J.-P. Thiran, and J. Solà. Accuracy and reliability of noninvasive stroke volume monitoring via ecg-gated 3d electrical impedance tomography in healthy volunteers. *PLOS ONE*, 13(1):1–19, 01 2018.

148. F. Braun, M. Proença, M. Rapin, M. Lemay, A. Adler, B. Grychtol, J. Solà, and J.-P. Thiran. Aortic blood pressure measured via EIT: investigation of different measurement settings. *Physiological Measurement*, 36(6):1147–1159, may 2015.

149. F. Braun, M. Proença, J. Solà, J.-P. Thiran, and A. Adler. A versatile noise performance metric for electrical impedance tomography algorithms. *IEEE Transactions on Biomedical Engineering*, 64(10):2321–2330, 2017.

150. F. Braun, M. Proença, M. Lemay, and A. Adler. Distribution of pulmonary pulse arrival in the healthy human lung. *Proceedings of the 20th International Conference on Biomedical Applications of Electrical Impedance Tomography*, page 64, 06 2019.

151. F. Braun, M. Proença, M. Sage, J.-P. Praud, M. Lemay, A. Adler, and É Fortin-Pellerin. EIT measurement of pulmonary artery pressure in neonatal lambs. In A. W.-W. Alistair Boyle, Kirill Aristovich and D. Holder, editors, *Proceedings of the 20th International Conference on Biomedical Applications of Electrical Impedance Tomography*, page 33, 06 2019.

152. F. Braun, M. Proença, A. Wendler, J. Solà, M. Lemay, J.-P. Thiran, N. Weiler, I. Frerichs, and T. Becher. Noninvasive measurement of stroke volume changes in critically ill patients by means of electrical impedance tomography. *Journal of Clinical Monitoring and Computing*, 34, 10 2020.

153. W. R. Breckon. *Image Reconstruction in Electrical Impedance Tomography, Ph*. PhD thesis, D. Thesis, Oxford Polytechnic, 1990.

154. W. R. Breckon and M. K. Pidcock. Mathematical aspects of impedance imaging. *Clinical Physics and Physiological Measurement*, 8(4A):77–84, nov 1987.

155. W. R. Breckon and M. K. Pidcock. Data errors and reconstruction algorithms in electrical impedance tomography. *Clin. Phys. Physiol. Meas*, 9(4):105–109, 1988.

156. L. Brochard, A. Slutsky, and A. Pesenti. Mechanical ventilation to minimize progression of lung injury in acute respiratory failure. *American Journal of Respiratory and Critical Care Medicine*, 195(4):438–442, Feb. 2017.

157. B. Brown. Electrical impedance tomography (EIT): a review. *Journal of Medical Engineering & Technology*, 27(3):97–108, 2003.

158. B. Brown, A. Leathard, A. Sinton, F. McArdle, R. Smith, and D. Barber. Blood flow imaging using electrical impedance tomography. *Clin Phys Physiol Meas.*, 13(Suppl A):175–179, 1993.

159. B. H. Brown, D. C. Barber, and A. D. Seagar. Applied potential tomography: possible clinical applications. *Clinical Physics and Physiological Measurement*, 6(2):109–121, May 1985.

160. B. H. Brown, R. Flewelling, H. Griffiths, N. D. Harris, A. D. Leathard, L. Lu, A. H. Morice, G. R. Neufeld, P. Nopp, and W. Wang. EITS changes following oleic acid induced lung water. *Physiological Measurement*, 17(4A):A117–A130, Nov. 1996.

161. B. H. Brown and A. D. Seagar. The sheffield data collection system. *Clinical Physics and Physiological Measurement*, 8(4A):91–97, nov 1987.

162. B. H. Brown, J. A. Tidy, K. Boston, A. D. Blackett, R. H. Smallwood, and F. Sharp. Relation between tissue structure and imposed electrical current flow in cervical neoplasia. *The Lancet*, 355(9207):892–895, Mar. 2000.

163. R. M. Brown and G. Uhlmann. Uniqueness in the inverse conductivity problem for nonsmooth conductivities in two dimensions. *Communications in Partial Differential Equations*, 22(5):1009–1027, 1997.

164. M. Brühl. Explicit characterization of inclusions in electrical impedance tomography, SIAM J. *Math Anal*, 32:1327–1341, 2001.

165. P. Brunner, R. Merwa, A. Missner, J. Rosell, K. Hollaus, and H. Scharfetter. Reconstruction of the shape of conductivity spectra using differential multi-frequency magnetic induction tomography. *Physiological measurement*, 27(5):S237–S248, 2006.

166. A. R. Brunoni, M. A. Nitsche, N. Bolognini, M. Bikson, T. Wagner, L. Merabet, D. J. Edwards, A. Valero-Cabre, A. Rotenberg, A. Pascual-Leone, et al. Clinical research with transcranial direct current stimulation (tdcs): challenges and future directions. *Brain stimulation*, 5(3):175–195, 2012.

167. S. Buehler, K. H. Wodack, S. H. Böhm, A. D. Waldmann, M. F. Graessler, S. Nishimoto, F. Thrk, E. Kaniusas, D. A. Reuter, and C. Trepte. Detection of the aorta in electrical impedance tomography images without the use of contrast agent. *Proceedings of the 18th International Conference on Biomedical Applications of Electrical Impedance Tomography*, page 18, 06 2017.

168. A. Bunse-Gerstner and R. St"over. On a conjugate gradient-type method for solving complex symmetric linear systems. *Linear Algebra Appl.*, 287:105–123, 1999.

169. S. Bureau, D. Bertrand, B. Jallais, P. Reling, B. Dekdouk, L. Marsh, M. O'Toole, D. W. Armitage, A. J. Peyton, and J. Alvarez-Garcia. Fruitgrading: Development of a fruit sorting technology based on internal quality parameters. In *16th International Conference on Near Infrared Spectroscopy*, 2013.

170. J. Bureš, O. Burešová, and J. Křivánek. *The mechanism and applications of Leao's spreading depression of electroencephalographic activity*. Academic Press, 1974.

171. M. Byars. Developments in electrical capacitance tomography. In *Proc. World Congress on Industrial Process Tomography*, pages 542–549. Hannover, 2001.

172. A.-P. Calderón. On an inverse boundary value problem. In *Seminar on Numerical Analysis and its Applications to Continuum Physics (Rio de Janeiro, 1980)*, pages 65–73. Soc. Brasil. Mat., Rio de Janeiro, 1980.

173. D. Calvetti, S. Nakkireddy, and E. Somersalo. Approximation of continuous EIT data from electrode measurements with bayesian methods. *Inverse Problems*, 35(4):045012, 2019.

174. D. Calvetti, L. Reichel, and A. Shuibi. Enriched krylov subspace methods for ill-posed problems. *Linear Algebra Appl.*, 362:257–273, 2003.

175. I. M. V. Caminiti, F. Ferraioli, A. Formisano, and R. Martone. Adaptive ablation treatment based on impedance imaging. *IEEE transactions on magnetics*, 46(8):3329–3332, 2010.

176. A. Campbell and D. Land. Dielectric properties of female human breast tissue measured in vitro at 3.2 ghz. *Physics in Medicine & Biology*, 37(1):193, 1992.

177. V. Candiani and M. Santacesaria. Neural networks for classification of strokes in electrical impedance tomography on a 3d head model. *arXiv preprint arXiv:2011.02852*, 2020.

178. L. Cao, H. Li, D. Fu, X. Liu, H. Ma, C. Xu, X. Dong, B. Yang, and F. Fu. Real-time imaging of infarction deterioration after ischemic stroke in rats using electrical impedance tomography. *Physiological Measurement*, 41(1):015004, 2020.

179. M. Capps and J. L. Mueller. Reconstruction of organ boundaries with deep learning in the d-bar method for electrical impedance tomography. *in press*, 2019.

180. R. Casañas, H. Scharfetter, A. Altes, A. Remacha, P. Sarda, J. Sierra, R. Merwa, K. Hollaus, and J. Rosell. Measurement of liver iron overload by magnetic induction using a planar gradiometer: preliminary human results. *Physiological Measurement*, 25(1):315–323, Feb. 2004.

181. L. R. Casanova, A. Silva, and A. Borges. A quantitative algorithm for parameter estimation in magnetic induction tomography. In *Proc. 3rd World Congress on Industrial Process Tomography*, 2003.

182. O. Casas, R. Bragos, P. J. Riu, J. Rosell, M. Tresanchez, M. Warren, A. Rodriguez-Sinovas, A. Carreno, and J. Cinca. In vivo and in situ ischemic tissue characterization using electrical impedance spectroscopya. *Annals of the New York Academy of Sciences*, 873(1 ELECTRICAL BI):51–58, Apr. 1999.

183. J. E. Chambers, D. Gunn, P. B. Wilkinson, P. Meldrum, E. Haslam, S. Holyoake, M. Kirkham, O. Kuras, A. Merritt, and J. Wragg. 4D electrical resistivity tomography monitoring of soil moisture dynamics in an operational railway embankment. *Near Surface Geophysics*, 12(1):61–72, Feb. 2014.

184. A. C. Charles and S. M. Baca. Cortical spreading depression and migraine. *Nature Reviews Neurology*, 9(11):637, 2013.

185. S. Chaudhary, R. Mishra, A. Swarup, and J. M. Thomas. Dielectric properties of normal & malignant human breast tissues at radiowave & microwave frequencies. *Indian journal of biochemistry & biophysics*, 21(1):76–79, 1984.

186. M. Chauhan, A. Indahlastari, A. K. Kasinadhuni, M. Schar, T. H. Mareci, and R. J. Sadleir. Low-frequency conductivity tensor imaging of the human head in vivo using DT-MREIT: First study. *IEEE Transactions on Medical Imaging*, 37(4):966–976, Apr. 2018.

187. X. D. Chen, D. X. Li, S. X. Lin, and N. Özkan. On-line fouling/cleaning detection by measuring electric resistance—-equipment development and application to milk fouling detection and chemical cleaning monitoring. *Journal of food engineering*, 61(2):181–189, 2004.

188. Y. Chen, C. C. Wong, T. S. Pui, R. Nadipalli, R. Weerasekera, J. Chandran, H. Yu, and A. R. Rahman. Cmos high density electrical impedance biosensor array for tumor cell detection. *Sensors and Actuators B: Chemical*, 173:903–907, 2012.

189. Y. Chen, M. Yan, D. Chen, M. Hamsch, H. Liu, H. Jin, M. Vauhkonen, C. H. Igney, J. Kahlert, and Y. Wang. Imaging hemorrhagic stroke with magnetic induction tomography: realistic simulation and evaluation. *Physiological Measurement*, 31(6):809–827, May 2010.

190. M. Cheney and D. Isaacson. Distinguishability in impedance imaging, *IEEE Trans Biomed Eng*, pages 852–860. Eng, 39, 1992.

191. M. Cheney, D. Isaacson, J. C. Newell, S. Simske, and J. Goble. Noser: An algorithm for solving the inverse conductivity problem. *Int. J. Imaging Systems & Technology*, 2:66–75, 1990.

192. K.-S. Cheng, S. J. Simske, D. Isaacson, J. C. Newell, and D. G. Gisser. Errors due to measuring voltage on current-carrying electrodes in electric current computed tomography. *IEEE transactions on biomedical engineering*, 37(1):60–65, 1990.

193. V. Cherepenin, Y. Gulyaev, A. Korjenevsky, S. Sapetsky, and T. Tuykin. An electrical impedance tomography system for gynecological application GIT with a tiny electrode array. *Physiological measurement*, 33(5):849, 2012.

194. V. Cherepenin, A. Karpov, A. Korjenevsky, V. Kornienko, A. Mazaletskaya, D. Mazourov, and D. Meister. A 3d electrical impedance tomography (EIT) system for breast cancer detection. *Physiological measurement*, 22(1):9, 2001.

195. V. A. Cherepenin, A. Y. Karpov, A. V. Korjenevsky, V. N. Kornienko, Y. S. Kultiasov, and D. Mazourov. Preliminary static EIT images of the thorax in health and disease. *Physiological Measurement*, 23:33–41, 2002.

196. V. A. Cherepenin, A. Y. Karpov, A. V. Korjenevsky, V. N. Kornienko, Y. S. Kultiasov, M. B. Ochapkin, O. V. Trochanova, and J. D. Meister. Three-dimensional EIT imaging of breast tissues: system design and clinical testing. *IEEE Transactions on Medical Imaging*, 21(6):662–667, 2002.

197. Y. Chiu, P. Arand, S. Shroff, T. Feldman, and J. Carroll. Determination of pulse wave velocities with computerized algorithms. *American heart journal*, 121(5):1460–1470, May 1991.

198. J. J. Cilliers, M. Wang, and S. J. Neethling. Measuring flowing foam density distributions using ERT. In *Proc. 1st World Congress on Industrial Process Tomography*, pages 108–112, Apr. 1999.

199. G. Cinnella, S. Grasso, P. Raimondo, D. D'Antini, L. Mirabella, M. Rauseo, and M. Dambrosio. Physiological effects of the open lung approach in patients with early, mild, diffuse acute respiratory distress syndrome. *Anesthesiology*, 123(5):1113–1121, Nov. 2015.

200. S. Ciulli, S. Ispas, M. K. Pidcock, and A. Stroian. On a mixed neumann-robin boundary value problem in electrical impedance tomography. *Z Angewandte Math Mech*, 80:681–696, 2000.

201. C. A. Clark, M. Hedehus, and M. E. Moseley. In vivo mapping of the fast and slow diffusion tensors in human brain. *Magnetic Resonance in Medicine: An Official Journal of the International Society for Magnetic Resonance in Medicine*, 47(4):623–628, 2002.

202. M. Clay and T. Ferree. Weighted regularization in electrical impedance tomography with applications to acute cerebral stroke. *IEEE Transactions on Medical Imaging*, 21(6):629–637, June 2002.

203. K. S. Cole and H. J. Curtis. Electric impedance of the squid giant axon during activity. *The Journal of general physiology*, 22(5):649–670, 1939.

204. Y. Colin de Verdière, I. Gitler, and D. Vertigan. Réseaux électriques planaires ii, comment. *Math. Helv.*, 71:144–167, 1996.

205. D. Colombo and D. Rovetta. Coupling strategies in multiparameter geophysical joint inversion. *Geophys. J. Int.*, 215(2):1171–1184, 2018.

206. D. Colton and R. Kress. *Inverse Acoustic and Electromagnetic Scattering Theory*. Springer, Berlin, 2nd edition, 1998.

207. J. Conway. Electrical impedance tomography for thermal monitoring of hyperthermia treatment: an assessment using in vitro and in vivo measurements. *Clinical Physics and Physiological Measurement*, 8(4A):141, 1987.

208. J. Conway, M. Hawley, Y. Mangnall, H. Amasha, and G. Van Rhoon. Experimental assessment of electrical impedance imaging for hyperthermia monitoring. *Clinical Physics and Physiological Measurement*, 13(A):185, 1992.

209. H. F. Cook. A comparison of the dielectric behaviour of pure water and human blood at microwave frequencies. *British Journal of Applied Physics*, 3(8):249–255, Aug. 1952.

210. R. D. Cook, G. J. Saulnier, D. G. Gisser, J. C. Goble, J. C. Newell, and D. Isaacson. ACT3: A high-speed, high-precision electrical impedance tomograph. *IEEE Transactions on Biomedical Engineering*, 41(8):713–722, August 1994.

211. H. Cornean, K. Knudsen, and S. Siltanen. Towards a *d*-bar reconstruction method for three-dimensional EIT. *Journal of Inverse and Ill-Posed Problems*, 14(2):111–134, 2006.

212. E. L. Costa, R. G. Lima, and M. B. Amato. Electrical impedance tomography. *Yearbook of Intensive Care and Emergency Medicine*, pages 394–404, 2009.

213. E. L. V. Costa, J. B. Borges, A. Melo, F. Suarez-Sipmann, C. Toufen, S. H. Bohm, and M. B. P. Amato. Bedside estimation of recruitable alveolar collapse and hyperdistension by electrical impedance tomography. *Intensive Care Medicine*, 35(6):1132–1137, Mar. 2009.

214. N. Coulombe, H. Gagnon, F. Marquis, Y. Skrobik, and R. Guardo. A parametric model of the relationship between EIT and total lung volume. *Physiological Measurement*, 26(4):401–411, apr 2005.

215. N. Coulter Jr and J. Pappenheimer. Development of turbulence in flowing blood. *American Journal of Physiology-Legacy Content*, 159(2):401–408, 1949.

216. J. D. Crowley and T. A. Rabson. Contactless method of measuring resistivity. *Review of Scientific Instruments*, 47(6):712–715, June 1976.

217. E. B. Curtis and J. A. Morrow. Inverse problems for electrical networks. *Series on Applied Mathematics -*, 13, 2000.

218. D. Murphy and P. Rolfe. Aspects of instrumentation design for impedance imaging. *Clin. Phys. Physiol. Meas.*, 9:5–14, 1988.

219. T. Dahlin and V. Leroux. Improvement in time-domain induced polarization data quality with multi-electrode systems by separating current and potential cables. *Near Surface Geophysics*, 10(6):545–565, Apr. 2012.

220. M. Dai, B. Li, S. Hu, C. Xu, B. Yang, J. Li, F. Fu, Z. Fei, and X. Dong. In vivo imaging of twist drill drainage for subdural hematoma: A clinical feasibility study on electrical impedance tomography for measuring intracranial bleeding in humans. *PLoS ONE*, 8(1):e55020, Jan. 2013.

221. M. Dai, L. Wang, C. Xu, L. Li, G. Gao, and X. Dong. Real-time imaging of subarachnoid hemorrhage in piglets with electrical impedance tomography. *Physiological Measurement*, 31(9):1229, 2010.

222. W. Daily and A. Ramirez. Environmental process tomography in the united states. *The Chemical Engineering Journal and the Biochemical Engineering Journal*, 56(3):159–165, Feb. 1995.

223. W. Daily and A. Ramirez. The role of electrical resistance tomography in the us nuclear waste site characterization program. In *Proc. 1st World Congress on Industrial Process Tomography*, pages 2–5, Apr. 1999.

224. OPA860 wide bandwidth operational transconductance amplifier and buffer. Texas Instruments, 2008.

225. M. de Hoop, M. Lassas, M. Santacesaria, S. Siltanen, and J. P. Tamminen. Positive-energy d-bar method for acoustic tomography: a computational study. *Inverse Problems*, 32(2):025003, 2016.

226. C. Deans, L. Marmugi, and F. Renzoni. Sub-picotesla widely tunable atomic magnetometer operating at room-temperature in unshielded environments. *Review of Scientific Instruments*, 89(8):083111, Aug. 2018.

227. J. Deceuster, O. Kaufmann, and M. V. Camp. Automated identification of changes in electrode contact properties for long-term permanent ERT monitoring experiments. *Geophysics*, 78(2):E79–E94, 2013.

228. J. M. Deibele, H. Luepschen, and S. Leonhardt. Dynamic separation of pulmonary and cardiac changes in electrical impedance tomography. *Physiological Measurement*, 29(6):S1–S14, jun 2008.

229. F. Delbary, P. C. Hansen, and K. Knudsen. Electrical impedance tomography: 3D reconstructions using scattering transforms. *Applicable Analysis*, 91(4):737–755, 2012.

230. F. Delbary and K. Knudsen. Full numerical implementation of the scattering transform algorithm for the 3d calderón problem. *Inverse Problems and Imaging*, 8:991, 2014.

231. F. Delbary, H. P.C., and K. Knudsen. A direct numerical reconstruction algorithm for the 3D Calderón problem. In *Journal of Physics: Conference Series*, volume 290, page 0120003. IOP Publishing, 2011.

232. R. Deloughry, M. Young, E. Pickup, and L. Barratt. Variable density flowmeter for loading road tankers using process tomography. In H. McCann and D. M. Scott, editors, *Process Imaging for Automatic Control*. SPIE, Feb. 2001.

233. E. Demidenko, A. Hartov, and K. Paulsen. Statistical estimation of resistance/conductance by electrical impedance tomography measurements. *IEEE transactions on medical imaging*, 23(7):829–838, 2004.

234. E. Demidenko, A. Hartov, N. Soni, and K. Paulsen. *On optimal current patters for electrical impedance tomography*. submitted to IEEE Trans Medical Imaging, 2004.

235. E. Demidenko, A. Hartov, N. Soni, and K. D. Paulsen. On optimal current patterns for electrical impedance tomography. *IEEE Transactions on Biomedical Engineering*, 52(2):238–248, February 2005.

236. X. Deng, F. Dong, L. Xu, X. Liu, and L. Xu. The design of a dual-plane ert system for cross correlation measurement of bubbly gas/liquid pipe flow. *Measurement Science and Technology*, 12(8):1024, 2001.

237. C. W. Denyer, F. J. Lidgey, C. N. McLeod, and Q. S. Zhu. Current source calibration simplifies high-accuracy current source measurement. *Innov. Tech. Biol. Med.*, 15:48–55, 1994.

238. A. Dey and H. Morrison. Resistivity modelling for arbitrarily shaped two-dimensional structures. *Geophys. Prospect.*, 27(1):106–136, 1979.

239. I. Dietzel, U. Heinemann, G. Hofmeier, and H. Lux. Stimulus-induced changes in extracellular na+ and cl- concentration in relation to changes in the size of the extracellular space. *Experimental brain research*, 46(1):73–84, 1982.

240. D. C. Dobson and F. Santosa. An image enhancement technique for electrical impedance tomography. *Inverse Problems*, 10:1994, 1994.

241. M. Dodd and J. Mueller. A real-time d-bar algorithm for 2-d electrical impedance tomography data. *Inverse Problems and Imaging*, 8(4):1013–1031, 2014.

242. Y. Dong. Regularization techniques for tomography problems. In P. C. Hansen, J. S. Jørgensen, and W. R. B. Lionheart, editors, *Computed Tomography: Algorithms, Insight and Just Enough Theory*, chapter 12. SIAM, Philadelphia, 2021.

243. O. Dorn, M. El, and C. M. Rappaport. A shape reconstruction method for electromagnetic tomography using adjoint fields and level sets. *Inverse Problems*, 16:1119–1156, 2000.

244. O. Dorn and D. Lesselier. Level set methods for structural inversion and image reconstruction. In *Handbook of Mathematical Methods in Imaging*. 2015.

245. T. Dowrick, C. Blochet, and D. Holder. In vivo bioimpedance measurement of healthy and ischaemic rat brain: implications for stroke imaging using electrical impedance tomography. *Physiological Measurement*, 36(6):1273–1282, May 2015.

246. T. Dowrick, C. Blochet, and D. Holder. In vivobioimpedance changes during haemorrhagic and is-chaemic stroke in rats: towards 3d stroke imaging using electrical impedance tomography. *Physiological Measurement*, 37(6):765–784, May 2016.

247. T. Dowrick and D. Holder. Phase division multiplexed EIT for enhanced temporal resolution. *Physiological Measurement*, 39(3):034005, 2018.

248. E. Dunne, M. O'Halloran, D. Craven, P. Puri, P. Frehill, S. Loughney, and E. Porter. Detection of vesicoureteral reflux using electrical impedance tomography. *IEEE Transactions on Biomedical Engineering*, 66(8):2279–2286, 2018.

249. T. Dyakowski, T. York, M. Mikos, D. Vlaev, R. Mann, G. Follows, A. Boxman, and M. Wilson. Imaging nylon polymerisation processes by applying electrical tomography. *Chemical Engineering Journal*, 77(1-2):105–109, Apr. 2000.

250. T. York (ed). *Proceedings of the 1st World Congress on Industrial Process Tomography*. VCIPT, Leeds, 1999.

251. J. F. Edd and B. Rubinsky. Detecting cryoablation with EIT and the benefit of including ice front imaging data. *Physiological measurement*, 27(5):S175, 2006.

252. P. M. Edic, G. J. Saulnier, J. C. Newell, and D. Isaacson. A real-time electrical impedance tomograph. *IEEE Trans. Biomed. Eng.*, 42(9):849–859, 1995.

253. R. Eichardt, C. H. Igney, J. Kahlert, M. Hamsch, M. Vauhkonen, and J. Haueisen. Sensitivity comparisons of cylindrical and hemi-spherical coil setups for magnetic induction tomography. In *IFMBE Proceedings*, pages 269–272. Springer Berlin Heidelberg, 2009.

254. Z. Elazar, R. Kado, and W. Adey. Impedance changes during epileptic seizures. *Epilepsia*, 7(4):291–307, 1966.

255. M. Elenkov, F. Thrk, A. Waldmann, K. Wodack, D. Reuter, S. Böhm, and E. Kaniusas. Localisation of pixels representing the aorta in electrical impedance tomography images based on time and frequency domain features. *Proceedings of the 19th International Conference on Biomedical Applications of Electrical Impedance Tomography*, page 27, 06 2018.

256. G. Elke, M. K. Fuld, A. F. Halaweish, B. Grychtol, N. Weiler, E. A. Hoffman, and I. Frerichs. Quantification of ventilation distribution in regional lung injury by electrical impedance tomography and xenon computed tomography. *Physiological Measurement*, 34(10):1303–1318, Sept. 2013.

257. J. G. Elmore, M. B. Barton, V. M. Moceri, S. Polk, P. J. Arena, and S. W. Fletcher. Ten-year risk of false positive screening mammograms and clinical breast examinations. *New England Journal of Medicine*, 338(16):1089–1096, 1998.

258. H. W. Engl, M. Hanke, and A. Neubauer. Regularization of inverse problems. *Kluwer, Dordrecht, 1996*, 1996.

259. N. Eronia, T. Mauri, E. Maffezzini, S. Gatti, A. Bronco, L. Alban, F. Binda, T. Sasso, C. Marenghi, G. Grasselli, G. Foti, A. Pesenti, and G. Bellani. Bedside selection of positive end-expiratory pressure by electrical impedance tomography in hypoxemic patients: a feasibility study. *Annals of Intensive Care*, 7(1), July 2017.

260. B. M. Eyüboğlu. Magnetic resonance electrical impedance tomography. *Wiley Encyclopedia of Biomedical Engineering*, 2006.

261. B. M. Eyüboğlu and T. C. Pilkington. Comment on distinguishability in electrical-impedance imaging, *IEEE Trans Biomed Eng*, pages 1328–1330. Eng, 40, 1993.

262. B. M. Eyboğlu, B. H. Brown, and D. C. Barber. In vivo imaging of cardiac related impedance changes. *IEEE Engineering in Medicine and Biology Magazine*, 8(1):39–45, March 1989.

263. L. Fabrizi, A. McEwan, T. Oh, E. Woo, and D. Holder. A comparison of two EIT systems suitable for imaging impedance changes in epilepsy. *Physiological measurement*, 30(6):S103, 2009.

264. L. Fabrizi, A. McEwan, T. Oh, E. Woo, and D. Holder. An electrode addressing protocol for imaging brain function with electrical impedance tomography using a 16-channel semi-parallel system. *Physiological measurement*, 30(6):S85, 2009.

265. L. Fabrizi, M. Sparkes, L. Horesh, J. P.-J. Abascal, A. McEwan, R. Bayford, R. Elwes, C. D. Binnie, and D. S. Holder. Factors limiting the application of electrical impedance tomography for identification of regional conductivity changes using scalp electrodes during epileptic seizures in humans. *Physiological measurement*, 27(5):S163, 2006.

266. L. D. Faddeev. Increasing solutions of the Schrödinger equation. *Soviet Physics Doklady*, 10:1033–1035, 1966.

267. T. J. C. Faes, H. A. van der Meij, J. C. de Munck, and R. M. Heethaar. The electric resistivity of human tissues (100 Hz–10 MHz): a meta-analysis of review studies. *Physiological Measurement*, 20(4):R1–R10, Nov. 1999.

268. A. Fagerberg, O. Stenqvist, and A. Åneman. Electrical impedance tomography applied to assess matching of pulmonary ventilation and perfusion in a porcine experimental model. *Critical Care*, 13(2):R34, Mar 2009.

269. A. Fagerberg, O. Stenqvist, and A. Åneman. Monitoring pulmonary perfusion by electrical impedance tomography: an evaluation in a pig model. *Acta Anaesthesiologica Scandinavica*, 53(2):152–158, 2009.

270. M. Faulkner, S. Hannan, K. Aristovich, J. Avery, and D. Holder. Characterising the frequency response of impedance changes during evoked physiological activity in the rat brain. *Physiological measurement*, 39(3):034007, 2018.

271. M. Faulkner, S. Hannan, K. Aristovich, J. Avery, and D. Holder. Feasibility of imaging evoked activity throughout the rat brain using electrical impedance tomography. *NeuroImage*, 178:1–10, 2018.

272. U. FDA. T-scan 2000, 1999.

273. E. C. Fear, S. C. Hagness, P. M. Meaney, M. Okoniewski, and M. A. Stuchly. Enhancing breast tumor detection with near-field imaging. *IEEE Microwave magazine*, 3(1):48–56, 2002.

274. J. R. Feldkamp. Single-coil magnetic induction tomographic three-dimensional imaging. *Journal of Medical Imaging*, 2(1):013502, Mar. 2015.

275. J. R. Feldkamp and S. Quirk. Optically tracked, single-coil, scanning magnetic induction tomography. *Journal of Medical Imaging*, 4(2):023504, June 2017.

276. M. Fernndez-Corazza, S. Turovets, P. Luu, N. Price, C. H. Muravchik, and D. Tucker. Skull modeling effects in conductivity estimates using parametric electrical impedance tomography. *IEEE Transactions on Biomedical Engineering*, 65(8):1785–1797, 2018.

277. F. Ferraioli, A. Formisano, and R. Martone. Effective exploitation of prior information in electrical impedance tomography for thermal monitoring of hyperthermia treatments. *IEEE Transactions on Magnetics*, 45(3):1554–1557, 2009.

278. D. Ferrario, A. Adler, J. Solà, S. Böhm, and M. Bodenstein. Unsupervised localization of heart and lung regions in EIT images: a validation study. In *Proc 12th Int conf Biomedical Applications of EIT (EIT 2011)*, pages 185–188, 05 2011.

279. D. Ferrario, B. Grychtol, A. Adler, J. Sola, S. H. Bohm, and M. Bodenstein. Toward morphological thoracic EIT: Major signal sources correspond to respective organ locations in CT. *IEEE Transactions on Biomedical Engineering*, 59(11):3000–3008, Nov. 2012.

280. R. Fletcher and Reeves C. Function minimization by conjugate gradients. *Computer J.*, 7:149–154, 1964.

281. G. B. Folland. *Introduction to Partial Differential Equations*. Univ Press, Princeton, second edition, 1995.

282. M. Forsman. Intragastric movement assessment by measuring magnetic field decay of magnetised tracer particles in a solid meal. *Medical & Biological Engineering & Computing*, 38(2):169–174, Mar. 2000.

283. G. Forte, A. Albano, M. J. Simmons, H. E. Stitt, E. Brunazzi, and F. Alberini. Assessing blending of non-newtonian fluids in static mixers by planar laser-induced fluorescence and electrical resistance tomography. *Chemical Engineering & Technology*, 42(8):1602–1610, 2019.

284. G. Forte, F. Alberini, M. J. Simmons, and E. H. Stitt. Measuring gas hold-up in gas–liquid/gas–solid–liquid stirred tanks with an electrical resistance tomography linear probe. *AIChE Journal*, 65(6):e16586, 2019.

285. K. R. Foster and H. P. Schwan. Dielectric properties of tissues – a review. In C. Polk and E. Postow, editors, *CRC Handbook of Biological Effects of Electromagnetic Fields*, pages 27–98. CRC Press, Boca Raton, FL, 1996.

286. C. Fox and G. Nicholls. Sampling conductivity images via MCMC. In K. Mardia, R. Ackroyd, and C. Gill, editors, *The Art and Science of Bayesian Image Analysis*, pages 91–100. Leeds Annual Statistics Research Workshop, University of Leeds, 1997.

287. M. A. Franceschini, K. T. Moesta, S. Fantini, G. Gaida, E. Gratton, H. Jess, W. W. Mantulin, M. Seeber, P. M. Schlag, and M. Kaschke. Frequency-domain techniques enhance optical mammography: initial clinical results. *Proceedings of the National Academy of Sciences*, 94(12):6468–6473, 1997.

288. G. Franchineau, N. Bréchot, G. Lebreton, G. Hekimian, A. Nieszkowska, J.-L. Trouillet, P. Leprince, J. Chastre, C.-E. Luyt, A. Combes, and M. Schmidt. Bedside contribution of electrical

impedance tomography to setting positive end-expiratory pressure for extracorporeal membrane oxygenation–treated patients with severe acute respiratory distress syndrome. *American Journal of Respiratory and Critical Care Medicine*, 196(4):447–457, Aug. 2017.

289. E. Francini. Recovering a complex coefficient in a planar domain from Dirichlet-to-Neumann map. *Inverse Problems*, 16:107–119, 2000.

290. S. Franco. *Design with Operational Amplifiers and Analog Integrated Circuits*. McGraw-Hill, 2014.

291. I. L. Freeston and R. C. Tozer. Impedance imaging using induced currents. *Physiological Measurement*, 16(3A):A257–A266, Aug. 1995.

292. D. Freimark, M. Arad, R. Sokolover, S. Zlochiver, and S. Abboud. Monitoring lung fluid content in CHF patients under intravenous diuretics treatment using bio-impedance measurements. *Physiological Measurement*, 28(7):S269–S277, June 2007.

293. I. Frerichs, M. B. P. Amato, A. H. van Kaam, D. G. Tingay, Z. Zhao, B. Grychtol, M. Bodenstein, H. Gagnon, S. H. Böhm, E. Teschner, O. Stenqvist, T. Mauri, V. Torsani, L. Camporota, A. Schibler, G. K. Wolf, D. Gommers, S. Leonhardt, and A. Adler. Chest electrical impedance tomography examination, data analysis, terminology, clinical use and recommendations: consensus statement of the TRanslational EIT developmeNt stuDy group. *Thorax*, 72(1):83–93, Jan. 2017.

294. I. Frerichs and T. Becher. Chest electrical impedance tomography measures in neonatology and paediatrics—a survey on clinical usefulness. *Physiological Measurement*, 40(5):054001, June 2019.

295. I. Frerichs, T. Becher, and N. Weiler. Electrical impedance tomography imaging of the cardiopulmonary system. *Current opinion in critical care*, 20(3):323–332, 2014.

296. I. Frerichs, M. Bodenstein, T. Dudykevych, J. Hinz, G. Hahn, and G. Hellige. Effect of lower body negative pressure and gravity on regional lung ventilation determined by EIT. *Physiological Measurement*, 26(2):S27–S37, Mar. 2005.

297. I. Frerichs, P. Braun, T. Dudykevych, G. Hahn, D. Genée, and G. Hellige. Distribution of ventilation in young and elderly adults determined by electrical impedance tomography. *Respiratory Physiology & Neurobiology*, 143(1):63–75, Oct. 2004.

298. I. Frerichs, T. Dudykevych, J. Hinz, M. Bodenstein, G. Hahn, and G. Hellige. Gravity effects on regional lung ventilation determined by functional EIT during parabolic flights. *Journal of Applied Physiology*, 91(1):39–50, July 2001.

299. I. Frerichs, G. Hahn, W. Golisch, M. Kurpitz, H. Burchardi, and G. Hellige. Monitoring perioperative changes in distribution of pulmonary ventilation by functional electrical impedance tomography. *Acta Anaesthesiologica Scandinavica*, 42(6):721–726, July 1998.

300. I. Frerichs, G. Hahn, and G. Hellige. Gravity-dependent phenomena in lung ventilation determined by functional EIT. *Physiological Measurement*, 17(4A):A149–A157, Nov. 1996.

301. I. Frerichs, J. Hinz, P. Herrmann, G. Weisser, G. Hahn, T. Dudykevych, M. Quintel, and G. Hellige. Detection of local lung air content by electrical impedance tomography compared with electron beam CT. *Journal of Applied Physiology*, 93(2):660–666, Aug. 2002.

302. I. Frerichs, J. Hinz, P. Herrmann, G. Weisser, G. Hahn, M. Quintel, and G. Hellige. Regional lung perfusion as determined by electrical impedance tomography in comparison with electron beam ct imaging. *IEEE transactions on medical imaging*, 21(6):646–652, 2002.

303. I. Frerichs, S. Pulletz, G. Elke, B. Gawelczyk, A. Frerichs, and N. Weiler. Patient examinations using electrical impedance tomography—sources of interference in the intensive care unit. *Physiological Measurement*, 32(12):L1–L10, Oct. 2011.

304. I. Frerichs, S. Pulletz, G. Elke, F. Reifferscheid, D. Schadler, J. Scholz, and N. Weiler. Assessment of changes in distribution of lung perfusion by electrical impedance tomography. *Respiration; international review of thoracic diseases*, 77(3):282–291, 2009.

305. I. Frerichs, H. Schiffmann, G. Hahn, and G. Hellige. Non-invasive radiation-free monitoring of regional lung ventilation in critically ill infants. *Intensive Care Medicine*, 27(8):1385–1394, July 2001.

306. I. Frerichs, H. Schiffmann, R. Oehler, T. Dudykevych, G. Hahn, J. Hinz, and G. Hellige. Distribution of lung ventilation in spontaneously breathing neonates lying in different body positions. *Intensive Care Medicine*, 29(5):787–794, May 2003.

307. I. Frerichs, B. Vogt, J. Wacker, R. Paradiso, F. Braun, M. Rapin, L. Caldani, O. Chételat, and N. Weiler. Multimodal remote chest monitoring system with wearable sensors: a validation study in healthy subjects. *Physiological Measurement*, 41(1):015006, Feb. 2020.

308. I. Frerichs, Z. Zhao, and T. Becher. Simple electrical impedance tomography measures for the assessment of ventilation distribution. *American Journal of Respiratory and Critical Care Medicine*, 201(3):386–388, Feb. 2020.

309. I. Frerichs, Z. Zhao, T. Becher, P. Zabel, N. Weiler, and B. Vogt. Regional lung function determined by electrical impedance tomography during bronchodilator reversibility testing in patients with asthma. *Physiological Measurement*, 37(6):698–712, May 2016.

310. W. Freygang Jr and W. M. Landau. Some relations between resistivity and electrical activity in the cerebral cortex of the cat. *Journal of Cellular and Comparative Physiology*, 45(3):377–392, 1955.

311. H. Fricke. A mathematical treatment of the electric conductivity and capacity of disperse systems i. the electric conductivity of a suspension of homogeneous spheroids. *Physical Review*, 24(5):575–587, Nov. 1924.

312. H. Fricke. The theory of electrolytic polarization. *The London, Edinburgh, and Dublin Philosophical Magazine and Journal of Science*, 14(90):310–318, Aug. 1932.

313. H. Fricke. The complex conductivity of a suspension of stratified particles of spherical or cylindrical form. *The Journal of Physical Chemistry*, 59(2):168–170, Feb. 1955.

314. H. Fricke and S. Morse. The electric resistance and capacity of blood for frequencies between 800 and $4^1/_2$ million cycles. *Journal of General Physiology*, 9(2):153–167, Nov. 1925.

315. H. Fricke and S. Morse. An experimental study of the electrical conductivity of disperse systems. i. cream. *Physical Review*, 25(3):361–367, Mar. 1925.

316. F. Fu, B. Li, M. Dai, S.-J. Hu, X. Li, C.-H. Xu, B. Wang, B. Yang, M.-X. Tang, X.-Z. Dong, Z. Fei, and X.-T. Shi. Use of electrical impedance tomography to monitor regional cerebral edema during clinical dehydration treatment. *PLoS ONE*, 9(12):e113202, Dec. 2014.

317. A. J. Gabor, A. G. Brooks, R. P. Scobey, and G. H. Parsons. Intracranial pressure during epileptic seizures. *Electroencephalography and clinical neurophysiology*, 57(6):497–506, 1984.

318. C. Gabriel. Compilation of the dielectric properties of body tissues at RF and microwave frequencies. Technical report, Jan. 1996.

319. C. Gabriel, T. Y. A. Chan, and E. H. Grant. Admittance models for open ended coaxial probes and their place in dielectric spectroscopy. *Physics in Medicine and Biology*, 39(12):2183–2200, Dec. 1994.

320. C. Gabriel, S. Gabriel, and y. E. Corthout. The dielectric properties of biological tissues: I. literature survey. *Physics in Medicine and Biology*, 41(11):2231–2249, 1996.

321. C. Gabriel, A. Peyman, and E. H. Grant. Electrical conductivity of tissue at frequencies below 1 MHz. *Physics in medicine & biology*, 54(16):4863, 2009.

322. S. Gabriel, R. Lau, and C. Gabriel. The dielectric properties of biological tissues: Ii. measurements in the frequency range 10 hz to 20 ghz. *Physics in medicine & biology*, 41(11):2251, 1996.

323. S. Gabriel, R. Lau, and C. Gabriel. The dielectric properties of biological tissues: Iii. parametric models for the dielectric spectrum of tissues. *Physics in Medicine & Biology*, 41(11):2271, 1996.

324. P. Gaggero, A. Adler, and B. Grychtol. Using real data to train GREIT improves image quality. In *Proc 15th Int conf Biomedical Applications of EIT (EIT 2014)*, page 39, Apr. 2014.

325. P. O. Gaggero, A. Adler, J. Brunner, and P. Seitz. Electrical impedance tomography system based on active electrodes. *Physiological Measurement*, 33:831–847, 2012.

326. H. Gagnon, M. Cousineau, A. Adler, and A. E. Hartinger. A resistive mesh phantom for assessing the performance of EIT systems. *IEEE transactions on biomedical engineering*, 57(9):2257–2266, 2010.

327. F. Gamache Jr, G. Dold, and R. Myers. Changes in cortical impedance and EEG activity induced by profound hypotension. *American Journal of Physiology-Legacy Content*, 228(6):1914–1920, 1975.

328. J. Gance, J. Malet, R. Supper, P. Sailhac, D. Ottowitz, and B. Jochum. Permanent electrical resistivity measurements for monitoring water circulation in clayey landslides. *J. Applied Geophys.*, 126:98–115, 2016.

329. J. Gance, P. Sailhac, and J. Malet. Corrections of surface fissure effect on apparent resistivity measurements. *Geophys. J. Int.*, 200(2):1118–1135, 2015.

330. N. Gao, S. Zhu, and B. He. Estimation of electrical conductivity distribution within the human head from magnetic flux density measurement. *Physics in Medicine & Biology*, 50(11):2675, 2005.

331. D. Garmatter and B. Harrach. Magnetic resonance electrical impedance tomography (MREIT): Convergence and reduced basis approach. *SIAM Journal on Imaging Sciences*, 11(1):863–887, 2018.

332. M. Gasulla, J. Jordana, and R. Palls-Areny. 2d and 3d subsurface resistivity imaging using a constrained least-squares algorithm. In *Proc. 1st World Congress on Industrial Process Tomography*, pages 20–27, Apr. 1999.

333. R. L. Gaw. *The effect of red blood cell orientation on the electrical impedance of pulsatile blood with implications for impedance cardiography*. PhD thesis, Queensland University of Technology, 2010.

334. L. A. Geddes. *Electrodes and the measurement of bioelectric events*. Wiley-Interscience, New York, USA, 1972.

335. L. A. Geddes and L. E. Baker. The specific resistance of biological material–a compendium of data for the biomedical engineer and physiologist. *Medical and biological engineering*, 5(3):271–293, 1967.

336. N. Gencer, Y. Ider, and S. Williamson. Electrical impedance tomography: induced-current imaging achieved with a multiple coil system. *IEEE Transactions on Biomedical Engineering*, 43(2):139–149, 1996.

337. N. Gencer and M. Tek. Electrical conductivity imaging via contactless measurements. *IEEE Transactions on Medical Imaging*, 18(7):617–627, July 1999.

338. A. George and J. Liu. The evolution of the minimum degree ordering algorithm. *SIAM Review*, 31:1–19, 1989.

339. A. K. Gerke and G. A. Schmidt. Physiology of heart-lung interactions. In S. P. Bhatt, editor, *Cardiac Considerations in Chronic Lung Disease*, pages 149–160. Humana Press, 2020.

340. D. B. Geselowitz. An application of electrocardiographic lead theory to impedance plethysmography. *IEEE Transactions on biomedical Engineering*, (1):38–41, 1971.

341. M. Gharibi and L. Bentley. Resolution of 3-D electrical resistivity images from inversions of 2-D orthogonal lines. *Journal of Environmental and Engineering Geophysics*, 10(4):339–343, 2005.

342. S.-M. Gho, J. Shin, M.-O. Kim, and D.-H. Kim. Simultaneous quantitative mapping of conductivity and susceptibility using a double-echo ultrashort echo time sequence: Example using a hematoma evolution study. *Magnetic resonance in medicine*, 76(1):214–221, 2016.

343. A. P. Gibson, J. Riley, M. Schweiger, J. C. Hebden, S. R. Arridge, and D. T. Delpy. A method for generating patient-specific finite element meshes for head modelling, phys. In *Med*, pages 481–495. Biol. 48, 2003.

344. R. Giguère, L. Fradette, D. Mignon, and P. Tanguy. ERT algorithms for quantitative concentration measurement of multiphase flows. *Chemical Engineering Journal*, 141(1-3):305–317, 2008.

345. O. Gilad, A. Ghosh, D. Oh, and D. S. Holder. A method for recording resistance changes non-invasively during neuronal depolarization with a view to imaging brain activity with electrical impedance tomography. *Journal of neuroscience methods*, 180(1):87–96, 2009.

346. O. Gilad and D. Holder. Impedance changes recorded with scalp electrodes during visual evoked responses: Implications for electrical impedance tomography of fast neural activity. *NeuroImage*, 47(2):514–522, Aug. 2009.

347. O. Gilad, L. Horesh, and D. Holder. A modelling study to inform specification and optimal electrode placement for imaging of neuronal depolarization during visual evoked responses by electrical and magnetic detection impedance tomography. *Physiological measurement*, 30(6):S201, 2009.

348. O. Gilad, L. Horesh, and D. S. Holder. Design of electrodes and current limits for low frequency electrical impedance tomography of the brain. *Medical & biological engineering & computing*, 45(7):621–633, 2007.

349. D. G. Gisser, D. Isaacson, and J. C. Newell. Current topics in impedance imaging. *Clin. Phys. Physiol. Meas*, 8:39–46, 1987.

350. D. G. Gisser, D. Isaacson, and J. C. Newell. Electric current computed-tomography and eigenvalues, *SIAM J Appl Math*, 50:1623–1634, 1990.

351. Y. A. Glickman, O. Filo, U. Nachaliel, S. Lenington, S. Amin-Spector, and R. Ginor. Novel EIS postprocessing algorithm for breast cancer diagnosis. *IEEE transactions on medical imaging*, 21(6):710–712, 2002.

352. M. E. Glidewell and K. T. Ng. Anatomically constrained electrical impedance tomography for three-dimensional anisotropic bodies, *IEEE Trans Med Imaging*, 16:572–580, 1997.

353. S. Gloning, K. Pieper, M. Zoellner, and A. Meyer-Lindenberg. Electrical impedance tomography for lung ventilation monitoring of the dog. *Tierärztliche Praxis Ausgabe K: Kleintiere / Heimtiere*, 45(01):15–21, 2017.

354. P. Glover. Geophysical properties of the near surface earth: Electrical properties. In *Treatise on Geophysics*, volume 11, pages 89–137. Elsevier, Oxford, UK, 2 edition, 2015.

355. J. Goble. *The three-dimensional inverse problem in electric current computed tomography*. PhD thesis, Rensselaer Polytechnic Institute, NY, USA, 1990.

356. J. Goble and D. Isaacson. Fast reconstruction algorithms for three-dimensional electrical impedance tomography. In *Proc. IEEE-EMBS Conf*, pages 100–101. 12(1), 1990.

357. C. Göksu, L. G. Hanson, H. R. Siebner, P. Ehses, K. Scheffler, and A. Thielscher. Human in-vivo brain magnetic resonance current density imaging (MRCDI). *NeuroImage*, 171:26–39, 2018.

358. G. H. Golub and C. F. Van Loan. *Matrix Computations*. Johns Hopkins University Press, 1996.

359. S. Goncalve, J. C. de Munck, J. P. A. Verbunt, R. M. Heethaar, and F. H. L. da Silva. In vivo measurement of the brain and skull resistivities using an EIT-based method and the combined analysis of SEF/SEP data. *IEEE Transactions on Biomedical Engineering*, 50(9):1124–1127, 2003.

360. S. Goncalves, J. de Munck, J. Verbunt, F. Bijma, R. Heethaar, and F. L. da Silva. In vivo measurement of the brain and skull resistivities using an EIT-based method and realistic models for the head. *IEEE Transactions on Biomedical Engineering*, 50(6):754–767, June 2003.

361. S. Gonçalves, J. C. de Munck, R. M. Heethaar, F. H. L. da Silva, and B. W. van Dijk. The application of electrical impedance tomography to reduce systematic errors in the EEG inverse problem - a simulation study. *Physiological Measurement*, 21(3):379–393, Aug. 2000.

362. L. Gong, K. Q. Zhang, and R. Unbehauen. 3-d anisotropic electrical impedance imaging. *IEEE Trans Magnetics*, 33:2120–2122, 1997.

363. C. A. Gonzalez, L. Horowitz, and B. Rubinsky. In vivo inductive phase shift measurements to detect intraperitoneal fluid. *IEEE Transactions on Biomedical Engineering*, 54(5):953–956, 2007.

364. C. A. González and B. Rubinsky. The detection of brain oedema with frequency-dependent phase shift electromagnetic induction. *Physiological Measurement*, 27(6):539–552, Apr. 2006.

365. C. A. González and B. Rubinsky. A theoretical study on magnetic induction frequency dependence of phase shift in oedema and haematoma. *Physiological Measurement*, 27(9):829–838, July 2006.

366. C. A. Gonzalez, J. A. Valencia, A. Mora, F. Gonzalez, B. Velasco, M. A. Porras, J. Salgado, S. M. Polo, N. Hevia-Montiel, S. Cordero, and B. Rubinsky. Volumetric electromagnetic phase-shift spectroscopy of brain edema and hematoma. *PLoS ONE*, 8(5):e63223, May 2013.

367. G. González, J. Huttunen, V. Kolehmainen, A. Seppänen, and M. Vauhkonen. Experimental evaluation of 3d electrical impedance tomography with total variation prior. *Inverse Problems in Science and Engineering*, 24(8):1411–1431, 2016.

368. C. A. González-Correa, B. H. Brown, R. H. Smallwood, D. C. Walker, and K. D. Bardhan. Electrical bioimpedance readings increase with higher pressure applied to the measuring probe. *Physiological Measurement*, 26(2):S39–S47, Mar. 2005.

369. N. Goren. *Clinical Applications of Electrical Impedance Tomography in Stroke and TBI Patients*. PhD thesis, University College London, London, UK, 2020.

370. N. Goren, J. Avery, T. Dowrick, E. Mackle, A. Witkowska-Wrobel, D. Werring, and D. Holder. Multi-frequency electrical impedance tomography and neuroimaging data in stroke patients. *Scientific data*, 5:180112, 2018.

371. D. Goss, R. O. Mackin, E. Crescenzo, H. S. Tapp, and A. J. Peyton. Development of electromagnetic inductance tomography (EMT) hardware for determining human body composition. In *Proc. 3rd World Congress on Industrial Process Tomography*, pages 377–383, 2003.

372. D. Goss, R. O. Mackin, E. Crescenzo, H. S. Tapp, and A. J. Peyton. Understanding the coupling mechanisms in high frequency EMT. In *Proc. 3rd World Congress on Industrial Process Tomography*, pages 354–369, 2003.

373. W. Gough. Wave propagation faster than light. *European Journal of Physics*, 23(1):17–19, Dec. 2001.

374. W. Gough. Circuit for the measurement of small phase delays in MIT. *Physiological Measurement*, 24(2):501–507, Apr. 2003.

375. W. Gough, H. Griffiths, and A. Morris. Wave propagation delays in magnetic induction tomography. In *Proc of 3rd EPSRC Engineering Network Meeting on Biomedical Applications of EIT*, Apr. 2001.

376. B. Graham and A. Adler. Objective selection of hyperparameter for EIT. *Physiological measurement*, 27(5):S65, 2006.

377. B. Graham and A. Adler. Electrode placement configurations for 3d EIT. *Physiological measurement*, 28(7):S29, 2007.

378. C. A. Grant, T. Pham, J. Hough, T. Riedel, C. Stocker, and A. Schibler. Measurement of ventilation and cardiac related impedance changes with electrical impedance tomography. *Critical Care*, 15(1):R37, 2011.

379. A. Greenleaf, M. Lassas, M. Santacesaria, S. Siltanen, and G. Uhlmann. Propagation and recovery of singularities in the inverse conductivity problem. *Anal. PDE*, 11(8):1901–1943, 2018.

380. B. D. Grieve, Q. Smit, R. Mann, and T. A. York. The application of electrical resistance tomography to a large volume production pressure filter. In *Proc 2nd World Congress on Process Tomography*, pages 175–182, Aug. 2001.

381. H. Griffiths. A cole phantom for EIT. *Physiological measurement*, 16(3A):A29, 1995.

382. H. Griffiths. Magnetic induction tomography. *Measurement Science and Technology*, 12(8):1126–1131, July 2001.

383. H. Griffiths and A. Ahmed. A dual-frequency applied potential tomography technique: computer simulations. *Clinical Physics and Physiological Measurement*, 8(4A):103, 1987.

384. H. Griffiths, W. Gough, S. Watson, and R. J. Williams. Residual capacitive coupling and the measurement of permittivity in magnetic induction tomography. *Physiological Measurement*, 28(7):S301–S311, June 2007.

385. H. Griffiths, W. R. Stewart, and W. Gough. Magnetic induction tomography: A measuring system for biological tissues. *Annals of the New York Academy of Sciences*, 873:335–345, Apr. 1999.

386. H. Griffiths, M. G. Tucker, J. Sage, and W. G. Herrenden-Harker. An electrical impedance tomography microscope. *Physiological Measurement*, 17(4A):A15–A24, Nov. 1996.

387. S. Grimnes and O. G. Martinsen. History of bioimpedance and bioelectricity. In *Bioimpedance and Bioelectricity Basics*, pages 313–319. Elsevier, 2000.

388. B. Grychtol and A. Adler. FEM electrode refinement for electrical impedance tomography. In *2013 35th Annual International Conference of the IEEE Engineering in Medicine and Biology Society (EMBC)*, pages 6429–6432. IEEE, 2013.

389. B. Grychtol, G. Elke, P. Meybohm, N. Weiler, I. Frerichs, and A. Adler. Functional validation and comparison framework for EIT lung imaging. *PLoS One*, 9(8):e103045, 2014.

390. B. Grychtol, B. Müller, and A. Adler. 3D EIT image reconstruction with GREIT. *Physiological measurement*, 37(6):785, 2016.

391. B. Grychtol, J. P. Schramel, F. Braun, T. Riedel, U. Auer, M. Mosing, C. Braun, A. D. Waldmann, S. H. Böhm, and A. Adler. Thoracic EIT in 3D: experiences and recommendations. *Physiological Measurement*, 40(7):074006, Aug. 2019.

392. R. Guardo, C. Boulay, B. Murray, and M. Bertrand. An experimental study in electrical impedance tomography using backprojection reconstruction. *IEEE transactions on biomedical engineering*, 38(7):617–627, 1991.

393. R. Guardo, J. Jehanne-Lacasse, A. Moumbe, and H. Gagnon. System front-end design for concurrent acquisition of electroencephalograms and EIT data. In *Journal of Physics: Conference Series*, volume 224, page 012012. IOP Publishing, 2010.

394. M. Guermandi, R. Cardu, E. F. Scarselli, and R. Guerrieri. Active electrode IC for EEG and electrical impedance tomography with continuous monitoring of contact impedance. *IEEE transactions on biomedical circuits and systems*, 9(1):21–33, 2014.

395. N. Gurler and Y. Z. Ider. Gradient-based electrical conductivity imaging using MR phase. *Magnetic resonance in medicine*, 77(1):137–150, 2017.

396. D. Gursoy, Y. Mamatjan, A. Adler, and H. Scharfetter. Enhancing impedance imaging through multimodal tomography. *IEEE Transactions on Biomedical Engineering*, 58(11):3215–3224, Nov. 2011.

397. D. Gürsoy and H. Scharfetter. The effect of receiver coil orientations on the imaging performance of magnetic induction tomography. *Measurement Science and Technology*, 20(10):105505, Sept. 2009.

398. D. Gürsoy and H. Scharfetter. Anisotropic conductivity tensor imaging using magnetic induction tomography. *Physiological Measurement*, 31(8):S135–S145, July 2010.

399. E. Haber and U. M. Ascher. Preconditioned all-at-once methods for large, sparse parameter estimation problems. *Inverse Problems*, 17:1847–1864, 2001.

400. D. Haemmerich, O. R. Ozkan, J. Z. Tsai, S. T. Staelin, S. Tungjitkusolmun, D. M. Mahvi, and J. G. Webster. Changes in electrical resistivity of swine liver after occlusion and postmortem. *Medical & Biological Engineering & Computing*, 40(1):29–33, Jan. 2002.

401. G. Hahn, I. Frerichs, M. Kleyer, and G. Hellige. Local mechanics of the lung tissue determined by functional EIT. *Physiological Measurement*, 17(4A):A159–A166, Nov. 1996.

402. G. Hahn, A. Just, T. Dudykevych, I. Frerichs, J. Hinz, M. Quintel, and G. Hellige. Imaging pathologic pulmonary air and fluid accumulation by functional and absolute EIT. *Physiological Measurement*, 27(5):S187–S198, Apr. 2006.

403. G. Hahn, I. Sipinkova, F. Baisch, and G. Hellige. Changes in the thoracic impedance distribution under different ventilatory conditions. *Physiological Measurement*, 16(3A):A161–A173, Aug. 1995.

404. R. Halter, A. Hartov, J. Heaney, K. Paulsen, and A. Schned. Electrical impedance spectroscopy of the human prostate. *IEEE Transactions on Biomedical Engineering*, 54(7):1321–1327, July 2007.

405. R. Halter, A. Hartov, and K. D. Paulsen. Design and implementation of a high frequency electrical impedance tomography system. *Physiological measurement*, 25(1):379, 2004.

406. R. J. Halter, A. Hartov, and K. D. Paulsen. A broadband high-frequency electrical impedance tomography system for breast imaging. *IEEE Transactions on biomedical engineering*, 55(2):650–659, 2008.

407. R. J. Halter, A. Hartov, K. D. Paulsen, A. Schned, and J. Heaney. Genetic and least squares algorithms for estimating spectral eis parameters of prostatic tissues. *Physiological measurement*, 29(6):S111, 2008.

408. R. J. Halter, A. Hartov, S. P. Poplack, W. A. Wells, K. M. Rosenkranz, R. J. Barth, P. A. Kaufman, K. D. Paulsen, et al. Real-time electrical impedance variations in women with and without breast cancer. *IEEE transactions on medical imaging*, 34(1):38–48, 2014.

409. R. J. Halter, A. Mahara, E. Hyams, and J. Pettus. Towards surgical margin assessment with microendoscopic electrical impedance sensing. In *Proc Annual Meeting of the Biomedical Engineering Society (BMES)*, Oct. 2017.

410. R. J. Halter, A. Schned, J. Heaney, A. Hartov, and K. D. Paulsen. Electrical properties of prostatic tissues: I. single frequency admittivity properties. *the Journal of Urology*, 182(4):1600–1607, 2009.

411. R. J. Halter, A. Schned, J. Heaney, A. Hartov, and K. D. Paulsen. Electrical properties of prostatic tissues: II. spectral admittivity properties. *The Journal of urology*, 182(4):1608–1613, 2009.

412. R. J. Halter, A. Schned, J. Heaney, A. Hartov, S. Schutz, and K. D. Paulsen. Electrical impedance spectroscopy of benign and malignant prostatic tissues. *The Journal of urology*, 179(4):1580–1586, 2008.

413. R. J. Halter, A. R. Schned, J. A. Heaney, and A. Hartov. Passive bioelectrical properties for assessing high-and low-grade prostate adenocarcinoma. *The Prostate*, 71(16):1759–1767, 2011.

414. R. J. Halter, T. Zhou, P. M. Meaney, A. Hartov, R. J. Barth Jr, K. M. Rosenkranz, W. A. Wells, C. A. Kogel, A. Borsic, E. J. Rizzo, et al. The correlation of in vivo and ex vivo tissue dielectric properties to validate electromagnetic breast imaging: initial clinical experience. *Physiological measurement*, 30(6):S121, 2009.

415. S. Hamilton and A. Hauptmann. Deep D-bar: Real time electrical impedance tomography imaging with deep neural networks. *IEEE Transactions on Medical Imaging*, 37:2367–2377, 2018.

416. S. Hamilton, C. Herrera, J. L. Mueller, and A. VonHerrmann. A direct D-bar reconstruction algorithm for recovering a complex conductivity in 2-D. *Inverse Problems*, 28:095005, 2012.

417. S. Hamilton, M. Lassas, and S. Siltanen. A direct reconstruction method for anisotropic electrical impedance tomography. *Inverse Problems*, 30:075007, 2014.

418. S. Hamilton, J. Mueller, and M. Alsaker. Incorporating a spatial prior into nonlinear D-bar EIT imaging for complex admittivities. *IEEE T. Med. Imaging*, 36(2):457–466, 2017.

419. S. Hamilton and S. Siltanen. Nonlinear inversion from partial EIT data: Computational experiments. *Contemporary Mathematics*, 615, 2014.

420. S. J. Hamilton, W. R. B. Lionheart, and A. Adler. Comparing D-bar and common regularization-based methods for electrical impedance tomography. *Physiological Measurement*, 40(4):044004, apr 2019.

421. S. J. Hamilton and J. L. Mueller. Direct EIT reconstructions of complex admittivities on a chest-shaped domain in 2-D. *IEEE Transactions on Medical Imaging*, 32(4):757–769, 2013.

422. S. J. Hamilton, J. L. Mueller, and T. Santos. Robust computation of 2D EIT absolute images with D-bar methods. *Physiological Measurement*, 39:064005, 2018.

423. S. J. Hamilton, J. M. Reyes, S. Siltanen, and X. Zhang. A hybrid segmentation and d-bar method for Electrical Impedance Tomography. *SIAM J. on Imaging Sciences*, 9(2):770–793, 2016.

424. E. Hammer, E. Abro, E. Cimpan, and G. Yan. High-frequency magnetic field probe for determination of interface levels in separation tanks. In H. McCann and D. M. Scott, editors, *Process Imaging for Automatic Control*, volume 4188, pages 294–299. SPIE, Feb. 2001.

425. E. Hammer and G. Fossdal. A new water-in-oil monitor cased on high frequency magnetic field excitation. In *Proc. 2nd International Symposium Process Tomography*, pages 9–16, Sept. 2002.

426. E. Hammer, F. Pettersen, and A. Nødseth. Numerical simulation of eddy current losses in high frequency magnetic field water fraction meters. In *Proc. 3rd World Congress on Industrial Process Tomography*, pages 347–351, 2003.

427. M. Hanke. *Conjugate gradient type methods for ill-posed problems*. Pitmannresearch notes in mathematics, Longman, Harlow Essex, 1995.

428. M. Hanke. On real-time algorithms for the location search of discontinuous conductivities with one measurement. *Inverse Problems*, 24(4):045005, 2008.

429. S. Hannan, M. Faulkner, K. Aristovich, J. Avery, and D. Holder. Frequency-dependent characterisation of impedance changes during epileptiform activity in a rat model of epilepsy. *Physiological measurement*, 39(8):085003, 2018.

430. S. Hannan, M. Faulkner, K. Aristovich, J. Avery, and D. Holder. Investigating the safety of fast neural electrical impedance tomography in the rat brain. *Physiological measurement*, 40(3):034003, 2019.

431. S. Hannan, M. Faulkner, K. Aristovich, J. Avery, M. C. Walker, and D. S. Holder. In vivo imaging of deep neural activity from the cortical surface during hippocampal epileptiform events in the rat brain using electrical impedance tomography. *NeuroImage*, 209:116525, 2020.

432. S. Hannan, D. S. Holder, K. Aristovich, M. Faulkner, J. Avery, and M. C. Walker. *Imaging slow brain activity during neocortical and hippocampal epileptiform events with electrical impedance tomography*, 42:014001, 2021.

433. A. J. Hansen. Effect of anoxia on ion distribution in the brain. *Physiological reviews*, 65(1):101–148, 1985.

434. Hansen and C.E. Olsen. Brain extracellular space during spreading depression and ischemia. *Acta Physiologica Scandinavica*, 108(4):355–365, 1980.

435. P. C. Hansen. *Rank-deficient and discrete ill-posed problems: numerical aspects of linear inversion*. SIAM, Philadelphia, 1998.

436. P. C. Hansen. *Discrete inverse problems: insight and algorithms*. SIAM, 2010.

437. P. C. Hansen, J. S. Jørgensen, and W. R. B. Lionheart. Computed tomography: Algorithms, insight and just enough theory. SIAM, Philadelphia, 2021.

438. B. Harrach. Uniqueness and lipschitz stability in electrical impedance tomography with finitely many electrodes. *Inverse problems*, 35(2):024005, 2019.

439. B. Harrach and M. Ullrich. Monotonicity-based shape reconstruction in electrical impedance tomography. *SIAM Journal on Mathematical Analysis*, 45(6):3382–3403, 2013.

440. A. E. Hartinger, H. Gagnon, and R. Guardo. A method for modelling and optimizing an electrical impedance tomography system. *Physiological measurement*, 27(5):S51, 2006.

441. A. E. Hartinger, H. Gagnon, and R. Guardo. Accounting for hardware imperfections in EIT image reconstruction algorithms. *Physiological measurement*, 28(7):S13, 2007.

442. A. E. Hartinger, R. Guardo, A. Adler, and H. Gagnon. Real-time management of faulty electrodes in electrical impedance tomography. *IEEE Transactions on Biomedical Engineering*, 56(2):369–377, 2008.

443. A. Hartov, P. LePivert, N. Soni, and K. Paulsen. Using multiple-electrode impedance measurements to monitor cryosurgery. *Medical physics*, 29(12):2806–2814, 2002.

444. A. Hartov, R. A. Mazzarese, F. R. Reiss, T. E. Kerner, K. S. Osterman, D. B. Williams, and K. D. Paulsen. A multichannel continuously selectable multifrequency electrical impedance spectroscopy measurement system. *IEEE transactions on biomedical engineering*, 47(1):49–58, 2000.

445. R. C. Hartwell and L. N. Sutton. Mannitol, intracranial pressure, and vasogenic edema. *Neurosurgery*, 32(3):444–450, Mar. 1993.

446. A. Hauptmann. Approximation of full-boundary data from partial-boundary electrode measurements. *Inverse Problems*, 33(12):125017, 22, 2017.

447. A. Hauptmann, M. Santacesaria, and S. Siltanen. Direct inversion from partial-boundary data in electrical impedance tomography. *Inverse Problems*, 33(2):025009, 2017.

448. K. Hayley, A. Pidlisecky, and L. Bentley. Simultaneous time-lapse electrical resistivity inversion. *J. Applied Geophys.*, 75(2):401–411, 2011.

449. W. M. Haynes, D. R. Lide, and T. J. Bruno. *CRC Handbook of Chemistry and Physics*. CRC Press, Boca Raton, FL, USA, 2014.

450. H. He, Y. Chi, Y. Long, S. Yuan, I. Frerichs, K. Möller, F. Fu, and Z. Zhao. Influence of overdistension/recruitment induced by high positive end-expiratory pressure on ventilation–perfusion matching assessed by electrical impedance tomography with saline bolus. *Critical Care*, 24(1):1–11, Sept. 2020.

451. H. He, Y. Chi, Y. Long, S. Yuan, R. Zhang, I. Frerichs, K. Möller, F. Fu, and Z. Zhao. Bedside evaluation of pulmonary embolism by saline contrast electrical impedance tomography method: a prospective observational study. *American Journal of Respiratory and Critical Care Medicine*, 202(10):1464–1468, 2020.

452. H. He, Y. Long, I. Frerichs, and Z. Zhao. Detection of acute pulmonary embolism by electrical impedance tomography and saline bolus injection. *American Journal of Respiratory and Critical Care Medicine*, 202(6):881–882, 2020.

453. L. M. Heikkinen, T. Vilhunen, R. M. West, and M. Vauhkonen. Simultaneous reconstruction of electrode contact impedances and internal electrical properties: II. *Laboratory experiments Meas*, 13:1855–1861, 2002.

454. B. Heincke, M. Jegen, M. Moorkamp, R. Hobbs, and J. Chen. An adaptive coupling strategy for joint inversions that use petrophysical information as constraints. *J. Applied Geophys.*, 136:279–297, 2017.

455. S. J. H. Heines, U. Strauch, M. C. G. van de Poll, P. M. H. J. Roekaerts, and D. C. J. J. Bergmans. Clinical implementation of electric impedance tomography in the treatment of ARDS: a single centre experience. *Journal of Clinical Monitoring and Computing*, 33(2):291–300, May 2018.

456. J. Heinitz and O. Minet. Dielectric properties of female breast tumors. In *Proc 9th Int Conference on Electrical Bio-Impedance*, pages 356–359.

457. S. Heinrich, H. Schiffmann, A. Frerichs, A. Klockgether-Radke, and I. Frerichs. Body and head position effects on regional lung ventilation in infants: an electrical impedance tomography study. *Intensive Care Medicine*, 32:1392–1398, June 2006.

458. R. Henderson, J. Webster, and D. Swanson. A thoracic electrical impedance camera. In *Proc 29th Annual Conference on Engineering in Medicine and Biology*, 1976.

459. R. P. Henderson and J. G. Webster. An impedance camera for spatially specific measurements of the thorax. *IEEE Transactions on Biomedical Engineering*, (3):250–254, 1978.

460. G. M. Henkin and M. Santacesaria. On an inverse problem for anisotropic conductivity in the plane. *Inverse Problems*, 26:095011, 2010.

461. M. Henningsson, M. Regner, K. Östergren, C. Trägrdh, and P. Dejmek. CFD simulation and ERT visualization of the displacement of yoghurt by water on industrial scale. *Journal of Food Engineering*, 80(1):166–175, 2007.

462. C. N. L. Herrera, M. F. M. Vallejo, J. L. Mueller, and R. Lima. Direct 2-D reconstructions of conductivity and permittivity from EIT data on a human chest. *IEEE Transactions on Medical Imaging*, 34(1):1–8, 2015.

463. N. Herteman, M. Mosing, K. Blaszczyk, C. Graubner, S. Lanz, and V. Gerber. Distribution of ventilation in equine pulmonary diseases measured by electrical impedance tomography: A case-series. In *Congress of the American College of Veterinary Internal Medicine*, 2019.

464. N. Herteman, M. Mosing, K. Blaszczyk, C. Graubner, S. Lanz, and V. Gerber. Effect of exercise on electrical impedance tomography derived flow variables in horses with equine asthma. In *Congress of the American College of Veterinary Internal Medicine*, 2019.

465. F. Hettlich and W. Rundell. The determination of a discontinuity in a conductivity from a single boundary measurement. *Inverse Problems*, 14:67–82, 1998.

466. N. J. Higham. *Accuracy and stability of numerical algorithms*. SIAM, Philadelphia, 1996.

467. S. Higson, P. Drake, D. W. Stamp, A. Peyton, A. Binns, A. Lyons, and W. Lionheart. Development of a sensor for visualisation of steel flow in the continuous casting nozzle. In *Proc 21st Int. ATS Steelmaking Conf*, Dec. 2002.

468. C. Hilbich, C. Fuss, and C. Hauck. Automated time-lapse ERT for improved process analysis and monitoring of frozen ground. *Permafrost and Periglacial Processes*, 22(4):306–319, Oct. 2011.

469. J. Hinz, G. Hahn, P. Neumann, M. Sydow, P. Mohrenweiser, G. Hellige, and H. Burchardi. End-expiratory lung impedance change enables bedside monitoring of end-expiratory lung volume change. *Intensive Care Medicine*, 29(1):37–43, Jan. 2003.

470. R. Höber. Messungen der inneren leitfähigkeit von zellen. *Pflüger's Archiv für die Gesamte Physiologie des Menschen und der Tiere*, 150(1-2):15–45, Feb. 1913.

471. A. L. Hodgkin and A. F. Huxley. A quantitative description of membrane current and its application to conduction and excitation in nerve. *The Journal of Physiology*, 117(4):500–544, Aug. 1952.

472. R. Hoekema, G. Wieneke, F. Leijten, C. van Veelen, P. van Rijen, G. Huiskamp, J. Ansems, and A. van Huffelen. *Brain Topography*, 16(1):29–38, 2003.

473. A. E. Hoerl. Application of ridge analysis to regression problems. *Chemical Engineering Progress*, 58:54–59, 1962.

474. A. Hoetink, T. Faes, K. Visser, and R. Heethaar. On the flow dependency of the electrical conductivity of blood. *IEEE Transactions on Biomedical Engineering*, 51(7):1251–1261, July 2004.

475. D. Holder. Feasibility of developing a method of imaging neuronal activity in the human brain: a theoretical review. *Medical and Biological Engineering and Computing*, 25(1):2, 1987.

476. D. Holder. Impedance changes during evoked nervous activity in human subjects: implications for the application of applied potential tomography (apt) to imaging neuronal discharge. *Clinical Physics and Physiological Measurement*, 10(3):267, 1989.

477. D. Holder. Detection of cerebral ischaemia in the anaesthetised rat by impedance measurement with scalp electrodes: implications for non-invasive imaging of stroke by electrical impedance tomography. *Clinical Physics and Physiological Measurement*, 13(1):63, 1992.

478. D. Holder. Electrical impedance tomography with cortical or scalp electrodes during global cerebral ischaemia in the anaesthetised rat. *Clinical Physics and Physiological Measurement*, 13(1):87, 1992.

479. D. Holder. *Clinical and physiological applications of electrical impedance tomography*. CRC Press, 1993.

480. D. Holder, C. Binnie, and C. Polkey. Cerebral impedance changes during seizures in human subjects: implications for non-invasive focus detection by electrical impedance tomography (EIT). *Brain Topography*, 5:331, 1993.

481. D. Holder, Y. Hanquan, and A. Rao. Some practical biological phantoms for calibrating multifrequency electrical impedance tomography. *Physiological measurement*, 17(4A):A167, 1996.

482. D. Holder and A. Khan. Use of polyacrylamide gels in a saline-filled tank to determine the linearity of the sheffield mark 1 electrical impedance tomography (EIT) system in measuring impedance disturbances. *Physiological measurement*, 15(2A):A45, 1994.

483. D. Holder, A. Rao, and Y. Hanquan. Imaging of physiologically evoked responses by electrical impedance tomography with cortical electrodes in the anaesthetized rabbit. *Physiological measurement*, 17(4A):A179, 1996.

484. D. S. Holder. *Electrical impedance tomography: methods, history and applications*. CRC Press, 2004.

485. K. Hollaus, C. Magele, R. Merwa, and H. Scharfetter. Fast calculation of the sensitivity matrix in magnetic induction tomography by tetrahedral edge finite elements and the reciprocity theorem. *Physiological Measurement*, 25(1):159–168, Feb. 2004.

486. H. Hong, J. Lee, J. Bae, and H. J. Yoo. A 10.4 mW electrical impedance tomography SoC for portable real-time lung ventilation monitoring system. *IEEE Journal of Solid State Circuits*, 50(11):2501–2512, 2015.

487. L. Horesh, S. Arridge, and D. Holder. Some novel approaches large scale algorithms for multi-frequency electrical impedance tomography of the human head, 2006.

488. R. Hou, P. J. Martin, H. J. Uppal, and A. J. Kowalski. An investigation on using electrical resistance tomography (ert) to monitor the removal of a non-newtonian soil by water from a cleaning-in-place (cip) circuit containing different pipe geometries. *Chemical Engineering Research and Design*, 111:332–341, 2016.

489. Y. Y. Hou, M. Wang, R. Holt, and R. A. Williams. A study of the mixing characteristics of a liquid magnetically stabilised fluidised bed using electrical resistance tomography. In *Proc 2nd World Congress on Process Tomography*, pages 315–323, 2001.

490. J. Hough, A. Trojman, and A. Schibler. Effect of time and body position on ventilation in premature infants. *Pediatric Research*, 80(4):499–504, May 2016.

491. B. S. Hoyle, X. Jia, F. J. W. Podd, H. I. Schlaberg, H. S. Tan, M. Wang, R. M. West, R. A. Williams, and T. A. York. Design and application of a multi-modal process tomography system. *Measurement Science and Technology*, 12(8):1157–1165, July 2001.

492. P. Hua, E. Woo, J. Webster, and W. Tompkins. Finite element modeling of electrode-skin contact impedance in electrical impedance tomography. *IEEE Transactions on Biomedical Engineering*, 40(4):335–343, Apr. 1993.

493. J.-J. Huang, Y.-H. Hung, J.-J. Wang, and B.-S. Lin. Design of wearable and wireless electrical impedance tomography system. *Measurement*, 78:9–17, 2016.

494. A. Hunt, J. D. Pendleton, and R. B. White. A novel tomographic flow analysis system. In *Proc. 3rd World Congress on Industrial Process Tomography*, Sept. 2003.

495. N. Hyvönen, L. Päivärinta, and J. P. Tamminen. Enhancing D-bar reconstructions for electrical impedance tomography with conformal maps. *Inverse Probl. Imaging*, 12(2):373–400, 2018.

496. Y. Z. Ider and S. Onart. Algebraic reconstruction for 3D magnetic resonance–electrical impedance tomography (MREIT) using one component of magnetic flux density. *Physiological measurement*, 25(1):281, 2004.

497. C. H. Igney, S. Watson, R. J. Williams, H. Griffiths, and O. Dössel. Design and performance of a planar-array MIT system with normal sensor alignment. *Physiological Measurement*, 26(2):S263–S278, Mar. 2005.

498. T. Ingeman-Nielsen, S. Tomaškovičová, and T. Dahlin. Effect of electrode shape on grounding resistances — part 1: The focus-one protocol. *Geophysics*, 81(1):1JF–Z7, 2016.

499. D. Ingerman and J. A. Morrow. On a characterization of the kernel of the dirichlet-to-neumann map for a planar region, *SIAM J Math Anal*, 29:106–115, 1998.

500. D. Isaacson. Distinguishability of conductivities by electric current computed tomography. *IEEE Transactions on Medical Imaging*, MI-5:92–95, 1986.

501. D. Isaacson, J. Mueller, J. Newell, and S. Siltanen. Imaging cardiac activity by the D-bar method for electrical impedance tomography. *Physiological Measurement*, 27:S43–S50, 2006.

502. D. Isaacson, J. L. Mueller, J. C. Newell, and S. Siltanen. Reconstructions of chest phantoms by the D-bar method for electrical impedance tomography. *IEEE Transactions on Medical Imaging*, 23:821–828, 2004.

503. V. Isakov. *Inverse Problems for Partial Differential Equations*. Springer, 1997.

504. Medical electrical equipment – part 1: General requirements for basic safety and essential performance. Standard, International Organization for Standardization, Geneva, CH, Dec. 2012.

505. J. J. Sikora, S. S. R. Arridge, R. R. H. Bayford, and L. Horesh. The application of hybrid BEM/FEM methods to solve electrical impedance tomography forward problem for the human head. *Proc. Proc X ICEBI and V EIT"* pages 503-506, 2004.

506. J. L. Jahn, E. L. Giovannucci, and M. J. Stampfer. The high prevalence of undiagnosed prostate cancer at autopsy: implications for epidemiology and treatment of prostate cancer in the prostate-specific antigen-era. *International journal of cancer*, 137(12):2795–2802, 2015.

507. G.-H. Jahng, M. B. Lee, H. J. Kim, E. J. Woo, and O.-I. Kwon. Low-frequency dominant electrical conductivity imaging of in vivo human brain using high-frequency conductivity at larmor-frequency and spherical mean diffusivity without external injection current. *NeuroImage*, 225:117466, 2020.

508. A. Jamshed, M. Cooke, Z. Ren, and T. L. Rodgers. Gas–liquid mixing in dual agitated vessels in the heterogeneous regime. *Chemical Engineering Research and Design*, 133:55–69, 2018.

509. A. Jamshed, M. Cooke, and T. L. Rodgers. Effect of zoning on mixing and mass transfer in dual agitated gassed vessels. *Chemical Engineering Research and Design*, 142:237–244, 2019.

510. G. Y. Jang, Y. J. Jeong, T. Zhang, T. I. Oh, R.-E. Ko, C. R. Chung, G. Y. Suh, and E. J. Woo. Noninvasive, simultaneous, and continuous measurements of stroke volume and tidal volume using EIT: feasibility study of animal experiments. *Scientific Reports*, 10(1), July 2020.

511. A. Javaherian, M. Soleimani, K. Moeller, A. Movafeghi, and R. Faghihi. An accelerated version of alternating direction method of multipliers for TV minimization in EIT. *Applied Mathematical Modelling*, 40(21-22):8985–9000, 2016.

512. A. J. Jaworski and T. Dyakowski. Application of electrical capacitance tomography for measurement of gas-solids flow characteristics in a pneumatic conveying system. *Measurement Science and Technology*, 12(8):1109–1119, July 2001.

513. M. Jehl, K. Aristovich, M. Faulkner, and D. Holder. Are patient specific meshes required for EIT head imaging? *Physiological Measurement*, 37(6):879, 2016.

514. M. Jehl, J. Avery, E. Malone, D. Holder, and T. Betcke. Correcting electrode modelling errors in EIT on realistic 3d head models. *Physiological Measurement*, 36(12):2423, 2015.

515. M. Jehl and D. Holder. Correction of electrode modelling errors in multi-frequency EIT imaging. *Physiological Measurement*, 37(6):893, 2016.

516. D. Jiang, Y. Wu, and A. Demosthenous. Hand gesture recognition using three-dimensional electrical impedance tomography. *IEEE Transactions on Circuits and Systems II: Express Briefs*, 67(9):1554–1558, 2020.

517. Y. Jiang and M. Soleimani. Capacitively coupled electrical impedance tomography for brain imaging. *IEEE transactions on medical imaging*, 38(9):2104–2113, 2019.

518. J.K Seo, O. Kwon, H. Ammari and E.J. Woo. A mathematical model for breast cancer lesion estimation: electrical impedance technique using ts2000 commercial system. *IEEE Transactions on Biomedical Engineering*, 51(11):1898–1906, 2004.

519. T. Johnson and J. Thomle. 3-D decoupled inversion of complex conductivity data in the real number domain. *Geophys. J. Int.*, 212(1):284–296, 2018.

520. J. Jordana, M. Gasulla, and R. Pallàs-Areny. Electrical resistance tomography to detect leaks from buried pipes. *Measurement Science and Technology*, 12(8):1061–1068, July 2001.

521. J. Jossinet. Variability of impedivity in normal and pathological breast tissue. *Medical and biological engineering and computing*, 34(5):346–350, 1996.

522. J. Jossinet. The impedivity of freshly excised human breast tissue. *Physiological Measurement*, 19(1):61–75, Feb. 1998.

523. J. Jossinet, A. Lobel, C. Michoudet, and M. Schmitt. Quantitative technique for bio-electrical spectroscopy. *Journal of biomedical engineering*, 7(4):289–294, 1985.

524. J. Jossinet and M. Schmitt. A review of parameters for the bioelectrical characterization of breast tissue. *Annals of the New York Academy of Sciences*, 873:30–41, Apr. 1999.

525. M. Joy, G. Scott, and M. Henkelman. In vivo detection of applied electric currents by magnetic resonance imaging. *Magnetic resonance imaging*, 7(1):89–94, 1989.

526. M. Jozwiak, J.-L. Teboul, and X. Monnet. Extravascular lung water in critical care: recent advances and clinical applications. *Annals of Intensive Care*, 5(1):38, 2015.

527. Y. M. Jung and S. Yun. Impedance imaging with first-order tv regularization. *IEEE transactions on medical imaging*, 34(1):193–202, 2014.

528. E. Kaden, N. D. Kelm, R. P. Carson, M. D. Does, and D. C. Alexander. Multi-compartment microscopic diffusion imaging. *NeuroImage*, 139:346–359, 2016.

529. J. P. Kaipio, V. Kolehmainen, E. Somersalo, and M. Vauhkonen. Statistical inversion and monte carlo sampling methods in electrical impedance tomography. *Inverse problems*, 16(5):1487, 2000.

530. J. P. Kaipio, A. Seppänen, E. Somersalo, and H. Haario. Posterior covariance related optimal current patterns in electrical impedance tomography. *Inverse Problems*, 20:919–936, 2004.

531. M. Kallio, A.-S. van der Zwaag, A. D. Waldmann, M. Rahtu, M. Miedema, T. Papadouri, A. H. van Kaam, P. C. Rimensberger, R. Bayford, and I. Frerichs. Initial observations on the effect of repeated surfactant dose on lung volume and ventilation in neonatal respiratory distress syndrome. *Neonatology*, 116(4):385–389, 2019.

532. T.-J. Kao, G. Boverman, B. S. Kim, D. Isaacson, G. J. Saulnier, J. C. Newell, M. H. Choi, R. H. Moore, and D. B. Kopans. Regional admittivity spectra with tomosynthesis images for breast cancer detection: preliminary patient study. *IEEE transactions on medical imaging*, 27(12):1762–1768, 2008.

533. C. Karagiannidis, A. D. Waldmann, P. L. Róka, T. Schreiber, S. Strassmann, W. Windisch, and S. H. Böhm. Regional expiratory time constants in severe respiratory failure estimated by electrical impedance tomography: a feasibility study. *Critical Care*, 22(1), Sept. 2018.

534. M. Karaoulis, P. Tsourlos, J. Kim, and A. Revil. 4D time-lapse ERT inversion: introducing combined time and space constraints. *Near Surface Geophysics*, 12(1):25–34, 2014.

535. B. Karbeyaz and N. Gencer. Electrical conductivity imaging via contactless measurements: an experimental study. *IEEE Transactions on Medical Imaging*, 22(5):627–635, May 2003.

536. J. Karsten, C. Grusnick, H. Paarmann, M. Heringlake, and H. Heinze. Positive end-expiratory pressure titration at bedside using electrical impedance tomography in post-operative cardiac surgery patients. *Acta Anaesthesiologica Scandinavica*, 59(6):723–732, Apr. 2015.

537. J. Karsten, K. Krabbe, H. Heinze, K. Dalhoff, T. Meier, and D. Drömann. Bedside monitoring of ventilation distribution and alveolar inflammation in community-acquired pneumonia. *Journal of Clinical Monitoring and Computing*, 28(4):403–408, Jan. 2014.

538. J. Karsten, T. Stueber, N. Voigt, E. Teschner, and H. Heinze. Influence of different electrode belt positions on electrical impedance tomography imaging of regional ventilation: a prospective observational study. *Critical Care*, 20(1), Dec. 2016.

539. N. Katoch, B. K. Choi, S. Z. Sajib, E. A. Lee, H. J. Kim, O. I. Kwon, and E. J. Woo. Conductivity tensor imaging of in vivo human brain and experimental validation using giant vesicle suspension. *IEEE Transactions on Medical Imaging*, 2018.

540. U. Katscher, T. Voigt, C. Findeklee, P. Vernickel, K. Nehrke, and O. Doessel. Determination of electric conductivity and local SAR via B1 mapping. *IEEE transactions on medical imaging*, 28(9):1365–1374, 2009.

541. J. M. Kay. Blood vessels of the lung. In *Comparative Biology of the Normal Lung*, pages 759–768. Elsevier, 2015.

542. C. P. Kellner, E. Sauvageau, K. V. Snyder, K. M. Fargen, A. S. Arthur, R. D. Turner, and A. V. Alexandrov. The vital study and overall pooled analysis with the vips non-invasive stroke detection device. *Journal of Neurointerventional Surgery*, 10(11):1079–1084, 2018.

543. K. M. Kelly, J. Dean, S.-J. Lee, and W. S. Comulada. Breast cancer detection: radiologists' performance using mammography with and without automated whole-breast ultrasound. *European radiology*, 20(11):2557–2564, 2010.

544. A. Kemna, A. Binley, G. Cassiani, E. Niederleithinger, A. Revil, L. Slater, K. Williams, F. Orozco, F. Haegel, A. Hrdt, S. Kruschwitz, K. Leroux, K. Titov, and E. Zimmermann. An overview of the spectral induced polarization method for near-surface applications. *Near Surface Geophysics*, 10(6):453–468, 2012.

545. J. B. Kendrick, A. D. Kaye, Y. Tong, K. Belani, R. D. Urman, C. Hoffman, and H. Liu. Goal-directed fluid therapy in the perioperative setting. *Journal of anaesthesiology, clinical pharmacology*, 35(Suppl 1):S29–S34, 2019.

546. K. Kerlikowske, D. Grady, J. Barclay, E. A. Sickles, and V. Ernster. Effect of age, breast density, and family history on the sensitivity of first screening mammography. *JAMA*, 276(1):33–38, 1996.

547. T. E. Kerner. *Electrical impedance tomography for breast imaging*. PhD thesis, Dartmouth College, Hannover, NH, USA, 2001.

548. T. E. Kerner, D. B. Williams, K. S. Osterman, F. R. Reiss, A. Hartov, and K. D. Paulsen. Electrical impedance imaging at multiple frequencies in phantoms. *Physiological measurement*, 21(1):67, 2000.

549. N. Kerrouche, C. McLeod, and W. Lionheart. Time series of EIT chest images using singular value decomposition and fourier transform. *Physiological Measurement*, 22(1):147–157, 2 2001.

550. T. K. A. Khairuddin and W. R. B. Lionheart. Characterization of objects by electrosensing fish based on the first order polarization tensor. *Bioinspiration & biomimetics*, 11(5):055004, 2016.

551. S. Khan, A. Mahara, E. S. Hyams, A. R. Schned, and R. J. Halter. Prostate cancer detection using composite impedance metric. *IEEE transactions on medical imaging*, 35(12):2513–2523, 2016.

552. H. Ki and D. Shen. Numerical inversion of discontinuous conductivities. *Inverse Problems*, 16:33–47, 2000.

553. B. S. Kim, D. Isaacson, H. Xia, T.-J. Kao, J. C. Newell, and G. J. Saulnier. A method for analyzing electrical impedance spectroscopy data from breast cancer patients. *Physiological measurement*, 28(7):S237, 2007.

554. D.-H. Kim, N. Choi, S.-M. Gho, J. Shin, and C. Liu. Simultaneous imaging of in vivo conductivity and susceptibility. *Magnetic resonance in medicine*, 71(3):1144–1150, 2014.

555. M. Kim, J. Jang, H. Kim, J. Lee, J. Lee, J. Lee, K.-R. Lee, K. Kim, Y. Lee, K. J. Lee, , and H.-J. Yoo. A 1.4-mΩ-sensitivity 94-dB dynamic-range electrical impedance tomography SoC and 48-channel hub-SoC for 3-D lung ventilation monitoring system. *IEEE Journal of Solid State Circuits*, 52(11):2829–2842, 2017.

556. M. N. Kim, T. Y. Ha, E. J. Woo, and O. I. Kwon. Improved conductivity reconstruction from multi-echo MREIT utilizing weighted voxel-specific signal-to-noise ratios. *Physics in Medicine & Biology*, 57(11):3643, 2012.

557. Y. T. Kim, P. J. Yoo, T. I. Oh, and E. J. Woo. Magnetic flux density measurement in magnetic resonance electrical impedance tomography using a low-noise current source. *Measurement Science and Technology*, 22(10):105803, 2011.

558. C. Klein and K. D. T.-R. McLaughlin. Spectral approach to D-bar problems. *Comm. Pure Appl. Math.*, 70(6):1052–1083, 2017.

559. K. A. Klivington and R. Galambos. Resistance shifts accompanying the evoked cortical response in the cat. *Science*, 157(3785):211–213, 1967.

560. K. A. Klivington and R. Galambos. Rapid resistance shifts in cat cortex during click-evoked responses. *Journal of Neurophysiology*, 31(4):565–573, 1968.

561. K. Knudsen, M. Lassas, J. Mueller, and S. Siltanen. D-bar method for electrical impedance tomography with discontinuous conductivities. *SIAM Journal on Applied Mathematics*, 67(3):893, 2007.

562. K. Knudsen, M. Lassas, J. Mueller, and S. Siltanen. Regularized D-bar method for the inverse conductivity problem. *Inverse Problems and Imaging*, 3(4):599–624, 2009.

563. K. Knudsen and J. Mueller. The born approximation and Calderón's method for reconstructions of conductivities in 3-D. *Discrete and Continuous Dynamical Systems*, pages 884–893, 2011.

564. K. Knudsen, J. Mueller, and S. Siltanen. Numerical solution method for the dbar-equation in the plane. *Journal of Computational Physics*, 198:500–517, 2004.

565. K. Knudsen and A. Tamasan. Reconstruction of less regular conductivities in the plane. *Communications in Partial Differential Equations*, 29:361–381, 2004.

566. L. Koessler, S. Colnat-Coulbois, T. Cecchin, J. Hofmanis, J. P. Dmochowski, A. M. Norcia, and L. G. Maillard. In-vivo measurements of human brain tissue conductivity using focal electrical current injection through intracerebral multicontact electrodes. *Human brain mapping*, 38(2):974–986, 2017.

567. R. V. Kohn and M. Vogelius. Determining conductivity by boundary measurements. II. interior results. *Comm. Pure Appl. Math*, 38:643–667, 1985.

568. A. Koksal, M. Eyuboglu, and M. Demirbilek. A quasi-static analysis for a class of induced-current EIT systems using discrete coils. *IEEE Transactions on Medical Imaging*, 21(6):688–694, June 2002.

569. A. Köksal and B. M. Eyüboğlu. Determination of optimum injected current patterns in electrical impedance tomography. *Physiol Meas*, 16:A99–A109, 1995.

570. K. Kolehmainen, S. R. Arridge, L. Wrb, M. Vauhkonen, and J. P. Kaipio. Recovery of region boundaries of piecewise constant coefficients of elliptic pde from boundary data. *Inverse Problems*, 15:1375–1391, 1999.

571. V. Kolehmainen. *Novel Approaches to Image Reconstruction in Diffusion Tomography*. PhD thesis, Department of Applied Physics Kuopio University, 2002.

572. V. Kolehmainen, M. Vauhkonen, J. P. Kaipio, and S. R. Arridge. Recovery of piecewise constant coefficients in optical diffusion tomography. *Optics Express*, 7:468–480, 2000.

573. A. Korjenevsky. Solving inverse problems in electrical impedance and magnetic induction tomography by artificial neural networks. *J Radioelectronics*, (12), 2001.

574. A. Korjenevsky, V. Cherepenin, and S. Sapetsky. Magnetic induction tomography: experimental realization. *Physiological Measurement*, 21(1):89–94, Feb. 2000.

575. A. Korjenevsky, V. Cherepenin, and S. Sapetsky. Magnetic induction tomography – new imaging method in biomedicine. In *Proc 2nd World Congress on Process Tomography*, pages 240–246, 2001.

576. A. Korzhenevskii and V. Cherapenin. Magnetic induction tomography. *J. Comm. Tech. Electronics*, 42(4):469–474, 1997.

577. A. Korzhenevsky and S. Sapetsky. Visualization of the internal structure of extended conducting objects by magnetoinduction tomography. *Bulletin of the Russian Academy of Sciences-Physics*, 65(12):1945–1949, 2001.

578. J. D. Kosterich, K. R. Foster, and S. R. Pollack. Dielectric permittivity and electrical conductivity of fluid saturated bone. *IEEE Transactions on Biomedical Engineering*, BME-30(2):81–86, Feb. 1983.

579. C. J. Kotre. A sensitivity coefficient method for the reconstruction of electrical impedance tomograms. *Clin. Phys. Physiol. Meas*, 10:275–81, 1989.

580. R. Kotzé, A. Adler, A. Sutherland, and C. Deba. Evaluation of electrical resistance tomography imaging algorithms to monitor settling slurry pipe flow. *Flow Measurement and Instrumentation*, 68:101572, 2019.

581. P. Krack and L. Vercueil. Review of the functional surgical treatment of dystonia. *European journal of neurology*, 8(5):389–399, 2001.

582. J. K. Krauss, J. Yianni, T. J. Loher, and T. Z. Aziz. Deep brain stimulation for dystonia. *Journal of Clinical Neurophysiology*, 21(1):18–30, 2004.

583. A. Krivoshei, M. Min, T. Parve, and A. Ronk. An adaptive filtering system for separation of cardiac and respiratory components of bioimpedance signal. In *IEEE International Workshop on Medical Measurement and Applications, MeMeA 2006*, volume 2006, pages 10 – 15, 02 2006.

584. S. Krueger-Ziolek, B. Schullcke, J. Kretschmer, U. Müller-Lisse, K. Möller, and Z. Zhao. Positioning of electrode plane systematically influences EIT imaging. *Physiological Measurement*, 36(6):1109–1118, May 2015.

585. S. Krueger-Ziolek, B. Schullcke, Z. Zhao, B. Gong, S. Naehrig, U. Müller-Lisse, and K. Moeller. Multi-layer ventilation inhomogeneity in cystic fibrosis. *Respiratory Physiology & Neurobiology*, 233:25–32, Nov. 2016.

586. C. Ktistis, D. W. Armitage, and A. J. Peyton. Calculation of the forward problem for absolute image reconstruction in MIT. *Physiological Measurement*, 29(6):S455–S464, June 2008.

587. G. Kühnel, G. Hahn, I. Frerichs, T. Schröder, and G. Hellige. Neue Verfahren zur Verbesserung der Ab bildungsqualität bei funktionellen EIT-tomogrammen der lung. *Biomedizinische Technik/Biomedical Engineering*, 42(s2):470–471, 1997.

588. E. Kuligowska, M. A. Barish, H. M. Fenlon, and M. Blake. Predictors of prostate carcinoma: accuracy of gray-scale and color doppler us and serum markers. *Radiology*, 220(3):757–764, 2001.

589. V. Kulkarni, J. Hutchison, and J. Mallard. The aberdeen impedance imaging system. *Biomedical sciences instrumentation*, 25:47–58, 1989.

590. V. Kulkarni, J. Hutchison, I. Ritchie, and J. Mallard. Impedance imaging in upper arm fractures. *Journal of biomedical engineering*, 12(3):219–227, 1990.

591. M.-F. Kuo, W. Paulus, and M. A. Nitsche. Therapeutic effects of non-invasive brain stimulation with direct currents (tdcs) in neuropsychiatric diseases. *Neuroimage*, 85:948–960, 2014.

592. Kuo-Sheng Cheng, D. Isaacson, J. C. Newell, and D. G. Gisser. Electrode models for electric current computed tomography. *IEEE Transactions on Biomedical Engineering*, 36(9):918–924, 1989.

593. O. Kuras, D. Beamish, P. I. Meldrum, and R. D. Ogilvy. Fundamentals of the capacitive resistivity technique. *Geophysics*, 71(3):G135–G152, May 2006.

594. O. Kuras, P. I. Meldrum, D. Beamish, R. D. Ogilvy, and D. Lala. Capacitive resistivity imaging with towed arrays. *Journal of Environmental and Engineering Geophysics*, 12(3):267–279, Sept. 2007.

595. O. Kuras, J. Pritchard, P. Meldrum, J. E. Chambers, P. B. Wilkinson, R. Ogilvy, and G. Wealthall. Monitoring hyrdaulic processes with automated time-lapse electrical resistivity tomography (ALERT). *Comptes Rendus Geoscience*, 341(10-11):868–885, Nov. 2009.

596. O. Kuras, P. B. Wilkinson, P. Meldrum, L. Oxby, S. Uhlemann, J. Chambers, A. Binley, J. Graham, N. Smith, and N. Atherton. Geoelectrical monitoring of simulated subsurface leakage to support high-hazard nuclear decommissioning at the Sellafield Site, UK. *Science of the Total Environment*, 566–567:350–359, 2016.

597. O. Kuras, P. B. Wilkinson, P. I. Meldrum, R. T. Swift, S. S. Uhlemann, J. E. Chambers, F. C. Walsh, J. A. Wharton, and N. Atherton. Performance assessment of novel electrode materials for long-term ERT monitoring. In *Near Surface Geoscience 2015*, Turin, Italy, Sept. 2015.

598. O. Kwon, E. J. Woo, J.-R. Yoon, and J. K. Seo. Magnetic resonance electrical impedance tomography (MREIT): simulation study of j-substitution algorithm. *IEEE Transactions on Biomedical Engineering*, 49(2):160–167, 2002.

599. O. I. Kwon, W. C. Jeong, S. Z. Sajib, H. J. Kim, E. J. Woo, and T. I. Oh. Reconstruction of dual-frequency conductivity by optimization of phase map in MREIT and MREPT. *Biomedical engineering online*, 13(1):24, 2014.

600. O. I. Kwon, S. Z. Sajib, I. Sersa, T. I. Oh, W. C. Jeong, H. J. Kim, and E. J. Woo. Current density imaging during transcranial direct current stimulation using DT-MRI and MREIT: algorithm development and numerical simulations. *IEEE Transactions on Biomedical Engineering*, 63(1):168–175, 2015.

601. D. LaBrecque, M. Miletto, W. Daily, A. Ramirez, and E. Owen. The effects of noise on Occam's inversion of resistivity tomography data. *Geophysics*, 62(2):538–548, 1996.

602. D. LaBrecque and X. Yang. Difference inversion of ert data: a fast inversion method for 3-d in situ monitoring. *Journal of Environmental and Engineering Geophysics*, 6(2):83–89, 2001.

603. J. Larsson. Electromagnetics from a quasistatic perspective. *American Journal of Physics*, 75(3):230–239, 2007.

604. L. Lasarow, B. Vogt, Z. Zhao, L. Balke, N. Weiler, and I. Frerichs. Regional lung function measures determined by electrical impedance tomography during repetitive ventilation manoeuvres in patients with COPD. *Physiological Measurement*, 42(1):015008, 2021.

605. M. Lassas, M. Taylor, and G. Uhlmann. The dirichlet-to-neumann map for complete riemannian manifolds with boundary, comm. *Comp. Anal. Geom*, 11:207–222, 2003.

606. J. Latikka, J. Hyttinen, T. Kuurne, H. Eskola, and J. Malmivuo. The conductivity of brain tissues: comparison of results in vivo and in vitro measurements. In *2001 Conference Proceedings of the 23rd Annual International Conference of the IEEE Engineering in Medicine and Biology Society*, volume 1, pages 910–912. IEEE, 2001.

607. S. K. Law. Thickness and resistivity variations over the upper surface of the human skull. *Brain Topography*, 6(2):99–109, Dec. 1993.

608. D. Le Bihan. Intravoxel incoherent motion perfusion MR imaging: a wake-up call. *Radiology*, 249(3):748–752, 2008.

609. B. R. Lee, W. W. Roberts, D. G. Smith, H. W. Ko, J. I. Epstein, K. Lecksell, and A. W. Partin. Bioimpedance: novel use of a minimally invasive technique for cancer localization in the intact prostate. *The Prostate*, 39(3):213–218, 1999.

610. J. Lee, S. Gweon, K. Lee, S. Um, K.-R. Lee, K. Kim, J. Lee, and H.-J. Yoo. A 9.6 mW/Ch 10 mHz wide-bandwidth electrical impedance tomography ic with accurate phase compensation for breast cancer detection. In *2020 IEEE Custom Integrated Circuits Conference (CICC)*, pages 1–4. IEEE, 2020.

611. J. Lee, J. Shin, and D. H. Kim. MR-based conductivity imaging using multiple receiver coils. *Magnetic resonance in medicine*, 76(2):530–539, 2016.

612. M. H. Lee, G. Y. Jang, Y. E. Kim, P. J. Yoo, H. Wi, T. I. Oh, and E. J. Woo. Portable multi-parameter electrical impedance tomography for sleep apnea and hypoventilation monitoring: Feasibility study. *Physiological measurement*, 39(12):124004, 2018.

613. C. D. Lehman, R. F. Arao, B. L. Sprague, J. M. Lee, D. S. Buist, K. Kerlikowske, L. M. Henderson, T. Onega, A. N. Tosteson, G. H. Rauscher, et al. National performance benchmarks for modern screening digital mammography: update from the breast cancer surveillance consortium. *Radiology*, 283(1):49–58, 2017.

614. S. Lehmann, S. Leonhardt, C. Ngo, L. Bergmann, I. Ayed, S. Schrading, and K. Tenbrock. Global and regional lung function in cystic fibrosis measured by electrical impedance tomography. *Pediatric Pulmonology*, 51(11):1191–1199, Apr. 2016.

615. S. Leonhardt, R. Pikkemaat, O. Stenqvist, and S. Lundin. Electrical impedance tomography for hemodynamic monitoring. *Conf IEEE Eng Med Biol Soc*, 2012:122–125, 2012.

616. N. Lesparre, A. Boyle, B. Grychtol, J. Cabrerra, J. Marteau, and A. Adler. Electrical resistivity imaging in transmission between the surface and underground tunnel for fault characterization. *J. Applied Geophys.*, 128(1):163–178, May 2016.

617. N. Lesparre, F. Nguyen, A. Kemna, T. Robert, T. Hermans, M. Daoudi, and A. Flores-Orozco. A new approach for time-lapse data weighting in electrical resistivity tomography. *Geophysics*, 82(6):E325–E333, 2017.

618. S. Levy, D. Adam, and Y. Bresler. Electromagnetic impedance tomography (EMIT): a new method for impedance imaging. *IEEE Transactions on Medical Imaging*, 21(6):676–687, June 2002.

619. C.-L. Li, G. Mathews, and A. F. Bak. Action potential of somatic and autonomic nerves. *Experimental neurology*, 56(3):527–537, 1977.

620. G. Li, K. Ma, J. Sun, G. Jin, M. Qin, and H. Feng. Twenty-four-hour real-time continuous monitoring of cerebral edema in rabbits based on a noninvasive and noncontact system of magnetic induction. *Sensors*, 17(3):537, Mar. 2017.

621. H. Li, R. Chen, C. Xu, B. Liu, M. Tang, L. Yang, X. Dong, and F. Fu. Unveiling the development of intracranial injury using dynamic brain EIT: an evaluation of current reconstruction algorithms. *Physiological measurement*, 38(9):1776, 2017.

622. H. Li, H. Ma, B. Yang, C. Xu, L. Cao, X. Dong, and F. Fu. Automatic evaluation of mannitol dehydration treatments on controlling intracranial pressure using electrical impedance tomography. *IEEE Sensors Journal*, 20(9):4832–4839, 2020.

623. Y. Li, R. Liu, and X. Dong. A magnetic induction tomography system using fully synchronous phase detection. In *3rd International Conference on Bioinformatics and Biomedical Engineering*, volume 1–11, pages 2271–2274. IEEE, June 2009.

624. F. Liebold, M. Hamsch, and C. Igney. Contact-less human vital sign monitoring with a 12 channel synchronous parallel processing magnetic impedance measurement system. In *IFMBE Proceedings*, pages 1070–1073. Springer Berlin Heidelberg, 2009.

625. P. Limousin, P. Krack, P. Pollak, A. Benazzouz, C. Ardouin, D. Hoffmann, and A.-L. Benabid. Electrical stimulation of the subthalamic nucleus in advanced parkinson's disease. *New England Journal of Medicine*, 339(16):1105–1111, 1998.

626. P. Linderholm, L. Marescot, M. H. Loke, and P. Renaud. Cell culture imaging using microimpedance tomography. *IEEE Transactions on Biomedical Engineering*, 55(1):138–146, Jan. 2008.

627. B. Lionheart. Reconstruction algorithms for permittivity and conductivity imaging. In *Proc 2nd World Congress on Process Tomography*, pages 4–11, 2001.

628. W. R. B. Lionheart. EIT reconstruction algorithms: pitfalls, challenges and recent developments. *Physiological Measurement*, 25(1):125–142, Feb. 2004.

629. W. R. B. Lionheart, J. Kaipio, and C. N. McLeod. Generalized optimal current patterns and electrical safety in EIT. *Physiological measurement*, 22(1):85, 2001.

630. W. R. B. Lionheart and K. Paridis. Finite elements and anisotropic EIT reconstruction. *Journal of Physics: Conference Series*, 224:012022, apr 2010.

631. W. R. B. Lionheart, M. Soleimani, and A. J. Peyton. Sensitivity analysis of 3d magnetic induction tomography (MIT). In *Proc. 3rd World Congress on Industrial Process Tomography*, pages 239–244, 2003.

632. A. Liston, R. Bayford, and D. Holder. A cable theory based biophysical model of resistance change in crab peripheral nerve and human cerebral cortex during neuronal depolarisation: implications for electrical impedance tomography of fast neural activity in the brain. *Medical & Biological Engineering & Computing*, 50(5):425–437, Apr. 2012.

633. A. D. Liston, R. Bayford, and D. S. Holder. The effect of layers in imaging brain function using electrical impedance tomograghy. *Physiological Measurement*, 25(1):143, 2004.

634. A. D. Liston, R. Bayford, A. Tidswell, and D. S. Holder. A multi-shell algorithm to reconstruct EIT images of brain function. *Physiological measurement*, 23(1):105, 2002.

635. B. Liu, G. Wang, Y. Li, L. Zeng, H. Li, Y. Gao, Y. Ma, Y. Lian, and C.-H. Heng. A 13-channel 1.53-mW 11.28-mm 2 electrical impedance tomography soc based on frequency division multiplexing for lung physiological imaging. *IEEE transactions on biomedical circuits and systems*, 13(5):938–949, 2019.

636. B. Liu, G. Wang, Y. Li, L. Zeng, H. Li, Y. Gao, Y. Ma, Y. Lian, and C.-H. Heng. A 13-channel 1.53-mW 11.28-mm^2 electrical impedance tomography SoC based on frequency division multiplexing for lung physiological imaging. *IEEE Transactions on Biomedical Circuits and Systems*, 2019.

637. D. Liu, Y. Zhao, A. K. Khambampati, A. Seppänen, and J. Du. A parametric level set method for imaging multiphase conductivity using electrical impedance tomography. *IEEE Transactions on Computational Imaging*, 4(4):552–561, 2018.

638. J. Liu, X. Zhang, S. Schmitter, P.-F. V. de Moortele, and B. He. Gradient-based electrical properties tomography (gEPT): A robust method for mapping electrical properties of biological tissues in vivo using magnetic resonance imaging. *Magnetic Resonance in Medicine*, 74(3):634–646, Sept. 2015.

639. Q. Liu, T. I. Oh, H. Wi, E. J. Lee, J. K. Seo, and E. J. Woo. Design of a microscopic electrical impedance tomography system using two current injections. *Physiological Measurement*, 32(9):1505–1516, Aug. 2011.

640. S. Liu, Y. Huang, H. Wu, C. Tan, and J. Jia. Efficient multitask structure-aware sparse bayesian learning for frequency-difference electrical impedance tomography. *IEEE Transactions on Industrial Informatics*, 17(1):463–472, 2020.

641. X. Liu, H. Li, H. Ma, C. Xu, B. Yang, M. Dai, X. Dong, and F. Fu. An iterative damped least-squares algorithm for simultaneously monitoring the development of hemorrhagic and secondary ischemic lesions in brain injuries. *Medical & Biological Engineering & Computing*, 57(9):1917–1931, June 2019.

642. N. K. Logothetis, C. Kayser, and A. Oeltermann. In vivo measurement of cortical impedance spectrum in monkeys: Implications for signal propagation. *Neuron*, 55(5):809–823, Sept. 2007.

643. W. W. Loh, R. C. Waterfall, J. Cory, and G. P. Lucas. Using ert for multi-phase flow monitoring. In *Proc. 1st World Congress on Industrial Process Tomography*, pages 47–53, Apr. 1999.

644. M. H. Loke and R. Barker. Rapid least-squares inversion of apparent resistivity pseudosections by a quasi-Newton method. *Geophys. Prospect.*, 44(1):131–152, Jan. 1996.

645. M. H. Loke and R. D. Barker. Practical techniques for 3d resistivity surveys and data inversion. *Geophysical Prospecting*, 44:499–523, 1996.

646. M. H. Loke, J. E. Chambers, D. Rucker, O. Kuras, and P. B. Wilkinson. Recent developments in the direct-current geoelectrical imaging method. *J. Applied Geophys.*, 95:135–156, Aug. 2013.

647. M. H. Loke, T. Dahlin, and D. Rucker. Smoothness-constrained time-lapse inversion of data from 3D resistivity surveys. *Near Surface Geophysics*, 12(1):5–24, 2014.

648. M. H. Loke, H. Kiflu, P. B. Wilkinson, D. Harro, and S. Kruse. Optimized arrays for 2D resistivity surveys with combined surface and buried arrays. *Near Surface Geophysics*, 13(5):505–517, 2015.

649. M. H. Loke, P. B. Wilkinson, J. E. Chambers, and P. Meldrum. Rapid inversion of data from 2D resistivity surveys with electrode displacements. *Geophys. Prospect.*, 66(3):579–594, 2018.

650. M. H. Loke, P. B. Wilkinson, J. E. Chambers, S. Uhlemann, and J. P. R. Sorensen. Optimized arrays for 2-D resistivity survey lines with a large number of electrodes. *Journal of Applied Geophysics*, 112:136–146, Jan. 2015.

651. Y. Long, D.-W. Liu, H.-W. He, and Z.-Q. Zhao. Positive end-expiratory pressure titration after alveolar recruitment directed by electrical impedance tomography. *Chinese Medical Journal*, 128(11):1421–1427, June 2015.

652. H. A. Lorentz. The theorem of poynting concerning the energy in the electromagnetic field and two general propositions concerning the propagation of light. *Amsterdammer Akademie der Wetenschappen*, 4:176–187, 1896.

653. M. Lourens, B. van den Berg, J. Aerts, A. Verbraak, H. Hoogsteden, and J. Bogaard. Expiratory time constants in mechanically ventilated patients with and without COPD. *Intensive Care Medicine*, 26(11):1612–1618, Oct. 2000.

654. J. R. Lovell. *Finite Element Methods in Resistivity Logging*. PhD thesis, Delft University of Technology, 1993.

655. K. Lowhagen, S. Lundin, and O. Stenqvist. Regional intratidal gas distribution in acute lung injury and acute respiratory distress syndrome–assessed by electric impedance tomography. *Minerva Anesthesiol*, 76:1024–1035, 2010.

656. H. Luepschen, S. Leonhardt, and C. Putensen. Measuring stroke volume using electrical impedance tomography. In J.-L. Vincent, editor, *Yearbook of Intensive Care and Emergency Medicine 2010*, pages 46 – 55. Springer Berlin Heidelberg, 2010.

657. M. Lukaschewitsch, P. Maass, and M. Pidcock. Tikhonov regularization for electrical impedance tomography on unbounded domains. *Inverse Problems*, 19:585–610, 2003.

658. S. Lundin and O. Stenqvist. Electrical impedance tomography: potentials and pitfalls. *Current opinion in critical care*, 18(1):35–41, 2012.

659. A. R. Lupton-Smith, A. C. Argent, P. C. Rimensberger, and B. M. Morrow. Challenging a paradigm: Positional changes in ventilation distribution are highly variable in healthy infants and children. *Pediatric Pulmonology*, 49(8):764–771, Sept. 2014.

660. H. Lux, U. Heinemann, and I. Dietzel. Ionic changes and alterations in the size of the extracellular space during epileptic activity. *Advances in neurology*, 44:619–639, 1986.

661. R. G. Lyons. *Understanding Digital Signal Processing (2nd Edition)*. Prentice Hall PTR, Upper Saddle River, NJ, USA, 2004.

662. G. Lytle, P. Perry, and S. Siltanen. Nachman's reconstruction method for the Calderón problem with discontinuous conductivities. *Inverse Problems*, 36:035018, 2020.

663. L. Ma, D. McCann, and A. Hunt. Combining magnetic induction tomography and electromagnetic velocity tomography for water continuous multiphase flows. *IEEE Sensors Journal*, 17(24):8271–8281, Dec. 2017.

664. L. Ma and M. Soleimani. Hidden defect identification in carbon fibre reinforced polymer plates using magnetic induction tomography. *Measurement Science and Technology*, 25(5):055404, Apr. 2014.

665. X. Ma, A. J. Peyton, R. Binns, and S. R. Higson. Imaging the flow profile of molten steel through a submerged pouring nozzle. In *Proc. 3rd World Congress on Industrial Process Tomography*, Sept. 2003.

666. T. D. Machin, H.-Y. Wei, R. W. Greenwood, and M. J. Simmons. In-pipe rheology and mixing characterisation using electrical resistance sensing. *Chemical Engineering Science*, 187:327–341, 2018.

667. A. Mahara, S. Khan, E. K. Murphy, A. R. Schned, E. S. Hyams, and R. J. Halter. 3D microendoscopic electrical impedance tomography for margin assessment during robot-assisted laparoscopic prostatectomy. *IEEE transactions on medical imaging*, 34(7):1590–1601, 2015.

668. R. Mahdavi, P. Hosseinpour, F. Abbasvandi, S. Mehrvarz, N. Yousefpour, H. Ataee, M. Parniani, A. Mamdouh, H. Ghafari, and M. Abdolahad. Bioelectrical pathology of the breast; real-time diagnosis of malignancy by clinically calibrated impedance spectroscopy of freshly dissected tissue. *Biosensors and Bioelectronics*, 165:112421, 2020.

669. S. E. Maier, S. Vajapeyam, H. Mamata, C.-F. Westin, F. A. Jolesz, and R. V. Mulkern. Biexponential diffusion tensor analysis of human brain diffusion data. *Magnetic Resonance in Medicine: An Official Journal of the International Society for Magnetic Resonance in Medicine*, 51(2):321–330, 2004.

670. S. Maisch, S. H. Bohm, J. Solà, M. S. Goepfert, J. C. Kubitz, H. P. Richter, J. Ridder, A. E. Goetz, and D. A. Reuter. Heart-lung interactions measured by electrical impedance tomography. *Critical care medicine*, 39(9):2173–2176, 2011.

671. J. Makkonen, L. Marsh, J. Vihonen, A. Järvi, D. Armitage, A. Visa, and A. Peyton. KNN classification of metallic targets using the magnetic polarizability tensor. *Measurement Science and Technology*, 25(5):055105, 2014.

672. A. Malehmir, L. Socco, M. Bastani, C. Krawczyk, A. Pfaffhuber, R. Miller, H. Maurer, R. Frauenfelder, K. Suto, S. Bazin, K. Merz, and T. Dahlin. Chapter two — near-surface geophysical characterization of areas prone to natural hazards: A review of the current and perspective on the future. *Advances in Geophysics*, 57:51–146, 2016.

673. A. Malich, T. Böhm, M. Facius, M. G. Freesmeyer, M. Fleck, R. Anderson, and W. A. Kaiser. Differentiation of mammographically suspicious lesions: evaluation of breast ultrasound, MRI mammography and electrical impedance scanning as adjunctive technologies in breast cancer detection. *Clinical radiology*, 56(4):278–283, 2001.

674. A. Malich, T. Böhm, M. Facius, M. Freessmeyer, M. Fleck, R. Anderson, and W. Kaiser. Additional value of electrical impedance scanning: experience of 240 histologically-proven breast lesions. *European Journal of Cancer*, 37(18):2324–2330, 2001.

675. A. Malich, T. Fritsch, R. Anderson, T. Boehm, M. Freesmeyer, M. Fleck, and W. Kaiser. Electrical impedance scanning for classifying suspicious breast lesions: first results. *European radiology*, 10(10):1555–1561, 2000.

676. J. Malmivuo and R. Plonsey. *BioelectromagnetismPrinciples and Applications of Bioelectric and Biomagnetic Fields*. Oxford University Press, Oct. 1995.

677. E. Malone, G. S. dos Santos, D. Holder, and S. Arridge. Multifrequency electrical impedance tomography using spectral constraints. *IEEE transactions on medical imaging*, 33(2):340–350, 2014.

678. E. Malone, G. S. dos Santos, D. Holder, and S. Arridge. A reconstruction-classification method for multifrequency electrical impedance tomography. *IEEE transactions on medical imaging*, 34(7):1486–1497, 2015.

679. E. Malone, M. Jehl, S. Arridge, T. Betcke, and D. Holder. Stroke type differentiation using spectrally constrained multifrequency EIT: evaluation of feasibility in a realistic head model. *Physiological measurement*, 35(6):1051, 2014.

680. D. Malonek, U. Dirnagl, U. Lindauer, K. Yamada, I. Kanno, and A. Grinvald. Vascular imprints of neuronal activity: relationships between the dynamics of cortical blood flow, oxygenation, and volume changes following sensory stimulation. *Proceedings of the National Academy of Sciences*, 94(26):14826–14831, 1997.

681. Y. Mamatjan, A. Borsic, D. Gürsoy, and A. Adler. Experimental/clinical evaluation of EIT image reconstruction with $\ell 1$ data and image norms. In *Proc. 15th Int. Conf. Electrical Bioimpedance and 14th Int. Conf. Biomedical Applications of EIT*, Apr. 2013.

682. R. Mann. Augmented-reality visualization of fluid mixing in stirred chemical reactors using electrical resistance tomography. *Journal of Electronic Imaging*, 10(3):620, July 2001.

683. P. Mansfield. Multi-planar image formation using NMR spin echoes. *Journal of Physics C: Solid State Physics*, 10(3):L55, 1977.

684. E. N. Marieb and K. Hoehn. *Human Anatomy & Physiology 9th Edition*. Pearson, 2015.

685. D. Marquardt. An algorithm for least squares estimation of nonlinear parameters. *SIAM J. Appl. Math*, 11:431–441, 1963.

686. F. Marquis, N. Coulombe, R. Costa, H. Gagnon, R. Guardo, and Y. Skrobik. Electrical impedance tomography's correlation to lung volume is not influenced by anthropometric parameters. *Journal of Clinical Monitoring and Computing*, 20(3):201–207, May 2006.

687. L. A. Marsh, C. Ktistis, A. Järvi, D. W. Armitage, and A. J. Peyton. Three-dimensional object location and inversion of the magnetic polarizability tensor at a single frequency using a walk-through metal detector. *Measurement Science and Technology*, 24(4):045102, Mar. 2013.

688. G. Martín, R. Martín, M. Brieva, and L. Santamaría. Electrical impedance scanning in breast cancer imaging: correlation with mammographic and histologic diagnosis. *European radiology*, 12(6):1471–1478, 2002.

689. O. G. Martinsen and S. Grimnes. *Bioimpedance and bioelectricity basics.* Academic press, 2011.

690. S. S. Marven, A. R. Hampshire, R. H. Smallwood, B. H. Brown, and R. A. Primhak. Reproducibility of electrical impedance tomographic spectroscopy (EITS) parametric images of neonatal lungs. *Physiological Measurement*, 17(4A):A205–A212, Nov. 1996.

691. N. Matoorian. Dental electromagnetic tomography: properties of tooth tissues. In *IEE Colloquium on Innovations in Instrumentation for Electrical Tomography*, pages 3/1–3/7. IEE, 1995.

692. Y. Matsuoka and K.-A. Hossmann. Cortical impedance and extracellular volume changes following middle cerebral artery occlusion in cats. *Journal of Cerebral Blood Flow & Metabolism*, 2(4):466–474, 1982.

693. T. Mauri, L. Alban, C. Turrini, B. Cambiaghi, E. Carlesso, P. Taccone, N. Bottino, A. Lissoni, S. Spadaro, C. A. Volta, L. Gattinoni, A. Pesenti, and G. Grasselli. Optimum support by high-flow nasal cannula in acute hypoxemic respiratory failure: effects of increasing flow rates. *Intensive Care Medicine*, 43(10):1453–1463, July 2017.

694. T. Mauri, G. Bellani, A. Confalonieri, P. Tagliabue, M. Turella, A. Coppadoro, G. Citerio, N. Patroniti, and A. Pesenti. Topographic distribution of tidal ventilation in acute respiratory distress syndrome. *Critical Care Medicine*, 41(7):1664–1673, July 2013.

695. T. Mauri, N. Eronia, C. Abbruzzese, R. Marcolin, A. Coppadoro, S. Spadaro, N. Patroniti, G. Bellani, and A. Pesenti. Effects of sigh on regional lung strain and ventilation heterogeneity in acute respiratory failure patients undergoing assisted mechanical ventilation. *Critical Care Medicine*, 43(9):1823–1831, Sept. 2015.

696. T. Mauri, N. Eronia, C. Turrini, M. Battistini, G. Grasselli, R. Rona, C. A. Volta, G. Bellani, and A. Pesenti. Bedside assessment of the effects of positive end-expiratory pressure on lung inflation and recruitment by the helium dilution technique and electrical impedance tomography. *Intensive Care Medicine*, 42(10):1576–1587, Aug. 2016.

697. T. Mauri, E. Spinelli, E. Scotti, G. Colussi, M. Basile, S. Crotti, D. Tubiolo, P. Tagliabue, A. Zanella, G. Grasselli, and A. Pesenti. Potential for lung recruitment and ventilation-perfusion mismatch in patients with the acute respiratory distress syndrome from coronavirus disease 2019. *Critical Care Medicine*, 48(8), Aug. 2020.

698. S. F. McCormick and J. G. Wade. Multigrid solution of a linearized regularized least-squares problem in electrical impedance tomography. *Inverse Problems*, 9(697):697–713, 1993.

699. B. McDermott, A. Elahi, A. Santorelli, M. O'Halloran, J. Avery, and E. Porter. Multi-frequency symmetry difference electrical impedance tomography with machine learning for human stroke diagnosis. *Physiological Measurement*, 41(7):075010, 2020.

700. K. T. McDonald. Dielectric (and magnetic) image methods, Mar. 2020.

701. A. McEwan, G. Cusick, and D. Holder. A review of errors in multi-frequency EIT instrumentation. *Physiological measurement*, 28(7):S197, 2007.

702. A. McEwan, A. Romsauerova, R. Yerworth, L. Horesh, R. Bayford, and D. Holder. Design and calibration of a compact multi-frequency EIT system for acute stroke imaging. *Physiological measurement*, 27(5):S199, 2006.

703. A. L. McEwan, M. Hamsch, S. Watson, C. H. Igney, and J. Kahlert. A comparison of two phase measurement techniques for magnetic impedance tomography. In *IFMBE Proceedings*, pages 4–6. Springer Berlin Heidelberg, 2009.

704. C. N. McLeod, C. W. Denyer, F. J. Lidgey, W. R. B. Lionheart, K. S. Paulson, M. K. Pidcock, and Y. Shi. High speed in vivo chest imaging with OXBACT III. In *Proceedings of the 18th Annual International Conference of the IEEE Engineering in Biology Society*, pages 770–771, 1996.

705. T. Meier, H. Luepschen, J. Karsten, T. Leibecke, M. Großherr, H. Gehring, and S. Leonhardt. Assessment of regional lung recruitment and derecruitment during a PEEP trial based on electrical impedance tomography. *Intensive Care Medicine*, 34(3):543–550, July 2008.

706. C. Meira, F. B. Joerger, A. P. Kutter, A. Waldmann, S. K. Ringer, S. H. Böhm, S. Iff, and M. Mosing. Comparison of three continuous positive airway pressure (CPAP) interfaces in healthy beagle dogs during medetomidine–propofol constant rate infusions. *Veterinary Anaesthesia and Analgesia*, 45(2):145–157, Mar. 2018.

707. M. M. Mellenthin, J. L. Mueller, E. D. L. B. de Camargo, F. S. de Moura, T. B. R. Santos, R. G. Lima, S. J. Hamilton, P. A. Muller, and M. Alsaker. The ACE1 electrical impedance tomography system for thoracic imaging. *IEEE Transactions on Instrumentation and Measurement*, 68(9):3137–3150, 2019.

708. R. Merwa, K. Hollaus, O. Biró, and H. Scharfetter. Detection of brain oedema using magnetic induction tomography: a feasibility study of the likely sensitivity and detectability. *Physiological Measurement*, 25(1):347–354, Feb. 2004.

709. R. Merwa, K. Hollaus, B. Brandstätter, and H. Scharfetter. Numerical solution of the general 3D eddy current problem for magnetic induction tomography (spectroscopy). *Physiological Measurement*, 24(2):545–554, Apr. 2003.

710. R. Merwa and H. Scharfetter. Magnetic induction tomography: evaluation of the point spread function and analysis of resolution and image distortion. *Physiological Measurement*, 28(7):S313–S324, June 2007.

711. R. Merwa and H. Scharfetter. Magnetic induction tomography: comparison of the image quality using different types of receivers. *Physiological Measurement*, 29(6):S417–S429, June 2008.

712. P. Metherall. *Three Dimensional Electrical Impedance Tomography of the Human Thorax*. PhD thesis, University of Sheffield, 1998.

713. P. Metherall, D. C. Barber, R. H. Smallwood, and B. H. Brown. Three dimensional electrical impedance tomography. *Nature*, 380:509–512, 1996.

714. M. Miedema, A. Adler, K. E. McCall, E. J. Perkins, A. H. van Kaam, and D. G. Tingay. Electrical impedance tomography identifies a distinct change in regional phase angle delay pattern in ventilation filling immediately prior to a spontaneous pneumothorax. *Journal of Applied Physiology*, 127(3):707–712, Sept. 2019.

715. M. Miedema, F. H. de Jongh, I. Frerichs, M. B. van Veenendaal, and A. H. van Kaam. Regional respiratory time constants during lung recruitment in high-frequency oscillatory ventilated preterm infants. *Intensive care medicine*, 38(2):294–299, 2012.

716. M. Miedema, K. E. McCall, E. J. Perkins, M. Sourial, S. H. Böhm, A. Waldmann, A. H. van Kaam, and D. G. Tingay. First real-time visualization of a spontaneous pneumothorax developing in a preterm lamb using electrical impedance tomography. *American Journal of Respiratory and Critical Care Medicine*, 194(1):116–118, July 2016.

717. S. Milne, J. Huvanandana, C. Nguyen, J. M. Duncan, D. G. Chapman, K. O. Tonga, S. C. Zimmermann, A. Slattery, G. G. King, and C. Thamrin. Time-based pulmonary features from electrical impedance tomography demonstrate ventilation heterogeneity in chronic obstructive pulmonary disease. *Journal of Applied Physiology*, 127(5):1441–1452, Nov. 2019.

718. R. Minns and J. Brown. Intracranial pressure changes associated with childhood seizures. *Developmental Medicine & Child Neurology*, 20(5):561–569, 1978.

719. S. A. Mitchell and S. A. Vavasis. Quality mesh generation in higher dimensions. *SIAM Journal on Computing*, 29(4):1334–1370, Jan. 2000.

720. A. Modiri, S. Goudreau, A. Rahimi, and K. Kiasaleh. Review of breast screening: Toward clinical realization of microwave imaging. *Medical physics*, 44(12):e446–e458, 2017.

721. Y. Moens, J. P. Schramel, G. Tusman, T. D. Ambrisko, J. Solà, J. X. Brunner, L. Kowalczyk, and S. H. Böhm. Variety of non-invasive continuous monitoring methodologies including electrical impedance tomography provides novel insights into the physiology of lung collapse and recruitment – case report of an anaesthetized horse. *Veterinary Anaesthesia and Analgesia*, 41(2):196–204, Mar. 2014.

722. M. Molinari. *High Fidelity Imaging in Electrical Impedance Tomography*. PhD thesis, University of Southampton, 2003.

723. M. Molinari, S. J. Cox, B. H. Blott, and G. J. Daniell. Comparison of algorithms for non-linear inverse 3d electrical tomography reconstruction. *Physiol Meas*, 23:95–104, 2002.

724. P. H. Möller, K.-G. Tranberg, B. Blad, P. Henriksson, L. Lindberg, L. Weber, and B. R. Persson. EIT for measurement of temperature distribution in laser thermotherapy (laserthermia). 1993.

725. G. Montante and A. Paglianti. Gas hold-up distribution and mixing time in gas–liquid stirred tanks. *Chemical Engineering Journal*, 279:648–658, 2015.

726. M. Moorkamp. Integrating electromagnetic data with other geophysical observations for enhanced imaging of the earth: a tutorial and review. *Surveys in Geophysics*, 38(5):935–962, 2017.

727. A. Morega, A. Dobre, and M. Morega. Electrical cardiometry simulation for the assessment of circulatory parameters. *Proceedings of the Romanian Academy - Series A: Mathematics, Physics, Technical Sciences, Information Science*, 17:259–266, 07 2016.

728. T. Morimoto, Y. Kinouchi, T. Iritani, S. Kimura, Y. Konishi, N. Mitsuyama, K. Komaki, and Y. Monden. Measurement of the electrical bio-impedance of breast tumors. *European surgical research*, 22(2):86–92, 1990.

729. A. Morris and H. Griffiths. A comparison of image reconstruction in EIT and MIT by inversion of the sensitivity matrix. In *Proc of 3rd EPSRC Engineering Network Meeting on Biomedical Applications of EIT*, Apr. 2001.

730. A. Morris, H. Griffiths, and W. Gough. A numerical model for magnetic induction tomographic measurements in biological tissues. *Physiological Measurement*, 22(1):113–119, Feb. 2001.

731. D. F. Morrison. *Applied linear statistical methods*. Prentice Hall, New Jersey, 1983.

732. M. Morrow, R. Schmidt, B. Cregger, C. Hassett, and S. Cox. Preoperative evaluation of abnormal mammographic findings to avoid unnecessary breast biopsies. *Archives of Surgery*, 129(10):1091–1096, 1994.

733. M. Mosing, U. Auer, P. MacFarlane, D. Bardell, J. P. Schramel, S. H. Böhm, R. Bettschart-Wolfensberger, and A. D. Waldmann. Regional ventilation distribution and dead space in anaesthetized horses treated with and without continuous positive airway pressure: novel insights by electrical impedance tomography and volumetric capnography. *Veterinary Anaesthesia and Analgesia*, 45(1):31–40, Jan. 2018.

734. M. Mosing, S. H. Böhm, A. Rasis, G. Hoosgood, U. Auer, G. Tusman, R. Bettschart-Wolfensberger, and J. P. Schramel. Physiologic factors influencing the arterial-to-end-tidal CO2 difference and the alveolar dead space fraction in spontaneously breathing anesthetised horses. *Frontiers in Veterinary Science*, 5, Mar. 2018.

735. M. Mosing, C. Marly-Voquer, P. MacFarlane, D. Bardell, S. H. Böhm, R. Bettschart-Wolfensberger, and A. D. Waldmann. Regional distribution of ventilation in horses in dorsal recumbency during spontaneous and mechanical ventilation assessed by electrical impedance tomography: a case series. *Veterinary Anaesthesia and Analgesia*, 44(1):127–132, Jan. 2017.

736. M. Mosing, M. Sacks, S. A. Tahas, E. Ranninger, S. H. Böhm, I. Campagnia, and A. D. Waldmann. Ventilatory incidents monitored by electrical impedance tomography in an anaesthetized orangutan (pongo abelii). *Veterinary anaesthesia and analgesia*, 44(4):973–976, 2017.

737. M. Mosing, M. Sacks, S. Wenger, S. H. Böhm, P. Buss, and D. V. Cooper. Distribution of ventilation in anaesthetised southern white rhinoceroses evaluated by electrical impedance tomography (EIT). In *Science Week of the Australian and New Zealand College of Veterinary Scientists*, 2017.

738. M. Mosing, A. D. Waldmann, P. MacFarlane, S. Iff, U. Auer, S. H. Bohm, R. Bettschart-Wolfensberger, and D. Bardell. Horses auto-recruit their lungs by inspiratory breath holding following recovery from general anaesthesia. *PLOS ONE*, 11(6):e0158080, June 2016.

739. M. Mosing, A. D. Waldmann, A. Raisis, S. H. Böhm, E. Drynan, and K. Wilson. Monitoring of tidal ventilation by electrical impedance tomography in anaesthetised horses. *Equine Veterinary Journal*, 51(2):222–226, Aug. 2019.

740. M. Mosing, A. D. Waldmann, M. Sacks, P. Buss, J. M. Boesch, G. E. Zeiler, G. Hosgood, R. D. Gleed, M. Miller, L. C. R. Meyer, and S. H. Böhm. What hinders pulmonary gas exchange and changes distribution of ventilation in immobilized white rhinoceroses (ceratotherium simum) in lateral recumbency? *Journal of Applied Physiology*, 129(5):1140–1149, Nov. 2020.

741. M. Moskowitz, T. Ryan, K. Paulsen, and S. Mitchell. Clinical implementation of electrical impedance tomography with hyperthermia. *International journal of hyperthermia*, 11(2):141–149, 1995.

742. T. Muders, H. Luepschen, J. Zinserling, S. Greschus, R. Fimmers, U. Guenther, M. Buchwald, D. Grigutsch, S. Leonhardt, C. Putensen, and H. Wrigge. Tidal recruitment assessed by electrical impedance tomography and computed tomography in a porcine model of lung injury. *Critical Care Medicine*, 40(3):903–911, Mar. 2012.

743. J. Mueller, P. Muller, M. Mellenthin, E. DeBoer, R. Murthy, M. Capps, M. Alsaker, R. Deterding, and S. Sagel. A method of estimating regions of air trapping from electrical impedance tomography data. *Physiological Measurement*, 39(5):05NT01, 2018.

744. J. Mueller and S. Siltanen. Direct reconstructions of conductivities from boundary measurements. *SIAM Journal on Scientific Computing*, 24(4):1232–1266, 2003.

745. J. Mueller and S. Siltanen. *Linear and Nonlinear Inverse Problems with Practical Applications*. SIAM, 2012.

746. J. L. Mueller, P. Muller, M. Mellenthin, R. Murthy, M. Capps, M. Alsaker, R. Deterding, S. D. Sagel, and E. DeBoer. Estimating regions of air trapping from electrical impedance tomography data. *Physiological Measurement*, 39(5):05NT01, May 2018.

747. J. L. Mueller, S. Siltanen, and D. Isaacson. A direct reconstruction algorithm for electrical impedance tomography. *IEEE Transactions on Medical Imaging*, 21(6):555–559, 2002.

748. B. G. Muller, J. H. Shih, S. Sankineni, J. Marko, S. Rais-Bahrami, A. K. George, J. J. de la Rosette, M. J. Merino, B. J. Wood, P. Pinto, et al. Prostate cancer: interobserver agreement and accuracy with the revised prostate imaging reporting and data system at multiparametric mr imaging. *Radiology*, 277(3):741–750, 2015.

749. P. A. Muller. *Numerical Methods for Electrical Impedance Tomography*. PhD thesis, Rensselaer Polytechnic Institute, Troy, NY, 2014.

750. P. A. Muller, D. Isaacson, J. C. Newell, and G. J. Saulnier. Calderón's method on an elliptical domain. *Physiological Measurement*, 32:609–622, 2013.

751. P. A. Muller, J. L. Mueller, and M. Mellenthin. Real-time implementation of Calderón's method on subject-specific domains. *IEEE Transactions on Medical Imaging*, 36(9):1868–1875, 2017.

752. P. A. Muller, J. L. Mueller, M. Mellenthin, R. Murthy, M. Capps, B. D. Wagner, M. Alsaker, R. Deterding, S. D. Sagel, and J. Hoppe. Evaluation of surrogate measures of pulmonary function derived from electrical impedance tomography data in children with cystic fibrosis. *Physiological Measurement*, 39(4):045008, Apr. 2018.

753. J. P. Mullin, M. Shriver, S. Alomar, I. Najm, J. Bulacio, P. Chauvel, and J. Gonzalez-Martinez. Is SEEG safe? a systematic review and meta-analysis of stereo-electroencephalography–related complications. *Epilepsia*, 57(3):386–401, 2016.

754. T. Murai and Y. Kagawa. Electrical impedance computed tomography based on a finite element model. *IEEE Transactions on Biomedical Engineering*, (3):177–184, 1985.

755. E. Murphy. *2-D D-bar Conductivity Reconstructions on Non-circular Domains*. PhD thesis, Colorado State University, Fort Collins, CO, 2007.

756. E. K. Murphy, J. Amoh, S. H. Arshad, R. J. Halter, and K. Odame. Noise-robust bioimpedance approach for cardiac output measurement. *Physiological Measurement*, 40(7):074004, 07 2019.

757. E. K. Murphy, A. Mahara, and R. J. Halter. A novel regularization technique for microendoscopic electrical impedance tomography. *IEEE transactions on medical imaging*, 35(7):1593–1603, 2016.

758. E. K. Murphy, A. Mahara, S. Khan, E. S. Hyams, A. R. Schned, J. Pettus, and R. J. Halter. Comparative study of separation between ex vivo prostatic malignant and benign tissue using electrical impedance spectroscopy and electrical impedance tomography. *Physiological measurement*, 38(6):1242, 2017.

759. E. K. Murphy, J. Skinner, M. Martucci, S. B. Rutkove, and R. J. Halter. Toward electrical impedance tomography coupled ultrasound imaging for assessing muscle health. *IEEE transactions on medical imaging*, 38(6):1409–1419, 2018.

760. E. K. Murphy, X. Wu, A. C. Everitt, and R. J. Halter. Phantom studies of fused-data TREIT using only biopsy-probe electrodes. *IEEE transactions on medical imaging*, 39(11):3367–3378, 2020.

761. E. K. Murphy, X. Wu, and R. J. Halter. Fused-data transrectal EIT for prostate cancer imaging. *Physiological measurement*, 39(5):054005, 2018.

762. J. Murton, O. Kuras, M. Krautblatter, T. Cane, D. Tschofen, S. Uhlemann, S. Schober, and P. Watson. Monitoring rock freezing and thawing by novel geoelectrical and acoustic techniques. *Journal of Geophysical Research: Earth Surface*, 121(12):2309–2332, 2016.

763. K. Mwakanyamale, L. Slater, A. Binley, and D. Ntarlagiannis. Lithologic imaging using complex conductivity: Lessons learned from the Hanford 300 area. *Geophysics*, 77(6):E397–E409, 2012.

764. A. I. Nachman. Global uniqueness for a two-dimensional inverse boundary value problem. *Annals of Mathematics*, 143:71–96, 1996.

765. H. S. Nam, B. I. Lee, J. Choi, C. Park, and O. I. Kwon. Conductivity imaging with low level current injection using transversal j-substitution algorithm in MREIT. *Physics in Medicine & Biology*, 52(22):6717, 2007.

766. H. S. Nam, C. Park, and O. I. Kwon. Non-iterative conductivity reconstruction algorithm using projected current density in MREIT. *Physics in Medicine & Biology*, 53(23):6947, 2008.

767. F. Natterer. *The Mathematics of Comuterized Tomogrpahy*. Wiley, 1982.

768. N. Neshatvar, P. Langlois, R. Bayford, and A. Demosthenous. Analog integrated current drivers for bioimpedance applications: A review. *Sensors (Basel)*, 19(4):756, 2019.

769. J. Netz, E. Forner, and S. Haagemann. Contactless impedance measurement by magnetic induction - a possible method for investigation of brain impedance. *Physiological Measurement*, 14(4):463–471, Nov. 1993.

770. M. Neukirch and N. Klitzsch. Inverting capacitive resistivity (line electrode) measurements with direct current inversion programs. *Vadose Zone Journal*, 9(4):882–892, 2010.

771. J. Newell, P. Edic, X. Ren, J. Larson-Wiseman, and M. Danyleiko. Assessment of acute pulmonary edema in dogs by electrical impedance imaging. *IEEE Transactions on Biomedical Engineering*, 43(2):133–138, 1996.

772. J. C. Newell, R. S. Blue, D. Isaacson, G. J. Saulnier, and A. S. Ross. Phasic three-dimensional impedance imaging of cardiac activity. *Physiological Measurement*, 23(1):203–209, Jan. 2002.

773. J. C. Newell, D. G. Gisser, and D. Isaacson. An electric current tomograph. *IEEE Trans. Biomed. Eng.*, 35(10):828–833, 1988.

774. C. Ngo, F. Dippel, K. Tenbrock, S. Leonhardt, and S. Lehmann. Flow-volume loops measured with electrical impedance tomography in pediatric patients with asthma. *Pediatric Pulmonology*, 53(5):636–644, Feb. 2018.

775. C. Ngo, S. Leonhardt, T. Zhang, M. Lüken, B. Misgeld, T. Vollmer, K. Tenbrock, and S. Lehmann. Linearity of electrical impedance tomography during maximum effort breathing and forced expiration maneuvers. *Physiological Measurement*, 38(1):77–86, Dec. 2017.

776. C. Ngo, S. Leonhardt, T. Zhang, M. Lken, B. Misgeld, T. Vollmer, K. Tenbrock, and S. Lehmann. Linearity of electrical impedance tomography during maximum effort breathing and forced expiration maneuvers. *Physiological Measurement*, 38(1):77–86, dec 2016.

777. D. T. Nguyen, A. Bhaskaran, W. Chik, M. A. Barry, J. Pouliopoulos, R. Kosobrodov, C. Jin, T. I. Oh, A. Thiagalingam, and A. L. McEwan. Perfusion redistribution after a pulmonary-embolism-like event with contrast enhanced EIT. *Physiological Measurement*, 36(6):1297–1309, May 2015.

778. D. T. Nguyen, C. Jin, A. Thiagalingam, and A. L. McEwan. A review on electrical impedance tomography for pulmonary perfusion imaging. *Physiological Measurement*, 33(5):695–706, apr 2012.

779. P. W. Nicholson. Specific impedance of cerebral white matter. *Experimental neurology*, 13(4):386–401, 1965.

780. J. D. Nielsen, K. H. Madsen, O. Puonti, H. R. Siebner, C. Bauer, C. G. Madsen, G. B. Saturnino, and A. Thielscher. Automatic skull segmentation from MR images for realistic volume conductor models of the head: Assessment of the state-of-the-art. *NeuroImage*, 174:587–598, July 2018.

781. M. A. Nitsche, P. S. Boggio, F. Fregni, and A. Pascual-Leone. Treatment of depression with transcranial direct current stimulation (tDCS): a review. *Experimental neurology*, 219(1):14–19, 2009.

782. M. A. Nitsche and W. Paulus. Excitability changes induced in the human motor cortex by weak transcranial direct current stimulation. *The Journal of physiology*, 527(3):633–639, 2000.

783. T. J. Noble, N. D. Harris, A. H. Morice, P. Milnes, and B. H. Brown. Diuretic induced change in lung water assessed by electrical impedance tomography. *Physiological Measurement*, 21(1):155–163, feb 2000.

784. M. Noel and B. Xu. Archaeological investigation by electrical resistivity tomography: a preliminary study. *Geophysical Journal International*, 107(1):95–102, Oct. 1991.

785. P. Nopp, N. D. Harris, T. X. Zhao, and B. H. Brown. Model for the dielectric properties of human lung tissue against frequency and air content. *Medical & Biological Engineering & Computing*, 35(6):695–702, Nov. 1997.

786. P. Nopp, E. Rapp, H. Pfutzner, H. Nakesch, and C. Rusham. Dielectric properties of lung tissue as a function of air content. *Physics in Medicine and Biology*, 38(6):699–716, June 1993.

787. S. Nordebo, M. Dalarsson, D. Khodadad, B. Müller, A. Waldmann, T. Becher, I. Frerichs, L. Sopho-cleous, D. Sjberg, N. Seifnaraghi, and R. Bayford. A parametric model for the changes in the complex valued conductivity of a lung during tidal breathing. *Journal of Physics D: Applied Physics*, 51, 01 2018.

788. R. Novikov. A multidimensional inverse spectral problem for the equation $-\delta\psi + (v(x) - eu(x))\psi = 0$. *Functional Analysis and Its Applications*, 22(4):263–272, 1988.

789. G. Nyman and G. Hedenstierna. Ventilation-perfusion relationships in the anaesthetised horse. *Equine Veterinary Journal*, 21(4):274–281, July 1989.

790. T. Oh, O. Gilad, A. Ghosh, M. Schuettler, and D. S. Holder. A novel method for recording neuronal depolarization with recording at 125–825 Hz: implications for imaging fast neural activity in the brain with electrical impedance tomography. *Medical & biological engineering & computing*, 49(5):593–604, 2011.

791. T. I. Oh, H. Wi, D. Y. Kim, P. J. Yoo, and E. J. Woo. A fully parallel multi-frequency EIT system with flexible electrode configuration: KHU Mark2. *Physiological Measurement*, 32:835–849, 2011.

792. T. I. Oh, E. J. Woo, and D. Holder. Multi-frequency EIT system with radially symmetricarchitecture: KHU Mark1. *Physiological Measurement*, 28:S183–S196, 2007.

793. Y. Ohmine, T. Morimoto, Y. Kinouchi, T. Iritani, M. Takeuchi, and Y. Monden. Noninvasive measurement of the electrical bioimpedance of breast tumors. *Anticancer research*, 20(3B):1941–1946, 2000.

794. J. O. Ollikainen, M. Vauhkonen, P. A. Karjalainen, and J. P. Kaipio. Effects of local skull inhomogeneities on EEG source estimation. *Medical Engineering & Physics*, 21(3):143–154, Apr. 1999.

795. T. Oostendorp, J. Delbeke, and D. Stegeman. The conductivity of the human skull: results of in vivo and in vitro measurements. *IEEE Transactions on Biomedical Engineering*, 47(11):1487–1492, Nov. 2000.

796. K. Osterman, T. Kerner, D. Williams, A. Hartov, S. Poplack, and K. Paulsen. Multifrequency electrical impedance imaging: preliminary in vivo experience in breast. *Physiological measurement*, 21(1):99, 2000.

797. K. Ostrowski, S. Luke, M. Bennett, and R. Williams. Application of capacitance electrical tomography for on-line and off-line analysis of flow pattern in horizontal pipeline of pneumatic conveyer. *Chemical Engineering Journal*, 77(1-2):43–50, Apr. 2000.

798. M. D. O'Toole, L. A. Marsh, J. L. Davidson, Y. M. Tan, D. W. Armitage, and A. J. Peyton. Non-contact multi-frequency magnetic induction spectroscopy system for industrial-scale bio-impedance measurement. *Measurement Science and Technology*, 26(3):035102, Feb. 2015.

799. D. M. Otten and B. Rubinsky. Cryosurgical monitoring using bioimpedance measurements-a feasibility study for electrical impedance tomography. *IEEE transactions on biomedical engineering*, 47(10):1376–1381, 2000.

800. S. Ozdemir and Y. Z. Ider. bSSFP phase correction and its use in magnetic resonance electrical properties tomography. *Magnetic Resonance in Medicine*, 81(2):934–946, 2019.

801. D. Pacho and G. Davies. Application of electrical capacitance measurements to study the collapse of oil foams. In *Proc 2nd World Congress on Process Tomography*, pages 618–627, 2001.

802. A. D. Pachowko, M. Wang, C. Poole, and D. Rhodes. The use of electrical resistance tomography (ERT) to monitor flow patterns in horizontal slurry transport pipelines. In *Proc. 3rd World Congress on Industrial Process Tomography*, Sept. 2001.

803. B. Packham, H. Koo, A. Romsauerova, S. Ahn, A. McEwan, S. Jun, and D. Holder. Comparison of frequency difference reconstruction algorithms for the detection of acute stroke using EIT in a realistic head-shaped tank. *Physiological measurement*, 33(5):767, 2012.

804. C. C. Pain, J. V. Herwanger, J. H. Saunders, M. H. Worthington, and C. R. E. de Oliveira. Anisotropic resistivity inversion. *Inverse Problems*, 19(5):1081–1111, Sept. 2003.

805. J. Palmer, A. De Crespigny, S.-P. Williams, E. Busch, and N. Van Bruggen. High-resolution mapping of discrete representational areas in rat somatosensory cortex using blood volume-dependent functional mri. *Neuroimage*, 9(4):383–392, 1999.

806. C. Park, B. I. Lee, O. Kwon, and E. J. Woo. Measurement of induced magnetic flux density using injection current nonlinear encoding (ICNE) in MREIT. *Physiological Measurement*, 28(2):117, 2006.

807. J. A. Park, K. J. Kang, I. O. Ko, K. C. Lee, B. K. Choi, N. Katoch, J. W. Kim, H. J. Kim, O. I. Kwon, and E. J. Woo. In vivo measurement of brain tissue response after irradiation: Comparison of T2 relaxation, apparent diffusion coefficient, and electrical conductivity. *IEEE transactions on medical imaging*, 2019.

808. R. L. Parke, A. Bloch, and S. P. McGuinness. Effect of very-high-flow nasal therapy on airway pressure and end-expiratory lung impedance in healthy volunteers. *Respiratory Care*, 60(10):1397–1403, Sept. 2015.

809. A. A. Pathiraja, R. A. Weerakkody, A. C. von Roon, P. Ziprin, and R. Bayford. The clinical application of electrical impedance technology in the detection of malignant neoplasms: a systematic review. *Journal of Translational Medicine*, 18:1–11, 2020.

810. R. Patz, S. Watson, C. Ktistis, M. Hamsch, and A. J. Peyton. Performance of a FPGA-based direct digitising signal measurement module for MIT. *Journal of Physics: Conference Series*, 224:012017, Apr. 2010.

811. K. Paulson, W. Breckon, and M. Pidcock. Electrode modelling in electrical impedance tomography. *SIAM Journal on Applied Mathematics*, 52(4):1012–1022, Aug. 1992.

812. K. S. Paulson, L. Wrb, and M. K. Pidcock. POMPUS - an optimized EIT reconstruction algorithm. *Inverse Problems*, 11:425–437, 1995.

813. J. L. Peake and K. E. Pinkerton. Gross and subgross anatomy of lungs, pleura, connective tissue septa, distal airways, and structural units. In *Comparative Biology of the Normal Lung*, pages 21–31. Elsevier, 2015.

814. S. Penz, H. Chauris, D. Donno, and C. Mehl. Resistivity modelling with topography. *Geophys. J. Int.*, 194(3):1486–1497, 2013.

815. S. M. Pereira, M. R. Tucci, C. C. A. Morais, C. M. Simões, B. F. F. Tonelotto, M. S. Pompeo, F. U. Kay, P. Pelosi, J. E. Vieira, and M. B. P. Amato. Individual positive end-expiratory pressure settings optimize intraoperative mechanical ventilation and reduce postoperative atelectasis. *Anesthesiology*, 129(6):1070–1081, Dec. 2018.

816. J. F. Perez-Juste Abascal. *The Anisotropic Inverse Conductivity Problem*. PhD thesis, MSc Thesis, University of Manchester, 2003.

817. A. Perrone, V. Lapenna, and S. Piscitelli. Electrical resistivity tomography technique for landslide investigation: A review. *Earth-Science Reviews*, 135:65–82, Aug. 2014.

818. A. Peyman, C. Gabriel, and E. Grant. Complex permittivity of sodium chloride solutions at microwave frequencies. *Bioelectromagnetics*, 28(4):264–274, 2007.

819. A. J. Peyton, A. R. Borges, J. de Oliveira, G. M. Lyon, Z. Z. Yu, M. W. Brown, and J. Ferreira. Development of electromagnetic tomography (emt) for industrial applications. part 1: Sensor design and instrumentation. In *Proc. 1st World Congress on Industrial Process Tomography*, pages 306–312, Apr. 1999.

820. A. J. Peyton, R. Mackin, D. Goss, E. Crescenzo, and H. S. Tapp. The development of high frequency electromagnetic inductance tomography for low conductivity materials. In *Proc. 2nd International Symposium Process Tomography*, pages 25–40, Sept. 2002.

821. A. J. Peyton, S. Watson, R. J. Williams, H. Griffiths, and W. Gough. Characterising the effects of the external electromagnetic shield on a magnetic induction tomography sensor. In *Proc. 3rd World Congress on Industrial Process Tomography*, pages 352–357, 2003.

822. A. J. Peyton, Z. Z. Yu, G. Lyon, S. Al-Zeibak, J. Ferreira, J. Velez, F. Linhares, A. R. Borges, H. L. Xiong, N. H. Saunders, and M. S. Beck. An overview of electromagnetic inductance tomography: description of three different systems. *Measurement Science and Technology*, 7(3):261–271, Mar. 1996.

823. H. Pfützner. Dielectric analysis of blood by means of a raster-electrode technique. *Medical and biological engineering and computing*, 22(2):142–146, 1984.

824. M. H. Pham, Y. Hua, and N. B. Gray. Eddy current tomography for metal solidification imaging. In *Proc. 1st World Congress on Industrial Process Tomography*, pages 451–458, Apr. 1999.

825. T. M. T. Pham, M. Yuill, C. Dakin, and A. Schibler. Regional ventilation distribution in the first 6 months of life. *European Respiratory Journal*, 37(4):919–924, July 2011.

826. D. L. Phillips. A technique for the numerical solution of certain integral equations of the first kind. *J Assoc Comput Mach*, 9:84–97, 1962.

827. A. Pidlisecky, R. Knight, and E. Haber. Cone-based electrical resistivity tomography. *Geophysics*, 71(4):G157–G167, 2006.

828. R. Pikkemaat, S. Lundin, O. Stenqvist, R.-D. Hilgers, and S. Leonhardt. Recent advances in and limitations of cardiac output monitoring by means of electrical impedance tomography. *Anesthesia and analgesia*, 119(1):76–83, 2014.

829. M. R. Pinsky. Functional haemodynamic monitoring. *Current opinion in critical care*, 20(3):288–293, 2014.

830. G. Piperno, E. Frei, and M. Moshitzky. Breast cancer screening by impedance measurements. *Frontiers of medical and biological engineering: the international journal of the Japan Society of Medical Electronics and Biological Engineering*, 2(2):111–117, 1990.

831. A. Plaskowski, T. Piotrowski, and M. Fraczak. Electrical process tomography application to industrial safety problems. In *Proc. 2nd International Symposium Process Tomography*, pages 63–72, 2002.

832. M. A. Player, J. van Weereld, A. R. Allen, and C. Dal. Truncated-newton algorithm for three-dimensional electrical impedance tomography. *Electronics Letters*, 35:2189–2191, 1999.

833. T. Pleyers, O. Levionnois, J. Siegenthaler, C. Spadavecchia, and M. Raillard. Investigation of selected respiratory effects of (dex)medetomidine in healthy beagles. *Veterinary Anaesthesia and Analgesia*, 47(5):667–671, Sept. 2020.

834. B. W. Pogue, S. P. Poplack, T. O. McBride, W. A. Wells, K. S. Osterman, U. L. Osterberg, and K. D. Paulsen. Quantitative hemoglobin tomography with diffuse near-infrared spectroscopy: pilot results in the breast. *Radiology*, 218(1):261–266, 2001.

835. L. Pointon. MRI breast screening study, 2003.

836. N. Polydorides. *Image Reconstruction Algorithms for Soft Field Tomography*. PhD thesis, UMIST, 2002.

837. N. Polydorides and H. Lionheart, W. R. B. McCann. Krylov subspace itemacserative techniques: on the detection of brain activity with electrical impedance tomography. *IEEE Trans Med Imaging*, 21:596–603, 2002.

838. N. Polydorides and W. R. B. Lionheart. A matlab toolkit for three-dimensional electrical impedance tomography: a contribution to the electrical impedance and diffuse optical reconstruction software project, meas. *Sci. Technol.*, 13:1871–1883, 2002.

839. S. P. Poplack, K. D. Paulsen, A. Hartov, P. M. Meaney, B. W. Pogue, T. D. Tosteson, M. R. Grove, S. K. Soho, and W. A. Wells. Electromagnetic breast imaging: average tissue property values in women with negative clinical findings. *Radiology*, 231(2):571–580, 2004.

840. S. P. Poplack, T. D. Tosteson, W. A. Wells, B. W. Pogue, P. M. Meaney, A. Hartov, C. A. Kogel, S. K. Soho, J. J. Gibson, and K. D. Paulsen. Electromagnetic breast imaging: results of a pilot study in women with abnormal mammograms. *Radiology*, 243(2):350–359, 2007.

841. M. Proença, F. Braun, J. Solà, A. Adler, M. Lemay, J.-P. Thiran, and S. F. Rimoldi. Non-invasive monitoring of pulmonary artery pressure from timing information by EIT: experimental evaluation during induced hypoxia. *Physiological Measurement*, 37(6):713–726, may 2016.

842. M. Proença, F. Braun, J. Solà, J.-P. Thiran, and M. Lemay. Noninvasive pulmonary artery pressure monitoring by EIT: a model-based feasibility study. *Medical & Biological Engineering & Computing*, 55(6):949–963, Jun 2017.

843. M. Proença, F. Braun, M. Rapin, J. Solà, A. Adler, B. Grychtol, S. Bohm, M. Lemay, and J.-P. Thiran. Influence of heart motion on cardiac output estimation by means of electrical impedance tomography: a case study. *Physiological measurement*, 36:1075–1091, 05 2015.

844. S. Pulletz, A. Adler, M. Kott, G. Elke, B. Gawelczyk, D. Schädler, G. Zick, N. Weiler, and I. Frerichs. Regional lung opening and closing pressures in patients with acute lung injury. *Journal of Critical Care*, 27(3):323.e11–323.e18, June 2012.

845. S. Pulletz, G. Elke, G. Zick, D. Schädler, F. Reifferscheid, N. Weiler, and I. Frerichs. Effects of restricted thoracic movement on the regional distribution of ventilation. *Acta Anaesthesiologica Scandinavica*, 54(6):751–760, Apr. 2010.

846. S. Pulletz, H. R. van Genderingen, G. Schmitz, G. Zick, D. Schädler, J. Scholz, N. Weiler, and I. Frerichs. Comparison of different methods to define regions of interest for evaluation of regional lung ventilation by EIT. *Physiological Measurement*, 27(5):S115–S127, apr 2006.

847. M. Radai, S. Zlochiver, M. Rosenfeld, and A. Abboud. Combined injected and induced current approaches in EIT – a simulation study. In *Proc 4th Int conf Biomedical Applications of EIT (EIT 2003)*, page 33, Apr. 2003.

848. R. W. Radcliffe, P. Morkel, A. Jago, A. A. Taft, P. Du Preez, M. A. Miller, D. Candra, D. V. Nydam, J. S. Barry, and R. D. Gleed. Pulmonary dead space in free-ranging immobilized black rhinoceroses (diceros bicornis) in namibia. *Journal of Zoo and Wildlife Medicine*, 45(2):263–271, 2014.

849. O. C. Radke, T. Schneider, A. R. Heller, and T. Koch. Spontaneous breathing during general anesthesia prevents the ventral redistribution of ventilation as detected by electrical impedance tomography. *Anesthesiology*, 116(6):1227–1234, June 2012.

850. A. R. A. Rahman, J. Register, G. Vuppala, and S. Bhansali. Cell culture monitoring by impedance mapping using a multielectrode scanning impedance spectroscopy system (CellMap). *Physiological Measurement*, 29(6):S227–S239, June 2008.

851. P. Rahmati, M. Soleimani, S. Pulletz, I. Frerichs, and A. Adler. Level-set-based reconstruction algorithm for EIT lung images: first clinical results. *Physiological measurement*, 33(5):739, 2012.

852. M. Rahtu, I. Frerichs, A. D. Waldmann, C. Strodthoff, T. Becher, R. Bayford, and M. Kallio. Early recognition of pneumothorax in neonatal respiratory distress syndrome with electrical impedance tomography. *American Journal of Respiratory and Critical Care Medicine*, 200(8):1060–1061, Oct. 2019.

853. V. Raicu, T. Saibara, H. Enzan, and A. Irimajiri. Dielectric properties of rat liver in vivo: analysis by modeling hepatocytes in the tissue architecture. *Bioelectrochemistry and Bioenergetics*, 47(2):333–342, Dec. 1998.

854. V. Raicu, T. Saibara, and A. Irimajiri. Dielectric properties of rat liver in vivo: a noninvasive approach using an open-ended coaxial probe at audio/radio frequencies. *Bioelectrochemistry and bioenergetics*, 47(2):325–332, 1998.

855. A. L. Raisis, M. Mosing, G. L. Hosgood, C. J. Secombe, A. Adler, and A. D. Waldmann. The use of cardiac related impedance changes measured with electrical impedance tomography (EIT) to evaluate pulse rate in anaesthetised horses. *submitted: The Veterinary Journal*, 2020.

856. S. Ramakrishna, L. Tian, and C. Wang. *Medical Devices: Regulations, Standards and Practices (Woodhead Publishing Series in Biomaterials*. Woodhead Publishing, 2015.

857. S. Ramli and A. J. Peyton. Feasibility study of planar-array electromagnetic inductance tomography. In *Proc. 1st World Congress on Industrial Process Tomography*, pages 502–510, 1999.

858. X. Ramus. Demystifying the operational transconductance amplifier. Texas Instruments Application Report SBOA117A, 2009 (revised 2013).

859. J. B. Ranck. Specific impedance of rabbit cerebral cortex. *Experimental Neurology*, 7(2):144–152, Feb. 1963.

860. J. B. Ranck Jr and S. L. BeMent. The specific impedance of the dorsal columns of cat: an anisotropic medium. *Experimental neurology*, 11(4):451–463, 1965.

861. A. Randazzo, E. Tavanti, M. Mikulenas, F. Boero, A. Fedeli, A. Sansalone, G. Allasia, and M. Pastorino. An electrical impedance tomography system for brain stroke imaging based on a lebesgue-space inversion procedure. *IEEE Journal of Electromagnetics, RF and Microwaves in Medicine and Biology*, 2020.

862. O. Raneta, V. Bella, L. Bellova, and E. Zamecnikova. The use of electrical impedance tomography to the differential diagnosis of pathological mammographic/sonographic findings. *Neoplasma*, 60(6):647–54, 2013.

863. A. Rao, A. Gibson, and D. Holder. EIT images of electrically induced epileptic activity in anaesthetised rabbits. *Medical and Biological Engineering and Computing*, 35(1):327, 1997.

864. A. Rao, E. K. Murphy, R. J. Halter, and K. M. Odame. A 1 mhz miniaturized electrical impedance tomography system for prostate imaging. *IEEE transactions on biomedical circuits and systems*, 14(4):787–799, 2020.

865. A. Rao, Y.-C. Teng, C. Schaef, E. K. Murphy, S. Arshad, R. J. Halter, and K. Odame. An analog front end ASIC for cardiac electrical impedance tomography. *IEEE Transactions on Biomedical Circuits and Systems*, 12(4):729–738, 2018.

866. A. J. Rao, E. K. Murphy, M. Shahghasemi, and K. M. Odame. Current-conveyor-based wide-band current driver for electrical impedance tomography. *Physiological Measurement*, 40(3), 2019.

867. S. Rauchenzauner, P. Thaler, B. Meldt, and H. Scharfetter. High resolution hardware and digital data acquisition for magnetic induction spectroscopy of biological tissue. In *Proc. Int. Fed. Med. Biol. Eng. (EMBEC02)*, pages 118–119, Dec. 2002.

868. E. Ravagli, S. Mastitkaya, N. Thompson, F. Iacoviello, P. R. Shearing, J. Perkins, A. V. Gourine, K. Aristovich, and D. Holder. Imaging fascicular organization of peripheral nerves with fast neural electrical impedance tomography (EIT). *bioRxiv*, 2020.

869. E. Ravagli, S. Mastitskaya, N. Thompson, K. Aristovich, and D. Holder. Optimization of the electrode drive pattern for imaging fascicular compound action potentials in peripheral nerve with fast neural electrical impedance tomography. *Physiological Measurement*, 40(11):115007, 2019.

870. F. Reifferscheid, G. Elke, S. Pulletz, B. Gawelczyk, I. Lautenschläger, M. Steinfath, N. Weiler, and I. Frerichs. Regional ventilation distribution determined by electrical impedance tomography: Reproducibility and effects of posture and chest plane. *Respirology*, 16(3):523–531, Mar. 2011.

871. Z. Ren, A. Kowalski, and T. Rodgers. Measuring inline velocity profile of shampoo by electrical resistance tomography (ert). *Flow Measurement and Instrumentation*, 58:31–37, 2017.

872. Z. Ren, L. Trinh, M. Cooke, S. C. De Hert, J. Silvaluengo, J. Ashley, I. E. Tothill, and T. L. Rodgers. Development of a novel linear ert sensor to measure surface deposits. *IEEE Transactions on Instrumentation and Measurement*, 68(3):754–761, 2018.

873. A. Renner, U. Marschner, and W.-J. Fischer. A new imaging approach for in situ and ex situ inspections of conductive fiber–reinforced composites by magnetic induction tomography. *Journal of Intelligent Material Systems and Structures*, 25(9):1149–1162, Oct. 2014.

874. D. A. Reuter, T. W. Felbinger, C. Schmidt, E. Kilger, O. Goedje, P. Lamm, and A. E. Goetz. Stroke volume variations for assessment of cardiac responsiveness to volume loading in mechanically ventilated patients after cardiac surgery. *Intensive care medicine*, 28(4):392–398, 2002.

875. D. A. Reuter, C. Huang, T. Edrich, S. K. Shernan, and H. K. Eltzschig. Cardiac output monitoring using indicator-dilution techniques: basics, limits, and perspectives. *Anesthesia and analgesia*, 110(3):799–811, 2010.

876. A. Revil, M. Karaoulis, T. Johnson, and A. Kemna. Review: Some low-frequency electrical methods for subsurface characterization and monitoring in hydrogeology. *Hydrogeology Journal*, 20(4):617–658, 2012.

877. F. Ricard, C. Brechtelsbauer, C. Lawrence, Y. Xu, and A. Pannier. Application of electrical resistance tomography technology to pharmaceutical processes. In *Proc. 3rd World Congress on Industrial Process Tomography*, Sept. 2003.

878. P. Ridd. Electric potential due to a ring electrode, *IEEE Journal of Oceanic Engineering*, 19(3):464–467, 1994.

879. C. H. Riedel and O. Dössel. Planar system for magnetic induction impedance measurement. In *Proc 4th Int conf Biomedical Applications of EIT (EIT 2003)*, page 32, Apr. 2003.

880. C. H. Riedel, M. A. Golombeck, M. von Saint-George, and O. Dössel. Data acquisition system for contact-free conductivity measurement of biological tissue. In *Proc. Int. Fed. Med. Biol. Eng. (EMBEC02)*, pages 86–87, Dec. 2002.

881. C. H. Riedel, M. Keppelen, S. Nani, and O. Dössel. Post mortem conductivity measurement of liver tissue using a contact free magnetic induction sensor. In *Proc 25th Int Conf IEEE EMBS*, volume 57, pages 3126–3129, 2003.

882. C. H. Riedel, M. Keppelen, S. Nani, R. D. Merges, and O. Dössel. Planar system for magnetic induction conductivity measurement using a sensor matrix. *Physiological Measurement*, 25(1):403–411, Feb. 2004.

883. T. Riedel, M. Kyburz, P. Latzin, C. Thamrin, and U. Frey. Regional and overall ventilation inhomogeneities in preterm and term-born infants. *Intensive Care Medicine*, 35(1):144–151, Sept. 2009.

884. T. Riedel, T. Richards, and A. Schibler. The value of electrical impedance tomography in assessing the effect of body position and positive airway pressures on regional lung ventilation in spontaneously breathing subjects. *Intensive Care Medicine*, 31(11):1522–1528, Sept. 2005.

885. J. Riera, P. Perez, J. Cortes, O. Roca, J. R. Masclans, and J. Rello. Effect of high-flow nasal cannula and body position on end-expiratory lung volume: A cohort study using electrical impedance tomography. *Respiratory Care*, 58(4):589–596, Apr. 2013.

886. B. Rigaud, Y. Shi, N. Chauveau, and J. P. Morucci. Experimental acquisition system for impedance tomography with active electrode approach. *Medical and Biological Engineering and Computing*, 31(6):593–599, 1993.

887. P. Ripka. *Magnetic Sensors and Magnetometers*. Artech House Inc., Northwood, MA, USA, 2001.

888. N. Robinson. Some functional consequences of species differences in lung anatomy. *Adv Vet Sci Comp Med*, 26:1–33, 1982.

889. N. Robitaille, R. Guardo, I. Maurice, A. E. Hartinger, and H. Gagnon. A multi-frequency EIT system design based on telecommunication signal processors. *Physiological measurement*, 30(6):S57, 2009.

890. A. Rocchi, R. Hagen, C. Rohrer, U. Auer, and M. Mosing. Comparison of three positions for the electric impedance tomography (EIT) belt in dogs. In *Association of Veterinary Anaesthetists, Spring Meeting*, 2014.

891. T. Rodgers and A. Kowalski. An electrical resistance tomography method for determining mixing in batch addition with a level change. *Chemical Engineering Research and Design*, 88(2):204–212, 2010.

892. T. L. Rodgers, M. Cooke, F. R. Siperstein, and A. Kowalski. Mixing and dissolution times for a cowles disk agitator in large-scale emulsion preparation. *Industrial & engineering chemistry research*, 48(14):6859–6868, 2009.

893. A. Romsauerova, A. McEwan, and D. Holder. Identification of a suitable current waveform for acute stroke imaging. *Physiological Measurement*, 27(5):S211, 2006.

894. A. Romsauerova, A. McEwan, L. Horesh, R. Yerworth, R. Bayford, and D. S. Holder. Multi-frequency electrical impedance tomography (EIT) of the adult human head: initial findings in brain tumours, arteriovenous malformations and chronic stroke, development of an analysis method and calibration. *Physiological measurement*, 27(5):S147, 2006.

895. L. Rondi and F. Santosa. Enhanced electrical impedance tomographyviathe mumford–shah functional. *ESAIM: Control, Optimisation and Calculus of Variations*, 6:517–538, 2001.

896. J. Rosell, R. Casañas, and H. Scharfetter. Sensitivity maps and system requirements for magnetic induction tomography using a planar gradiometer. *Physiological Measurement*, 22(1):121–130, Feb. 2001.

897. J. Rosell and P. Riu. Common-mode feedback in electrical impedance tomography. *Clinical Physics and Physiological Measurement*, 13:11–14, 1992.

898. J. Rosell-Ferrer, R. Merwa, P. Brunner, and H. Scharfetter. A multifrequency magnetic induction tomography system using planar gradiometers: data collection and calibration. *Physiological Measurement*, 27(5):S271–S280, Apr. 2006.

899. R. D. Rosenberg, W. C. Hunt, M. R. Williamson, F. D. Gilliland, P. W. Wiest, C. A. Kelsey, C. R. Key, and M. N. Linver. Effects of age, breast density, ethnicity, and estrogen replacement therapy on screening mammographic sensitivity and cancer stage at diagnosis: review of 183,134 screening mammograms in Albuquerque, New Mexico. *Radiology*, 209(2):511–518, 1998.

900. A. B. Rosenkrantz, A. Oto, B. Turkbey, and A. C. Westphalen. Prostate imaging reporting and data system (pi-rads), version 2: a critical look. *American Journal of Roentgenology*, 206(6):1179–1183, 2016.

901. A. S. Ross, G. J. Saulnier, J. C. Newell, and D. Isaacson. Current source design for electrical impedance tomography. *Physiological Measurement*, 24(2):509–516, May 2003.

902. C. J. Roth, A. Ehrl, T. Becher, I. Frerichs, J. C. Schittny, N. Weiler, and W. A. Wall. Correlation between alveolar ventilation and electrical properties of lung parenchyma. *Physiological Measurement*, 36(6):1211–1226, may 2015.

903. C. Rücker and T. Günther. The simulation of finite ERT electrodes using the complete electrode model. *Geophysics*, 76(4):F227–F238, 2011.

904. C. Rücker, T. Günther, and K. Spitzer. Three-dimensional modelling and inversion of DC resistivity data incorporating topography - I. Modelling. *Geophys. J. Int.*, 166(2):495–505, 2006.

905. S. Rush, J. Abildskov, and R. McFee. Resistivity of body tissues at low frequencies. *Circulation research*, 12(1):40–50, 1963.

906. S. Rush and D. A. Driscoll. Current distribution in the brain from surface electrodes. *Anesthesia & Analgesia*, 47(6):717–723, Nov. 1968.

907. T. Rymarczyk, G. Kłosowski, E. Kozłowski, and P. Tchórzewski. Comparison of selected machine learning algorithms for industrial electrical tomography. *Sensors*, 19(7):1521, 2019.

908. T. Rymarczyk and P. Tchórzewski. Implementation 3D level set method to solve inverse problem in EIT. In *2018 International Interdisciplinary PhD Workshop (IIPhDW)*, pages 159–161. IEEE, 2018.

909. Y. Saad and M. H. Schultz. GMRES: A generalized minimal residual algorithm for solving nonsymmetric linear systems, *SIAM J Sci Statist Comput*, 7:856–869, 1986.

910. R. Sadleir and R. Fox. Quantification of blood volume by electrical impedance tomography using a tissue-equivalent phantom. *Physiological measurement*, 19(4):501, 1998.

911. R. Sadleir, S. Grant, S. U. Zhang, B. I. Lee, H. C. Pyo, S. H. Oh, C. Park, E. J. Woo, S. Y. Lee, O. Kwon, et al. Noise analysis in magnetic resonance electrical impedance tomography at 3 and 11 T field strengths. *Physiological measurement*, 26(5):875, 2005.

912. R. J. Sadleir and A. Argibay. Modeling skull electrical properties. *Annals of Biomedical Engineering*, 35(10):1699–1712, July 2007.

913. R. J. Sadleir and R. A. Fox. Detection and quantification of intraperitoneal fluid using electrical impedance tomography. *IEEE transactions on biomedical engineering*, 48(4):484–491, 2001.

914. R. J. Sadleir, F. Fu, and M. Chauhan. Functional magnetic resonance electrical impedance tomography (fMREIT) sensitivity analysis using an active bidomain finite-element model of neural tissue. *Magnetic Resonance in Medicine*, 81(1):602–614, May 2019.

915. R. J. Sadleir, F. Fu, C. Falgas, S. Holland, M. Boggess, S. C. Grant, and E. J. Woo. Direct detection of neural activity in vitro using magnetic resonance electrical impedance tomography (MREIT). *NeuroImage*, 161:104–119, Nov. 2017.

916. S. Z. Sajib, N. Katoch, H. J. Kim, O. I. Kwon, and E. J. Woo. Software toolbox for low-frequency conductivity and current density imaging using MRI. *IEEE Transactions on Biomedical Engineering*, 64(11):2505–2514, 2017.

917. S. Z. Sajib, O. I. Kwon, H. J. Kim, and E. J. Woo. Electrodeless conductivity tensor imaging (cti) using mri: basic theory and animal experiments. *Biomedical Engineering Letters*, 8:273–282, 2018.

918. S. Z. Sajib, T. I. Oh, H. J. Kim, O. I. Kwon, and E. J. Woo. In vivo mapping of current density distribution in brain tissues during deep brain stimulation (DBS). *AIP Advances*, 7(1):015004, 2017.

919. S. G. Sakka, D. A. Reuter, and A. Perel. The transpulmonary thermodilution technique. *Journal of Clinical Monitoring and Computing*, 26(5):347–353, Oct 2012.

920. S. G. Sakka, C. C. Rühl, U. J. Pfeiffer, R. Beale, A. McLuckie, K. Reinhart, and A. Meier-Hellmann. Assessment of cardiac preload and extravascular lung water by single transpulmonary thermodilution. *Intensive Care Medicine*, 26(2):180–187, Mar 2000.

921. G. Salomon, T. Hess, A. Erbersdobler, C. Eichelberg, S. Greschner, A. N. Sobchuk, A. K. Korolik, N. A. Nemkovich, J. Schreiber, M. Herms, et al. The feasibility of prostate cancer detection by triple spectroscopy. *european urology*, 55(2):376–384, 2009.

922. S. A. Santos, M. Czaplik, J. Orschulik, N. Hochhausen, and S. Leonhardt. Lung pathologies analyzed with multi-frequency electrical impedance tomography: Pilot animal study. *Respiratory Physiology & Neurobiology*, 254:1–9, Aug. 2018.

923. T. B. R. Santos, R. M. Nakanishi, J. P. Kaipio, J. L. Mueller, and R. G. Lima. Introduction of sample based prior into the D-bar method through a Schur complement property. *IEEE Transactions on Medical Imaging*, 39(12):4085–4093, 2020.

924. T. B. R. Santos, R. M. Nakanishi, J. P. Kaipio, J. L. Mueller, and R. G. Lima. Introduction of sample based prior into the D-bar method through a Schur complement property. *IEEE Transactions on Medical Imaging*, 2020.

925. F. Santosa and M. Vogelius. A backprojection algorithm for electrical impedance imaging, siam j. *Appl. Math*, 50:216–243, 1991.

926. G. J. Saulnier, A. Abdelwahab, and O. R. Shishvan. Dsp-based current source for electrical impedance tomography. *Physiological Measurement*, 41(6):64002, June 2020.

927. G. J. Saulnier, A. S. Ross, and N. Liu. A high-precision voltage source for EIT. *Physiological Measurement*, 27(5):S221–S236, 2006.

928. I. Savukov, Y. J. Kim, V. Shah, and M. G. Boshier. High-sensitivity operation of single-beam optically pumped magnetometer in a kHz frequency range. *Measurement Science and Technology*, 28(3):035104, Feb. 2017.

929. B. Schappel. Electrical impedance tomography of the half space: Locating obstacles by electrostatic measurements on the boundary. In *Proceedings of the 3rd World Congress on Industrial Process Tomography*, pages 2–5, Canada, September, 788-793, 2003. Banff.

930. H. Scharfetter. Single-shot dual frequency excitation for magnetic induction tomography (MIT) at frequencies above 1 MHz. *Journal of Physics: Conference Series*, 224:012041, Apr. 2010.

931. H. Scharfetter, R. Casanas, R. Merwa, and J. Rosell. Magnetic induction spectroscopy of biological tissue with a conducting background: experimental demonstration within the β-dispersion. In *Proc. Int. Fed. Med. Biol. Eng. (EMBEC02)*, pages 88–89, Dec. 2002.

932. H. Scharfetter, R. Casanas, and J. Rosell. Biological tissue characterization by magnetic induction spectroscopy (MIS): requirements and limitations. *IEEE Transactions on Biomedical Engineering*, 50(7):870–880, July 2003.

933. H. Scharfetter, A. Köstinger, and S. Issa. Hardware for quasi-single-shot multifrequency magnetic induction tomography (MIT): the Graz Mk2 system. *Physiological Measurement*, 29(6):S431–S443, June 2008.

934. H. Scharfetter, H. K. Lackner, and J. Rosell. Magnetic induction tomography: hardware for multi-frequency measurements in biological tissues. *Physiological Measurement*, 22(1):131–146, Feb. 2001.

935. H. Scharfetter, R. Merwa, and K. Pilz. A new type of gradiometer for the receiving circuit of magnetic induction tomography (MIT). *Physiological Measurement*, 26(2):S307–S318, Mar. 2005.

936. H. Scharfetter, S. Rauchenzauner, R. Merwa, O. Biró, and K. Hollaus. Planar gradiometer for magnetic induction tomography (MIT): theoretical and experimental sensitivity maps for a low-contrast phantom. *Physiological Measurement*, 25(1):325–333, Feb. 2004.

937. H. Scharfetter, P. Riu, M. Populo, and J. Rosell. Sensitivity maps for low-contrast perturbations within conducting background in magnetic induction tomography. *Physiological Measurement*, 23(1):195–202, Jan. 2002.

938. U. Schaumloffel-Schulze, S. H. Heywang-Kobrunner, C. Alter, D. Lampe, and J. Buchmann. Diagnostische vakuumbiopsie der brust–ergebnisse von 600 patienten. *Fortschr. Roentgenstr.*, 72:S1–170, 1999.

939. A. Schibler, T. M. T. Pham, A. A. Moray, and C. Stocker. Ventilation and cardiac related impedance changes in children undergoing corrective open heart surgery. *Physiological Measurement*, 34(10):1319–1327, Sept. 2013.

940. A. Schibler, M. Yuill, C. Parsley, T. Pham, K. Gilshenan, and C. Dakin. Regional ventilation distribution in non-sedated spontaneously breathing newborns and adults is not different. *Pediatric Pulmonology*, 44(9):851–858, Aug. 2009.

941. C. Schlumberger. *Étude sur la Prospection Électrique du Sous-sol.* Gauthier-Villars et Cie, 1920.

942. S. Schnidrig, C. Casaulta, A. Schibler, and T. Riedel. Influence of end-expiratory level and tidal volume on gravitational ventilation distribution during tidal breathing in healthy adults. *European Journal of Applied Physiology*, 113(3):591–598, Aug. 2013.

943. J. Schöberl. NETGEN - an advancing front 2d/3d-mesh generator based on abstract rules, *Visual Sci Comput*, 1:41–52, 1997.

944. J. Schramel, C. Nagel, U. Auer, F. Palm, C. Aurich, and Y. Moens. Distribution of ventilation in pregnant shetland ponies measured by electrical impedance tomography. *Respiratory Physiology & Neurobiology*, 180(2-3):258–262, Mar. 2012.

945. B. Schullcke, S. Krueger-Ziolek, B. Gong, R. A. Jörres, U. Mueller-Lisse, and K. Moeller. Ventilation inhomogeneity in obstructive lung diseases measured by electrical impedance tomography: a simulation study. *Journal of Clinical Monitoring and Computing*, 32(4):753–761, Oct. 2017.

946. M. Schwarz, M. Jendrusch, and I. Constantinou. Spatially resolved electrical impedance methods for cell and particle characterization. *Electrophoresis*, 41(1-2):65–80, 2020.

947. D. M. Scott and O. W. Gutsche. ECT studies of bead fluidization in vertical mills. In *Proc. 1st World Congress on Industrial Process Tomography*, pages 90–95, Apr. 1999.

948. G. Scott, M. Joy, R. Armstrong, and R. Henkelman. Sensitivity of magnetic-resonance current-density imaging. *Journal of Magnetic Resonance (1969)*, 97(2):235–254, 1992.

949. A. D. Seagar. *Probing with low frequency electric current, PhD Thesis.* PhD thesis, University of Canterbury, Christchurch, NZ, 1983.

950. C. Secombe, A. Adler, A. Raisis, G. Hosgood, and M. Mosing. Bronchoconstriction and bronchodilation in horses can be verified using electrical impedance tomography (EIT). In *Proc. Asthma Workshop 2019*, 2019.

951. C. Secombe, A. D. Waldmann, G. Hosgood, and M. Mosing. Evaluation of histamine-provoked changes in airflow using electrical impedance tomography in horses. *Equine Veterinary Journal*, 52(4):556–563, Feb. 2020.

952. A. S. Sedra and K. C. Smith. A second-generation current conveyor and its applications. *IEEE Transactions on Circuit Theory*, 17(1), February 1970.

953. M. Sekino, K. Yamaguchi, N. Iriguchi, and S. Ueno. Conductivity tensor imaging of the brain using diffusion-weighted magnetic resonance imaging. *Journal of applied physics*, 93(10):6730–6732, 2003.

954. N. Sella, F. Zarantonello, G. Andreatta, V. Gagliardi, A. Boscolo, and P. Navalesi. Positive end-expiratory pressure titration in COVID-19 acute respiratory failure: electrical impedance tomography vs. PEEP/FiO2 tables. *Critical Care*, 24(1), Sept. 2020.

955. J. K. Seo, D.-H. Kim, J. Lee, O. I. Kwon, S. Z. Sajib, and E. J. Woo. Electrical tissue property imaging using mri at dc and larmor frequency. *Inverse Problems*, 28(8):084002, 2012.

956. J. K. Seo, M.-O. Kim, J. Lee, N. Choi, E. J. Woo, H. J. Kim, O. I. Kwon, and D.-H. Kim. Error analysis of nonconstant admittivity for mr-based electric property imaging. *IEEE transactions on medical imaging*, 31(2):430–437, 2011.

957. J. K. Seo, J. Lee, S. W. Kim, H. Zribi, and E. J. Woo. Frequency-difference electrical impedance tomography (fdEIT): algorithm development and feasibility study. *Physiological measurement*, 29(8):929, 2008.

958. J. K. Seo and E. J. Woo. Electrical tissue property imaging at low frequency using MREIT. *IEEE Transactions on Biomedical Engineering*, 61(5):1390–1399, 2014.

959. J. K. Seo, J.-R. Yoon, E. J. Woo, and O. Kwon. Reconstruction of conductivity and current density images using only one component of magnetic field measurements. *IEEE Transactions on Biomedical Engineering*, 50(9):1121–1124, 2003.

960. J. E. Serrallés, L. Daniel, J. K. White, D. K. Sodickson, R. Lattanzi, and A. G. Polimeridis. Global maxwell tomography: a novel technique for electrical properties mapping based on mr measurements and volume integral equation formulations. In *2016 IEEE International Symposium on Antennas and Propagation (APSURSI)*, pages 1395–1396. IEEE, 2016.

961. R. E. Serrano, B. de Lema, O. Casas, T. Feixas, N. Calaf, V. Camacho, I. Carrió, P. Casan, J. Sanchis, and P. J. Riu. Use of electrical impedance tomography (EIT) for the assessment of unilateral pulmonary function. *Physiological Measurement*, 23(1):211–220, Jan. 2002.

962. R. E. Serrano, P. J. Riu, B. de Lema, and P. Casan. Assessment of the unilateral pulmonary function by means of electrical impedance tomography using a reduced electrode set. *Physiological Measurement*, 25(4):803–813, July 2004.

963. M. Shalit. The effect of metrazol on the hemodynamics and impedance of the cat's brain cortex. *Journal of Neuropathology & Experimental Neurology*, 24(1):75–84, 1965.

964. E. Sharif, C. Bell, P. F. Morris, and A. J. Peyton. Imaging the transformation of hot strip steel using magnetic techniques. *Journal of Electronic Imaging*, 10(3):669, July 2001.

965. M. Sharifi and B. Young. Electrical resistance tomography (ERT) applications to chemical engineering. *Chemical Engineering Research and Design*, 91(9):1625–1645, 2013.

966. X. Shi, W. Li, F. You, X. Huo, C. Xu, Z. Ji, R. Liu, B. Liu, Y. Li, F. Fu, et al. High-precision electrical impedance tomography data acquisition system for brain imaging. *IEEE Sensors Journal*, 18(14):5974–5984, 2018.

967. K. Shimada and D. C. Gossard. Bubble mesh: automated triangular meshing of non-manifold geometry by sphere packing. *Proceedings of the third ACM symposium on Solid modeling and applications*, pages 409–419, 1995.

968. K. Shin and J. L. Mueller. An improved calderón's method for absolute images with a priori information. *Inverse Problem*, 36(12):124005, 2020.

969. A. Shono, N. Katayama, T. Fujihara, S. H. Böhm, A. D. Waldmann, K. Ugata, T. Nikai, and Y. Saito. Positive end-expiratory pressure and distribution of ventilation in pneumoperitoneum combined with steep trendelenburg position. *Anesthesiology*, 132(3):476–490, Mar. 2020.

970. W. Shuai, F. You, H. Zhang, W. Zhang, F. Fu, X. Shi, R. Liu, T. Bao, and X. Dong. Application of electrical impedance tomography for continuous monitoring of retroperitoneal bleeding after blunt trauma. *Annals of biomedical engineering*, 37(11):2373–2379, 2009.

971. R. L. Siegel, K. D. Miller, H. E. Fuchs, and A. Jemal. Cancer statistics, 2021. *CA: a Cancer Journal for Clinicians*, 71(1):7–33, 2021.

972. K. C. Siegmann, T. Xydeas, R. Sinkus, B. Kraemer, U. Vogel, and C. D. Claussen. Diagnostic value of mr elastography in addition to contrast-enhanced mr imaging of the breast–initial clinical results. *European radiology*, 20(2):318–325, 2010.

973. S. Siltanen, J. Mueller, and D. Isaacson. An implementation of the reconstruction algorithm of A. Nachman for the 2-D inverse conductivity problem. *Inverse Problems*, 16:681–699, 2000.

974. S. Siltanen and J. P. Tamminen. Reconstructing conductivities with boundary corrected D-bar method. *Journal of Inverse and Ill-posed Problems*, 22(6):847–870, 2014.

975. P. P. Silvester and R. L. Ferrari. *Finite Elements for Electrical Engineers*. Cambridge University Press, Cambridge, 1990.

976. B. Singh, C. Smith, and R. Hughes. In vivo dielectric spectrometer. *Medical and Biological Engineering and Computing*, 17(1):45–60, 1979.

977. K. Singha, F. Day-Lewis, T. Johnson, and L. Slater. Advances in interpretation of subsurface processes with time-lapse electrical imaging. *Hydrological Processes*, 29(6):1549–1576, 2015.

978. A. Sinton, B. Brown, D. Barber, F. McArdle, and A. Leathard. Noise and spatial resolution of a real-time electrical impedance tomograph. *Clinical Physics and Physiological Measurement*, 13(A):125, 1992.

979. R. Sirian and J. Wills. Physiology of apnoea and the benefits of preoxygenation. *Continuing Education in Anaesthesia Critical Care & Pain*, 9(4):105–108, 06 2009.

980. L. Slater. Near surface electrical characterization of hydraulic conductivity: From petrophysical properties to aquifer geometries — a review. *Surveys in Geophysics*, 28(2-3):169–197, 2007.

981. A. S. Slutsky and V. M. Ranieri. Ventilator-induced lung injury. *New England Journal of Medicine*, 369(22):2126–2136, Nov. 2013.

982. H. J. Smit, A. Vonk Noordegraaf, J. T. Marcus, A. Boonstra, P. M. de Vries, and P. E. Postmus. Determinants of pulmonary perfusion measured by electrical impedance tomography. *European Journal of Applied Physiology*, 92(1):45–49, Jun 2004.

983. R. Smith, I. Freeston, and B. Brown. A real-time electrical impedance tomography system for clinical use-design and preliminary results. *IEEE Transactions on Biomedical Engineering*, 42(2):133–140, 1995.

984. R. W. M. Smith, I. L. Freeston, B. H. Brown, and A. M. Sinton. Design of a phase sensitive detector to maximize signal-to-noise ratio in the presence of Gaussian wideband noise. *Meas. Sci. Technol.*, 3:1054–1062, 1992.

985. V. Sobota, M. Müller, and K. Roubík. Intravenous administration of normal saline may be misinterpreted as a change of end-expiratory lung volume when using electrical impedance tomography. *Scientific Reports*, 9(1), Apr. 2019.

986. V. Sobota and K. Roubik. Center of ventilation-methods of calculation using electrical impedance tomography and the influence of image segmentation. In *Proc IFBME*, volume 57, pages 1258–1263, 2016.

987. J. Solà, A. Adler, A. Santos, G. Tusman, F. S. Sipmann, and S. H. Bohm. Non-invasive monitoring of central blood pressure by electrical impedance tomography: first experimental evidence. *Medical & Biological Engineering & Computing*, 49(4):409, Mar 2011.

988. J. Solà, R. Vetter, P. Renevey, O. Chételat, C. Sartori, and S. F. Rimoldi. Parametric estimation of pulse arrival time: a robust approach to pulse wave velocity. *Physiological Measurement*, 30(7):603–615, jun 2009.

989. J. Solà i Caros and J. X. Brunner. Method for determining non-invasively a heart-lung interaction, 04 2014.

990. M. Soleimani. Electrical impedance tomography imaging using a priori ultrasound data. *Biomedical engineering online*, 5(1):1–8, 2006.

991. M. Soleimani, C. Gómez-Laberge, and A. Adler. Imaging of conductivity changes and electrode movement in EIT. *Physiological measurement*, 27(5):S103, 2006.

992. M. Soleimani and W. R. B. Lionheart. Absolute conductivity reconstruction in magnetic induction tomography using a nonlinear method. *IEEE Transactions on Medical Imaging*, 25(12):1521–1530, Dec. 2006.

993. B. D. Sollish, Y. Drier, E. Hammerman, E. H. Frei, and B. Man. Dielectric breast scanner. In *Proc. XII International Conference on Medicine and Biology Engineering/V Conference on Medical Physics II*, pages 30–33, 1979.

994. E. Somersalo, M. Cheney, D. Isaacson, and E. Isaacson. Layer stripping, a direct numerical method for impedance imaging. *Inverse Problems*, 7:899–926, 1991.

995. E. Somersalo, D. Isaacson, and M. Cheney. A linearized inverse boundary value problem for maxwell's equations. *J. Comput. Appl. Math*, 42:123–136, 1992.

996. E. Somersalo, J. P. Kaipio, M. Vauhkonen, and D. D. Baroudi. Impedance imaging and Markov chain monte carlo methods. In *Proc. SPIE's 42nd Annual Meeting*, pages 175–185, 1997.

997. N. K. Soni, A. Hartov, C. Kogel, S. P. Poplack, and K. D. Paulsen. Multi-frequency electrical impedance tomography of the breast: new clinical results. *Physiological measurement*, 25(1):301, 2004.

998. B. Souffaché, P. Cosenza, S. Flageul, J. Pencolé, S. Seladji, and A. Tabbagh. Electrostatic multipole for electrical resistivity measurements at the decimetric scale. *J. Applied Geophys.*, 71(1):6–12, 2010.

999. S. Spadaro, T. Mauri, S. H. Böhm, G. Scaramuzzo, C. Turrini, A. D. Waldmann, R. Ragazzi, A. Pesenti, and C. A. Volta. Variation of poorly ventilated lung units (silent spaces) measured by electrical impedance tomography to dynamically assess recruitment. *Critical Care*, 22(1), Jan. 2018.

1000. B. L. Sprague, R. F. Arao, D. L. Miglioretti, L. M. Henderson, D. S. Buist, T. Onega, G. H. Rauscher, J. M. Lee, A. N. Tosteson, K. Kerlikowske, et al. National performance benchmarks for modern diagnostic digital mammography: update from the breast cancer surveillance consortium. *Radiology*, 283(1):59–69, 2017.

1001. C. A. Stahl, K. Möller, S. Schumann, R. Kuhlen, M. Sydow, C. Putensen, and J. Guttmann. Dynamic versus static respiratory mechanics in acute lung injury and acute respiratory distress syndrome. *Critical Care Medicine*, 34(8):2090–2098, Aug. 2006.

1002. E. H. Starling and M. B. Visscher. The regulation of the energy output of the heart. *The Journal of physiology*, 62(3):243–261, 1927.

1003. M. Steffen, K. Heimann, N. Bernstein, and S. Leonhardt. Multichannel simultaneous magnetic induction measurement system (MUSIMITOS). *Physiological Measurement*, 29(6):S291–S306, June 2008.

1004. C. Stehning, T. Voigt, and U. Katscher. Real-time conductivity mapping using balanced ssfp and phase-based reconstruction. In *Proceedings of the 19th Scientific Meeting of the International Society of Magnetic Resonance in Medicine (ISMRM11)*, volume 128, 2011.

1005. J. Stelter, J. Wtorek, A. Nowakowski, A. Kopacz, and T. Jastrzembski. Complex permittivity of breast tumor tissue. In *Proc 10th Int Conference on Electrical Bio-Impedance*, 1998.

1006. A. Stojadinovic, O. Moskovitz, Z. Gallimidi, S. Fields, A. D. Brooks, R. Brem, R. N. Mucciola, M. Singh, M. Maniscalco-Theberge, H. E. Rockette, et al. Prospective study of electrical impedance scanning for identifying young women at risk for breast cancer. *Breast cancer research and treatment*, 97(2):179–189, 2006.

1007. A. Stojadinovic, A. Nissan, Z. Gallimidi, S. Lenington, W. Logan, M. Zuley, A. Yeshaya, M. Shimonov, M. Melloul, S. Fields, et al. Electrical impedance scanning for the early detection of breast cancer in young women: preliminary results of a multicenter prospective clinical trial. *Journal of Clinical Oncology*, 23(12):2703–2715, 2005.

1008. S. Stowe, A. Boyle, M. Sage, W. See, J.-P. Praud, É. Fortin-Pellerin, and A. Adler. Comparison of bolus- and filtering-based EIT measures of lung perfusion in an animal model. *Physiological Measurement*, 40(5):054002, June 2019.

1009. G. Strang. *Introduction to Linear Algebra, 3rd edition*, volume 3. Wellesley-Cambridge Press, 1988.

1010. G. Strang and G. J. Fix. *An Analysis of the Finite Element Method*. Prentice-Hall, New York, 1973.

1011. N. A. Stutzke, S. E. Russek, D. P. Pappas, and M. Tondra. Low-frequency noise measurements on commercial magnetoresistive magnetic field sensors. *Journal of Applied Physics*, 97(10):10Q107, 2005.

1012. B. Sun, S. Yue, Z. Hao, Z. Cui, and H. Wang. An improved tikhonov regularization method for lung cancer monitoring using electrical impedance tomography. *IEEE Sensors Journal*, 19(8):3049–3057, 2019.

1013. R. Supper, D. Ottowitz, B. Jochum, J. Kim, A. Römer, I. Baron, S. Pfeiler, M. Lovisolo, S. Gruber, and F. Vecchiotti. Geoelectrical monitoring: an innovative method to supplement landslide surveillance and early warning. *Near Surface Geophysics*, 12(1):133–150, Feb. 2014.

1014. A. Surowiec, S. Stuchly, J. Barr, and A. Swarup. Dielectric properties of breast carcinoma and the surrounding tissues. *IEEE Transactions on Biomedical Engineering*, 35(4):257–263, Apr. 1988.

1015. A. Surowiec, S. S. Stuchly, M. Keaney, and A. Swarup. In vivo and in vitro dielectric properties of feline tissues at low radiofrequencies. *Physics in Medicine and Biology*, 31(8):901–909, Aug. 1986.

1016. J. Sylvester and G. Uhlmann. A uniqueness theorem for an inverse boundary value problem in electrical prospection. *Communications on Pure and Applied Mathematics*, 39(1):91–112, 1986.

1017. J. Sylvester and G. Uhlmann. A global uniqueness theorem for an inverse boundary value problem. *Annals of Mathematics*, 125:153–169, 1987.

1018. M. Takhti, Y.-C. Teng, and K. Odame. A 10 mhz read-out chain for electrical impedance tomography. *IEEE transactions on biomedical circuits and systems*, 12(1):222–230, 2018.

1019. A. Tamburrino and G. Rubinacci. A new non-iterative inversion method in electrical resistancetomography. *Inverse Problems*, 18:2002, 2002.

1020. A. Tamburrino and G. Rubinacci. Fast methods for quantitative eddy-current tomography of conductive materials. *IEEE transactions on magnetics*, 42(8):2017–2028, 2006.

1021. A. Tamburrino, G. Rubinacci, M. Soleimani, and W. R. B. Lionheart. Non iterative inversion method for electrical resistance, capacitance and inductance tomography for two phase materials. In *Proc. 3rd World Congress on Industrial Process Tomography*, pages 233–238, 2003.

1022. J. Tamminen, T. Tarvainen, and S. Siltanen. The D-bar method for diffuse optical tomography: a computational study. *Experimental Mathematics*, 26(2):225–240, 2017.

1023. C. Tang, F. You, G. Cheng, D. Gao, F. Fu, G. Yang, and X. Dong. Correlation between structure and resistivity variations of the live human skull. *IEEE Transactions on Biomedical Engineering*, 55(9):2286–2292, 2008.

1024. L.-F. Tanguay, H. Gagnon, and R. Guardo. Comparison of applied and induced current electrical impedance tomography. *IEEE transactions on biomedical engineering*, 54(9):1643–1649, 2007.

1025. H. Tapp, A. Peyton, E. Kemsley, and R. Wilson. Chemical engineering applications of electrical process tomography. *Sensors and Actuators B: Chemical*, 92(1-2):17–24, July 2003.

1026. H. S. Tapp and A. J. Peyton. A state of the art review of electromagnetic tomography. In *Proc. 3rd World Congress on Industrial Process Tomography*, pages 340–346, Sept. 2003.

1027. A. Tarantola. *Inverse Problem Theory*. Elsevier, 1987.

1028. P. P. Tarjan and R. McFee. Electrodeless measurements of the effective resistivity of the human torso and head by magnetic induction. *IEEE Transactions on Biomedical Engineering*, BME-15(4):266–278, Oct. 1968.

1029. I. Tarotin, K. Aristovich, and D. Holder. Effect of dispersion in nerve on compound action potential and impedance change: a modelling study. *Physiological Measurement*, 40(3):034001, Mar. 2019.

1030. I. Tarotin, K. Aristovich, and D. Holder. Model of impedance changes in unmyelinated nerve fibers. *IEEE Transactions on Biomedical Engineering*, 66(2):471–484, Feb. 2019.

1031. D. Thomas, J. Siddall-Allum, I. Sutherland, and R. Beard. Correction of the non-uniform spatial sensitivity of electrical impedance tomography images. *Physiological measurement*, 15(2A):A147, 1994.

1032. F. Thuerk, A. Waldmann, K. H. Wodack, M. F. Grässler, S. Nishimoto, C. J. Trepte, D. Reuter, S. H. Böhm, S. Kampusch, and E. Kaniusas. Hypertonic saline injection to detect aorta in porcine EIT. In *Proceedings of the 17th International Conference on Electrical Impedance Tomography*, page 121, 06 2016.

1033. F. Thürk, S. Boehme, D. Mudrak, S. Kampusch, A. Wielandner, H. Prosch, C. Braun, F. P. R. Toemboel, J. Hofmanninger, and E. Kaniusas. Effects of individualized electrical impedance tomography and image reconstruction settings upon the assessment of regional ventilation distribution: Comparison to 4-dimensional computed tomography in a porcine model. *PLOS ONE*, 12(8):e0182215, Aug. 2017.

1034. F. Thürk, A. D. Waldmann, K. H. Wodack, C. J. Trepte, D. Reuter, S. Kampusch, and E. Kaniusas. Evaluation of reconstruction parameters of electrical impedance tomography on aorta detection during saline bolus injection. *Current Directions in Biomedical Engineering*, 2(1):394, 2016.

1035. G. Y. Tian, A. Al-Qubaa, and J. Wilson. Design of an electromagnetic imaging system for weapon detection based on GMR sensor arrays. *Sensors and Actuators A: Physical*, 174:75–84, Feb. 2012.

1036. A. Tidswell, A. P. Bagshaw, D. S. Holder, R. J. Yerworth, L. Eadie, S. Murray, L. Morgan, and R. Bayford. A comparison of headnet electrode arrays for electrical impedance tomography of the human head. *Physiological measurement*, 24(2):527, 2003.

1037. A. Tidswell, A. Gibson, R. Bayford, and D. S. Holder. Validation of a 3d reconstruction algorithm for EIT of human brain function in a realistic head-shaped tank. *Physiological measurement*, 22(1):177, 2001.

1038. T. Tidswell, A. Gibson, R. H. Bayford, and D. S. Holder. Three-dimensional electrical impedance tomography of human brain activity. *NeuroImage*, 13(2):283–294, 2001.

1039. A. N. Tikhonov. Solution of incorrectly formulated problems and the regularization method. *Soviet Math Dokl*, 4:1035–1038, 1963.

1040. D. D. G. Tingay, A. A. D. Waldmann, I. I. Frerichs, S. S. Ranganathan, and A. Adler. Electrical impedance tomography can identify ventilation and perfusion defects: A neonatal case. *American Journal of Respiratory and Critical Care Medicine*, 199(3):384–386, Feb. 2019.

1041. D. G. Tingay, R. Bhatia, G. M. Schmölzer, M. J. Wallace, V. A. Zahra, and P. G. Davis. Effect of sustained inflation vs. stepwise PEEP strategy at birth on gas exchange and lung mechanics in preterm lambs. *Pediatric Research*, 75(2):288–294, Nov. 2014.

1042. D. G. Tingay, O. Farrell, J. Thomson, E. J. Perkins, P. M. Pereira-Fantini, A. D. Waldmann, C. Rüegger, A. Adler, P. G. Davis, and I. Frerichs. Imaging the respiratory transition at birth: Unravelling the complexities of the first breaths of life. *American Journal of Respiratory and Critical Care Medicine*, 2021.

1043. D. G. Tingay, G. R. Polglase, R. Bhatia, C. A. Berry, R. J. Kopotic, C. P. Kopotic, Y. Song, E. Szyld, A. H. Jobe, and J. J. Pillow. Pressure-limited sustained inflation vs. gradual tidal inflations for resuscitation in preterm lambs. *Journal of Applied Physiology*, 118(7):890–897, Apr. 2015.

1044. D. G. Tingay, A. Togo, P. M. Pereira-Fantini, M. Miedema, K. E. McCall, E. J. Perkins, J. Thomson, G. Dowse, M. Sourial, R. L. Dellacà, P. G. Davis, and P. A. Dargaville. Aeration strategy at birth influences the physiological response to surfactant in preterm lambs. *Archives of Disease in Childhood - Fetal and Neonatal Edition*, 104(6):F587–F593, Feb. 2019.

1045. D. G. Tingay, M. J. Wallace, R. Bhatia, G. M. Schmölzer, V. A. Zahra, M. J. Dolan, S. B. Hooper, and P. G. Davis. Surfactant before the first inflation at birth improves spatial distribution of ventilation and reduces lung injury in preterm lambs. *Journal of Applied Physiology*, 116(3):251–258, Feb. 2014.

1046. K. Tomkiewicz, A. Plaskowski, M. S. Beck, and M. Byars. Testing of the failure of solid rocket propellant with tomography methods. In *Proc. 1st World Congress on Industrial Process Tomography*, pages 249–255, Apr. 1999.

1047. A. Torosyan and A. Willson. Exact analysis of DDS spurs and SNR due to phase truncation and arbitrary phase-to-amplitude errors. In *Proceedings of the 2005 IEEE International Frequency Control Symposium and Exposition*, pages 50–58, August 2005.

1048. J. C. Tozer, R. H. Ireland, D. C. Barber, and A. T. Barker. Magnetic impedance tomography. In *Proc of 10th Int. Conf. on Electrical Bioimpedance*, pages 369–372, Apr. 1998.

1049. A. Trakic, N. Eskandarnia, B. K. Li, E. Weber, H. Wang, and S. Crozier. Rotational magnetic induction tomography. *Measurement Science and Technology*, 23(2):025402, Jan. 2012.

1050. L. Traser, J. Knab, M. Echternach, H. Fuhrer, B. Richter, H. Buerkle, and S. Schumann. Regional ventilation during phonation in professional male and female singers. *Respiratory Physiology & Neurobiology*, 239:26–33, May 2017.

1051. C. J. C. Trepte, C. Phillips, J. Solà, A. Adler, B. Saugel, S. Haas, S. H. Bohm, and D. A. Reuter. Electrical impedance tomography for non-invasive assessment of stroke volume variation in health and experimental lung injury. *British journal of anaesthesia*, 118(1):68–76, 2017.

1052. C. J. C. Trepte, C. R. Phillips, J. Solà, A. Adler, S. A. Haas, M. Rapin, S. H. Böhm, and D. A. Reuter. Electrical impedance tomography (EIT) for quantification of pulmonary edema in acute lung injury. *Critical care (London, England)*, 20:18, 2016.

1053. I. F. Triantis, A. Demosthenous, M. Rahal, H. Hong, and R. Bayford. A multi-frequency bioimpedance measurement asic for electrical impedance tomography. In *2011 Proceedings of the ESSCIRC (ESSCIRC)*, pages 331–334. IEEE, 2011.

1054. O. Trokhanova, Y. Chijova, M. Okhapkin, A. Korjenevsky, and T. Tuykin. Possibilities of electrical impedance tomography in gynecology. In *Journal of Physics: Conference Series*, volume 434, page 012038, Apr. 2013.

1055. O. Trokhanova, M. Okhapkin, and A. Korjenevsky. Dual-frequency electrical impedance mammography for the diagnosis of non-malignant breast disease. *Physiological measurement*, 29(6):S331, 2008.

1056. B. J. Tromberg, B. W. Pogue, K. D. Paulsen, A. G. Yodh, D. A. Boas, and A. E. Cerussi. Assessing the future of diffuse optical imaging technologies for breast cancer management. *Medical physics*, 35(6Part1):2443–2451, 2008.

1057. C. Tso, O. Kuras, P. B. Wilkinson, S. Uhlemann, J. E. Chambers, P. Meldrum, J. Graham, E. Sherlock, and A. Binley. Improved characterisation and modelling of measurement errors in electrical resistivity tomography (ERT) surveys. *J. Applied Geophys.*, 146(1):103–119, 2017.

1058. M. Tšoeu and M. R. Inggs. Fully parallel electrical impedance tomography using code division multiplexing. *IEEE Transactions on Biomedical Circuits and Systems*, 10(3):556–566, 2015.

1059. G. Tsokas, P. Tsourlos, J. Kim, C. Papazachos, G. Vargemezis, and P. Bogiatzis. Assessing the condition of the rock mass over the Tunnel of Eupalinus in Samos (Greece) using both conventional geophysical methods and surface to tunnel electrical resistivity tomography. *Archaeological Prospection*, 21(4):277–291, 2014.

1060. A. S. Tucker, E. A. Ross, J. Paugh-Miller, and R. J. Sadleir. In vivo quantification of accumulating abdominal fluid using an electrical impedance tomography hemiarray. *Physiological measurement*, 32(2):151, 2010.

1061. S. Uhlemann, J. E. Chambers, P. B. Wilkinson, H. Maurer, A. Merritt, P. Meldrum, O. Kuras, D. Gunn, A. Smith, and T. Dijkstra. Four-dimensional imaging of moisture dynamics during landslide reactivation. *Journal of Geophysical Research: Earth Surface*, 122(1):398–418, 2017.

1062. S. Uhlemann, J. Sorensen, A. House, P. B. Wilkinson, C. Roberts, D. Gooddy, A. Binley, and J. E. Chambers. Integrated time-lapse geoelectrical imaging of wetland hydrological processes. *Water Resources Research*, 52(3):1607–1625, 2016.

1063. S. Uhlemann, P. B. Wilkinson, H. Maurer, F. Wagner, T. Johnson, and J. Chambers. Optimized survey design for electrical resistivity tomography: combined optimization of measurement configuration and electrode placement. *Geophys. J. Int.*, 214(1):108–121, 2018.

1064. A. Ukere, A. März, K. Wodack, C. Trepte, A. Haese, A. Waldmann, S. Böhm, and D. Reuter. Perioperative assessment of regional ventilation during changing body positions and ventilation conditions by electrical impedance tomography. *British Journal of Anaesthesia*, 117(2):228–235, Aug. 2016.

1065. B. Ulker and N. Gencer. Implementation of a data acquisition system for contactless conductivity imaging. *IEEE Engineering in Medicine and Biology Magazine*, 21(5):152–155, Sept. 2002.

1066. G. Vainikko. Fast solvers of the Lippmann-Schwinger equation. In *Direct and inverse problems of mathematical physics (Newark, DE, 1997)*, volume 5 of *Int. Soc. Anal. Appl. Comput.*, pages 423–440. Kluwer Acad. Publ., Dordrecht, 2000.

1067. M. E. V. Valkenburg. *Analog Filter Design*. Holt, Rinehart and Winston, 1982.

1068. P. S. van der Burg, F. H. de Jongh, M. Miedema, I. Frerichs, and A. H. van Kaam. The effect of prolonged lateral positioning during routine care on regional lung volume changes in preterm infants. *Pediatric Pulmonology*, 51(3):280–285, Aug. 2016.

1069. P. van der Zee, P. Somhorst, H. Endeman, and D. Gommers. Electrical impedance tomography for positive end-expiratory pressure titration in COVID-19–related acute respiratory distress syndrome. *American Journal of Respiratory and Critical Care Medicine*, 202(2):280–284, July 2020.

1070. A. van Harreveld, T. Murphy, and K. Nobel. Specific impedance of rabbit's cortical tissue. *American Journal of Physiology-Legacy Content*, 205(1):203–207, 1963.

1071. A. Van Harreveld and S. Ochs. Cerebral impedance changes after circulatory arrest. *American Journal of Physiology-Legacy Content*, 187(1):180–192, 1956.

1072. A. van Harreveld and J. Schadé. Changes in the electrical conductivity of cerebral cortex during seizure activity. *Experimental neurology*, 5(5):383–400, 1962.

1073. E. E. Van Houten, M. M. Doyley, F. E. Kennedy, J. B. Weaver, and K. D. Paulsen. Initial in vivo experience with steady-state subzone-based mr elastography of the human breast. *Journal of Magnetic Resonance Imaging: An Official Journal of the International Society for Magnetic Resonance in Medicine*, 17(1):72–85, 2003.

1074. H. Vargas, A. Hötker, D. Goldman, C. Moskowitz, T. Gondo, K. Matsumoto, B. Ehdaie, S. Woo, S. Fine, V. Reuter, et al. Updated prostate imaging reporting and data system (pirads v2) recommendations for the detection of clinically significant prostate cancer using multiparametric mri: critical evaluation using whole-mount pathology as standard of reference. *European radiology*, 26(6):1606–1612, 2016.

1075. M. Vauhkonen. *Electrical Impedance Tomography and Prior Information*. PhD thesis, University of Kuopio, 1997.

1076. M. Vauhkonen, M. Hamsch, and C. H. Igney. A measurement system and image reconstruction in magnetic induction tomography. *Physiological Measurement*, 29(6):S445–S454, June 2008.

1077. M. Vauhkonen, P. A. Karjalainen, and J. P. Kaipio. A kalman filter approach to track fast impedance changes in electrical impedance tomography. *IEEE Transactions on Biomedical Engineering*, 45(4):486–493, 1998.

1078. M. Vauhkonen, W. R. B. Lionheart, L. M. Heikkinen, P. J. Vauhkonen, and J. P. Kaipio. A MATLAB package for the EIDORS project to reconstruct two-dimensional EIT images. *Physiological Measurement*, 22(1):107–111, Feb. 2001.

1079. M. Vauhkonen, D. Vadasz, P. A. Karjalainen, E. Somersalo, and J. P. Kaipio. Tikhonov regularization and prior information in electrical impedance tomography. *IEEE Transactions on Medical Imaging*, 17(2):285–293, 1998.

1080. P. J. Vauhkonen. *Second order and Infinite Elements in Three-Dimensional Electrical Impedance Tomography*. PhD thesis, Department of Applied Physics, University of Kuopio, Finland, 1999.

1081. P. J. Vauhkonen, M. Vauhkonen, T. Savolainen, and J. P. Kaipio. Static three dimensional electrical impedance tomography. In *Proceedings of ICEBI'98*, pages 125–135, Spain, 41 PaiviInfVauhkonen

PJ, Vauhkonen M, Kaipio JP, 2000, Errors due to the truncation of the computational domain in static three-dimensional electrical impedance tomography. Physiol Meas, 21, 1998. Barcelona.

1082. R. Velluti, K. Klivington, and R. Galambos. Evoked resistance shifts in subcortical nuclei. *Biosystems*, 2(2):78–80, 1968.

1083. C. Verdet, Y. Anguy, C. Sirieix, R. Clément, and C. Gaborieau. On the effect of electrode finiteness in small-scale electrical resistivity imaging. *Geophysics*, 83(6):1ND–Z38, 2018.

1084. J.-L. Vincent, A. Rhodes, A. Perel, G. S. Martin, G. Della Rocca, B. Vallet, M. R. Pinsky, C. K. Hofer, J.-L. Teboul, W.-P. de Boode, S. Scolletta, A. Vieillard-Baron, D. de Backer, K. R. Walley, M. Maggiorini, and M. Singer. Clinical review: Update on hemodynamic monitoring–a consensus of 16. *Critical care*, 15(4):229, 2011.

1085. A. Vinciguerra, M. Aleardi, and P. Costantini. Full-waveform inversion of complex resistivity IP spectra: Sensitivity analysis and inversion tests using local and global optimization strategies on synthetic datasets. *Near Surface Geophysics*, 17(2):109–125, 2019.

1086. K. R. Visser. Electric conductivity of stationary and flowing human blood at low frequencies. *Medical & Biological Engineering & Computing*, 30(6):636–640, Nov. 1992.

1087. C. R. Vogel. *Computational Methods for Inverse Problems*. Society for Industrial and Applied Mathematics, Jan. 2002.

1088. B. Vogt, K. Deuß, V. Hennig, Z. Zhao, I. Lautenschläger, N. Weiler, and I. Frerichs. Regional lung function in nonsmokers and asymptomatic current and former smokers. *ERJ Open Research*, 5(3):00240–2018, July 2019.

1089. B. Vogt, S. Löhr, Z. Zhao, C. Falkenberg, T. Ankermann, N. Weiler, and I. Frerichs. Regional lung function testing in children using electrical impedance tomography. *Pediatric Pulmonology*, 53(3):293–301, Nov. 2018.

1090. B. Vogt, L. Mendes, I. Chouvarda, E. Perantoni, E. Kaimakamis, T. Becher, N. Weiler, V. Tsara, R. P. Paiva, N. Maglaveras, and I. Frerichs. Influence of torso and arm positions on chest examinations by electrical impedance tomography. *Physiological Measurement*, 37(6):904–921, May 2016.

1091. B. Vogt, S. Pulletz, G. Elke, Z. Zhao, P. Zabel, N. Weiler, and I. Frerichs. Spatial and temporal heterogeneity of regional lung ventilation determined by electrical impedance tomography during pulmonary function testing. *Journal of Applied Physiology*, 113(7):1154–1161, Oct. 2012.

1092. B. Vogt, Z. Zhao, P. Zabel, N. Weiler, and I. Frerichs. Regional lung response to bronchodilator reversibility testing determined by electrical impedance tomography in chronic obstructive pulmonary disease. *American Journal of Physiology-Lung Cellular and Molecular Physiology*, 311(1):L8–L19, July 2016.

1093. T. Voigt, H. Homann, U. Katscher, and O. Doessel. Patient-individual local sar determination: in vivo measurements and numerical validation. *Magnetic resonance in medicine*, 68(4):1117–1126, 2012.

1094. T. Voigt, U. Katscher, and O. Doessel. Quantitative conductivity and permittivity imaging of the human brain using electric properties tomography. *Magnetic Resonance in Medicine*, 66(2):456–466, 2011.

1095. A. Volkov, S. Paula, and D. Deamer. Two mechanisms of permeation of small neutral molecules and hydrated ions across phospholipid bilayers. *Bioelectrochemistry and bioenergetics*, 42(2):153–160, 1997.

1096. J. Vollmer-Haase, H. W. Folkerts, C. G. Haase, M. Deppe, and E. B. Ringelstein. Cerebral hemodynamics during electrically induced seizures. *Neuroreport*, 9(3):407–410, 1998.

1097. M. Vonach, B. Marson, M. Yun, J. Cardoso, S. Ourselin, and D. Holder. A method for rapid production of subject specific finite element meshes for electrical impedance tomography of the human head. *Physiological measurement*, 33(5):801, 2012.

1098. A. N. Vongerichten, G. S. dos Santos, K. Aristovich, J. Avery, A. McEvoy, M. Walker, and D. S. Holder. Characterisation and imaging of cortical impedance changes during interictal and ictal activity in the anaesthetised rat. *NeuroImage*, 124:813–823, Jan. 2016.

1099. A. Vonk Noordegraaf, T. J. C. Faes, A. Janse, J. T. Marcus, R. M. Heethaar, P. E. Postmus, and P. M. J. M. de Vries. Improvement of cardiac imaging in electrical impedance tomography by means of a new electrode configuration. *Physiological Measurement*, 17(3):179–188, aug 1996.

1100. A. Vonk Noordegraaf, A. Janse, J. T. Marcus, J. G. F. Bronzwaer, P. E. Postmus, T. J. C. Faes, and P. M. J. M. de Vries. Determination of stroke volume by means of electrical impedance tomography. *Physiological Measurement*, 21(2):285–293, may 2000.

1101. A. Vonk Noordegraaf, P. W. A. Kunst, A. Janse, R. A. Smulders, R. M. Heethaar, P. E. Postmus, T. J. C. Faes, and P. M. J. M. de Vries. Validity and reproducibility of electrical impedance tomography for measurement of calf blood flow in healthy subjects. *Medical and Biological Engineering and Computing*, 35(2):107–112, Mar 1997.

1102. J. G. Wade, K. Senior, and S. Seubert. Convergence of derivative approximations in the inverse conductivity problem. Technical Report No. 96-14, Bowling Green State University, 1996.

1103. F. M. Wagner, P. Bergmann, C. Rücker, B. Wiese, T. Labitzke, C. Schmidt-Hattenberger, and H. Maurer. Impact and mitigation of borehole related effects in permanent crosshole resistivity imaging: An example from the Ketzin CO_2 storage site. *J. Applied Geophys.*, 123(1):102–111, 2015.

1104. F. M. Wagner, T. Günther, C. Schmidt-Hattenberger, and H. Maurer. Constructive optimization of electrode locations for target-focused resistivity monitoring. *Geophysics*, 80(2):1MA–Z50, 2015.

1105. F. M. Wagner and B. Wiese. Fully coupled inversion on a multi-physical reservoir model — part II: The Ketzin CO_2 storage reservoir. *International Journal of Greenhouse Gas Control*, 75(1):273–281, 2018.

1106. A. Waldmann, C. Meira, U. Auer, S. Böhm, C. Braun, and S. Böhm. Finite element data base for animals. In *Proc 17th Int conf Biomedical Applications of EIT (EIT 2016)*, 2016.

1107. A. Waldmann, C. Meira, S. Böhm, M. Dennler, and M. Mosing. Construction of a robust beagle model for EIT applications. In *Proc 17th Int conf Biomedical Applications of EIT (EIT 2016)*, 2016.

1108. A. J. Walker, J. Ruzevick, A. A. Malayeri, D. Rigamonti, M. Lim, K. J. Redmond, and L. Kleinberg. Postradiation imaging changes in the cns: how can we differentiate between treatment effect and disease progression? *Future Oncology*, 10(7):1277–1297, 2014.

1109. Y. Wan, A. Borsic, A. Hartov, and R. Halter. Incorporating a biopsy needle as an electrode in transrectal electrical impedance imaging. In *2012 Annual International Conference of the IEEE Engineering in Medicine and Biology Society*, pages 6220–6223. IEEE, 2012.

1110. Y. Wan, A. Borsic, J. Heaney, J. Seigne, A. Schned, M. Baker, S. Wason, A. Hartov, and R. Halter. Transrectal electrical impedance tomography of the prostate: spatially coregistered pathological findings for prostate cancer detection. *Medical physics*, 40(6Part1):063102, 2013.

1111. Y. Wan, R. Halter, A. Borsic, P. Manwaring, A. Hartov, and K. Paulsen. Sensitivity study of an ultrasound coupled transrectal electrical impedance tomography system for prostate imaging. *Physiological measurement*, 31(8):S17, 2010.

1112. C. Wang, H. He, Z. Cui, Q. Cao, P. Zou, and H. Wang. A novel EMT system based on TMR sensors for reconstruction of permeability distribution. *Measurement Science and Technology*, 29(10):104008, Sept. 2018.

1113. J.-R. Wang, B.-U. Sun, H.-X. Wang, S. Pang, X. Xu, and Q. Sun. Experimental study of dielectric properties of human lung tissue in vitro. *Journal of Medical and Biological Engineering*, 34:598–604, 2014.

1114. J.-Y. Wang, T. Healey, A. Barker, B. Brown, C. Monk, and D. Anumba. Magnetic induction spectroscopy (MIS)—probe design for cervical tissue measurements. *Physiological Measurement*, 38(5):729–744, Apr. 2017.

1115. L. Wang and R. Patterson. Multiple sources of the impedance cardiogram based on 3-d finite difference human thorax models. *IEEE Transactions on Biomedical Engineering*, 42(2):141–148, 1 1995.

1116. M. Wang, S. Johnstone, W. J. N. Pritchard, and T. A. York. Modelling and mapping electrical resistance changes due to hearth erosion in a 'cold' model of a blast furnace. In *Proc. 1st World Congress on Industrial Process Tomography*, pages 161–166, Apr. 1999.

1117. S. Wang, D. Geldart, M. Beck, and T. Dyakowski. A behaviour of a catalyst powder flowing down in a dipleg. *Chemical Engineering Journal*, 77(1-2):51–56, Apr. 2000.

1118. S. Wang, T. Kalscheuer, M. Bastani, A. Malehmir, L. Pedersen, T. Dahlin, and N. Meqbel. Joint inversion of lake-floor electrical resistivity tomography and boat-towed radio-magnetotelluric data illustrated on synthetic data and an application from the äspö Hard Rock Laboratory site, Sweden. *Geophys. J. Int.*, 213(1):511–533, 2018.

1119. S. Wang and W. Yin. Monitoring cleaning-in-place by electrical resistance tomography with dynamic references. In *2016 IEEE International Conference on Imaging Systems and Techniques (IST)*, pages 300–305. IEEE, 2016.

1120. Y. Wang, P. Spincemaille, Z. Liu, A. Dimov, K. Deh, J. Li, Y. Zhang, Y. Yao, K. M. Gillen, A. H. Wilman, A. Gupta, A. J. Tsiouris, I. Kovanlikaya, G. C.-Y. Chiang, J. W. Weinsaft, L. Tanenbaum, W. Chen, W. Zhu, S. Chang, M. Lou, B. H. Kopell, M. G. Kaplitt, D. Devos, T. Hirai, X. Huang, Y. Korogi,

A. Shtilbans, G.-H. Jahng, D. Pelletier, S. A. Gauthier, D. Pitt, A. I. Bush, G. M. Brittenham, and M. R. Prince. Clinical quantitative susceptibility mapping (QSM): Biometal imaging and its emerging roles in patient care. *Journal of Magnetic Resonance Imaging*, 46(4):951–971, Mar. 2017.

1121. R. Ward, M. Joseph, A. Langley, S. Taylor, and J. C. Watson. Magnetic induction tomography of objects for security applications. In *Emerging Imaging and Sensing Technologies for Security and Defence II*, volume 10438, page 104380G. International Society for Optics and Photonics, 2017.

1122. R. C. Waterfall, R. He, P. Wolanski, and Z. Gut. Monitoring flame position and stability in combustion cans using ect. In *Proc. 1st World Congress on Industrial Process Tomography*, pages 35–38, Apr. 1999.

1123. S. Watson. *Instrumentation for low-conductivity Magnetic Induction Tomography*. PhD thesis, University of South Wales, Cardiff, United Kingdom, 2009.

1124. S. Watson, A. Morris, R. J. Williams, H. Griffiths, and W. Gough. A primary field compensation scheme for planar array magnetic induction tomography. *Physiological Measurement*, 25(1):271–279, Feb. 2004.

1125. S. Watson, H. C. Wee, H. Griffiths, and R. J. Williams. A highly phase-stable differential detector amplifier for magnetic induction tomography. *Physiological Measurement*, 32(7):917–926, June 2011.

1126. S. Watson, R. Williams, A. Morris, W. Gough, and H. Griffiths. The Cardiff magnetic induction tomography system. *Proc. Int. Fed. Med. Biol. Eng. EMBEC02*, 3:116–7, 2002.

1127. S. Watson, R. J. Williams, W. Gough, and H. Griffiths. A magnetic induction tomography system for samples with conductivities below 10 $\mathrm{S\,m^{-1}}$. *Measurement Science and Technology*, 19(4):045501, Feb. 2008.

1128. S. Watson, R. J. Williams, W. Gough, A. Morris, and H. Griffiths. Phase measurement in biomedical magnetic induction tomography. In *Proc 2nd World Congress on Process Tomography*, pages 517–524, Aug. 2001.

1129. S. Watson, R. J. Williams, H. Griffiths, W. Gough, and A. Morris. Frequency downconversion and phase noise in MIT. *Physiological Measurement*, 23(1):189–194, Jan. 2002.

1130. S. Watson, R. J. Williams, H. Griffiths, W. Gough, and A. Morris. Magnetic induction tomography: phase versus vector-voltmeter measurement techniques. *Physiological Measurement*, 24(2):555–564, Apr. 2003.

1131. M. Waxman and L. Smits. Electrical conductivities in oil-bearing shaly sands. *SPE Journal*, 8(2):107–122, June 1968.

1132. H.-Y. Wei and M. Soleimani. Hardware and software design for a national instrument-based magnetic induction tomography system for prospective biomedical applications. *Physiological Measurement*, 33(5):863–879, Apr. 2012.

1133. H.-Y. Wei and M. Soleimani. Theoretical and experimental evaluation of rotational magnetic induction tomography. *IEEE Transactions on Instrumentation and Measurement*, 61(12):3324–3331, 2012.

1134. K. Wei, C. Qiu, D. McCormack, and K. Primrose. Its densitometer: an electrical resistance tomography based densitometer. In *Proc. 8th World Congress on Industrial Process Tomography (WCIPT8)*, Sept. 2016.

1135. K. Wei, C.-H. Qiu, and K. Primrose. Super-sensing technology: Industrial applications and future challenges of electrical tomography. *Philosophical Transactions of the Royal Society A: Mathematical, Physical and Engineering Sciences*, 374(2070):20150328, 2016.

1136. J. C. Weinreb, J. O. Barentsz, P. L. Choyke, F. Cornud, M. A. Haider, K. J. Macura, D. Margolis, M. D. Schnall, F. Shtern, C. M. Tempany, et al. Pi-rads prostate imaging–reporting and data system: 2015, version 2. *European urology*, 69(1):16–40, 2016.

1137. R. M. West, R. G. Aykroyd, S. Meng, and R. A. Williams. Markov chain monte carlo techniques and spatial–temporal modelling for medical EIT. *Physiological Measurement*, 25(1):181–194, Feb. 2004.

1138. R. M. West, D. M. Scott, G. Sunshine, J. Kostuch, L. Heikkinen, M. Vauhkonen, B. S. Hoyle, H. I. Schlaberg, R. Hou, and R. A. Williams. In situ imaging of paste extrusion using electrical impedance tomography. *Measurement Science and Technology*, 13(12):1890–1897, Nov. 2002.

1139. M. Wettstein, L. Radlinger, and T. Riedel. Effect of different breathing aids on ventilation distribution in adults with cystic fibrosis. *PLoS ONE*, 9(9):e106591, Sept. 2014.

1140. R. B. White. Using electrical capacitance tomography to monitor gas voids in a packed bed of solids. In *Proc 2nd World Congress on Process Tomography*, pages 307–314, 2001.

1141. J. Whiteley, J. Chambers, S. Uhlemann, P. B. Wilkinson, and J. Kendall. Geophysical monitoring of moisture-induced landslides: A review. *Review of Geophysics*, 57(1):106–145, 2019.

1142. H. Wi, A. L. McEwan, V. Lam, H. J. Kim, E. J. Woo, and T. I. Oh. Real-time conductivity imaging of temperature and tissue property changes during radiofrequency ablation: An ex vivo model using weighted frequency difference. *Bioelectromagnetics*, 36(4):277–286, 2015.

1143. H. Wi, H. Sohal, A. L. McEwan, E. J. Woo, and T. I. Oh. Multi-frequency electrical impedance tomography system with automatic self-calibration for long-term monitoring. *IEEE Transactions on Biomedical Circuits and Systems*, 8(1):119–128, 2014.

1144. A. Wickenbrock, F. Tricot, and F. Renzoni. Magnetic induction measurements using an all-optical 87 Rb atomic magnetometer. *Applied Physics Letters*, 103(24):243503, Dec. 2013.

1145. M. Wiegel, S. Hammermüller, H. Wrigge, and A. W. Reske. Electrical impedance tomography visualizes impaired ventilation due to hemidiaphragmatic paresis after interscalene brachial plexus block. *Anesthesiology*, 125(4):807–807, Oct. 2016.

1146. P. B. Wilkinson, J. Chambers, O. Kuras, P. Meldrum, and D. Gunn. Long-term time-lapse geoelectrical monitoring. *First Break*, 29(8):77–84, 2011.

1147. P. B. Wilkinson, J. Chambers, P. Meldrum, R. Ogilvy, and S. Caunt. Optimization of array configurations and panel combinations for the detection and imaging of abandoned mineshafts using 3D cross-hole electrical resistivity tomography. *Journal of Environmental & Engineering Geophysics*, 11(3):161–224, 2006.

1148. P. B. Wilkinson, J. E. Chambers, M. Lelliot, G. Wealthall, and R. Ogilvy. Extreme sensitivity of cross-hole electrical resistivity tomography measurements to geometric errors. *Geophys. J. Int.*, 173(1):49–62, Apr. 2008.

1149. P. B. Wilkinson, J. E. Chambers, S. Uhlemann, P. Meldrum, A. Smith, N. Dixon, and M. H. Loke. Reconstruction of landslide movements by inversion of 4-D electrical resistivity tomography monitoring data. *Geophysical Research Letters*, 43(3):1166–1174, Feb. 2016.

1150. P. B. Wilkinson, M. H. Loke, P. I. Meldrum, J. E. Chambers, O. Kuras, D. A. Gunn, and R. D. Ogilvy. Practical aspects of applied optimized survey design for electrical resistivity tomography. *Geophys. J. Int.*, 189(1):428–440, Apr. 2012.

1151. P. B. Wilkinson, P. Meldrum, O. Kuras, J. Chambers, S. Holyoake, and R. Ogilvy. High-resolution electrical resistivity tomography monitoring of a tracer test in a confined aquifer. *J. Applied Geophys.*, 70(4):268–276, 2010.

1152. P. B. Wilkinson, S. Uhlemann, P. Meldrum, J. Chambers, S. Carrière, L. Oxby, and M. H. Loke. Adaptive time-lapse optimized survey design for electrical resistivity tomography monitoring. *Geophys. J. Int.*, 203(1):755–766, 2015.

1153. R. Williams, S. Luke, K. Ostrowski, and M. Bennett. Measurement of bulk particulates on belt conveyor using dielectric tomography. *Chemical Engineering Journal*, 77(1-2):57–63, Apr. 2000.

1154. R. A. Williams and M. S. Beck. *Process Tomography: Principles, Techniques and Applications*. Butterworth-Heinemann: Oxford, UK, 1995.

1155. R. A. Williams and T. A. York. Microtomographic sensors for microfactories. In *Proc Int Conf on Process Innovation and Intensification*, Oct. 1998.

1156. A. J. Wilson, P. Milnes, A. R. Waterworth, R. H. Smallwood, and B. H. Brown. Mk3.5: a modular, multi-frequency successor to the mk3a EIS/EIT system. *Physiological Measurement*, 22(1):49–54, Feb. 2001.

1157. A. Witkowska-Wrobel. *Imaging Physiological Brain Activity and Epilepsy with Electrical Impedance Tomography*. PhD thesis, University College London, 2020.

1158. A. Witkowska-Wrobel, K. Aristovich, M. Faulkner, J. Avery, and D. Holder. Feasibility of imaging epileptic seizure onset with EIT and depth electrodes. *NeuroImage*, 173:311–321, 2018.

1159. K. H. Wodack, S. Buehler, S. A. Nishimoto, M. F. Graessler, C. R. Behem, A. D. Waldmann, B. Mueller, S. H. Böhm, E. Kaniusas, F. Thürk, A. Maerz, C. J. C. Trepte, and D. A. Reuter. Detection of thoracic vascular structures by electrical impedance tomography: a systematic assessment of prominence peak analysis of impedance changes. *Physiological measurement*, 39(2):024002, 2018.

1160. C. F. Wojslaw and E. A. Moustakas. *Operational Amplifiers*. Wiley, New York, 1986.

1161. G. K. Wolf, B. Grychtol, I. Frerichs, D. Zurakowski, and J. H. Arnold. Regional lung volume changes during high-frequency oscillatory ventilation. *Pediatric Critical Care Medicine*, 11(5):610–615, Sept. 2010.

1162. T. Wondrak, U. Hampel, M. Ratajczak, I. Glavinic, F. Stefani, S. Eckert, D. van der Plas, P. Pennerstorfer, I. Muttakin, M. Soleimani, S. Abouelazayem, J. Hlava, A. Blishchik, and S. Kenjeres. Real-time control of the mould flow in a model of continuous casting in frame of the TOMOCON project. *IOP Conference Series: Materials Science and Engineering*, 424:012003, Oct. 2018.

1163. E. J. Woo, P. Hua, J. G. Webster, and W. J. Tompkins. Measuring lung resistivity using electrical impedance tomography. *IEEE transactions on biomedical engineering*, 39(7):756–760, 1992.

1164. E. J. Woo and M. Kranjc. Principles and use of magnetic resonance electrical impedance tomography in tissue electroporation. *Handbook of Electroporation*, pages 1–18, 2016.

1165. E. J. Woo and J. K. Seo. Magnetic resonance electrical impedance tomography (MREIT) for high-resolution conductivity imaging. *Physiological measurement*, 29(10):R1, 2008.

1166. A. J. Woolcock and P. T. Macklem. Mechanical factors influencing collateral ventilation in human, dog, and pig lungs. *Journal of Applied Physiology*, 30(1):99–115, Jan. 1971.

1167. W. R. B. Lionheart. Conformal uniqueness results in anisotropic electrical impedance imaging. *Inverse Problems*, 13:125–134, 1997.

1168. W. R. B. Lionheart, EIT reconstruction algorithms: pitfalls, challenges and recent developments, *Physiol. Meas.* 25:125, 2004

1169. H. Wrigge, J. Zinserling, T. Muders, D. Varelmann, U. Günther, C. von der Groeben, A. Magnusson, G. Hedenstierna, and C. Putensen. Electrical impedance tomography compared with thoracic computed tomography during a slow inflation maneuver in experimental models of lung injury. *Critical Care Medicine*, 36(3):903–909, Mar. 2008.

1170. C. Wu and M. Soleimani. Frequency difference EIT with localization: A potential medical imaging tool during cancer treatment. *IEEE Access*, 7:21870–21878, 2019.

1171. H. Wu, Y. Yang, P.-O. Bagnaninchi, and J. Jia. Calibrated frequency-difference electrical impedance tomography for 3d tissue culture monitoring. *IEEE Sensors Journal*, 19(18):7813–7821, 2019.

1172. H. Wu, W. Zhou, Y. Yang, J. Jia, and P. Bagnaninchi. Exploring the potential of electrical impedance tomography for tissue engineering applications. *Materials*, 11(6):930, 2018.

1173. Y. Wu, D. Jiang, A. Bardill, R. Bayford, and A. Demosthenous. A 122 fps, 1 MHz bandwidth multi-frequency wearable EIT belt featuring novel active electrode architecture for neonatal thorax vital sign monitoring. *IEEE Transactions on Biomedical Circuits and Systems*, 2019.

1174. Y. Wu, D. Jiang, A. Bardill, S. de Gelidi, R. Bayford, and A. Demosthenous. A high frame rate wearable EIT system using active electrode ASICs for lung respiration and heart rate monitoring. *IEEE Transactions on Circuits and Systems-I: Regular Papers*, 65(11), 2018.

1175. Y. Wu, D. Jiang, X. Liu, R. Bayford, and A. Demosthenous. A human–machine interface using electrical impedance tomography for hand prosthesis control. *IEEE transactions on biomedical circuits and systems*, 12(6):1322–1333, 2018.

1176. Z. Xiao, C. Tan, and F. Dong. Multi-frequency difference method for intracranial hemorrhage detection by magnetic induction tomography. *Physiological Measurement*, 39(5):055006, May 2018.

1177. Z. Xu, H. Luo, W. He, C. He, X. Song, and Z. Zahng. A multi-channel magnetic induction tomography measurement system for human brain model imaging. *Physiological Measurement*, 30(6):S175–S186, June 2009.

1178. D. A. Yablonskiy, J. J. Ackerman, and M. E. Raichle. Coupling between changes in human brain temperature and oxidative metabolism during prolonged visual stimulation. *Proceedings of the National Academy of Sciences*, 97(13):7603–7608, 2000.

1179. B. Yang, B. Li, C. Xu, S. Hu, M. Dai, J. Xia, P. Luo, X. Shi, Z. Zhao, X. Dong, et al. Comparison of electrical impedance tomography and intracranial pressure during dehydration treatment of cerebral edema. *NeuroImage: Clinical*, 23:101909, 2019.

1180. B. Yang, X. Shi, M. Dai, C. Xu, F. You, F. Fu, R. Liu, and X. Dong. Real-time imaging of cerebral infarction in rabbits using electrical impedance tomography. *Journal of international medical research*, 42(1):173–183, 2014.

1181. J. Yang, Y. Liu, and X. Wu. 3-D DC resistivity modelling with arbitrary long electrode sources using finite element method on unstructured grids. *Geophys. J. Int.*, 211(2):1162–1176, 2017.

1182. L. Yang, M. Dai, K. Möller, I. Frerichs, A. Adler, F. Fu, and Z. Zhao. Lung regions identified with ct improve the value of global inhomogeneity index measured with electrical impedance tomography. *in Press, Quantitative Imaging in Medicine and Surgery*, 2020.

1183. L. Yang, G. Zhang, J. Song, M. Dai, C. Xu, X. Dong, and F. Fu. Ex-vivo characterization of bioimpedance spectroscopy of normal, ischemic and hemorrhagic rabbit brain tissue at frequencies from 10 hz to 1 MHz. *Sensors*, 16(11):1942, Nov. 2016.

1184. W. Q. Yang, D. M. Spink, T. A. York, and H. McCann. An image-reconstruction algorithm based on landweber's iteration method for electrical-capacitance tomography. *Measurement Science and Technology*, 10(11):1065–1069, Sept. 1999.

1185. Y. Yang, J. Jia, S. Smith, N. Jamil, W. Gamal, and P.-O. Bagnaninchi. A miniature electrical impedance tomography sensor and 3-d image reconstruction for cell imaging. *IEEE Sensors Journal*, 17(2):514–523, 2016.

1186. J. Yao, H. Chen, Z. Xu, J. Huang, J. Li, J. Jia, and H. Wu. Development of a wearable electrical impedance tomographic sensor for gesture recognition with machine learning. *IEEE journal of biomedical and health informatics*, 24(6):1550–1556, 2019.

1187. R. J. Yerworth, R. Bayford, B. Brown, P. Milnes, M. Conway, and D. S. Holder. Electrical impedance tomography spectroscopy (EITS) for human head imaging. *Physiological measurement*, 24(2):477, 2003.

1188. R. J. Yerworth, R. H. Bayford, G. Cusick, M. Conway, and D. S. Holder. Design and performance of the UCLH mark 1b 64 channel electrical impedance tomography (EIT) system, optimized for imaging brain function. *Physiological Measurement*, 23(1):149–158, Jan. 2002.

1189. R. J. Yerworth, I. Frerichs, and R. Bayford. Analysis and compensation for errors in electrical impedance tomography images and ventilation-related measures due to serial data collection. *Journal of clinical monitoring and computing*, 31(5):1093–1101, 2017.

1190. W. Yin and A. J. Peyton. A planar EMT system for the detection of faults on thin metallic plates. *Measurement Science and Technology*, 17(8):2130–2135, July 2006.

1191. T. York. Status of electrical tomography in industrial applications. *Journal of Electronic Imaging*, 10(3):608, July 2001.

1192. T. York, Q. Smit, J. Davidson, and B. Grieve. An intrinsically safe electrical tomography system. In *2003 IEEE International Symposium on Industrial Electronics*, volume 2, pages 946–951. IEEE, 2003.

1193. T. York, L. Sun, C. Gregory, and J. Hatfield. Silicon-based miniature sensor for electrical tomography. *Sensors and Actuators A: Physical*, 110(1-3):213–218, Feb. 2004.

1194. T. J. Yorkey, J. G. Webster, and W. J. Tompkins. Comparing reconstruction algorithms for electrical impedance tomography. *IEEE Transactions on Biomedical Engineering*, BME-34(11):843–852, 1987.

1195. Z. Yu, A. Peyton, and M. Beck. Electromagnetic tomography (EMT), part i: Design of a sensor and a system with a parallel excitation field. In *Proc. European Concerted Action in Process Tomography*, pages 147–154, Mar. 1994.

1196. Z. Yu, A. Peyton, M. Beck, W. Conway, and L. Xu. Imaging system based on electromagnetic tomography (EMT). *Electronics Letters*, 29(7):625–626, 1993.

1197. Y. Yuan, J. Qiang, J. Tang, Z. Ren, and Z. Xiao. 2.5D direct-current resistivity forward modelling and inversion by finite-element—infinite-element coupled method. *Geophys. Prospect.*, 64(3):767–779, 2016.

1198. E. Yuen, D. Vlaev, R. Mann, T. Dyakowski, B. Grieve, and T. A. York. Applying electrical resistance tomography (ert) to solid-fluid filtration processes. In *Proc World Filtration Congress 8*, April 2000.

1199. A. D. Zacharopoulos, S. R. Arridge, O. Dorn, V. Kolehmainen, and J. Sikora. Three-dimensional reconstruction of shape and piecewise constant region values for optical tomography using spherical harmonic parametrization and a boundary element method. *Inverse Problems*, 22(5):1509, 2006.

1200. M. Zadehkoochak, B. Blott, T. K. Hames, and R. F. George. Pulmonary perfusion and ventricular ejection imaging by frequency domain filtering of EIT images. *Clinical Physics and Physiological Measurement*, 13:191–196, 01 1992.

1201. T. Zaehle, S. Rach, and C. S. Herrmann. Transcranial alternating current stimulation enhances individual alpha activity in human eeg. *PloS one*, 5(11):e13766, 2010.

1202. C. Zhang, M. Dai, W. Liu, X. Bai, J. Wu, C. Xu, J. Xia, F. Fu, X. Shi, X. Dong, F. Jin, and F. You. Global and regional degree of obstruction determined by electrical impedance tomography in patients with obstructive ventilatory defect. *PLOS ONE*, 13(12):e0209473, Dec. 2018.

1203. H. Zhang, T. Schneider, C. Wheeler-Kingshott, and D. Alexander. NODDI: practical in vivo neurite orientation dispersion and density imaging of the human brain. *Neuroimage*, 61(4):1000–1016, 2012.

1204. J. Zhang, B. Yang, H. Li, F. Fu, X. Shi, X. Dong, and M. Dai. A novel 3d-printed head phantom with anatomically realistic geometry and continuously varying skull resistivity distribution for electrical impedance tomography. *Scientific reports*, 7(1):1–9, 2017.

1205. Y. Zhang and C. Harrison. Tomo: Wearable, low-cost electrical impedance tomography for hand gesture recognition. In *Proceedings of the 28th Annual ACM Symposium on User Interface Software & Technology*, pages 167–173, 2015.

1206. Z. Zhao, R. Fischer, I. Frerichs, U. Müller-Lisse, and K. Möller. Regional ventilation in cystic fibrosis measured by electrical impedance tomography. *Journal of Cystic Fibrosis*, 11(5):412–418, Sept. 2012.

1207. Z. Zhao, H. He, J. Luo, A. Adler, X. Zhang, R. Liu, Y. Lan, S. Lu, X. Luo, Y. Lei, I. Frerichs, X. Huang, and K. Möller. Detection of pulmonary oedema by electrical impedance tomography: validation of previously proposed approaches in a clinical setting. *Physiological Measurement*, 40(5):054008, June 2019.

1208. Z. Zhao, K. Möller, D. Steinmann, I. Frerichs, and J. Guttmann. Evaluation of an electrical impedance tomography-based global inhomogeneity index for pulmonary ventilation distribution. *Intensive Care Medicine*, 35(11), Aug. 2009.

1209. Z. Zhao, U. Müller-Lisse, I. Frerichs, R. Fischer, and K. Möller. Regional airway obstruction in cystic fibrosis determined by electrical impedance tomography in comparison with high resolution CT. *Physiological Measurement*, 34(11):N107–N114, Oct. 2013.

1210. Z. Zhao, S.-Y. Peng, M.-Y. Chang, Y.-L. Hsu, I. Frerichs, H.-T. Chang, and K. Möller. Spontaneous breathing trials after prolonged mechanical ventilation monitored by electrical impedance tomography: an observational study. *Acta Anaesthesiologica Scandinavica*, 61(9):1166–1175, Aug. 2017.

1211. Z. Zhao, S. Pulletz, I. Frerichs, U. Müller-Lisse, and K. Möller. The EIT-based global inhomogeneity index is highly correlated with regional lung opening in patients with acute respiratory distress syndrome. *BMC Research Notes*, 7(1), Feb. 2014.

1212. Z. Zhao, D. Steinmann, I. Frerichs, J. Guttmann, and K. Möller. PEEP titration guided by ventilation homogeneity: a feasibility study using electrical impedance tomography. *Critical Care*, 14(1):R8, 2010.

1213. Z. Zhao, P.-J. Yun, Y.-L. Kuo, F. Fu, M. Dai, I. Frerichs, and K. Möller. Comparison of different functional EIT approaches to quantify tidal ventilation distribution. *Physiological Measurement*, 39(1):01NT01, Jan. 2018.

1214. L. Zhou, B. Harrach, and J. K. Seo. Monotonicity-based electrical impedance tomography for lung imaging. *Inverse Problems*, 34(4):045005, 2018.

1215. Z. Zhou, G. S. dos Santos, T. Dowrick, J. Avery, Z. Sun, H. Xu, and D. S. Holder. Comparison of total variation algorithms for electrical impedance tomography. *Physiological measurement*, 36(6):1193, 2015.

1216. Z. Zhu, W. R. B. Lionheart, F. J. Lidgey, C. N. McLeod, K. S. Paulson, and M. K. Pidcock. An adaptive current tomograph using voltage sources. *IEEE Transactions on Biomedical Engineering*, 40(2):163–168, 1993.

1217. G. Zick, G. Elke, T. Becher, D. Schädler, S. Pulletz, S. Freitag-Wolf, N. Weiler, and I. Frerichs. Effect of PEEP and tidal volume on ventilation distribution and end-expiratory lung volume: A prospective experimental animal and pilot clinical study. *PLoS ONE*, 8(8):e72675, Aug. 2013.

1218. S. Zlochiver, M. M. Radai, S. Abboud, M. Rosenfeld, X.-Z. Dong, R.-G. Liu, F.-S. You, H.-Y. Xiang, and X.-T. Shi. Induced current electrical impedance tomography system: experimental results and numerical simulations. *Physiological Measurement*, 25(1):239–255, Feb. 2004.

1219. M. Zolgharni, H. Griffiths, and P. D. Ledger. Frequency-difference MIT imaging of cerebral haemorrhage with a hemispherical coil array: numerical modelling. *Physiological Measurement*, 31(8):S111–S125, July 2010.

1220. M. Zolgharni, P. D. Ledger, D. W. Armitage, D. S. Holder, and H. Griffiths. Imaging cerebral haemorrhage with magnetic induction tomography: numerical modelling. *Physiological Measurement*, 30(6):S187–S200, June 2009.

1221. Y. Zou and Z. Guo. A review of electrical impedance techniques for breast cancer detection. *Medical engineering & physics*, 25(2):79–90, 2003.

Index

9 781032 161174